HUMAN SCALE

ALSO BY KIRKPATRICK SALE

THE LAND AND PEOPLE OF GHANA

SDS

POWER SHIFT: THE RISE OF THE SOUTHERN RIM
AND ITS CHALLENGE TO THE EASTERN ESTABLISHMENT

HUMAN SCALE

Kirkpatrick Sale

A PERIGEE BOOK

Perigee Books
are published by
G. P. Putnam's Sons
200 Madison Avenue
New York, New York 10016

Published simultaneously in Canada
by General Publishing Co. Limited, Toronto.

The author gratefully acknowledges the following for granting permission to quote from copyrighted materials:

Houghton Mifflin Co. for an excerpt from "America Was Promises" from *New & Collected Poems, 1917–1976* by Archibald MacLeish, reprinted by permission of Houghton Mifflin Co., copyright © 1976 by Archibald MacLeish.

Rutgers University Press for an excerpt from "The Size of Song" by John Ciardi from *Person to Person,* reprinted by permission of Rutgers University Press, copyright © 1964 by Rutgers, the State University.

Library of Congress Cataloging in Publication Data

Sale, Kirkpatrick.
Human scale.

1. United States—Economic conditions—1971–
2. United States—Politics and government—1945–
3. Decentralization in government—United States.
4. Environmental policy—United States. 5. Technology—Social aspects—United States. I. Title.
HC106.7.S24 1982 306'.0973 81-17843
ISBN 0-399-50621-7 AACR2

First Perigee printing, 1982
Printed in the United States of America

AN OLD CAUTIONARY TALE has it that there once was a kingdom in which all of the grain crop one exceptional year somehow became poisoned, causing anyone who ate its products to go insane. That posed a terrible dilemma for the king and his advisors, for the stores of grain from previous years were very modest, not nearly enough to feed the entire population of the land, and there was no way to procure food from without. The kingdom would face either widespread famine and starvation, if the harvest were destroyed, or widespread madness and chaos. After much deliberation, the king reluctantly decided to have the people go ahead and eat the grain, hoping its effects would be temporary, that at the very least human lives would be preserved. "But," he added, "we must at the same time keep a few people apart and feed them on an unpoisoned diet of the grain from previous years. That way there will at least be a few among us who will remember that the rest of us are insane."

It is to those few that this book is dedicated:

Steve Baer
Tom Bender
C. George Benello
Wendell Berry
Murray Bookchin
Ralph Borsodi
Scott Burns
Ernest Callenbach
Noam Chomsky
Joseph Collins
Richard Cornuelle
Herman E. Daly
René Dubos
Edgar Z. Friedenberg
Paul Goodman
Percival Goodman
Hazel Henderson
Karl Hess
John Holt
Ivan Illich
Judson Jerome
Lee Johnson
Rosabeth Moss Kanter
Leopold Kohr
Milton Kotler
Ursula LeGuin
Frances Moore Lappé
Mildred Loomis
Amory Lovins
Michael Marien
John McClaughry
Ian McHarg
Margaret Mead
Arthur Morgan
Griscom Morgan
David Morris
Lewis Mumford
Carol Pateman
Theodore Roszak
E. F. Schumacher
Neil Seldman
L. S. Stavrianos
Barry Stein
Bob Swann
Lee Swenson
Gordon Rattray Taylor
Frederick Thayer
William I. Thompson
John Todd
Nancy Todd
Peter van Dresser

CONTENTS

PART ONE: TOWARD THE HUMAN SCALE

PART TWO: THE BURDEN OF BIGNESS

PART THREE: SOCIETY ON A HUMAN SCALE

PART FOUR: ECONOMY ON A HUMAN SCALE

PART FIVE: POLITICS ON A HUMAN SCALE

PART SIX: CONCLUSION

HUMAN SCALE

PART ONE

TOWARD THE HUMAN SCALE

Units of measure are the first condition of all. The builder takes as his measure what is easiest and most constant: his pace, his foot, his elbow, his finger. He has created a unit which regulates the whole work. . . . It is in harmony with him. That is the main point.

LE CORBUSIER
Towards a New Architecture, 1958

The proper size of a bedroom has not changed in thousands of years. Neither has the proper size of a door nor the proper size of a community. . . .
Scale: by that we mean that buildings and their components are related harmoniously to each other and to human beings. In urban design we also mean that a city and its parts are interrelated and also related to people and their abilities to comprehend their surroundings.

PAUL D. SPREIREGEN
Urban Design: The Architecture of Towns and Cities, 1965

The most important balance of all the elements in space is that of the human scale.

CONSTANTINE DOXIADIS
Ekistiks, 1968

1

Parthenothanatos

FOR 2,400 YEARS the Parthenon has stood atop the Acropolis, an enduring monument to the imagination and craft of humankind and to the complex civilization that gave it birth. Artfully placed against the backdrop of two dramatic mountains, on a large stone outcrop 500 feet above the Aegean's Saronic Gulf, it was purposefully built at an angle to the entrance gate so that you see it first not head on but in perspective, the columns receding in order and harmony, their delicately fluted lines etching a series of shadows in the Attic light against the bright, creamy stone. As you approach, the temple seems almost to float, massive and assertive though it is, for it rests on a slight hill and it was crafted without any true verticals whatsoever, the columns bending inward from base to capital with infinite subtlety and precision, the flutes so carefully measured that each one had to be carved individually like a jewel, the whole effect pulling the eye imperceptibly upward. Close to, the building's decorations, or what is left of them, immediately draw attention: the exterior sculptures display an extraordinary concern for the varieties of the human form, in motion and at rest, clothed and naked, while the interior friezes of the Panathenaic procession convey the energy and centrality of the workaday civic life of the city below. Within, where the measured classical spaces, even in their ruined condition, suggest the kind of monumentality that befits a temple, the sense of the human measure is again reflected in the dimensions of the columns and remnant forms, and the rational, humanistic spirit that originally informed it is unmistakable still.

To architects a model, to archaeologists a treasure, to classicists a palimpsest, to historians a time chamber, to humanists an inspiration, the Parthenon has no equal, on any continent, from any age. "Earth proudly wears the Parthenon," Emerson wrote, "as the best gem upon her zone." It has been the object of pilgrimages for many peoples of the world, but for the West it is even more: the seat of the civilization that has done more than any other to shape our own, our arts and our sciences, our politics and our governments, our culture and our most basic perceptions of the world. During the course of twenty-four centuries—longer than any single civilization has lasted since the dawn of time—the Parthenon has stood as the embodiment of our heritage. It has suffered much, to be sure, in the course of human warfare and human greed—the Byzantines looted it in the fifth century, the Venetians bombed it in the seventeenth, the British pried off part of its treasures in the nineteenth—but it has always endured, always seemed to be possessed, as Plutarch had written, of "a living and incorruptible breath, a spirit impervious to age."

Now the Parthenon is literally crumbling away, and it may not survive into the next century, certainly not as we know it now.

The Athens of Periclean times, a city of perhaps 50,000 people, has grown now to a sprawling metropolitan area of 2.5 million. It has chosen as its primary means of transportation the automobile, its primary system of heating the oil furnace, its primary source of electricity the coal-burning power plant, and its primary engine of production the industrial factory. All these for the last thirty years or so have been gushing out a huge spew of pollutants that have slowly eaten into the very stones of the buildings of the Acropolis, particularly the soft Doric marble of the Parthenon and its smaller neighbor, the Erechtheum; of special noxiousness is sulfur dioxide, which pours out of Athenian smokestacks at the rate of some 2,000 tons a day and out of Athenian car exhausts at some 5,000 pounds a day and which combines with droplets of water vapor in the air to become sulfuric acid, a chemical that is quite literally and methodically melting the stones of antiquity. Today the faces of the relief figures all along the Parthenon's entablature are falling away, the sculptures are being worn to indistinguishable blobs, the horseman on the west side is practically obliterated, the columnar flutes are fading, and hands and arms and horses' legs in the Panathenaic frieze have disappeared entirely. The monument that has been the pride of centuries is now being irretrievably destroyed in the space of half a lifetime. As the UNESCO director general has put it, "After resisting the onslaughts of weather and human assailants for 2,400 years, this magnificent monument is threatened with destruction as a result of the damage which industrial civilization has increasingly inflicted on it." How wry, how ironic, that phrase "industrial civilization."

Measures have naturally been taken to try to moderate the disaster. Tourists—more visitors in a single summer now than in all the years of ancient Athens—have been excluded from the interior of the Parthenon and made to stay behind barriers some twenty-five feet away, lest the incessant reverberations of their feet compound the pollutants' effect; jet airplanes, too, are normally routed away from the Acropolis area to minimize the effect of their vibrations. UNESCO has begun a fund-raising campaign, hoping to find at least $15 million for protection and restoration of the various buildings, though much of that money will be spent actually to undo an earlier industrial calamity caused when misguided restorers bored through the center of the Parthenon columns and reinforced them with long iron bars that have now rusted and corroded. The Greek government has removed several sculptures (and plans to remove all the Acropolis figures in time), replacing them with fiberglass replicas of the originals; it even has plans to remove the smaller Erechtheum and put it into a still-to-be-built museum, simply bulldozing over the spot that for countless generations even before Pericles was a holy shrine. And lately the experts have been seriously considering a method, during the winter when the tourist flow abates, of covering the entire Acropolis in plastic.

But none of it will make a difference. The desperate attempts to patch up the ravages of huge industrial systems with money and technology—always the favored solution of our contemporary age—cannot restore what has been lost, cannot prevent the ongoing devastation. The Parthenon that was, the shrine that even in imperfect form excited centuries, will never—*never*—be the same. And even if the current fixers succeed in all their schemes, at best they can only transform the temple into a lifeless picture postcard, its sculptures no more than "authentic reproductions" (as the art world's contradiction has it), its reliefs turned into machine-molded panels of drabness, its columns shaped to some restorer's plan, its interiors barred to public experience. Pilgrimages have not been made these many centuries so that the inheritors of Greece could stand off in mid-distance and gaze upon the miracles of molded fiberglass.

NOR IS THIS awesome devastation confined to the Parthenon, or to the Acropolis, or to Greece alone. The monuments of the world's civilizations in every country are being obliterated wherever they have the misfortune of being located in the vicinity of a large industrial city; the United Nations International Symposium on the Deterioration of Building Stones, held in September 1976, identified at least 500 important buildings suffering this destruction. The Roman statues in Italy are disintegrating, losing noses and limbs to the pollutions of the air. The Sphinx and the Pyramids of Egypt are being eroded, the Carthaginian remains in Tunisia, the Inca temples in Mexico. Cathedrals throughout Europe are losing their statuary and decorations to the relentless chemicals: St. Sophia, St. Paul's, Salisbury, Chartres, Notre Dame, St. Peter's, St. Mark's, the churches whose very names tell much of the story of our heritage, all are decaying daily. The windows of Chartres, some of the finest stained-glass art known to the world, and of course unreproduceable, have had to be plastic-coated to preserve them from air pollution, and they now have turned yellow and lusterless and appear to be somehow fake, like paste replicas in a golden crown.

No one is designing this catastrophe, and surely no one wants it. Yet just as surely, as long as the priorities of industrial systems pertain, as long as private cars and fossil-fuel heating plants and chemical-based industries occupy such central parts of our societies, there can never be, however great a will we muster, and however many millions, a reversal of this process.

THE ONSLAUGHT ON the Parthenon is not of course the worst offense of the contemporary world. But it is a symbol, for me a haunting one: as the Parthenon so fittingly embodies the heritage of Western civilization, so it displays as well the condition to which that civilization has been brought over the last few decades and the crises with which, I think one can say without hyperbole, it is now imperilled.

Its tragedy suggests at least four hard truths. That the crisis of the contemporary world is *real,* not some temporary aberration or media contrivance, and as palpable and perceptible as the sulfur dioxide on the Parthenon. That it cannot be solved, though it may for some time be ameliorated, by the devices of modern technology, by some combination of plastics and chemicals that will somehow emerge if enough laboratories are endowed with enough grants. That it can be dealt with only by a reordering of priorities, a rethinking of values, a reorganization of our systems and institutions so that we can begin to remove the pollutants from the economic and political environments as well as from the natural one. And that if we do not perform some such reordering and reworking we will almost certainly find our cities, our cultures, our ecologies, and perhaps our very lives eroding and disintegrating just as surely and as irretrievably as the Parthenon. These are the truths that guide this book throughout.

But there is more, happily, to the Parthenon than just that, for it can also symbolize for us the direction of that reordering and reorganization. For the Parthenon is a building carefully, gracefully, designed on the human scale, measured by the human thumb and pace, celebrating the human form, created in its every detail with the principle of "man the measure."[1] It was built in a land whose society was governed by the

1. The great French architect Le Corbusier was once sent a set of documents giving the exact measurements of each of the marble blocks used in the building of the Parthenon—ledges, columns, entablatures—and from them he determined that the Athenians had used a human-scale dimension of the height of a man throughout their design; he calculated the height as close to his "Modulor I" figure of 1.75 meters, or roughly 5 feet 9 inches, "with the help of conviction and a few inches (or millimetres) suggested by pure faith" *(The Modulor,* MIT Press, 1954). That in fact would be a little large for the ancient Greeks, whose stature was smaller than that of the modern European, but something closer to 5 feet 7½ inches would be very like the average male height, and therefore undoubtedly the architectural measure, of the Athenians. And I have calculated that, allowing for minuscule variations in the original construction and in the settlement of the building over the years, it

human scale, whose economic relations were ordered by the human scale, whose government was determined by the human scale. When we appreciate that, we can begin to see something of what we are lacking in our contemporary world, something of what it might be pertinent to be striving for.

Not that Periclean Athens was ideal by any means—the subjugation of women and slaves would alone have made it repugnant to the modern soul—nor would anyone today knowing of the numerable advantages of the present possibly advocate recreating such a remote past. Yet for all its ills, that city at that time had such an appreciation of the central role of the human within the society, of individual worth mixed with communal value, of civic participation and reward, that a contemporary American could not observe it without some sense of what has been lost in the intervening years.

Unlike the peoples of the preceding empires—Babylonia, Mesopotamia, Assyria, Egypt—the Greeks did not worship omnipotent gods, did not serve almighty kings, did not cluster themselves into faceless urban multitudes. They evolved, for the first time, philosophies and organizations built on the quite remarkable notion of the free-born citizen, an individual with an inalienable equality within the community, or *polis,* who was expected to participate in its arts, sciences, athletics, politics, discourses, and games not only for the betterment of the self but for that of the entire population.

In Athens, the daily life evidenced that human principle. In the *agora,* the public square that was at once the marketplace and the meeting place, there would be an amorphous and spontaneous movement of people and goods and ideas from dawn to sunset, a social axis on which the rest of the city's life spun. In the *ecclesia,* the democratic assembly of the citizens, the free men (or at least the more purposeful among them) would meet to formulate the decisions of the community on the principles of open participation and individual rights, and the offices of the city would be held by various of them chosen by lot throughout the year. In the sports-grounds and parks and at the periodic games and dances, the human body would be celebrated with a pagan zeal, and at the schools and *gymnasia* a similar passion, at least among those who had the leisure, was devoted to the development of the human mind. For every citizen there would be a full range of activity through

---

is in exact multiples of that measure that the major dimensions were built. The full height is 540.12 inches (5′7½″ × 8), the width 1215.3 inches (5′7½″ × 18), and the length 2734.03 inches (5′7½″ × 40.5); in addition, the interior columns are 202.5 inches high (5′7½″ × 3), the distance between the architraves is 405 inches (5′7½″ × 6), and the statue of Athena herself, the goddess to whom the city and temple were consecrated, was said to be 540 inches tall (5′7½″ × 8). The multiples of the three outside dimensions when the 5′7½″ measure is used match exactly the 4:9 ratio that has been regarded as the governing ratio of the Acropolis: 4:9 = 8:18 = 18:40.5.

the year, a rotation of economic and political and even artistic function, so that each would be able to participate in all parts of urban life and none would grow to exert undue dominance. And though there would be toil for most, as likely as not with artisan and laborer and slave and free standing shoulder-to-shoulder, Athenian life was meant, and the day was organized, as much as possible for the individual's intellectual, aesthetic, sexual, social, and athletic satisfactions.

In short, Athens was, in the words of the great urban historian Lewis Mumford, a city "cut closer to the human measure."

THE PERILS AND the promise, then, coexist in the singular shrine that is the Parthenon, as fitting an exemplification of our own age as of Pericles's. Its present plight makes manifest our crises, its past glories suggest the direction of our remedies.

What follows is an extrapolation from just that duality, meant to be as definitive and judicious as I can possibly make it but, dealing as it does with much that is speculative and not widely practiced, capable of refinement and development. In this first part, I want to indicate, briefly, the nature and seriousness of our predicament, the unique challenge it offers, and the responses it is already provoking among most of the industrial nations of the West. In the second part, I try to isolate the malady that has brought us to this pass—it is, not to try for suspense, the idea that *bigger is better*—and to show that this fallacy is not only dangerous and absurd but flatly contradicted by the considerable evidence showing that smaller systems—smaller buildings, communities, cities, offices, factories, farms, economic networks, and societies—are both more efficient and more humane.

In the next three parts, the heart of the book, I hope to demonstrate for the three main sectors of our lives—our society, economy, and polity—that smaller and more people-sized institutions and arrangements are not simply necessary and desirable but flat-out possible, using examples from other cultures, other ages, and around our own country to show that we have the *means* to achieve the desirable future as soon as we can apply the will. In the final part, I touch briefly on the possibilities of such a future coming about.

It takes some time to pursue all of this, and you will not have failed to notice that this, a book about the virtues of the small and human-scaled, is quite unusually large. The easy answer is that, in books as in other artifacts or systems, I am no advocate of the needlessly small, rather only of the appropriate size, kept within ecological and humanitarian limits. The more elaborate answer is that I have found that to present so many ideas running against the current thought, not only of this epoch but of the past few centuries, to reassess not merely one aspect of our present difficulties but a whole range of them, and to survey the serious and workable alternatives that have been tried and proven within that range throughout history, has inevitably taken a goodly number of pages—as it has of travels and interviews and researches, and years.

I do not ask that you agree with me as we begin. Only that you keep an open mind, and heart, and remember the sulfuric acid eating into the figures on the Panathenaic frieze.

2

Crises and Double Binds

WILLIAM SLOANE COFFIN tells the story of a scientist from Harvard flying on an experimental mission in a private plane over the lake country of northern Alabama, measuring with elaborate instruments the fish populations of the various lakes. Sighting two fishermen out at some remote lake he had just surveyed, the scientist figured that as a favor he would land his plane on the water nearby and tell them that his instruments had discovered there were no fish to speak of in those waters and they would have better luck if they went on to another lake. So despite the delay, he landed near the anglers and explained the bad news to them, expecting their grateful thanks. They were outraged, instantly, and told the scientist in rich Southern expletives where he could take his plane and his instruments and what he could do with them, whereupon they baited their lines once again and kept on fishing. The scientist flew off, much abashed and much puzzled. "I expected their disappointment," he said later, "but not their anger."

But of course we all react that way to unpleasant truth much of the time: it upsets our preconceptions and our comforting illusions and therefore angers us, and often as not we choose to ignore it. None of us wants to be told, even though deep down we may know it, that there are no fish.

Still, the scientist has his obligation to the truth, and there is ultimately no real point in turning a deaf ear to what he says. So, at the risk of exciting anger or producing instant deafness, I feel it is necessary for us to begin with one unpleasant truth of great importance: *in the last half of the twentieth century, particularly in advanced industrial countries, we are witnessing a series of crises beyond any yet experienced in the procession of Western civilization.*

Now, dire predictions of universal crisis have been common to all ages at least since the Sumerians and Egyptians first settled into urban societies 5,000 years ago. And I do not mean to seem foolishly pessimistic, a professional Chicken Little, when I speak of this series of crises, as if I thought the ingenuity of the human species had reached its limits or the cockroaches were about to inherit the earth. Nonetheless, there is enough evidence around us, and affirmations from enough different kinds of people in enough different disciplines, to prove that our current predicament is quite real and quite unique.[1]

I need only touch on the crises briefly to suggest both their magnitude and their scope:

An imperilled ecology, irremediable pollution of atmosphere and oceans, overpopulation, world hunger and starvation, the depletion of resources, environmental diseases, the vanishing wilderness, uncontrolled technologies, chemical toxins in water, air, and foods, and endangered species on land and sea.

A deepening suspicion of authority, distrust of established institutions, breakdown of family ties, decline of community, erosion of religious commitment, contempt for law, disregard for tradition, ethical and moral confusion, cultural ignorance, artistic chaos, and aesthetic uncertainty.

Deteriorating cities, megalopolitan sprawls, stifling ghettoes, overcrowding, traffic congestion, untreated wastes, smog and soot, budget insolvency, inadequate schools, mounting illiteracy, declining university standards, dehumanizing welfare systems, police brutality, overcrowded hospitals, clogged court calendars, inhuman prisons, racial injustice, sex discrimination, poverty, crime and vandalism, and fear.

The growth of loneliness, powerlessness, insecurity, anxiety, anomie, boredom, bewilderment, alienation, rudeness, suicide, mental illness, alcoholism, drug usage, divorce, violence, and sexual dysfunction.

Political alienation and discontent, bureaucratic rigidification, administrative inefficiency, legislative ineptitude, judicial inequity, bribery and corruption, inadequate government regulations and enforcement, the use of repressive machinery, abuses of power, ineradicable national debt, collapse of the two-party system, defense overspending, nuclear proliferation, the arms race and arms sales, and the threat of nuclear annihilation.

Economic uncertainty, unemployment, inflation, devaluation and displacement of the dollar, capital shortages, the energy crisis, absentee-

---

1. I just glance down my bookshelf: *Mankind at the Turning Point, The Domesday Book, The Limits of Growth, The Coming Dark Age, The Promise of the Coming Dark Age, The Twilight of Capitalism, The Environmental Crisis, The Transformation, The Biological Time-Bomb, Awakening from the American Dream, The Poverty of Power, The Stalled Society, Our Synthetic Environment, Future Shock, Blueprint for Survival, Nightmare, The Myth of the Machine, The End of the American Future, The End of the American Era.*

ism, employee sabotage and theft, corporate mismanagement, industrial espionage, business payoffs and bribes, white-collar criminality, shoddy goods, waste and inefficiency, planned obsolescence, fraudulent and incessant advertising, mounting personal debt, and maldistribution of wealth.

International instability, worldwide inflation, national and civil warfare, arms buildups, nuclear reactors, plutonium stockpiles, disputes over laws of the sea, inadequate international law, the failure of the United Nations, multinational exploitation, Third World poverty and unrepayable debt, and the end of the American imperial arrangement.

Or to put it another way:

Vietnam, Watergate, New York City bankruptcy, gas lines, Mirex, Equity Funding, ITT, riots, Medicaid fraud, redlining, CIA drug-testing, hostages, price fixing, Vesco, nursing homes, coffee prices, product recalls, assassinations, heroin, the Middle East, Rio Rancho, Kepone, skyjacking, the SLA, *Hustler,* Spiro Agnew, saccharin, the square tomato, Harlequin books, Los Angeles, OPEC, Wilbur Mills, power failures, My Lai, Charles Manson, PCB, the SST, Andy Warhol, Appalachia, organized crime, Three Mile Island, Valium, the Wilmington 10, REITs, TV violence, strip-mining, FBI break-ins, the Sahel, microwaves, McDonald's, Kent State, Penn Central, Attica, the *Torrey Canyon,* psychosurgery, mercury, Chile . . .

But that is too fast. Put that way, they come to seem as unreal as the evening news programs, where death and fires and ball scores and tomorrow's temperature all have the same hue and value. And we all become inured to the crises, like the frog in the laboratory experiment who jumps out of the frying pan immediately if you put him over a high flame all at once but who stays on unaware until he is fried to a crisp if you start with a tiny flame and increase it only gradually. But the crises are real nonetheless, and they reflect the condition of America with shocking aptness.

LET US ISOLATE a few at a slower tempo:

▪ In the United States today, seventy people every day commit suicide and another thousand or so attempt it. That is a truly frightening fact, for suicide, as Durkheim argued so long ago, is the ultimate form of alienation from society, the ultimate evidence that life is meaningless. That a thousand people *every day*—365,000 every year, or 15 million people in my lifetime so far—should be moved to such a statement, even if some of them indeed only want to attract attention or revenge a slight and never actually succeed, surely says something about the society we have created and the life we have offered. And then to learn that the suicide rate is increasing fastest among the young—nearly 300 percent increase for people 15 to 24 in the last twenty years, and teenage suicides at something like a dozen a day—is to realize that some kind of serious social pathology is at work.

▪ A lesser symptom of desperation is alcohol, but its prevalent and increasing use, and the sharply increasing number of alcoholics, suggest that it is symptomatic of the same maladies in the nation. Depending on who is doing the defining, there are anywhere from 11 million to 20 million full-scale alcoholics in the U.S. today, people who regularly must drink themselves to stupefaction to get through their days, and each year there are an additional 270,000 or so. Again, it is wrenching to learn that teenagers, the very people who should have the most optimistic view of life, are turning more and more to alcohol, some 1.3 million of them now regarded as out-and-out alcoholics—that's equivalent to the entire high-school population of Pennsylvania—and a 1978 study by the Department of Health, Education, and Welfare found that 60 percent of all 16-year-old boys, and 25 percent of *all* 12-year-olds, can be classified as moderate or heavy drinkers.

▪ The number of murders in the U.S. has increased steadily over the last twenty years, up by nearly 60 percent since 1970, with some 20,000 people a year now driven to this ultimate cruelty, giving the U.S. a greater murder rate than any nation on earth, perhaps greater than any nation known to history. To be sure, many of these killings are crimes of passion, still others committed in the course of another crime, and hence are not expressions of straight-out social hostility and vengeance. But the fact that this form of death has become so automatic in our country and so commonplace (in New York City, for example, which averages more than four homicides a day, the great majority of cases are never even reported in the newspapers) suggests a stark failure of the society to nourish among its citizens even the most basic social value, the respect for human life.

▪ The number of people with mental disorders, a reflection of an inability to cope with the surrounding world, has also increased steadily every year for the past twenty. Nearly 6 million individuals in the U.S. are treated *each year* in mental facilities, and according to the 1978 President's Commission on Mental Disorders approximately 32 million people received some treatment for mental disorders in 1976—in all, it estimated, 40 million Americans have "diagnosable disturbances" and are in need of professional care, and another 55 million suffer "severe emotional distress."

▪ In a nation that prides itself on being the richest in history (though in truth there are a number of others with greater per capita wealth), at least 30 million people live in deplorable poverty. Washington's standards are altered every few years to cosmeticize the bitter facts, but even using current Federal data it is clear that some 25 million people—more than the combined populations of Sweden, Norway, Denmark, and Finland—are living beneath an acceptable standard and another 15 million are barely above it. Considering that the percentage of those in poverty is greater here than in most other industrialized nations, and that governments here spend only about half as much as those in other countries, one might justifiably conclude that the nation's anti-poverty commitment is abjectly minimal, and apparently cruel and unfeeling to boot. If a society is to be judged by the way it treats its poor, as has been said, then American society, in letting nearly a fifth of its people languish in destitution, surely is to be found wanting.

These particular evils, just these few, should be enough, I would think, to cause us all some trepidation and to prompt a reassessment, both personal and national, of the society we have built. Can it be right to organize a nation this way? Even if we can point to some other ages or some other countries with worse performances, as no doubt we can, does that forgive our own record, these ominous and growing crises?

Suppose a Martian were to descend tomorrow and ask how efficiently the United States had gone about its business in the last twenty years or so, how effective it had been in using its obvious riches, its large universities, its mighty bureaucracies to solve the social problems of its people. What would we be able to say? That Sweden has a higher suicide rate and there seems to be about the same percentage of mental disorders in Nova Scotia? That many well-intentioned books had been written on all these subjects and many good-hearted people have given them thought? That there seemed to be other things to attend to and for some reason other priorities for institutions both public and private?

We would avert our eyes.

OF COURSE, SOCIAL DISJUNCTIONS of this magnitude do not exist without their deep and pervasive effect on the populace, no matter how immune we think we may be. The evening news may not mention alcoholism, but we all know about the woman in the apartment upstairs and the broken marriage down the street; there are no banner headlines about mental disorders, but we all have friends or children or colleagues who have tortured themselves into psychic knots. Increasingly, the existence of all these crises, and the inability of any of our institutions to alleviate them, has created an awareness among the people that, as Miss Clavell puts it so well, something is not right.

Though certainly not definitive, the clearest indicators of this national spirit are the public opinion polls. Every poll and survey taken in the last decade to gauge the mood of the country indicates that Americans are troubled, unhappy, distrustful, disgruntled, alienated, you name it, and that these attitudes are shared by more and more people with each passing year.

Some 53 percent of the citizens recently agreed that there is "something deeply wrong in America," and some 45 percent declared that the "quality of life had deteriorated in the last 10 years." Well over half say every year that "most people with power only try to take advantage of you" and "what you think doesn't count" and a consistent three-quarters of the population is resigned to the proposition that "the rich get richer and the poor get poorer." Not one of the institutions that make up the daily fabric of the country inspires the confidence of a majority: only 43 percent of Americans have a "great deal" of confidence in doctors, 30 percent in the press, 20 percent in the U.S. Congress, and 19 percent in the major corporations.

As for the organs of government, public trust and allegiance have been eroding every year since 1958, when the pollsters first thought to ask about it, and by now every single survey shows that a clear majority is totally disillusioned, a most astonishing fact. Recent polls have counted 58 percent who are "alienated and disenchanted by the government," 72 percent who believe the government is run "on behalf of a few special interests," and 55 percent who feel that "public officials don't care much about what people like me think." From 1966 to 1977, the Harris polls show that public confidence in the presidency dropped from 41 percent to 23 percent, in the Congress from 42 percent to 17 percent. Only 53 percent of the eligible voters bothered to go to the polls in the 1976 presidential election, and the percentage has been declining since 1960, although the number of eligible voters has been increased by 17 percent; 41 million voted for Carter, 39 million voted for Ford, and 66 million stayed home.

"A central fact," the Harris organization concluded in our bicentennial year, "is that in our nation, our people, disaffection and disenchantment abound at every turn. *That disaffection has now reached majority proportions.*"

NOW IT CAN BE fairly objected here that every age has its crises and so far the ingenuity of the human brain, often the scientific brain, has been capable of solving, or appearing to solve, them all. Here we are, after all, in a large and successful country, with lots of comforts and luxuries, and however numerous the problems have been in the past they obviously haven't done us in.

But that lesson from the past disguises one important fact of the present: our crises proceed, like the very growth of our systems, *exponentially*. "During the last two centuries," in the words of Dr. M. King Hubbert, a geophysicist from the U.S. Geological Survey with a worldwide reputation for vision and acumen, "we have known nothing but exponential growth, and we have evolved what amounts to an exponential-growth culture, a culture so heavily dependent upon the continuance of exponential growth for its stability that it is incapable of reckoning with problems of non-growth."

What that means is best expressed in the ancient fable of the Arab potentate who offered to give one of his subjects, in return for some well-received favor, any gift he desired. The humble subject asked only for some grains of wheat on a chessboard, one grain on the first square, two on the second, four on the third, eight on the fourth, and so on, doubling each time until all the squares were filled. Naturally the potentate was willing to comply with this modest request, and the granary managers were called forth to begin distributing. Much to everyone's amazement, there wasn't enough wheat in the entire country to supply the subject. Before they reached the thirty-third square he was owed 8,000 bushels of wheat, by the fiftieth square 1 billion bushels (which is equal to the modern annual output of the United States), and by the sixty-fourth square some 7.4 trillion bushels, or 2,000 times as much as the annual production of the entire modern world. That's exponential growth.

The Club of Rome and others in recent years have shown with sobering clarity that exponential growth will, quite literally, devastate the society that practices it. Mineral resources and fossil fuels used exponentially—as we are now doing—will dry up, and quite quickly, too, while pollution and population growing exponentially—as they are now doing—will overstretch the ecological limits. "Our conclusion from these extrapolations," the original Club of Rome researchers wrote, "is one that many perceptive people have already realized—that the short doubling times of many of man's activities, combined with the immense quantities being doubled, will bring us close to the limits of growth of those activities surprisingly soon." Since then some experts have challenged their figures and their calculations, but none have seriously disputed that conclusion.

In simplest terms, this means we are now living through a time that is markedly different from that of the past. Our crises are not only different in degree, they are also, *because of that,* different in kind. Never before have nations grown so large, never have corporations become so powerful, never have governments swollen to such sizes, never have the instruments, the factories, the farms, the technologies been so huge—hence, never before have the crises been so acute. It's as if humankind were living in a huge unbreakable bottle, into which is placed a small drop of exponential water that doubles in size once a day. One day we discover that the water has filled up half the bottle, and we say, well, it's taken an awful long time for it to get up to here, we probably have no reason to worry just yet, surely with our technology we'll be able to adjust to it as we have so far. The next day we drown.

The crises of the present, in other words, have now grown so large, so interlocked, so exponential, that they pose a threat unlike that ever known. It has come to the point where we cannot solve one problem, or try to, without creating some other problem, or a score of problems, usually unanticipated. Then we are suddenly faced with the task of coming up with new solutions without enough time to figure out their consequences, and when we hastily put that solution into effect, it goes on and creates another set of problems.

That is the double bind.

THE DOUBLE BIND is the paradigmatic condition of our age. The term was invented by the psychologist Gregory Bateson some twenty years ago to describe a central dilemma he perceived in his schizophrenic patients and their families: the patient, needing love and familial reinforcement, wants to believe that his parents make sense at all times and doesn't want to disagree with them, so even when they say something wrong and he *knows* it to be wrong he agrees with them, thus producing the situation that when he's right he's wrong and when he's wrong he's right. Bateson amplifies this by describing a hypothetical five-person game in which each person tries to form alliances with the others for some personal gain. If A aligns with B, a correct decision, he will then be outnumbered three-to-two, which is incorrect, so he must try for an alliance also with C, a correct decision, but which may offend B and therefore forces A to ally also with D so as to prevent the counter-coalition BDE from taking shape; but that, too, is incorrect because at that point E, left alone, will try for an alliance with any of the others and each one will hope to forestall his partners so as to block a new counter-coalition, but if A thus allies with E, a correct decision, he is likely to find himself abandoned by his partners and stuck again in a minority of two, an incorrect decision, and if he does *not* join with E, a correct decision, he is likely to be outnumbered by a majority of four, an incorrect decision. As Bateson notes:

> Every move which he makes is the common-sense move in the situation as he correctly sees it at that moment, but his every move is subsequently demonstrated to have been wrong by the moves which other members of the system make in response to his "right" move. The individual is thus caught in a perpetual sequence of what we have called double bind experiences.

In other words, you can't win.

It doesn't take much reflection to realize that the series of crises that I have been describing so far has reached the double-bind point.

Let's say that America wants to alleviate the crisis of domestic hunger, not simply for humanitarian reasons but because feeding 20 million underfed citizens turns them into better workers, better consumers, and better taxpayers and prevents them from turning to social unrest. But given the nature of corporate agriculture, a decision to grow more food means a far greater use of energy for farm equipment, fertilizers, pesticides, and transportation to markets, thus adding to the energy crisis, driving up energy prices, and making the price of growing and distributing food even more expensive, thus ultimately putting it out of the price-range of the needy. It means increased use of pesticides, some of which in the air, soil, or food will cause additional disease and debilitation, especially among the poor, thus putting them out of work and limiting the amount of money they can spend on food. It means increased use of chemical fertilizers, the mining of which adds radioactivity to the air and can cause further sickness, and the fertilizers will eventually leach even more into surrounding water systems to damage the marine life, curtailing the supply of fish for food. It means the expansion of the larger farms with greater capital, thus driving out small and marginal farmers who will be forced into the cities and either join the ranks of the underfed or get on the welfare rolls, adding to governmental spending and thus to inflation, driving up food prices. With increased inflation and abundant agricultural supplies, farmers will be getting less money for their crops, so either they will have to be given subsidies from the Federal treasury, increasing inflation still further, particularly for the poor, or they will have to cut back on production to force prices up, thus making less food available for the underfed.

Whichever way you look at it: double bind.

Or let's say that it is decided, as many politicians and criminologists have urged, that the country solve its problem of soaring crime rates by beefing up state and local police forces, expanding and computerizing Federal agencies, increasing the number of prosecutors and judges at all levels, building more courts and jails, and expanding budgets for prisons and rehabilitation services. This of course places an enormous extra burden on already stretched governmental budgets, particularly those of the larger cities where crime is greatest, which must then draw funds away from other services, including schools, hospitals, welfare, community development, mass transit, housing, and job-training programs, and this inevitably leads to increased unemployment, poverty, urban deterioration, addiction, and prostitution. That in turn inevitably means more crime. Double bind.

Or let's say that the crises of resource depletion and misallocation become so acute within a few years that the nation finally decides to follow the recommendations of such varied people as John Kenneth Galbraith and Herman Kahn for greater centralized planning, so that rationing and conservation can be enforced, allocation and distribution made rational, and the full range of technology brought to bear. Since only the central government can accomplish all this, that will mean an expanded bureaucracy in Washington that, whatever its other virtues, must by its very nature be cumbersome, unresponsive, time-wasting, and inefficient. Planning at this level would have to mean more reliance on high-technology and centralized systems, mechanical and human, which are inherently prone to greater error and breakdown (and are harder to fix) as they grow in size. Greater control of how resources are obtained and used would lead to greater roles for the government in extractive businesses, distribution services, the market, and even the home itself (checking, for example, on the amount of gas and oil being used), substantially increasing government interference in hitherto private areas and creating resentment and resistance among many citizens. All of that—bureaucracy, complex systems, citizen resistance—adds up to a high probability of overall inefficiency and waste, leading to misallocation and misuse of resources, and thus the likelihood of their earlier depletion. Double bind.

But we don't have to be quite so ethereal about it all, so futuristic. Double binds are all around us.

To build up heart muscle and ward off coronary diseases that affect people who don't exercise, people living in cities, where natural forms of exercise have been pretty much done away with, have taken to jogging and cycling. But jogging and cycling along city streets exposes the lungs to about ten times as much air pollution as normal and the activity itself leads to hyperventilation and the inhalation of even greater quantities of pollutants, many of which are known to cause heart disease. So if you do *not* exercise you risk coronary illness one way and if you *do* you risk it another way.

To ease traffic congestion on overcrowded highways, cities and states spend millions of dollars creating new multi-lane superhighways. But every time a bigger highway is built more people, even those who used mass transit before, choose to ride on it because it is faster and easier and safer, so that it very quickly becomes as overcrowded as the original road. So if a city has old highways it will have a traffic congestion problem and if it replaces them with new highways it will have a traffic congestion problem.

Double bind.

Nor is this all esoterica: the double-bind crises are the stuff of our daily headlines.

Take the energy crisis. The nation has wrestled with the problem of how to power its generating stations for five years now, and each year the mess only becomes murkier. We can't use oil, of course, because it is increasingly expensive to produce domestically, it makes our economy dependent on the OPEC nations, it is a non-renewable resource that we are fast running out of, and it is polluting. Nuclear power is no alternative because conventional plants pose grave risks of accidents and air and thermal pollution, and they depend on uranium, another fast-diminishing non-renewable resource; breeder reactors might be used to make plutonium instead, but that is even more toxic and radioactive, it can easily be made into bombs, it can't be efficiently recycled, and plutonium power plants are even riskier and more expensive than conventional ones. Coal offers the advantage of a resource that won't run out for a few more centuries, if you can take comfort from that, but it has its own problems: high-sulfur coal produces a deadly pollutant (that's where much of Athens's sulfuric acid is coming from), and smokestack "scrubber" devices used to lessen its sulfur content are enormously expensive, needing an investment of some $20 billion over the next decade, and produce acres and acres of sludge that nobody knows what to do with; low-sulfur coal is less polluting, but it would have to be provided by strip-mining great sections of the American West, probably turning a sizeable portion into a desert wasteland no matter how much reclamation is attempted, and it provides fewer BTUs of energy so it would be necessary to use much more of it. Solar energy, as we shall see later, is a real alternative, but because of government foot-dragging there is no technology that can turn it into electricity cheaply enough for massive power-plant use, and all the huge government projects for turning it to citywide heating use have so far been abject failures. So what's left—candles?

Or take Tris. That is the chemical that clothing manufacturers decided to put into children's sleepwear when, after more than a decade of pressure, the Department of Commerce ordered some steps to be taken to fireproof those garments. This increased the price of sleepwear by a third, and it forced the manufacturers to use petroleum-based synthetics instead of cotton because they took the chemical better, but it did lessen the number of burn incidents. Until, as you remember, it was discovered that Tris is a mutagen, gets absorbed by skin cells, and is a likely source of cancer, after which it was banned from the market—although there were 20 million garments with the stuff already in it and it was impossible to tell which polyester nightclothes were contaminated and which weren't. Then it turned out that Tris was also used in dolls, toys, car seats, and draperies, so that even the child who slept in bare skin stood a good chance of being exposed to it. And when scientists began a search for other flame-retardants to replace Tris, they discovered that all of the other chemicals suggested as flame-retardants were also—what else?—highly suspect carcinogens.

Or take the Aswan Dam. Egypt built that with much fanfare and vast expense in order to provide electricity for its people, increase agricultural production through controlled irrigation, increase fish production by providing a new lake, and thus improve the general standard of living. But the dam has blocked off the Nile waters so that the millions of tons of natural fertilizers end up in the lake behind it and never get either to the farmlands downstream, severely harming agricultural production, or to the marine life of the delta, severely curtailing fish production. So the government planners were forced to use much of the electricity from the dam not for home or industry but to make artificial fertilizers for the farmers and, someday they hope, artificial chemicals for the delta fishermen, thus using electricity to solve the problem created by the dam that was built to solve the problem of electricity. But since the artificial fertilizers so far have been strange to the soil and don't work as well as the natural ones, and since the delta waters, stagnant now for much of the year, have bred a variety of diseases, the overall standard of living has in fact been lowered.

Or take—a last example—mass transit. It is commonly agreed that the nation needs mass transit systems to save on energy use and stop traffic pollution and congestion in the big cities. But in the years between 1971 and 1977, the deficit of all the country's urban transit systems went from $300 million to $1.8 billion, a rise of 600 percent, while the number of riders went up only 3.2 percent, thus putting great strains on city budgets, creating taxpayer resentment and revolts, and putting future generations of city dwellers deep into debt, but not lessening traffic by more than an iota. Here is the pattern: a city builds a mass transit system, which is correct, but then has a bigger municipal debt, which is incorrect, so it cuts back the frequency and quality of the service, which is correct, but then finds itself with fewer riders, which is incorrect, so it has to charge more for the service, which is correct, but then fewer people ride and the deficit goes up more, which is incorrect. So building a mass transit system, which is correct, ends up simply enlarging city debts year after year and affecting pollution and congestion not at all, which is incorrect.

Double bind.

I hope the point is clear, though we shall have occasion to return to it again: the crises of the present are of such magnitude and synergy that they cannot be solved by any of the conventional solutions we have applied, and indeed the solutions themselves turn out to be problems more often than not. In virtually all our systems, all our conventions, all our institutions, we are caught in a double bind, and we apply our solutions, one after another, in vain. "In vain" in Latin, we might remember, is *frustrere,* from which, of course, we get our word "frustration."

There are no fish.

3

Turning Point

THAT THIS MULTIPLICITY of crises has grown upon us in these last few decades is not an accident, a celestial happenstance. It is rather a signal that, in some way we don't yet fully understand, the arrangements of the past are coming unstuck and the systems that have sustained us for centuries past are approaching their end. It is not hyperbolic to say that the present era is as surely a transition period in human affairs as was the Renaissance or the fall of Rome. Indeed, the urban planner E. A. Gutkind goes further: "We are at one of the decisive turning points in the history of humanity," he argues, "comparable to the domestication of animals, the invention of the earliest tools, the foundation of the first cities, and the conception of the heliocentric universe."

All of the basic characteristics that, taken in the broad, have marked what we like to call the "modern age"—that is, the 500-year period of Western civilization beginning with the Renaissance and lasting to the middle years of this century—are in the process of change.

That period was marked by the exploration and settlement of every geographic corner of the world. But now for the first time in human history the global limits have been reached and there are no more frontiers to which surplus people and capital can move and from which new agricultural and mineral wealth can be drawn. The great British historian Arnold Toynbee, who has called this frontier development "the dominant movement in world history," is only one of many to observe that by the middle of the twentieth century "it was evident that the West's expansion of every kind and in every continent was coming to an end."

That "modern" period was marked by the triumph of capitalism and its total penetration into all European (and, later, world) systems, including those of the "socialist" states. But now the capitalist world economy is in a decline more precipitous than even that of the seventeenth century, and its most basic elements—the free market, for example, and the play of supply and demand—are being transformed or discarded everywhere, even in the United States. Michael Harrington, who as a socialist may be presumed to be biased but as an economic philosopher has proven to be indisputably accurate, has noted that "it is now clear that the Western capitalist hegemony over the world market, which has its origins as far back as the thirteenth century, is beginning to come to an end."

That period was marked by the rise of the political nation-state and its evolution into the large centralized government that has become the most important institution (replacing the church and even the family) in contemporary societies. But now the state, particularly in the West, is at the point of serious trouble, without even the ability to claim either the automatic allegiance of its citizens (indeed, as we have seen, often blamed by them as the main source of evil and discontent) or their unquestioned obedience (indeed, often challenged in recent decades by protest, unrest, resistance, and violence). The conservative sociologist Robert Nisbet, seeing in all this a "twilight of authority," has written that we are witnessing "the seeming incapacity of the political order any longer to sustain the lives and hopes of its citizens," meaning that "today we are present . . . at the commencement of the retreat of the state as we have known this institution for some five centuries."

That period was marked by the exploitation of a great portion of the world's people for the well-being of a succession of Western nations, most recently the United States. But now, particularly with the American defeat in Vietnam and the rise of the OPEC nations, the imperial era has come to an end, even the neo-colonial patchwork is falling apart, and the formerly subject nations are moving toward totally new relationships with the West. Hannah Arendt, the brilliant historical philosopher, remarking on "the swift decline in power of the United States" during the 1970s, saw in long perspective that "we may very well stand at one of those decisive turning points of history which separate whole eras from each other."

That period was marked by the full and often reckless use of the whole range of the earth's resources, many of them irreplaceable, in service to the West's growing technology and industrialism. But now for the first time the world is able to perceive that these resources are finite and to foresee the time when they could be used up entirely. Even the most moderate recent work by the Club of Rome suggests that "a new ethic in the use of material resources must be developed which will result in a style of life compatible with the oncoming age of scarcity."

That period was marked by a remarkably even plateau in the world economy, with prices for food, fuel, and clothing holding fairly steady except for a brief surge in the middle of the nineteenth century. But now over the last forty years or so prices have shot up along an unbroken inflationary incline and the economic stability we have known for centuries seems to be at an end. *Forbes* magazine, only one of the business press to have drawn attention to this process, says this adds up to a modern "price revolution" and adds, "The last time anything like *that* happened, and on such a scale as this, was in the 16th century."

That period was marked by the ascendance of certain agreed-upon perceptions and philosophies, at least in the West—including a belief in progress, the work ethic, materialism, romantic love, patriarchy, static sex roles, social and economic hierarchies, Judeo-Christian morals, and patriotism. But now all of these values are being questioned and challenged in greater or lesser degree in every country, and for many, particularly among the younger generations, they have already lost their hold. Gordon Rattray Taylor, the perceptive British science writer, is among those who have seen that this "reconsideration of basic assumptions is underway in many spheres—domestic, economic, industrial, political, and religious," and as he sees it, "the Western world is engaged in a massive rethink which cannot fail to prove a turning point in its development."

And that period was marked by the hegemony of the cultural ideas of the Renaissance and their expression throughout all European artforms. But now there are no more common assumptions or standards in any of the arts, the cultural precepts of the past are no longer shared either among or within disciplines, and the aesthetics of the Euro-American tradition are generally regarded as having lost their hold. Jacques Barzun, the scholar who was one of the most intelligent defenders of this dying culture, was not alone in believing that "what we are witnessing in all the arts, and in all that the arts refer to, is the liquidation of 500 years of civilization—the entire modern age dating from the Renaissance," and at the end of his life he concluded that we were at the point where "every act and all its opposites converge to bring about the leveling of the ground, the wiping clean of the cultural slate."

There is no mistaking it: the evidence comes from too many directions, too many thoughtful people of all kinds of political persuasions in all sorts of disciplines—and I am citing here only the most cautious and sober observers, not the professional doom-sayers or the overexcited journalists. Human civilization, particularly that of the West and more particularly still that of the United States, is at a momentous turning point. It is not in simply one or two dimensions that our world is changing, but in all of them, and synergistically. It seems clear that future historians will mark a new age beginning somewhere within our lifetimes.

THE QUESTION THEN IS: what kind of a new age will it be?

There are, in truth, only two answers to that.

It could be an age of bigness, continuing certain obvious trends of the present toward large-scale institutions, multinational corporations, centralized governments, high-technology machinery, large cities, high-rise buildings, luxury cars, and all that is implied in the American ethic of unimpeded growth. That would seem to have to entail the expansion of the present corporate-governmental alliance, leading to a fully mixed system of state and private capitalism, government regulation of scarce resources, increased corporate conglomeration, some greater degree of social regulation by the organs of government, further consolidation of political power within the executive branch, and corporate-governmental encouragement of the arts. Allowing, as always, for a few pockets of discontent, big would be better, progress our most important product.

Essential to this future is a belief in what I would call *technofix:* that is, that our present crises can be solved, or at least meliorated, by the application of modern technology and its attendant concentrations of science, government, and capital. "We must pursue the idea," former Atomic Energy Commission chairman Glenn T. Seaborg said not long ago, "that it is more science, better science, more wisely applied that is going to free us from [our] predicaments . . . and to set the underlying philosophy for a rationale for the future handling of our technological and social development." A shortage of oil? The technofix solution is to develop and perfect the processes for liquefaction of coal, and if that fuel ends up costing more money, the unimpeded workings of corporations in the marketplace will simply develop more efficient cars and furnaces that will require less of it. More food? All you have to do is run water into the Sahara, which could be done with large nuclear-power desalination plants along the Mediterranean and Atlantic coasts and develop a few new kinds of crops such as a single-cell protein that could be made into an all-purpose supercereal. Civil unrest? It has been seriously proposed that enormous geodesic domes be built over large American cities that have populations likely to riot or rebel, especially in summer, and in that way you could control the temperatures so that there would be no "long hot summers" and control the air so that you could create oxygen mixes to induce lethargy when needed. So, just as the nation has developed the technofix of Valium for anxiety, or chlorine for water pollution, or the Law Enforcement Assistance Administration for crime, so we can surely find technofixes for the other, admittedly difficult, problems of the future.

Of course, a serious commitment to the technofix future will require far greater concentrations of capital than at present, larger scientific and technological centers, bigger corporations to raise capital and sponsor research, and expanded national bureaucracies to coordinate the varied activities, particularly those on an international scale. And, many admit, it will doubtless require some measure of greater secrecy (as, for example, with the Manhattan Project) and hence greater security, some degree of independence of the technologists from interference by legislative or public bodies, and some kind of social and political controls to persuade citizens along the most efficient technological paths in spite of their doubts and fears and ignorance.

This sort of technofix world is a real possibility, particularly because it does not require much change in the psychological sets or economic systems of the present and of course because so many of the larger forces in American society—corporate, governmental, academic—are leading us in this direction. It would not surprise me all that much, if I should live to the next century, to find that this in fact is the future that we have chosen—or, rather, that we have been given.

Still, it is not such a comforting future, either, considering that it has been these larger forces—as we will see in the next section—that have brought us to the critical pass we now find ourselves in. There are some among us, and not a radical fringe either, who might feel that the facts that Valium is the most widely abused drug in the country, chlorine is reliably suspected of being a carcinogen, and the LEAA has spent $7 billion in ten years and the crime rate has *increased* do not inspire all that much confidence in technofixability. Not everything, we might point out, can always be solved by technology: the cost of all the elaborate techniques now used to try to control air pollution in this country comes to more than $95 million *a day*—and yet in most places the air is not getting appreciably better, and in some places it is getting steadily, unalterably worse. Nor can we always be sure of the effects technology will have: the technofix in the 1940s and 1950s for acne, tonsilitis, adenoids, and ringworm among children was high-dosage X-ray treatment, but now the National Cancer Institute says that may have given thyroid cancer to as many as 4 million people, one-third of those who were exposed.

I tend to feel somewhat skeptical about the unremitting optimism of the technofix people—their sanguinity tempered only by occasional flashes of judicious concern—no matter where they come from on the political spectrum. Herman Kahn, the Hudson Institute analyst on the conservative side, has suggested that the energy problem in a technofix America would be solved with nuclear power from ten large "nuclear reservations" scattered around the country, each a *thousand* times bigger than present sites. But since *present* nuclear plants take enormous amounts of public money to build and operate, pose problems of radiation leakage, thermal pollution, undisposable wastes, sabotage, and theft, have safety records that are spotty at best, and use incredibly complex systems that must be 100 percent "fail-safe," and yet—as Three Mile Island showed—can never be, it makes one reluctant to place a whole lot of trust in *future* plants that are supposed to be so much bigger. And Aurelio Peccei, the Club of Rome leader from the liberal side, has urged more "global cooperation" and regularized relations among governments to provide the international planning necessary for a rational technofix future. But the rather pathetic record of world cooperation (for example, on laws of the sea, arms proliferation, whale fishing), not to mention the dismal records of all kinds of governmental planning (for example, the Soviet five-year plans or housing development in the U.S.), suggests that this sort of solution may be the kind that ends up creating more problems. The Aswan Dam, after all, was a technofix.

It is hard to welcome any of the large-scale technofix ideas with very much enthusiasm, given their histories. The technofixers' optimism, even if it can in some way be justified, is a cold one—like British economist Wilfred Beckerman's jaunty prediction that "even though it may be impossible at present to mine to a depth of one mile at every point in the earth's crust, by the time we reach the year A.D. 100,000,000 I am sure we will think up something." Of considerable comfort for our posterity 2,500,000 generations hence, no doubt, but not the sort of news that makes those of us in the present sigh with much relief.

THE OTHER POSSIBILITY for the new age to which we are moving lies in exactly the opposite direction: toward the decentralization of institutions and the devolution of power, with the slow dismantling of all the large-scale systems that one way or another have created or perpetuated the current crises, and their replacement by smaller, more controllable, more efficient, people-sized units, rooted in local circumstances and guided by local citizens.

In short, the *human-scale* alternative.

In the search for the proper order of things and societies, a search that has inspired humankind since its earliest sentient days, no better guide has been found than the human form, no better measure than the human scale. "Man the measure"—has that not been the standard, or at least the goal, for the greatest number of human societies since they first began gathering collectively some 15,000 years ago? Was that not the explicitly stated principle of Pericles, of Leonardo, of Jefferson, of Corbusier, and hundreds of others of our most capable planners and thinkers? Indeed, is it not the essential spirit within Jewish mythology, Christian ethics, Anglo-Saxon law, Renaissance humanism, Protestant separatism, merchant capitalism, the French Revolution, the American Republic, Marxism, Darwinism, Freudianism, of so many of the basic currents—though by no means all, of course—of Western history? So in any search for a desirable future, for the ways in which tools, buildings, communities, cities, shops, offices, factories, meeting places, forums, and legislatures should be constructed, I see no reason to go beyond this simple rule: they should all be built to human scale.

"Human scale" is originally an architectural term, used to describe the components of a building in relation to the people who use it. A cottage door, for example, is of necessity built to human scale, high enough and wide enough so that a body can move through it comfortably, located at a place convenient for the body to use it, in some harmonious relation with the other elements of the building; a hangar door, by contrast, is not, for it has nothing to do with the human form, and it is outsized and disproportionate to the human body. From earliest times until quite recent eras, most conscious building has been a reflection of human scale, for in every society the measurements most convenient and most constant were those of the finger, the hand, the arm, the stride, and the height of the builder—a tradition we honor today in the English system, in which an inch is based on the length of the first joint of the thumb, the foot on the length of the forearm, and the yard on the length of a normal pace or an extended arm from fingertip to nose.[1] Even buildings intended to evoke awe and inspiration, such as the Parthenon or Peking's Temple of Heaven, were, when successful, built on these human measures and in proportions that took account of the human body.

But the idea of human scale can also be used to govern the design of communities and towns, indeed of whole cities. It means buildings that can be easily taken in by the human eye, in harmonious relations that do not engulf or dwarf the individual; streets that can be comfortably walked, parks and arenas for habitual human contact, places for work and play and sleep within easy distance of each other; the natural world brought into daily life, with grass and trees and flowers in every part, open spaces to experience scenery by day and the stars by night, woods and farms and grazing grounds somewhere within walking distance. And all of this of such a size as can be comprehended by a single individual, known at least by acquaintance to all others, where the problems of life are thus kept to manageable proportions, and where security is the natural outcome of association. Cities, too, with their overlays of urbanity, can arise from an amalgam of such communities, with interlocking networks and cross-neighborhood relationships of all kinds, providing only that the cities themselves do not lose the human scale, either in their buildings or their total size, and do not smother their separate parts.

And if buildings and cities to human scale, then surely it is not so difficult to imagine all the other aspects of human life, by extension, governed by the same principle. Social arrangements, economic conditions, and political structures could all be designed so that individuals can take in their experience whole and coherently, relate with other

---

1. A tradition that is being threatened by current attempts to get the United States to adopt a metric system, based not on anything human at all but on a measure that the French Convention, in 1799, chose quite arbitrarily by taking 1/40-millionth of the meridian of the earth.

people freely and honestly, comprehend all that goes on in their working and civic lives, share in the decisions that make it all function, and not feel intimidated or impotent in the face of large hidden forces beyond their control or reckoning. The same sense of security and self-worth that a person inevitably feels within an effective community, the family member can feel within the home, the worker on the job, the citizen at the town meeting, and all for precisely parallel reasons. What it takes is a scale at which one can feel a degree of *control* over the processes of life, at which individuals become neighbors and lovers instead of just acquaintances and ciphers, makers and creators instead of just users and consumers, participants and protagonists instead of just voters and taxpayers.

That scale is the human scale.

This alternative kind of new age would certainly not be without its problems, some considerable, and would likely face crises of its own in the course of its development—a development that, even in the best of circumstances, would take place over several decades. But at a minimum it does suggest something in the way of obvious relief from the imperilment brought on by the large-scale institutions of the present. For after all, such an age would not be congenial to centralized bureaucracies and high-energy machinery and corporate conglomerates, it would not require multi-billion-dollar investments in nuclear-power plants or military gadgetry. It would not have need for the 8 million polluting cars turned out every year, or complicated electric grids covering the entire continent, or large standing armies, or cities bulging with two, three, seven million people. It would survive without the military-industrial complex, the agribusiness giants, the real-estate speculators. It would eliminate the convoluted systems whereby, at present, the citizens of New York City are governed by 1,487 different governments, agencies, and boards, and the citizens of California pay an accumulation of 454 taxes on a single loaf of bread.

This alternative future is somewhat less likely than the technofix one, I should think, since it calls for somewhat broader changes over time and for the dislocation of powers that, despite being caught in recurrent double binds, retain considerable momentum. But it is by no means a utopian pipe-dream, and there are many reasons to imagine its coming about.

It accords with some of the deepest instincts of the human animal, possibly encoded in our DNA, such as the drive for individual expression, for tribal and community sustenance, for harmony with the natural world of stars and trees and songbirds, for companionship and cooperation. It accords with the experience of by far the greatest part of human history, from the earliest settlements right down to many parts of the world today, during which people lived in compact villages and self-contained towns and cities, crafting and farming for themselves, knit to other settlements through travel and trade without any sacrifice of sovereignty. And it accords with much that is rooted in the American experience, such as the anti-authoritarian beliefs of the Pilgrim settlers, the traditions of cooperation and self-sufficiency that grew up in the early towns, the town-meeting democracies that extended at one time from New England to Virginia, the rural and agrarian values among the Founding Fathers, their suspicion of authority and centralism, and the tenets of individualism that for generations drove people from the cities to the frontier.

Moreover, the human-scale future can take advantage of current technology, which for the first time in human development permits the creation of smaller units without sacrificing any of the clear benefits of modern living—in fact with healthier and more congenial settings, one could reasonably expect, and with greater access to material comforts for more people. Provided it was kept on a small scale and made simple, safe, and controllable, modern technology could be used without all the perils that accompany it on a large scale; though both are the products of contemporary advances, there is a qualitative difference between a desktop calculator and the space shuttle, or between solar collectors and nuclear plants. This allows for the first time, for example, an answer to the nagging question that has been thrown up to proponents of egalitarian societies since the time of Plato—*But who will collect the garbage?*—the implication being that the trash-removers are always accorded the lowest ranks of any society. Simple. Putting already developed technology to appropriate use, it is now possible to have community garbage-collection systems that channel wastes from homes and shops on pipelines feeding into a community recycling center, where they may be automatically sorted according to type and sent to separate compartments to be processed by simple machines and made available again in the form of usable raw materials. And not only is that an egalitarian solution, it is a communitarian and ecological one as well.

Finally, the evidence continues to mount that such a human-scale future is, at least in its major aspects, *proven.* Models for almost every part of such a future already exist now, or have existed in the very recent past, in many different nations of the world, including our own: the worker-owned plywood companies of the Pacific Northwest; the ecologically sound and solar-powered community of Davis, California; the consensual democracy practiced by Quaker meetings across the country; the worker self-management schemes of Yugoslavian factories; the non-hierarchical societies of Eastern Africa; the self-sufficient intentional communities, from Twin Oaks, Virginia, to Cerro Gordo, Oregon; the generations-old communes of the Jews in Israel, the Amana colonies in Iowa, the Bruderhof in New York, and others; the direct democracy of various New England communities; the non-authoritarian work programs of many American corporations; the cooperative movements of the United States, Canada, and Britain; and countless other pieces of evidence that we shall be exploring in later sections. None of these models is perfect, by any means; some are deeply flawed, and none of them exhibits all the elements of a human-scale life as it might be in optimum form. Yet taken together, the successful innovation of this one with the most durable practice of that, they show that all the elements of such a life, far from being utopian, are practical and possible, should we wish to pursue them.

Inuit ("Eskimo") children are given a puzzle at a fairly early age that most of them have no difficulty in solving after a few minutes: given a square of nine dots, how can you connect all the dots with only four straight lines, never taking your pencil off the paper?

It is a problem that the most sophisticated children in other parts of the world have failed to solve, and it stumps most adults as well. They are accustomed by their culture to certain ways of thinking that are difficult to break out of, and they do not have the same sense of space that Inuit children naturally come by. But in fact new ways of thinking may be perfectly logical and efficient, once you can get your mind around them, no more fantastic or impractical than the idea of extending the straight lines *beyond* the nine-dot square:

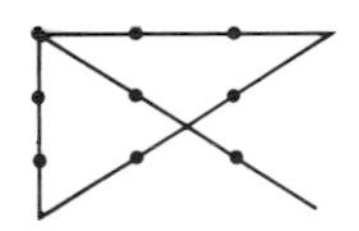

There is much in the human-scale alternative that at first seems strange, romantic maybe, visionary and impractical. But that is largely because our culture has conditioned us in myriad ways over the last several generations to thinking of certain kinds of solutions and to disregarding—in fact to not even seeing—others. But they are there.

THERE IS NO doubt that the next twenty-five years or so will bring major changes to our present systems, our present styles of life. Whether doomsayer or professional optimist, most of our contemporary thinkers seem to agree that the American people, indeed all Western peoples, are going to have to make thoroughgoing readjustments, in one direction or the other.

Our choices, then, would seem to be clear: large scale or human scale, continual technofix or ecological balance, governance by bureaucracy or governance by democracy, increasing authoritarianism or increasing libertarianism, complex mass systems or individual self-sufficiency, opportunities for chaos or opportunities for community. I hope that the course of this book will convince the fair-minded reader that there is only one rational selection between them.

4

Trend

UNTIL 1972, BOB LIGHT, then 30, and his wife Lee, 28, lived a rather ordinary suburban life in Upper Saddle River, New Jersey, one of those plush redwood-deck-and-blacktop-driveway places outside of New York City. He worked for his father in a successful textile machinery plant, she did freelance work, together they raised two children, and much of their free time they spent in the chic nightspots of New York—all the stuff of statistical tables and Hollywood movies. Then, that year, they sold their split-level house and moved to a 25-acre farm just outside the tiny village of Plainfield, Vermont.

The Lights put most of their life savings into restoring the 150-year-old farmhouse and broken-down barn, with the help of neighbors with whom they would swap work-shifts. Around the wood stove in winter they would plan for the summer, calculate what lands to clear, what crops to grow, what repairs to make, and then, with the first thaw of spring, working from dawn to sunset, they would set about making those winter dreams a reality—or nearly. They sought out the locals for help on how to make a go of the place, carefully amassed some cattle, a few hogs, some chickens, turned several acres over to corn for the livestock, established a vegetable garden in back of the house, and planted a field of strawberries and some fruit trees. Within a few years they were able to raise 80 percent of the food they needed to live on, and with a neat sign on their lawn reading, "Milk, Butter, Eggs" they were able to sell some of their dairy products for an extra $3,000 a year. Bob Light figures that with another few thousand dollars, maybe $7,000 or so from the milk-butter-eggs trade, he and Lee will be fully separated from most of the market economy and be able to live within a network of their friends and neighbors.

"What we want," Bob Light says, "is self-sufficiency—to make and produce as much as we can so we have to have as few dollars as possible. We don't want to support the system. We don't want to support the factories belching out the smoke. It's not our goal to make money."

The Lights, of course, are not "typical" of anything; among other things, they started with some savings and some money from the sale of their New Jersey house and have been able to put $60,000 or so into buying and fixing up their property. Nor is there any suggestion of one-swallowism; the Lights are part of a notable new development in America, but they are as yet only a tiny minority in the American landscape compared to all those at workaday jobs and familiar professions.

Still, there is no gainsaying the importance of what they represent. The *New York Times* has described their transformation as one part of "a full-scale back-to-the-land movement" that "has attracted tens of thousands of people across the United States," people who aim "to find ways to live simply but well on the land, outside the economic institutions that dominate the United States . . . rely on their own personal resources and labor, especially for their food and shelter," using "more wind and solar power" and "less machinery, less technology, less everything that comes from and depends on big business." They are a movement "deeply antagonistic to the American economic system, whose adherents it sees as controlled by unbridled corporate power, corrupted with surfeit and crazed by an impulse to consume and throw away more and more faster and faster."

To have "tens of thousands" of Americans in such a sweeping movement is consequential enough, but the back-to-the-landers are in turn part of a major new population shift that has taken the demographers by surprise and may signal something of a transformation in America's basic living styles. In the last decade there has been an abrupt reversal in the pattern of urbanization that has characterized the United States since the 1820s, and now more Americans are moving *out* of the big cities than *into* them—in fact, for every 100 that moved into metropolitan areas in the 1970–75 period, according to a Rand Corporation study, 131 moved out. For the first time in decades, small towns under 2,500 people are growing at a faster rate than any others, and rural areas all over the country, from Vermont to Arkansas, Oregon to Texas, are experiencing unprecedented population influxes.

And *that,* in turn, is part of a still larger demographic movement that has taken place over the last thirty years, the migrations out of the big cities into the suburban and exurban areas and, in a parallel move, out of the older cities into the warmer and younger regions of the Sunbelt.

All of these demographic changes reflect what I suspect is that innate desire for a little clean air, some green grass, a home one could call one's own, a town where a family could sink its roots, and a bit of room to move around in. They add up to an extraordinary phenomenon of people "voting with their feet" for a more natural world and a more communitarian setting than they could find in the choking and overgrown cities, a phenomenon of which Bob and Lee Light are only the most recent and most dramatic examples.

THERE IS AN unmistakable trend in American society—indeed in much of the Western world—that has been gaining momentum at least since World War II but most particularly in the last few years, the trend, one may call it broadly, toward human-scale values. I don't for one minute suggest that this trend will necessarily prevail. The forces arrayed against it are obviously considerable, and it runs smack against the entrenched messages of our culture that say big is better. But I do suggest that this is something quite undeniable, something that seems to be growing year after year pretty much on its own, something that anyone concerned with our future must reckon with. Indeed, one would not have any substantive reason to believe in the possibility of a human-scale future of any kind if there was not already a very real, very significant trend in that direction.

Examine just these few items:

▪ At least 18 million adults now are allied to the alternate religions and spiritual movements of the last fifteen years that Theodore Roszak—who has managed to enumerate no fewer than 140 of them—says make up "the biggest introspective binge any society has undergone." A *New York Times* survey in 1977 indicated these movements "are gaining a foothold for an enduring future" and "appear to represent serious challenges to Western thought that have caused many people to turn away from material gain, competition and the success ethic."

▪ The consumer movement that began with a few cranky protesters in the 1960s has blossomed into a full-scale political force of, according to pollster Louis Harris, "staggeringly" large proportions. Interestingly, his polls have found, Americans do not believe that business can police itself or that government can regulate it, and in overwhelming numbers they opt for some kind of grass-roots regulatory powers.

▪ Patrick Caddell, the professional pollster and one of Jimmy Carter's closest advisors, explained the President's "moral-equivalent-of-war" campaign against energy waste in 1977 by saying, "The idea that big is bad and that there is something good to smallness is something that the country has come to accept much more today than it did ten years ago. This has been one of the biggest changes in America over the past decade."

▪ The back-to-nature spirit in America has become so intense that it has outpaced the leisure boom of which it is a part: visits to national parks have grown from 80 million in 1960 to 283 million in 1978, an increase three times greater than that in general sports attendance in those years. There are estimated to be 8.6 million backpackers and 7.7 million wilderness campers in the U.S. today.

▪ Individuals operating on their own outside of the standard market economy, doing business by barter and cold cash and keeping free of the IRS, have created an "underground economy" that by the late 1970s was thought to amount to hundreds of *billions* of dollars. According to *U.S. News,* as many as 4.5 million Americans get all their income from this hidden, non-corporate economy and 15 million more get some of their support from it.

▪ New York City, which drove itself to near bankruptcy through reckless giantism, gave itself a new decentralized City Charter in 1977 designed (according to the *New York Times)* to provide "a sharply increased role for local communities in the conduct of city government and the gradual assumption of responsibility for the delivery of most services." (The performance may not live up to this promise—that would be par for New York—but the direction of the reform is important in itself.)

▪ At least 71 million Americans belong to one kind of cooperative or another, obtaining credit, health care, food, housing, car repairs, electric power, agricultural supplies, and other goods and services outside the profit-centered business system; credit unions have grown from 17 million members in 1965 to 32 million in 1977, health co-ops from 3.4 million to 6.1 million, electricity co-ops from 4.9 million to 6.6 million. The food co-op movement has burgeoned so much in so many places that there is no way to be sure just how many various outlets there are in existence at any given moment, but the latest figure contrived by a food co-op newspaper (yes, they have their own newspaper) is 10,000, which works out to more than two in every city or town over 5,000 across the land.

▪ Some 32 million American households are estimated to do backyard and city-lot gardening nowadays, perhaps twice as many as ever before, producing goods reckoned to be worth more than $13 billion on the market. "There has been nothing like it since the patriotic Victory Gardens of World War II," says the *New York Times.*

▪ It is impossible to measure the total number of people who have turned their backs on traditional American society over the last decade or so, but the Stanford Research Institute in 1977 reported that 4 to 5 million adults have actively "dropped out" into a life of—as they so quaintly put it—"voluntary simplicity."

▪ "Do-it-yourself," according to the Bureau of Building Marketing Research, had become a $16.5-billion-a-year business by 1977, a 200 percent increase from 1974.

Very disparate, all these nuggets, but I don't have to emphasize that they are all pointing in a single general direction, toward individual self-worth, community cooperation, harmony with nature, decentralization of power, self-sufficiency—the human-scale values.

Indeed, I think we can find this same direction in the progress of every significant movement for social and political change of the past two decades. The civil-rights and black-power movements were searching for self-identity and pride, community power, equality of opportunity, and meaningful participation; so too the later "ethnic revivals" among descendants of European immigrants. The women's movement was—and is—a movement for individual self-worth, for communality among women, dignity and equality in the workplace, and full participation in political and economic matters; similarly for the homosexual-rights movements. The New Left was an expression of a whole generation's demand that the individual "share in those social decisions determining the quality and direction of his life" (as SDS's *Port Huron Statement* put it) and that power be redistributed from centralized systems to the grass roots; it grew—or at least the healthier parts of it grew—into a movement in the 1970s that rooted itself in communities, worked on community organizing and development, became concerned with local rather than national issues, and is found today working in alternative local organizations and institutions of all kinds. The counterculture that began in that same era was a movement in the main for individual expression, sexual freedom, and communal living, and proclaimed a denial of old standards like the work ethic, Puritanism, success, materialism, the profit motive, and authority. The consumer movement, though begun as a simple protest against high prices and shoddy goods, developed into a broad demand for consumer participation, individual rights, an end to corporate exploitation, and in recent years has set itself explicitly against the large monopolistic corporations and the government agencies that perform as their handmaidens. And, finally, the environmental and alternate-technology movements have been impelled by a concern for the natural world and its despoliation, for the proper role of humans within the ecosystem, and for the preservation of animal and plant species, which has broadened out in recent years to include alternative sources of energy, simple living, intentional ecological communities, natural foods, self-sufficiency, community health and communications services, and the like.

BUT I THINK it is possible to locate even broader manifestations of the inchoate trend toward a human-scale future, profounder currents so large and slow as to be almost imperceptible in our daily lives, yet of the sort that future historians may well mark as the dominant characteristics of our time. Which citizens of fifth-century Athens actually realized they were living in the Golden Age of Greece? How many Europeans of the twelfth century thought they were in the Dark Ages? Did the people of the nineteenth century know they were going through something that would be the Industrial Revolution? Just so, how many of us ever think to stop and measure the subcutaneous pulses that will be read as the heartbeat of this era by ages still to come? Yet at the risk of presumption, and certainly of oversimplification, let me try to isolate a few of these pulses that seem most particularly to define this period.

Feminism. In the broad view of history, at no time since the Romantic Age at the turn of the nineteenth century have the values associated with feminism in the largest sense been so triumphant. I don't mean simply the rise of the women's movement and its quick political and economic consequences all across this country, important though those have been; nor just the changing ideas in many sectors about women's roles, child-rearing, the patriarchal family, sexual stereotypes, and all that we now mean by sexism, radical though those have been. I mean the ascendance of what have been regarded, in our culture, as the womanly virtues: spontaneity, permissiveness, sexuality, emotion, softness, cooperation, lovingness, participation, as against, say, analysis, authoritarianism, denial, reason, hardness, competition, sternness, elitism, what might be called the manly virtues. Obviously both of these sets of elements play a part in our current society, as in all societies, but who could look at the current revolt against and distrust of authority, the free expression of sex in all ages and quarters of the population, the rise of communal and communitarian arrangements, the growing permissiveness in homes, schools, offices, even boot camps, the upsurge in the nurturing skills of gardening and husbandry, the relative status of the scholar as against the soldier, the growing acceptance of homosexuality, the rise of pacifist, anti-war, and anti-nuclear sentiments, the revival of fundamentalist and transcendental religions—who could look at these alone and deny the ascendance of the feminine virtues over the last decades? Indeed, if masculine cultures are associated with sky-gods (as in ancient Egypt, Judaism, Christianity) and feminine cultures with earth-gods (Phoenicia, Greece, Phrygia), the present widespread interest in ecological and environmental matters suggests that we may be experiencing a profound feminist alteration in our culture.

Naturalism. This new concern for the natural world, the unprecedented concern for everything from renewable resources to endangered species, may be a very deep biological reaction to the damage we are inflicting on our environment, a primordial human impulse toward self-preservation. But it is an extraordinary phenomenon, whatever the roots, and it finds expression just now in every Western nation, among all segments of society. In the U.S. the most dramatic manifestation has been the emergence, and the clout, of the environmental movement, which, as former *New York Times* columnist William Shannon has noted, "is one of the few big popular movements that continues to enlist volunteers, excite idealism, and evoke steadfast and unselfish commitment . . . in every region and virtually every community." Along with it has come the swift development of the alternate-technology crusade, with British economist E. F. Schumacher at its head and literally thousands of solar and compost-toilet entrepreneurs in his wake, and the closely allied drive against nuclear-power plants in almost every state of the union. Today the spirit of naturalism is seen in everything from the boom in health foods (one New England company in 1977 reported selling, it's hard to believe, twelve *tons* of granola a month), through the development of a $200-million-a-year houseplant business, never before known in our economy (it is now so big that it has inspired a plant-napping profession in the Sunbelt states where most of the greenery is grown), to the mania for backyard gardening, now carried to rooftops and corner lots (born no doubt of the desire to taste, before it is too late, a real, non-square, non-gassed, non-mealy tomato).

Localism. Part of the disintegration of national loyalties in the West—witness the sharp decline in patriotism in the United States—and the rise of separatist and regional movements—among American Indians, blacks, Inuits, Quebecois, Scots, Welsh, Basques—has to do with a resurgence of local interests and local allegiances that has probably not been so intense for more than a century. René Dubos, the eminent biologist who has been studying the human animal for some fifty years now, believes that "the most interesting and powerful force in our time is that people are getting more and more interested in regional and local affairs," and he has noted another remarkable fact: the American people's great mobility has produced an "extraordinary social trend—for the first time in human history a large and increasing number of people can *select* their place of residence" rather than simply take it as a matter of birth. This undoubtedly accounts for the sweeping population migrations noted above, and for the fierce passions that people tend to exhibit over the places they have selected to live. Hence the spontaneous growth of block associations and neighborhood associations over the last few years, of community organizations and communes and intentional communities; hence also, in a darker vein, the growth of neighborhood vigilante groups and "voluntary civilian patrols" in the 1970s as local responses to the increase in crime, a development favored, according to one survey, by at least 62 percent of urban dwellers. Dubos, again, makes the point that localism can only increase in the future: first, because of increased mobility; second, because of the drying up of shippable fuel sources (such as oil and gas) and the turn to unshippable local ones (solar, wind, wave); third, because of the need for societies, buildings, and technology to adapt to local conditions to harness local energy and preserve local ecosystems; and fourth, because of the need for greater regional self-sufficiency to avoid overdependence on foreign resources and to supply food for growing populations. All of this, he predicts, "will have enormous effects on the evolution of this country."

Populism. It is safe to say that not since the late nineteenth century have the goals and ideas broadly associated with populism been so pervasive as they are today. The current mood of disillusionment with big government, big business, and big labor to which the polls continually attest is reminiscent of nothing so much as that period when, as historian Mark Sullivan put it, "the average American had the feeling he was being 'put upon' by something he couldn't quite see or get his fingers on . . . he felt his freedom of action was being frustrated . . . his economic freedom and his capacity to direct his political liberty . . . was being circumscribed in a tightening ring." The classical populist response is the one we are witnessing now: the rise of alternative institutions, the growth of consumer and citizen protest, and the emergence of mainstream political activity aimed at reducing government interference rather than applying government remedies. The list of alternative organizations and networks that have grown up over the past fifteen years is almost unlimited, but for a start there are free schools, free clinics, storefront law offices, food co-ops and suppliers, community newspapers, counterculture magazines, alternative publishing houses, "underground" churches, independent credit unions, community day-care centers, communal houses, garage and tag sales, and health-food restaurants, and they are common now in every large city and most college towns, coast to coast. A similar people-oriented opposition to the established systems has produced the remarkable number of citizen-action groups on all fronts just in the last decade or so: Common Cause, the Sierra Club, Friends of the Earth, the National Organization for Women, Ralph Nader's Public Citizen and local Public Interest Research Groups, the National Association of Neighborhoods, the Peoples Business Commission, California's Campaign for Economic Democracy, the Congress of Racial Equality, and literally dozens more, plus another hundred or so statewide action groups organized around consumer protection, utility regulation, taxation, the environment, and other political issues. Taken altogether, these alternative groups have at least twice as many contributors as the two major political parties (about 1.4 million to 520,000), and they get even more money in donations and dues ($16.6 million to $13.6 million in 1976, a presidential election year), which gives a good nut-cutting idea of their relative importance in the eyes of the populace. Widespread support for certain mainstream politicians is part of this same impulse and accounts both for the strong anti-Washington, anti-Establishment trend in recent elections and for the victory of so many campaigners—most notably Jerry Brown in California, Tom Cahill in Oregon, Gary Hart in Colorado, David Pryor in Arkansas, Thomas Salmon in Vermont, and of course denim-wearing, down-home-sounding Jimmy Carter himself—who have taken a populist or quasi-populist line. To veteran Washington correspondent Tris Coffin, all this signals that "the U.S. is turning left to neo-populism"; to the conservative academic Daniel Bell it means that "the Populist revolt . . . has already begun at the very outset of the post-industrial society."

Individualism. Probably as a result of the proven inadequacy of so many governmental, corporate, and academic remedies, and possibly in biological response to the increasing pressures of depersonalization and homogenization in our society, Americans are asserting a new kind of individualism, a claim for self-identity and self-worth. It can be seen in the fashions of our times, the physical-fitness boom, personalized gifts and clothing, "self-assertiveness" training for women, do-it-yourselfers, individualistic styles of dancing and dress, at-home "gourmet" cooking, the search for "roots" and family trees, the demands of women to control their own bodies, even the demands of cancer victims for laetrile. It is there in the new religious movements, which, in the words of San Francisco State sociologist Jacob Needleman, "can no longer be taken as a transitory cultural aberration but rather as a central feature of the profound change through which the American civilization is now passing." And it is the wellspring of the contemporary currents toward "flextime," workplace democracy, self-management, worker participation in decision-making, employee-ownership plans, and the like, currents that (as we shall later see) are gradually changing the nature of many offices and factories in the industrialized world. In a larger sense, this spirit of individualism is what people like Daniel Bell have been denouncing for the last decade or so, a spirit in which "the self is taken as the touchstone of cultural judgments," the individual asserts claims above those of society, and large segments of the populace ask for equality, not simply of opportunity, but of results, with the redistribution of wealth and power from a few individuals to many. This "sociologizing mode," in Bell's phrase, replacing traditional economic values with a new set of social ones—not how much you make but how well you live—is today the dominant mode in all Western societies.

THESE FIVE PULSES of our era—feminism, naturalism, localism, populism, and individualism—must be reckoned, if not as the only defining characteristics of the present period, at least as among the most important. They are all durable, indeed all have existed in one form or another through most of history, now waxing, now on the wane, but never really extinguished, as if they described at least one set of priorities and passions basic to the human psyche. They are all connected, all part of some complicated reticular pattern—so the nurturing of feminism relates directly to the rootedness of localism, and both to the ecological harmony of naturalism—that expresses similar concerns in different but reinforcing ways. And they have all become resurgent in a fairly short span of time, roughly in the years since 1960, despite the obvious array of static and traditional styles that oppose them. That suggests to me that the trend they reflect is a real and powerful one and that it is in their direction that the entire societal pendulum may perhaps be swinging.

It is that which can give one some hard-nosed optimism about the potential, though hardly inevitable, emergence of a new period of human history in the not-too-distant future in which small communities and self-sustaining cities, locally rooted and ecologically sound, cooperatively managed and democratically based, developing new resources and technologies according to their new philosophies, might evolve a society built to the human scale.

It is said that his contemporaries laughed with scorn when Sophocles, as a young man, told them that Athens—just then engaged in a costly and protracted war with the Persian Empire—was about to enter into a Golden Age.

PART TWO

THE BURDEN OF BIGNESS

Everywhere Nature works true to scale, and everything has its proper size accordingly. Men and trees, birds and fishes, stars and star-systems, have their appropriate dimensions, and their more or less narrow range of absolute magnitudes.

D'ARCY WENTWORTH THOMPSON
On Growth and Form, 1915

On a small scale, everything becomes flexible, healthy, manageable, and delightful, even a baby's ferocious bite. On a large scale, on the other hand, everything becomes unstable and assumes the proportions of terror.

LEOPOLD KOHR
The Breakdown of Nations, 1957

Some rule of birds kills off the song
in any that begin to grow
much larger than a fist or so.
What happens as they move along
to power and size? Something goes wrong.

JOHN CIARDI
"The Size of Song," 1964

1

The Beanstalk Principle

IT OCCURRED TO ME one day, looking at the drawings in a book of fairy tales I was reading to one of my children, that the giant in "Jack and the Beanstalk" looked somehow more *fragile* than menacing, as if he weren't put together quite right. I sat down and made a few calculations. If the giant was, as he looked in the picture, about five times as big as Jack—five times taller, five times wider, five times thicker—then he would have to weigh not just five times as much but *five-to-the-third-power* times as much, five times in each direction—just as the volume of a box of five square feet is not just five times bigger than a box of one square foot, but five-to-the-third-power times as big, or 125 cubic feet. So if Jack weighed, say, 50 pounds, the giant would have to weigh 50 × 50 × 50 pounds, or 125,000 pounds—more than 60 tons. Naturally lugging around all that weight would be something of a problem, an immensely greater problem than the human form has been developed for. Particularly so because the giant's leg bones, though indeed they were 125 times as big as Jack's in all, were only 25 times as big in the dimensions where it counts in carrying weight, that is, in a cross-section through the bone, which takes in only width and breadth. In short, the giant would have been trying to support 125 times the weight on bones only 25 times as strong—like an average man trying to carry a ton of bricks—and he would have cracked his legs in two and fallen flat on his face if he were even to stand up, much less try to chase after Jack. No wonder he didn't seem very menacing.

Some time later I happened upon a forgotten essay by the British biologist J. B. S. Haldane entitled "On Being the Right Size," quite a delightful little exercise in comparative anatomy. And there, sure enough, Haldane had made the same sorts of computations I had, though he had figured them out from a copy of *Pilgrim's Progress* showing Giant Pope and Giant Pagan towering over tiny Christian, and had come to the same conclusions: "As the human thigh bone breaks under about ten times the human weight, Pope and Pagan would have broken their thighs every time they took a step. This is doubtless why they were sitting down in the picture I remember." But he went beyond Bunyan's morality tale to show the importance of size in the entire biological world.

Haldane shows, for example, that size matters when you consider the fate of various animals when they fall any distance. Since smaller animals tend to have greater surfaces proportional to their weight than larger ones, their air-resistance is greater; at the same time, having less weight, their gravitational impulse is smaller. So if a mouse were to fall from a ten-story building, it would be only slightly bruised, get up, and scamper away. A rat would be temporarily stunned and dazed, but probably not seriously hurt. A man would be killed, a horse splattered.

Size matters, too, he shows, in the form an animal takes. A gazelle, for example, could not become larger without putting so much weight on its skinny little legs that they would shatter—unless it adapted like two other members of the same general family: "It may make its legs short and thick, like the rhinoceros, so that every pound of weight has still about the same area of bone to support it. Or it can compress its body and stretch out its legs obliquely to gain stability like the giraffe." Similarly, a human being cannot ever be able to fly because he would need a totally different form to supply enough power to keep his body continually in the air. Thus the notion of an angel—or at least a human-shaped angel—is patently ridiculous: "An angel whose muscles developed no more power, weight for weight, than those of an eagle or a pigeon would require a breast projecting for about four feet to house the muscles engaged in working its wings, while to economize in weight, its legs would have to be reduced to mere stilts."

Haldane limits his brief inquiry to animals, but he is not unmindful of the broader conclusion, one with implications important for our own inquiry: "Just as there is a best size for every animal," he says, "so the same is true for every human institution." And, of course, giants.

THE POINT CAN BE amplified a hundred ways: we have only to look around.

A rose by any other name would no doubt smell as sweet, but not by any other size. Imagine a rose five feet in diameter, say, or as big as an elephant. If its perfume were exuded at a rate proportional to its size, it would overwhelm anyone who came within twenty feet of it, something like falling into a vat of Chanel No. 5. As for beauty, we wouldn't even begin to know how to measure that, but the thing would undoubtedly have more in common with a Times Square billboard than a garden flower.

A committee meeting, similarly, has—or rather should have—a certain optimum size. We all know what it is like to be in interminable meetings, and it generally happens that the greater the number of participants, the more talking, the longer the meeting—and usually the less decided. There are several effects of size here, but the basic one is that the number of signals between participants increases exponentially once you get past a certain very small number. For example:

Primus is seated at the committee-room table, next to Secundus. He has only one other person with whom to relate, one other source of signals, words, expressions, gestures, body language, all that goes into the communication by which to arrive at intelligent decisions, and the same is true for Secundus. When they are joined by Tertius, there are suddenly *nine* possible ways of sending and receiving signals—from Primus to Secundus or Tertius, from Secundus to Primus or Tertius, from Tertius to Primus or Secundus, and from each one of them to any other pair. (And of course any of these signals may be taking place simultaneously—as when Primus's proposal produces a grin from Secundus, who is at the same time signalling acute nervousness by drumming his fingers, and produces a scowl from Tertius, who at the same time is doodling dollar signs on a pad that only Secundus can see.) Now when Quadrius joins them, the elemental signals are multiplied again, this time to twenty-eight, since each participant then has three other individuals, three possible pairs, and one trio to relate to ($7 \times 4 = 28$). A fifth person continues the process, producing a total of seventy-five signals (four individuals, six pairs, four trios, and one quartet for each, $15 \times 5 = 75$), a sixth produces no fewer than 186, and so on (the formula is $N [2^{n-1} - 1] = x$)—and by the time ten people are sitting around the table, not an unusual number for a committee meeting, there are a total of 5,110 ways for all the participants to relate to each other:

2 people	2 signals
3 people	9 signals
4 people	28 signals
5 people	75 signals
6 people	186 signals
7 people	441 signals
8 people	1,016 signals
9 people	2,295 signals
10 people	5,110 signals

No wonder that a committee of ten takes so long to achieve anything at all, and that's usually only when most of the participants are so worn down that they cease to participate or give off any signals whatsoever. Committees, like roses and giants, should not get too big.

Or take a university. Obviously it is possible to build such an institution to any size one wants (as many of our present state universities seem determined to demonstrate), providing only that you make the buildings large enough and the amplification systems loud enough. There are university systems with a quarter of a million students in them, and single campus systems with 40,000 or 50,000, even 75,000 students. But just as obviously, there is a size beyond which real *teaching* and *learning* can no longer take place, except in isolated and happenstantial cases. A university of 77,000 students, say, taught by 1,200 professors (the figures happen to be those of Indiana University, though they are comparable to many in the U.S.) in which students daily go to class after class of 300 and 400 people and have only the most limited personal contact with their teachers—such a place might be said hardly to be a university at all, in the sense that knowledge is imparted and minds are developed and the higher learning is undertaken. As Cardinal Newman pointed out a century ago, long before institutions of this size were even imagined, "A university is an Alma Mater, knowing her children one by one, not a foundry, or a mint, or a treadmill."

No doubt there are certain "economies of scale" in these large places—after all, why have a high-paid English professor teach only six students when he could just as easily be saying the same things to 60 or 600 students for the same amount of money? And some tests indicate that learning can indeed take place irrespective of classroom size, providing it is measured by retention of facts rather than by growth, creativity, or intellectual innovation. As long as what matters is "efficiency"—giving the maximum number of degrees to the maximum number of students in the minimum amount of time and with the minimum amount of expense—then American universities have certainly performed their function over the last two decades. But if it is *education* that's wanted, or the creation of acquisitive and logical minds, or even the production of well-read and well-equipped individuals, then the large institutions that characterize this country may be said for the most part to have failed. "The growth of the campus in size," summarizes the definitive *Higher Education in America,* "inhibits the education of the individual."

It is of course true that contemporary "audiovisual mechanisms" and suchlike can sometimes be used to overcome some difficulties of size. But I wonder how useful in fact these tools can really be. A visiting professor at a large Eastern university a few years ago swears the following actually happened. Knowing he would be on a trip out of town during an upcoming class, a professor announced he would put his lecture on tape and have an assistant play the tape during the assigned period. His trip unexpectedly cancelled, the professor decided to go down to the lecture hall and see how many students were actually there listening. He was not prepared for the sight, on walking into the back of the room, of nothing but empty seats, not a single live student present. Instead, ranked all around the tape recorder on the table at the front were the students' *own* tape recorders, copying the lecture.

It's very simple. When the intimate teacher-student relationship is distended and broken, the student can no longer be individually probed and guided and nurtured and stretched, nor does the teacher have much incentive to do so. Studies confirm the point. One project in the 1960s found "a negative relationship between size, and individual participation, involvement, and satisfaction": "As schools increase in size, the number of persons increases much faster than either the number of learning settings or the varieties of settings. . . . Students in small schools were involved in more activities than those in large schools and had more satisfying experiences related to developing competence, being challenged, and engaging in important activities." Alison R. Bernstein, one of the few educators to have confronted this issue, similarly has concluded that "the massive size of a great many institutions stifles not only student development but the entire process of innovation." "The process of change itself is often hindered at large institutions," she found, and even communication, the very basis of an education system, becomes "a massive problem in logistics." And the faculty is generally "more cynical" at large universities, senses its powerlessness, and feels little sense of loyalty, and at big institutions "the frustrations, cynicism, and alienation of the faculty are reflected in the students."

THE SIMPLE CONCLUSION: size matters, in human institutions as well as human forms, and it has its limits. We may formulate this more precisely as a principle—let's call it the Beanstalk Principle in honor of the medium that after all brought Jack and the Giant to the point of comparison—that holds:

For every animal, object, institution, or system, there is an optimal limit beyond which it ought not to grow.

To which might be added the Beanstalk Corollary:

Beyond this optimal size, all other elements of an animal, object, institution, or system will be affected adversely.

I am not saying that size—and its companion notion of scale, or sizes in relation to each other—is the *only* measurement to make in judging something. But it does make sense that it should be the first, and the central, consideration, inasmuch as it is likely to affect, in one degree or another, all other considerations.

As the biologist knows. If an earthworm were ten times bigger, its weight would be a thousand times greater and its need for air a thousand times greater, but the surface area through which it absorbs oxygen would be only a hundred times greater, so it would get only a tenth of the air it needed and would immediately die.

As the architect knows. To build a skyscraper 110 stories tall, you have to make special allowances for sinkage, wind stress, tensile strength, and movement, and work out special problems of heating and cooling, lighting, plumbing, transportation, congestion, and maintenance that are completely different from those when you are doing a building of two or three stories.

As the city planner knows. Doubling a city's population means vastly increasing its area—since as a population increases arithmetically its space tends to increase geometrically—and it means completely reorganizing its systems—since as population grows, urban services become increasingly complex, interdependent, rigid, and vulnerable, while levels of citizen participation and bureaucratic job performance decline.

As we all know. A big mansion is not simply a bungalow with more rooms, a big party is not simply an intimate dinner with more people, a big metropolitan hospital is not simply a clinic with more beds and more doctors, a big corporation is not simply a family firm with more employees and products, a big government is not simply a town council with more branches.

Size, indeed, might well be regarded as *the* crucial variable in anything. More important than, say, ideology—for a large disciplined party like the Communist Party in the Soviet Union is no better than a large undisciplined one like the Christian Democrats in Italy; both are unwieldy, unrepresentative, undemocratic, and inefficient, not because of their politics but because of their size. More important than, say, public or private governance—a large public system like the Tennessee Valley Authority is just as autocratic and impersonal as a large private one like AT&T, the Post Office (in either public or quasi-public form) is as inefficient as Penn Central, and the American Stock Exchange is just as chaotic as the Philadelphia school system; all fail to the degree that they are oversized. More important even than wealth—an enormously wealthy company like General Motors is as inept at regularly making good cars as a poor and tottering one like British Leyland, and the wonderfully affluent city of Houston is no better at halting the pollution of its air than a tattered and impoverished city like Newark; they are all of such a size that they can no longer keep control over their disparate parts.

The lesson would seem to be, as we shall see in the course of this section, that *size governs.* That is what the Beanstalk Principle is all about.

THE BEANSTALK PRINCIPLE has one particularly poignant exemplar: the dinosaur. As the various dinosaur species grew larger and larger, mostly to permit them to forage higher and to store their food longer, they grew increasingly incapable of dealing with their world. They were unable to regulate their own internal conditions—they were, presumably, cold-blooded and hence sluggish at night, when small warm-blooded predators were on the prowl; and they were unable to regulate their external conditions—particularly defenses against a species of small, hairy, shrew-like animals that came along and fed upon their unattended eggs. Eventually the dinosaurs died out, giving way to animals of more moderate size, marsupials and primitive mammals, superior because they did not—not even, later on, the elephant or the whale—grow to a size beyond which they lost control.

2

The Condition of Bigness

BIGNESS IS PERVASIVE. It is as much a part of the American system, the American way of thinking, and that part of the world that America has touched as the capillaries are a part of the vascular system. Consider:

▪ In the U.S. there are eleven different sizes of olives. The smallest size is "jumbo."

▪ The producer of Easter extravaganzas at New York's Radio City Music Hall has declared his theater to be the "greatest theater in the history of the world." Not because it has produced any art of distinction, not because it has nurtured a culture in the way that the Dionysian theater of Athens or the Globe Theatre of London did—but for the single reason that it has had a bigger cumulative audience, some 250 million people, than any other previous theatrical building.

▪ Big cars have been so quintessentially American that, even after successive gasoline shortages and continuing price rises, they continue to make up well over half of the American automobile market. For three decades Americans bought nothing but the biggest models Detroit could turn out, even though it was clear enough that this ran counter both to the public interest—because the two-ton machines are so wasteful of fuel (in one week of summer travel American cars use as much gasoline as all of the U.S. Armed Forces in 1944)—and to private interests—because they cost about a quarter of the average family income to buy and maintain. Even when Detroit was forced by the government to cut back on its mastodon production, it was still able to sell plenty of outsized, gas-eating vehicles to people like John E. Fix of Blaine, Washington, who bought five huge Eldorado convertibles and built a special garage for them just because "I love them big cars": "Sometimes," he says, "I just go in the garage and turn the lights on and just sit there and just look at 'em."

▪ The U.S. manufactures the largest newspapers in the world. The

New York Times, the largest of them all, for an average Sunday in November, *any* November, weighs about 10 pounds, but that seems to feel so right that most New Yorkers lug the whole thing home before they throw out the sections they don't intend to read. It takes 850 acres of Canadian trees to produce one average Sunday's edition, more than it would take to build 100 three-bedroom ranchhouses. It also costs New York City ten cents a copy in Sanitation Department expenses just to pick up the littered copies on Monday morning.

▪ In a nation that is noted for massive public-works projects, the Albany Mall is said to be the largest ever built: it consists of a bunch of odd-shaped ponderosities flanked by two quite gargantuan high-rise spires that so overpower the city they can be seen from 15 miles away. The project was scheduled to cost $450 million, which would have made it the most expensive public complex in New York history, and ended up costing $2 billion, making it the largest and most expensive state project in the world, and everyone in the state took that quite in stride.

▪ Texans, as everyone knows, typify American bigness. As John Bainbridge explains it in his account of *The Super-Americans:*

> It really means something to Texans that their San Jacinto Monument, just outside Houston, is a little higher than the Washington Monument and that their Capitol, in Austin, has a similar edge on the Capitol in Washington. They also take satisfaction in knowing that of all the states Texas has the most farms, the most churches per capita, the biggest state fair, the most airports, the most insurance companies, the most species of birds, the most banks, the most football teams, and the most holidays, among other things.

And since that time they have acquired the largest single-story convention hall in the country, the largest artificial seaport, and the largest airport.

▪ American advertising expresses its most important messages in superlatives about size: the biggest shopping and travel card . . . the nation's largest airline with the world's biggest fleet of widebodies . . . the biggest sale we've ever held . . . once you've had a taste of it, the only thing you'll want is more, more, more . . . the largest condominiums in Florida . . . the most spacious rooms in Boston . . . come to the widest beach in Aruba . . . we're the biggest little bank in Milwaukee . . . we're growing bigger every day . . . a growing concern for the nation's growing needs . . . everything you always wanted—and more . . . Big Macs, Whoppers, Jumbo Cakes, Green Giant . . . king-size, super-longs, extra-large. . . .

AND IT'S NOT simply that we are continually immersed in bigness in this country, we are conditioned to accept it as inevitable. I well remember my father once telling me when I was in high school that *progress*—which in contemporary culture is equated with growth, and growth with bigness—was not the inevitable fact of humankind, an automatic outcome of time and human endeavor, but rather just an invented concept, *one way* of thinking about and ordering the processes of history. I had no idea what he meant. Try as I might, I could not imagine another way of seeing the world except that it got better, gradually, every day, that there were newer and better inventions, more and better products, and that was progress. Look, I said, look at what we have today that they didn't have fifty years ago—for example, *cars,* isn't that a sign of progress? My father just smiled and said he wouldn't be so sure.

He was patient. A few years later he happened to mention to me that one of his colleagues, the historian Carl Becker, had written an essay on progress for the *Encyclopedia of Social Sciences,* showing that the concept was not eternal at all and was in fact mostly a product of the modern era and the Euro-American culture. I remained baffled, but I dutifully went to the library some weeks later to see what he was talking about, and I read the whole essay through, slowly, twice. I still didn't understand it. Can't this man see, I said, that all people, all cultures, are inevitably better off than their predecessors because they learn from them and improve upon their ways?

It was really not for another ten years or so that I began to get a glimmering of what my father had been trying to say, and one day in the library I took the opportunity to sit down and read that essay again. Suddenly I understood: history is not necessarily a straight-line inclination from less to better, as the theory of progress would have it. One could just as well think of history as a fluctuating curve, or peaks and valleys, or even as a wheel or a Möbius strip, in which some things are better at one time and worse at another. Progress is not the necessary path of the human animal but simply a contemporary, and essentially materialistic, way of redescribing the events of the past so as to make the present seem superior.

In the next instant there flashed in my mind a picture of my father's patient smile and the realization that cars were hardly much evidence of "progress," given the fact that they cause pollution, disease, traffic jams, sprawl, stress, fuel consumption, resource depletion, indebtedness, jealousy, and 50,000 deaths and 5 million injuries a year. But so ingrained in me had the standard American notion of progress been that I was well into my adulthood before I could even begin to question it.

The natural result of this sort of conditioning—and everyone has it to a greater or lesser degree—is that it has become part of the American character not only to accept bigness but actually to admire, respect, love at times even worship bigness. Size is the measure of excellence: in cars, tomatoes, cigarettes, houses, breasts, audiences, salaries, freeways, skyscrapers, muscles, children, penises, and fish. Where but in America could a con man like P. T. Barnum actually get people to pay money to see someone he billed as "The Biggest Midget in America" (a perfectly normal-sized man, as you found out after you paid your nickel)? Or could people successfully sell a cigarette called simply "More" or a tennis racket called "The Giant"? Or, in a national art form known as the tall story, could a giant lumberman with a big blue ox be considered an admirable folk hero? Every outsider from Tocqueville to Jean-François Revel has noted this peculiarity in the American psyche; the French writer Raoul de Roussy de Sales was bemused by the whole thing:

> When I listen to Americans talking on shipboard, or in a Paris restaurant, or here in New York, it is only a question of time before someone will come out with that favorite boast of yours—"the biggest in the world!" The New York skyline, or the Washington Monument, or the Chicago Merchandise Mart—the biggest in the world. You say it without thinking what it means.

Or as one nineteenth-century German journalist summed it up: "To say that something is large, massive, gigantic, is in America not a mere statement of fact but the highest commendation."

PART OF THE reason for this has to do with American uniqueness.

To begin with, we had such a gigantic continent that we didn't even know quite how big it really was until the Lewis and Clark expedition some 175 years after we began settling the place. Generations of settlers just assumed it was *there,* land to move to, to build on, whenever anyone wanted to; a kind of infinitude settled into our national psyche. And of course we did not have much else—no aristocracy, no fine castles, no institutions, no cathedrals, no traditions, no titled families, in short no History—so it was natural enough, as a young country, to emphasize what we *did* have: size. When the United States began, even though it was a quarter of the size it is today, it was already far bigger than any European country, indeed larger than any other nation in the world except Russia, India, and China.

Gradually, too, blessed as we were by the riches that this vast territory conferred on us, we came to use its resources with a prodigious energy, proving to the people back in Europe that we were doing all right, thanks, and could build our buildings taller, our cities bigger, our roads farther, our fields wider. And for more than three centuries we grew accustomed to living in an ever-expanding boom society, with free land, rich soil, abundant timber, animals for hunting and trapping, gold and iron ore and coal and oil there for the taking, all in an abundance quite unlike anything known to the world before.

The pace with which all this happened also had its effect. America became instantly rich, within a few generations of its first settlement, and established its independence only a century and a half later, hardly much time at all in the eyes of Europeans—the Greeks, say, who had to wait 2,000 years for their independence. The speed with which the country grew in the nineteenth century surpassed even that of Britain, at the time the most powerful nation on earth, and the rate of growth by the end of that century, particularly in the economic realm, was truly dizzying, as E. L. Doctorow suggests so hauntingly in both the style and the story of his novel, *Ragtime*. Remarkably, that pace has accelerated in this century, particularly within the last few decades, a helter-skelter of both extreme and rapid social and scientific change that Alvin Toffler has demonstrated amounts to something like a "future shock."

The result of all this accumulating change is that Americans have tended to notice and admire that which is easily perceptible and most obvious—which is to say, bigness. When there is no time to pause and judge, to reflect, to measure, *quantitative* values must replace *qualitative* values; as Margaret Mead has noted, when a speeding car rushes past us it is only the size that we can judge; we have no way of knowing how well the engine works.

Finally, the fact that the United States has been the most exemplary capitalist nation during most of this period has contributed to its love of bigness as well. Capitalism is a system that depends, quite simply, on growth: minor cycles of boom and bust there may be, but overall there must be expanding markets, and therefore greater production, and therefore greater income and more expansion, and therefore more workers and more salaries, and therefore expanding markets. When capitalism does not grow it does not work, a lesson that Lord Keynes forcefully bore in upon the politicians of our era in persuading them that something always had to be priming the pumps and stoking the furnaces of the economy, and that if business couldn't do it the government must. Long schooling in this system of continual growth has taught nothing so well as the lesson that more is better, size is value, and big is beautiful.

BIGNESS IS THE condition of America also because ever since World War II it has been the function of the national government—fulfilling, it is presumed, the will of the people—to foster and promote it.

That is the process that British economist Ezra Mishan has called "growthmania," by which governments take it as their proper responsibility to measure everything in terms of economic development. "Although economic growth was not unheard of before this century," Mishan says, "it was not until the recent postwar recovery turned into a period of sustained economic advance for the West, and the latest products of technological innovation were everywhere visible, and audible, that countries rich and poor became aware of a new phenomenon in the calendar of events, since watched everywhere with intentness and anxiety, the growth index." None has been more purposeful in this adventure than the American government, which also enjoyed the benefits of an undamaged infrastructure after World War II and a consequent worldwide hegemony for the next thirty years.

On every front, U.S. government regulations and policies have worked, both deliberately and accidentally, to create and sustain the large corporations that are the underpinnings of our economic system. Federal subsidies encourage large agribusiness farms, regardless of how much they may actually produce—in fact, often in inverse ratio to their productivity, for the more land the farm can afford to keep idle, the higher the subsidy payments. Federal subsidies sustain American shipbuilding companies and pay more than half the costs of the construction of new supertankers and giant container ships. Federal tax law favors large oil companies and even allows them large accumulations of cash that they may use to buy up smaller companies and thus increase their size and control; between 1956 and 1968, for example, large oil companies took over no fewer than 226 smaller companies, all with the approval of the government. Federal trade policies for many years allowed American steel companies to increase in size even beyond the point of economic efficiency, without fear of foreign competition, thereby giving, in the words of Michigan State economics professor Walter Adams, "legitimacy and endurance to a cartel that could not survive without government succor." Federal construction of large water systems with low-cost irrigation encouraged the growth of larger and larger agricultural holdings, while Federal construction of the massive interstate highway system allowed these and other large corporations to greatly extend their markets and drive out local competitors. Federal subsidization of nuclear-power research and development has led to the construction of large nuclear-power plants by private utilities that otherwise would not have been able to afford them. Federal research funds and purchases of massive defense and space systems—not to mention occasional direct government handouts—have encouraged the growth of giant aerospace corporations. Federal laws permitting tax-free exchanges of stock between merging companies have encouraged conglomeratization in all industries, since it is the one sure way of acquiring assets without paying taxes on them; one economic consultant in the *New York Times* has showed how this "taxpayer's subsidy" has been "a major cause of the trend to bigness." Federal regulatory bodies serve to protect the large companies already operating, to encourage their further growth, and to discourage smaller, competitive firms; the Interstate Commerce Commission, one example among many, has set trucking rates and routes over the years so as to allow a few big firms to operate over wide sections of the country, has encouraged large truckers to buy up operating rights from smaller carriers, and has sharply restricted the number of new companies allowed to come in to compete with new routes.

The examples could be multiplied, but the point is clear enough. In the words of Sim Van der Ryn, the former California state architect and an expert on technological scale: "Public policy favors and often subsidizes large-scale standardized activity at the expense of diversified local enterprise." Or as economist Walter Adams sums it up: "As I read the industrial history of the United States, I find that the trend toward concentration of economic power is not a response to natural law or inexorable technological imperatives. Rather it is the result of institutional forces which are subject to control, change, and reversal . . . ; the inadequacy of existing antitrust laws and/or the desultory performance of the enforcement agencies; and the unwise, man-made, discriminatory, privilege-creating actions of Big Government."

But—I CAN HEAR the question beginning to form—*is there anything wrong with all this bigness?* Perhaps America does go overboard in its enthusiasm sometimes, but isn't bigness usually a positive and desirable goal? Aren't there a lot of things that are just plain better because they're bigger? What about the American corporation or the telephone system, or a library or hospital, or fireworks displays, telescopes, parks, supermarkets, sundaes, and dictionaries?

Well, yes, there is a point there. The Beanstalk Principle, after all, says only that there is a certain limit beyond which things ought not to grow, not that everything has to be miniaturized. The optimal size for one thing—a symphony orchestra, say—might not be the optimal size for another—a bridge game, for example. Each group or function or object will have its own most desirable size, and some are bound to be bigger than others.

Nonetheless, we should keep several important points in mind.

1. In the first place, there is no automatic or necessary connection between *size* and *value* in most instances. There are a few good big hospitals and a lot of appallingly bad big hospitals; there are some very modest hospitals with burn units and kidney machines and many very large ones without them. Most dictionaries of 3,000 pages are not worth the energy required to lift them, though a very few, like *Webster's Second,* are invaluable monitors of the English language—not so much because of their size, though that is obviously important in a work of comprehensiveness, but because of their skill and sensitivity in selection and definition. A large library may be preferable to a small one, but only if 90 percent of its books are not solely on embryopathology and if it

does not shut down for five days a week. Size, in other words, where pertinent at all, is often simply an ancillary, not an essential, component of worth.

2. Also, even when benefits *are* produced by size they are almost always accompanied—and in many cases outweighed—by drawbacks. This is very often true of the bigger institutions in our society: large hospitals (which provide more varied equipment, but less patient care per nurse and more chances for mechanical breakdowns and mistakes), large libraries (more volumes, but less personal attention, more pilferage and damage, more system breakdowns), and large parks (more room and more sights, but greater risks of mugging and robbery, more despoliation, more litter). It is certainly true of a variety of large machines—nuclear power plants, supersonic jets, oil supertankers, and massive cars, for example—where the purported benefits of size are well overbalanced by the manifold dangers. And it is increasingly true (as we shall see in some detail later on) even of the very cities of humanity, where the traditional advantages of size (greater diversity, spontaneity, anonymity, cultural opportunity, urbanity) are more and more overshadowed by the accreted disadvantages (pollution, congestion, crime, stress, powerlessness, higher prices and taxes).

3. Moreover, many big institutions and systems are large not because they are better that way but simply because they happened to grow that way for reasons other than need or value. Government bureaucracies are an obvious case in point, as Parkinson's Law reminds us: these institutions tend to grow by about 6 percent a year, regardless of the amount of work to be done, *or whether there is any work at all.* Parkinson's chief evidence was the British Admiralty Office, which expanded its office staff by no less than 768 percent between 1914 and 1967, although the actual number of ships had *decreased* by 475 percent in that period, with the result that "over 33,000 civil servants are barely sufficient to administer the navy we no longer possess." The American Navy, as no one yet seems to have discovered, offers a comparable example, growing by about 6 percent a year quite regardless of the number of ships there are to supervise. In 1915, before our entry into World War I, we had 22,978 civil servants in the Navy Department to administer to 324 ships; after the war the number of civil servants shot up to 90,000 but the number of ships actually declined to 300. Since then the size of the fleet and the size of the Navy Department have gone up and down depending on the wars we were engaged in, but at the end of the hostilities we have always ended up with more desk admirals than during wartime, when there was actually something for them to do: In 1950, for the Korean War, we had 293,000 workers to care for 4,297 ships—68 per ship—but in 1955, after the war, there were 410,000 civil servants and only 2,953 ships—139 per ship; by 1970, during the

Vietnam War, there were 3,146 ships and 376,000 civil servants—120 per ship—but by 1975, when we cut the fleet to 1,039, there were 320,000 Navy Department people whose job was to take care of them—an all-time high of 307 per ship. As Parkinson says, "The number of the officials and the quantity of the work are not related to each other at all." And one of them always gets bigger.

4. Finally, the virtues of size, as we have been taught to regard them, are mostly illusory. Americans of course are proffered a good many myths about the benefits of bigness, but on examination the greatest percentage of them turn out to be quite erroneous. We may be told that foot-long carrots and 15-pound radishes are better than the ordinary kind (there's a seed firm in New Jersey that guarantees you can grow such stuff), but that is willfully reckoning without our tastebuds. Captains of industry often assert that the American economic system is far more successful than those of smaller countries, but in fact Switzerland, Sweden, Denmark, Norway, and several tiny Mideast oil nations all have higher per capita incomes than the U.S. and industrial growth rates quite a bit healthier. Con Edison may spend a lot of advertising money to get New Yorkers to believe that because it is the largest electric utility in America it is the best, but the patent absurdity of that is made manifest once or twice a year. Supermarket chains repeatedly try to persuade us that their enormity enables them to offer us quality we could not expect from a smaller store on the corner, but a comparison of their tomatoes or their fish fillets will instantly refute the claim.

One class of myths, perhaps the commonest of all, has to do with the supposed superiority of large American corporations. It is a complex matter and we will have occasion to touch on it again when weighing the virtues of human-scale economic units, but for the moment it may suffice to examine a few of these myths. For example, there is the one that big corporations are more productive than small ones—but in fact, as Barry A. Stein, the leading scale-economist, has shown, "studies of productivity, as measured by value of shipments per manufacturing employee, show that the highest efficiencies tend to occur not in the largest size plant, but in those of moderate size," that is, *with fewer than forty-five people.* There is the myth that big companies make more profit than small ones—but in fact Stein's figures show that "there is a declining relationship between profitability and size," that "smaller firms are more efficient users of capital," and that as firms grow bigger they show a "decreasing efficiency of asset use." There is the myth that big companies can sell their goods more cheaply than small ones, which is why consumers benefit from the American system—but in fact, because of increased inefficiency, accident rates, strikes, sabotage, absenteeism, storage facilities, distribution networks, transportation costs, advertising, promotion, and product differentiation, large firms actually must charge more for their products. It seems to be, in sum, that many of the "advantages" of big business, as we have been taught them, have all the substantiality of an over-the-counter mining stock.

With those four cautionary points, then, we can conclude at the very least that "what's wrong with all this bigness" is that it leads us to assumptions that are often unsupportable and hence to practices and policies that are often detrimental. Bigness may have its place here or there, but it is the pervasiveness of it that is so alarming. Because bigness is by now so rooted in our culture, we have not really ever come to grips with the questions of size, of quantity, of extent, for our individual possessions or dwellings, for our organizations and workplaces, for our cities and systems. Because we do not really know how much is enough, we assume that bigger is better.

The danger in that should be evident, made all the more ominous by Lewis Mumford's sobering judgment after a lifetime of studying history: "In the repeated decay and breakdown of one civilization after another, after it has achieved power and centralized control, one may read the failure to reach an organic solution of the problems of quantity."

TO CELEBRATE THE Bicentennial of the United States, New York City's Triborough Bridge and Tunnel Authority decided to display the largest American flag ever made (indeed, presumably the largest flag of any kind ever made), a huge banner as big as a football field and a half, 18,000 yards of nylon taffeta sailcloth, enough to make something like 6,000 dresses.

Early on the morning of June 29, a crew of workmen supervised by a team of project engineers began to lower the huge flag from the upper framework of the Verrazano-Narrows Bridge, where it could be viewed by the gathering of "Tall Ships" as they sailed into the harbor on Independence Day. Almost as soon as the cloth was first unfurled it caught the high winds sweeping in from the Atlantic, coming in at a velocity somewhat greater than the engineers had expected. By the time the flag had been completely unrolled, the winds had increased from 6 miles an hour to 12 and shortly to 16, whipping the cloth out of the workmen's control and flinging it against the steel bars descending from the bridge's suspension cables; it hung there for several minutes, looking like the bottom side of a fat man on a lawn chair, as the crews desperately tried to retie their fastenings.

Gradually, under the wind's pressure, the horizontal stitching at one end began to give way, and even as the workmen struggled to roll up the banner again the other end began to unravel. "Then," as the chief administrative engineer recalled it later, "all hell broke loose": all the horizontal seams started to pull apart, strips of cloth whipped by the wind tore away from the flag, and within two hours the giant banner was in pieces, flapping so wildly over the roadway below that the traffic had to halt. An hour later even those strips had been blown into the harbor waters below, and all that was left were a few tattered strands of blue, dotted by stars, twitching against the cables like trapped butterflies.

When the first Bicentennial sailing ships came through five days later, the bridge was bare.[1]

---

1. In 1980, some corporations and advertising firms got together to sponsor the production of a new flag, even larger, to be fixed permanently to the Verrazano Bridge and "unfurled" on national holidays. It is to be called "The Great American Flag," is expected to cost $850,000, and will last, according to corporate publicists, ten years.

3

But Not Always

THE PERCEPTION THAT bigness held sway over human affairs has no doubt been fashioned for us by our historians, for just as it is the victors who write the histories of wars, it is the inhabitants of big institutions—university professors, corporate publicists, government researchers, and the like—who have generally written the histories of institutions. Thus we become trained to see history as a succession of empires, like the Hellenic and Roman and British, of giant cities, like Cairo and Peking and London, of outsized monuments like the Pyramids and the Great Wall and Versailles, of military achievements like Hannibal's elephants, Napoleon's campaigns, and Hitler's blitzkriegs, of inventions like the steam engine, the railroad, and the atomic bomb, of wars and kings and capitals and conquests. The broad periods of quietude and assimilation tend to be overlooked, the lives of ordinary people ignored, the accomplishments of small communities and organizations made secondary. We are given the castle but not the cottage, the parliamentarian but not the plowman; we learn of the Library at Alexandria but not the scrolls therein, of McCormick and his reaper but not the inventors of crop rotation. "The masses themselves," as the great French historian Ferdinand Braudel comments, "lie, as it were, outside the lively, garrulous chronicles of history."

But the fact is, as common sense would tell us, that for most of the human period on earth, small arrangements, small groups, small communities and cities have tended to predominate—even within our own recent historical period. The big institutions that did arise came fairly late in time and generally lasted for fairly limited periods.

Cities. Not until about 2000 B.C. did humans fashion anything resembling what we would regard as a city, and that was some tens of thousands of years after people all over the globe, from Africa to China, had grown accustomed to living in small communities. Our guesses are very rough, of course, but it is probable that the first recognizable cities did not have more than 5,000 or 10,000 people in them, and one of the earliest known town remains, Çatal Hüyük in Turkey, may have had the attributes of a city when it had no more than 2,500 people. When larger cities did grow up in later pre-Christian times—Eridu, Ur, Babylon, Thebes, Memphis in the Middle East, Harappa and Mohenjo-Daro in India, Anyang and Chengchou in China—there is no indication that they ever grew to sizes much beyond 35,000 or 50,000, perhaps 100,000 at the very most, definitely of urban proportions but nothing so large that they could not be taken in with a single view, circumnavigated with a single day's walk. The cities of the Christian era occasionally grew to massive sizes—Rome may have had as many as 300,000 at the time of Augustus, Constantinople perhaps 180,000 in the ninth century, Baghdad under the Caliphate something close to 300,000—but it is notable that such agglomerations seemed so inherently unstable that they typically lasted for less than a century.

It was only an eye-blink ago, so to speak, that any city reached the size of a million—that was London, in the 1820s. By 1900 there were only ten cities of a million or more in all the world. Even as late as 1800, only 2 percent of the world's people lived in urban areas over 20,000, by 1900 only 10 percent, and as of 1970 still only 20 percent. The giant metropolis, in other words, is not something eternally with us but, historically speaking, a brand-new phenomenon, and even now an arrangement that encompasses only a small minority of us.

Empires. Not until about 2300 B.C., with a certain Sargon of Akkad, was there anything we could reasonably call an empire, after four or five thousand years of settled independent communities—and it lasted for all of about fifty years before it reverted to a congeries of city-states. Indeed, for all the attention paid to them, it seems that empires don't really last very long as a rule; their periods of successful dominance rarely extend more than a century before some apparently inevitable decentralizing process breaks their hold. The Hellenic Empire of Alexander the Great, which comes down to us as a symbol of successful enormity, lasted barely twenty-six years, from the conquest of Gaugamela in 331 B.C. to the secession of Egypt in 305 B.C., leading on to full-scale dissolution with the civil wars of 281 B.C. The Roman Empire, so often celebrated by the historians, had an effective lifetime of only about two hundred years, from Augustus's accession in 27 B.C. to the death of Marcus Aurelius and the end of the Pax Romana in A.D. 180—a considerable stretch of years, to be sure, but nothing that should suggest eternalness. Even the British Empire, the largest imperium ever known, may be said to have started formally only with the colonization of Ceylon and Malta in 1815 and lasted to the transition to Commonwealth in 1931 (or alternatively to the beginning of national independence in 1947), a period of not much more than a hundred years.

And we should also note that even within the most extensive of these empires, by far the greatest number of people lived in small communities rather than the imperial capitals, and that always after the empires had crumbled it was the small communities that persisted, durable and solid, provincial and conservative to be sure but for that reason the repositories of culture and the sanctuaries of civilization. As Richard Goodwin has put it:

> Even during the great world empires—the Macedonian, the Roman, and the British—most of life was centered in a small community. From a remote capital came armies to devastate or protect, new governors, despoiling collectors of revenues, but though an ultimate power might reside in Rome or London, most aspects of everyday life were regulated by the community. Men ordered their affairs according to local custom and local law. Among neighbors and in a familiar landscape, they developed the identity that sustained them—found satisfaction in the knowledge that their acts and labors had human consequences.

Nations. Although the processes of nation-making began several centuries before, it was not until the nineteenth century that the full-fledged nation-state emerged in any part of the world; as late as 1815 there were only thirty-five nations or so that were recognized as sovereign, and of them only a handful had the centralized, bureaucratized form that was characteristically modern. Nations themselves were simply non-existent through the greater part of history, the notion of separate racial and ethnic groupings finding political expression through a single government being regarded as impractical if not absurd. The very idea of nationalism, in fact, far from being the universal and everlasting notion we sometimes assume it to be, is of comparatively recent origin, and the word does not even enter the English language until the middle of the nineteenth century.

Moreover, it was not until the rise of Britain after the Napoleonic Wars that any nation developed the characteristics we associate with contemporary nations: a large centralized government supported by a standing army and a civil service, an industrial substructure and world-wide commerce, and state-run systems of education, health care, prisons, police, welfare, communication, and international diplomacy. Much of what we now take for granted in national organizations, from income taxes and official police to national anthems and flags, was in fact created only in the last century and a half.

Nations, empires, cities: in the long run of history, we may thus conclude, the classic exemplars of bigness have been rather more the exception than the rule.

ALTHOUGH THERE IS no denying the contributions that some forms of bigness have made to our civilization—I am thinking not about the

Pyramids and the Great Wall, but such things as the cross-fertilization of ideas and the development of communications that empires encourage—nonetheless it has been in the small cities and city-states, and occasionally in quite remote areas, that many of what we regard as the high accomplishments of civilization have taken place.

The earliest and still the most fundamental manifestations of civilization were nurtured, of course, in the small communities where fire was first captured and then tended, speech and history and religion were created, tools and wheels and weapons were fashioned, and animals and foodcrops were domesticated. Later on, it was in the originally modest cities of Mesopotamia and the Indus Valley that additional elements of settled civilization, from astronomy and hydrology to brick-making and bronze-working, were fashioned, in populations that may have ranged from 2,000 to 20,000 but for the most part were probably little bigger than 5,000. Later still it was the independent cities of Byblos, Mycenae, and Troy, and the constricted islands of Cyprus and Crete—none of them probably with many more than 10,000 people, to judge from excavated sites—that developed the maritime trading cultures of the West, created navigation and geography, and gathered and transmitted new ideas and material around the Mediterranean world—principally the knowledge of how to mine and work iron, the development that was to transform Europe.

The Greek city-states that took form from the seventh century B.C. on are justly famous as the incubators of so much that is a part of the contemporary world, and they too were all of quite modest size, kept small both by the natural geography and the hard-scrabble farmlands of the Aegean regions, which could not support large populations, and later by a policy of colonization and deportation. Even at their height, the biggest of them—Athens, Syracuse, and Acragas—probably did not have over 50,000 people in their city cores, according to the reasoned estimates of Greek urbanologist Constantine Doxiadis, and the rest of the cities probably had no more than 10,000 on average. These little units, such a stark contrast to the conurbations that earlier Eastern empires had developed—Lord Ritchie-Calder goes so far as to call them of "insignificant size"—were, in Lewis Mumford's words,

> cut closer to the human measure, and were delivered from the paranoid claims of quasi-divine monarchs, with all the attending compulsions and regimentations of militarism and bureaucracy. . . . The result was not merely a torrential outpouring of ideas and images in drama, poetry, sculpture, painting, logic, mathematics, and philosophy; but a collective life more highly energized, more heightened in its capacity for esthetic expression and rational evaluation, than had ever been achieved before. Within a couple of centuries the Greeks discovered more about

the nature and potentialities of man than the Egyptians or the Sumerians seem to have discovered in as many millennia.

Though the Greek cities gave way to the Hellenic, and those to the Roman, with the fall of Rome it was once again the small unit that became the repository of civilization. The small cities of the Middle Ages, despite their reputation for "Dark Ages" barbarity, and with them the small monasteries and retreats, were in fact the nurturers of Western culture for more than a thousand years. Somewhat more ignorant and often far more parochial than their Greek counterparts, these cities were nonetheless almost equally skilled in the creation of both human institutions and arts. The guilds of merchants and craftsmen that gave them economic self-sufficiency, for example, were seminal and enduring organizations; the folkmotes and town councils that provided their political independence were, though male-dominated, the instruments by which Europe's traditions of political liberty and independence were fashioned. It was in this period, too, and in quite limited settlements, that the institution of the university took root—the first universities were in Bologna (c. 1100), a place of perhaps 35,000 people, Paris (1150), which probably had no more than 50,000, Oxford (1167) and Cambridge (1209), both with less than 20,000, and such small cities as Vicenza, Salamanca, Padua, Toulouse, Siena, Piacenza, Montpelier, and Avignon. Notable cathedrals were erected in almost every urban place of more than 10,000, and some of the very greatest—Chartres, Amiens, Rheims, Cologne, Canterbury, Salisbury—arose in the most modest and provincial cities. The extraordinary medieval painters of Northern Europe flourished in a region where most cities were under 15,000 and where Bruges, with 25,000, and Ghent, with not more than 60,000 at its height, were cultural meccas of continentwide influence. The Italy of Dante and Giotto had the largest cities in Europe at the time, yet none of them was more than 80,000 and most were far smaller:

Naples (in 1278)	27,000
Bologna (1371)	32,000
Pisa (1293)	38,000
Padua (1320)	41,000
Milan (13th c.)	52,000
Siena (1328)	52,000
Florence (1381)	55,000
Venice (1363)	78,000

And when the Middle Ages blossomed into the Renaissance, Europe was still a land of small cities, its burst of creativity taking place among very limited populations. The Rome of Michelangelo had perhaps 55,000 people, the Florence of Botticelli and Leonardo 40,000,

Padua and Genoa were probably under 50,000, and Pisa, Modena, Siena, and Vicenza may have had no more than 20,000. Fifteenth-century Germany was said to have 150 large cities worth enumerating, yet the largest did not have more than 35,000 people, and it was here only a few decades later that such giants as Dürer, Cranach, and Holbein were nurtured. The England that created Skelton and Spenser and Shakespeare had only one city with more than 40,000 inhabitants, and that was of course London, which had a population of only 93,000 as late at 1563 and no more than 150,000 at the end of the Elizabethan reign.

Closer to home, the genius and vision that created the American nation sprang from a land of very small cities, not a single one with even 30,000 people in it, and even they harbored no more than 5 percent of the total population. Even by the time of the first census in 1790, New York had only 33,131 people, Philadelphia 28,522, Boston 18,320, and Baltimore 13,603. In fact, the instruments of the American state were fashioned and tested for nearly fifty years before any city in this country reached the 100,000 mark—that was New York, in the 1820s.

Bertrand Russell once observed that "where art flourished in the past it has flourished as a rule amongst small communities"—and political institutions, too, we might add, seem to flourish in the same soil.

EVEN TODAY, in this country, smallness persists. We should not get the idea that, though the cult of bigness is unquestionably dominant, it is all-inclusive and irresistible; in many places and in many ways smaller institutions and units, so far at least, have continued to hold on successfully. Indeed, it is quite revealing to realize just how much smallness we have with us still.

Let us take two examples, places where the pervasiveness of bigness seems to be axiomatic: cities and corporations.

Hardly a day passes that we are not reminded of the enormity of our cities and their centrality in our lives, and it is common to read references to how "three-quarters" of the nation is now "urbanized." Yet the fact is that, as of the 1970 census, only ninety-seven cities had more than 100,000 people in them, barely 1 percent of all urban places in the land, and those cities represented no more than 28 percent of the total population. *Most* people in this nation—*64 percent*—live in small cities (19 percent in cities between 10,000 and 50,000) or in villages (11 percent in places under 10,000) or in rural areas (26 percent) or in unincorporated towns (8 percent). The myth of extensive urbanization is probably the result of the Census Bureau, for reasons I have never fathomed, choosing to call a "city" any place over 2,500 people or any settlement of any size located on the fringe of a metropolitan area.

Moreover, the notion that the United States is somehow dense with people, another common myth, is belied by a nice statistic that economist E. F. Schumacher liked to cite: if all the people of the entire world, all 4 billion of them, were squeezed into the space of the United States, some 3.6 million square miles, the population density here would still be only equal to that of England right now—and far, far less than that of such places as Malta, Singapore, Barbados, or Hong Kong.

Perhaps even more startling is the persistence of small units within corporate America, acknowledged to be the mightiest and most concentrated economic engine ever amassed. Here we need only consult a few pages of that most extraordinary fund of information, the Federal government's *Statistical Abstract.*

Despite the unquestioned predominance of a few thousand very large businesses, hundreds of thousands of individual firms do exist in every industry in the country, in fact nearly 11 million firms in all. Some of these, to be sure, are mere mom-and-pop affairs and out-of-the-garage dispensaries (though even so, their numbers are extraordinary), but 1.6 million of them are full-fledged corporations, the form that usually denotes size and durability. And though a small portion of these corporations are quite large—165,000 do more than a million dollars worth of business a year—843,000 of them, just over 50 percent, gross no more than $100,000, which is small business by any definition. To take just one example, the category of businesses commonly labeled "FIRE" (financial, insurance, and real-estate institutions), here, where we would expect to find the very largest organizations of corporate capitalism, there are indeed 16,000 corporations with receipts of over $1 million and another 11,000 middle-sized ones with receipts between $500,000 and $1 million, but there are 400,000 quite small ones that bring in less than $500,000. *Most* businesses in this country, in sum, are small.

But of course we're looking at all kinds of enterprises here, small corporate farms and local contractors and retail stores, so one might expect to find such a perseverance of limited size. To be fairer, let's confine the examination to manufacturing industries, where it is more reasonable to think that only large-scale operations could be successful, given modern technology and modern capital requirements. Remarkably, the same story holds. Of the 200,000 manufacturing corporations in the land, you could take away the *Fortune* 500, or the *Fortune* 1000, or ten times that, and you would still have 95 percent of all the manufacturers left. In this industry of expected giants, no fewer than 65 percent—128,000 manufacturers—bring in receipts of less than half a million, and that does not include 213,000 *non*-corporate manufacturers (partnerships and proprietorships) of the same small size. In the plastics industry there are no fewer than 4,500 companies, in machine manufacturing 15,000, in steel fabrication 1,900, in iron foundries 1,000, in printing 18,000, and all but a handful are small businesses by any standard.

Even in those industries where the familiar giants hold sway and three or four firms dominate the market, there are literally hundreds of other small businesses that manage to endure, apparently doing nicely on the scraps. For example, although four companies produce 60 percent of all auto parts and accessories, there are 1,420 others that supply the remaining 40 percent; though 68 percent of the market in aircraft engines and parts is controlled by four firms, another 200 or so vie profitably for the remaining portion; and though fully 70 percent of the soap industry is controlled by four giant companies, there are another 600 that stay in business and divide up the rest.

There's one more interesting fact among the statistics. It turns out that almost all American companies operate with a very small number of people. Taking businesses of all kinds, 98 percent have under a hundred employees, and those employees are actually divided up into separate units (factories, laboratories, offices, warehouses, and the like) of an average of just about *seven people each.* Even in manufacturing industries, where we have images of endless assembly lines and mammoth steel plants, 65 percent of the establishments employed *less than twenty people* as of 1967, a year typical enough of those since World War II, and the average number of workers per unit in all manufacturing was just 44.9.

I do not mean for one minute to minimize the importance of the big corporation, of bigness in general, in the American economy. The giant firms are responsible for many billions more dollars, many millions more people, than the small firms, and only 1 percent of the corporations manage to control something like 80 percent of all corporate assets. But I do think it is necessary to hold that in perspective. The endurance of just these few examples of smallness within the economy suggests that we ought not to let the advertisements and the media delude us. Corporate bigness, whatever its power, is still more like an aberration, or at least an exception, in the economic system, even now, even here.

AT THE PHILADELPHIA EXPOSITION of 1876, the World's Fair held to mark our centennial celebrations, the most popular and most publicized single object in all its 20 acres was inventor George Henry Corliss's "Centennial" steam engine. This was an enormous combination of boilers and pipes and pulleys and belts and flywheels and cogs and cranks and shafts, fully 40 feet high and 80 feet wide, weighing several thousand pounds and costing $100,000, more than any single exhibition to that date. It was a truly remarkable collection of steam-power technology, probably the finest pinnacle to which steam mechanics, the dominant energy form of the nineteenth century, could be brought, and it was hailed as such by the knowledgeable and fashionable alike.

Three years earlier, at another exposition, in Vienna, a far smaller machine for making energy had been exhibited. This was a small-scale dynamo, which generated electricity, combined with a motor, which converted the electricity into motion, the whole affair being less than a fiftieth of the size of the Corliss "Centennial." That exhibition was almost totally ignored. Nobody could figure out what a dynamo would be used for.

Within twenty years of its creation, the Corliss engine was an anomaly. The twentieth century, needless to say, was built upon the dynamo.

4

Beanstalk Violations

JUST OVER TWENTY YEARS AGO a little book called *The Breakdown of Nations* was published in England. It was written by Leopold Kohr, an Austrian-born economist who was a recognized expert in international customs unions and who had taught economics at several prominent universities in America, but its thesis went far beyond run-of-the-mill economic matters, indeed beyond run-of-the-mill matters of any kind. It offered one of the most extraordinarily provocative arguments of contemporary political philosophy, right from its opening lines:

> As the physicists of our time have tried to elaborate an integrated single theory, capable of explaining not only some but all phenomena of the *physical* universe, so I have tried on a different plane to develop a single theory through which not only some but all phenomena of the *social* universe can be reduced to a common denominator. The result is a new and unified political philosophy centering in the *theory of size.* It suggests that there seems only one cause behind all forms of social misery: *bigness.* Oversimplified as this may seem, we shall find the idea more easily acceptable if we consider that bigness, or oversize, is really much more than just a social problem. It appears to be the one and only problem permeating all creation. Wherever something is wrong, something is too big.

So far as I can tell, this remarkable thesis was almost totally ignored at the time, though its pertinence then, in the aftermath of a second worldwide war and the rise of the giant superpowers, would seem to have been obvious. Its relevance today, in a world ever more plagued by bigness, is clearer still. For today we are surrounded by countless examples of difficulties and crises brought about by one form or another of overgrowth, excess, distention, enormity, or expansion, for which we are seemingly unable to contrive any solutions at all. As Kohr notes:

> Social problems have the unfortunate tendency to grow at a geometric ratio with the growth of the organism of which they are a part, while the

> ability of man to cope with them, if it can be extended at all, grows only at an arithmetic ratio. Which means that if a society grows beyond its optimum size, its problems must eventually outrun the growth of those human faculties which are necessary for dealing with them.

In short, we are a society plagued by violations of the Beanstalk Principle, the victim of institutions and systems grown beyond their optimal size to the point where they are literally out of control.

Examples of Beanstalk violations are so numerous I hardly know where to begin, but perhaps the best place is the day that I was sitting down to write this chapter—and all the lights went out.

THE GREAT BLACKOUT of 1977 in New York City has been analyzed up and down by everyone from journalists to electrical engineers, but everyone seems to have overlooked the obvious source of the problem: *bigness*. The Consolidated Edison Company runs the biggest private power system in the world, so big that, try as it might, it is unable to operate it efficiently—a point that you would assume was obvious in light of the fact that Con Edison dividends are the lowest for any utility company in North America although its rates are the highest. Yet far from trying to diminish that size or decentralize its operations or install smaller pieces of equipment, Con Edison has worked relentlessly ever since the first Great Blackout of 1965 to increase its system and enlarge its components.

First, it shut down all the "small, uneconomical" generating plants that once covered the area and decided instead to rely on eleven large generators, notably "Big Allis," the 1,000-megawatt installation in Ravenswood, Queens. Those large generators, of course, cost considerably more, since they couldn't very well be mass-produced; they needed elaborate fail-safe devices to protect their costly mechanisms at the first sign of trouble (that's for Con Ed's protection, you understand, not the public's); and they made the entire system more dependent on the unhindered operation of each one, though in fact each generator is of such a size that it loses in reliability what it gains in unit cost as it gets bigger.

Second, Con Ed created a system of large gas-turbine generators as a back-up for its main generators. Those machines, however, turned out to be so expensive to operate at that size that they were never used, even in emergencies, and so expensive to repair that they were rarely in operating order even if anyone had wanted to use them.

Third, Con Ed installed large computer banks to oversee its service and complicated sensors to detect failures and overloads. This meant a greater reliance on complex high technology and on the experts needed to monitor it, both of which are subject to greater liability of error than in simpler systems.

Finally, Con Ed put in large underground transmission cables throughout its service area, cooling and insulating them with great amounts of oil pumped through the lines. Again the result was higher costs, both for the equipment and for its installation, and costs went higher still when the price of oil started going up in 1973. And since the whole cable system was so large, no auxiliary pumping stations could be provided and all the pumping had to rely on standard electric power, which of course would be lost in the case of a blackout.

Now the first consequence of all this big-betterism was that Con Ed spent enormous sums of money, so much that it soon found that it was simply too expensive to generate electricity under those conditions and determined that *it was actually cheaper to buy its electricity from other systems*. Which would seem to defeat somewhat the idea of having its own generating capacity in the first place—rather like a person who buys and repairs a Cadillac but chooses to walk everywhere because it's cheaper. The second consequence was the creation of a brittle system totally dependent on a few large machines and a frail network of power lines, "a very complex system," according to Con Ed itself. The third consequence was the blackout.

On the night of July 13, nothing in this overbuilt system worked quite the way it was supposed to. The "shield wires" Con Ed used to protect its transmission lines from lightning failed during a severe thunderstorm that began about 8:30 that night, and several lines to the north of the city went out, five of which were bringing in the outside electricity Con Ed had come to depend on. Next a circuit breaker in the Buchanan plant in Westchester failed, and because of a faulty timer and a malfunctioning reclosing circuit, lines from there were not reopened as they should have been. Shortly thereafter, a mistakenly set protective relay for the Indian Point plant was knocked out, and a bent metal contact at the Millwood substation caused an incorrect response to the remaining power load, shutting that relay down. Responding to all this, operating engineers in charge at the Manhattan control center tried to patch things up by "load-shedding" some of the electricity—allowing selective blackouts and brownouts around the area—but the information coming in was complex, the decisions difficult, the communication haphazard, and what engineers like to call "the human factor" intervened; according to a subsequent inquiry, there was "a feeling of confusion in that room, a sense that they were not entirely sure of what they were doing." In fact, for about thirty minutes nothing very much did get done, at least not properly, and finally at 9:29 an automated control regulator on the line coming in from New Jersey shut down, and that last source of outside power was cut off. Then, just as these calamities were about to bring the system to its knees, Con Ed's "fail-safe" mechanisms went into operation, automatically shedding loads to protect the expensive machinery and prevent a total blackout. But as Dr. John Gall points out in his insightful review of contemporary technology, *Systemantics,* "When a fail-safe system fails, it fails by failing to fail safe": the Con Ed system shed so many loads that there was an instant *over-supply* of power in the remaining part of the system, and Big Allis's generator relays tripped within seconds to prevent a burnout, followed in succession by the other nine remaining generators. By 9:40, the night was black.

Start it up again, you say? Have you ever floated a beached whale? The system was so large and so complex, and depended so much on automated machinery that couldn't operate since there was no power to run it, that there was just no easy way to get things going again. The big generators, once shut down, could not be started up again without a long and difficult process, since the huge rotor blades weigh several tons each and even the slightest mistake would cost the company millions of dollars. The underground transmission lines, lacking oil in all the higher areas because the electrically run pumps no longer worked, had to get new oil pumped into them manually. The equipment throughout the system, all of it in all its complexity, had to be checked both by on-site inspection and, when alternate generators were functioning, by computer. All in all, as Con Ed President Arthur Hauspurg put it, "The system's complexity slowed recovery." Not only that: "The whole system concept was a design that said that we're not going to shut down," so a blackout had simply not been programmed for and no scheme of starting up again had ever been devised.

The result: perhaps a dozen deaths, numerous injuries, 3,000 arrests and weeks of kangaroo justice, something on the order of a quarter of a billion dollars in property losses, another quarter of a billion in city expenses, more than $50 million in unrecovered wage losses, $10 million for restarting costs and $100 million for new equipment, and a permanent imprint on the city psyche of a night of looting, fear, chaos, and crime proving unmistakably thc fragility of the social and legal conventions.

The lessons from this night would seem to have been obvious. You cannot *design* such a large system as Con Ed's for infallibility, since that requires too great a need for technical perfection. You cannot *build* such a large system to work, continuously and indefinitely, without fail. You cannot *operate* such a large system without mistakes and the intervention of the unforeseen. You cannot *protect* such a large system from breakdowns and a multitude of human and mechanical errors. And you cannot *repair* and restart such a large system before serious damage has been done. The very size of the system magnifies all of the dangers.

Yet these very simple lessons seem not to have been learned. Con Ed has changed nothing but a few computers in the years since the blackout, added a few new pieces of equipment here and there, and made repairs and patches to one of its New Jersey connections; it has made its oil-cooling systems even larger than they were before and even more complex, and with the hopes of eliminating all human error it has installed even more labyrinthian computers to take care of load-shedding automatically. It has even gone blithely along in its campaign to be allowed to add a few more nuclear-powered plants to the system, to supply enough power, they assure us, so that such a blackout will never happen again. The prospect of this company, with its history of two total blackouts in twelve years and a half-dozen smaller ones in between, with its complexity of equipment, with its acknowledged fallibility of personnel, with its record of inefficiency, being put in charge of more nuclear plants and breeder reactors that could quite literally wipe out much of the New York region, does not invite sanguinity.

There are some who say that the blame for the Great Blackouts shouldn't really be placed on Con Ed, for there was nothing else the company could have done, given the fact that they have to supply electric power for 9 million people who depend on it for almost all their everyday needs, and they have to do it within environmental and financial restraints. That makes a certain sense, except that there *were* other things for Con Ed to have done, and there still are: make the system smaller, simpler, cheaper, and safer. That is not so farfetched. The British, for example, and some of the systems operating in this country rely on a network of smaller generators and more of them: the British system intertwines 137 separate power plants for about 50 million people, an average of 365,000 per generator, whereas Con Edison has eleven plants for 9 million people, an average of more than 800,000 per generator. With the right kind of thinking, Con Ed could even now work toward knitting itself into a wider grid system made up of far more, but smaller, generating plants, sharing and cooperating with other systems as the need arises; according to *Newsweek,* current thinking is that "utilities will have to abandon the old ideal of huge, centralized power grids in favor of smaller, more efficient plants." And Con Ed could certainly work toward creating a truly decentralized power system within a decade or so, using solar energy in small-scale community plants, industrial co-generation of electricity from wasted factory energy, hydraulic windmill systems along the shorelines, and watermill generators on nearby streams and rivers—all technologically possible *right now,* if the will existed to put them into practice.

But ultimately the question is not really whether there is an alternative to the present method but whether the people of America's cities want to go on depending on such overextended and unstable power systems as Con Ed's. And whether they wish to risk going through more of these nights of madness, waiting for the one that will perhaps be accompanied by another calamity—a failure of the telephone system as in 1969 and 1975, or a crippling transit strike as in 1966, or periods of street protest as in the 1960s—leading finally to the systems breakdown that may not be remedied so quickly and the social chaos that may not be contained at all?

THE CAUSE OF the Great Blackout of 1977 is only one spectacular example of violations of the Beanstalk Principle. In the wide range of this society of Kohrian excess we can find many other institutions and systems that have similarly been extended beyond their optimum size to the detriment of the individuals they were theoretically meant to serve. A short survey will suggest their extent.

THE GREATEST SALES RECORDS in all retail trade in recent years have been made by the incredibly proliferous fast-food stores, owned or licensed by large corporations or conglomerates that are able to buy in large quantities, mass-produce both containings and containers, and advertise incessantly in all markets. (It has been calculated that more people have heard the McDonald's advertising jingle in the last decade than have heard Beethoven's *Eroica* since it was first played in 1805.) Fast-food operations now account for 40 cents out of every food dollar Americans spend, and the take is increasing by about 2 cents a year; in 1978 it amounted to $20 billion, more money than Americans spend for all personal-care products or for private education and research.

Of course the quality of the food produced is pretty generally deplorable—their "milk" shakes, for example, contain no milk or ice cream but are made rather with vegetable oil, casein, nonfat milk solids, emulsifiers, flavorings, and sugar (which is why they are called "thick" shakes or somesuch, instead of milk shakes). And even when they start out with food of adequate, occasionally high, quality, the scale at which it must be processed to reach several billion customers a year throughout the country is so vast that it is not possible for the end product to be anything but stomach-threatening. Even such a fast-food figure as Colonel Harland Sanders himself has acknowledged this sad truth. After selling out control of his Kentucky Fried Chicken business in 1964 (though staying on for public-relations stints), he saw it go through various corporate hands until it ended up controlled by Heublein, a conglomerate that knows more about liquor—and conglomeratization—than it does about chicken, fried or otherwise. A few years ago Colonel Sanders went into a Kentucky Fried Chicken outlet in New York City with a *New York Times* reporter and pronounced the product there "the worst fried chicken I've ever seen," likened the mashed potatoes to "wallpaper paste" (with the gravy added it was more like "sludge"), and concluded to the startled store manager, "You're just working for a company that doesn't know what it's doing." Later he added remorsefully, "You know, that company is just too big to control now. I'm sorry I sold it back in 1964. It would have been smaller now, but a lot better. People see me doing those commercials and they wonder how I could ever let such products bear my name. It's downright embarrassing." The executives at Heublein unwittingly confirmed the Colonel's judgment: "We're very grateful to have the colonel around to keep us on our toes, but he is a purist and his standards were all right when he was operating just a few stores. But we have over 5,500 now and that means more than 10,000 fry cooks of all ages and abilities." And 10,000 gradations of chicken.

But the consequence of the giant fast-food operations for the neighborhoods in which they are located is far greater than anything the fast-food products may do to one's innards. In social terms, the fast-food invasions are almost always devastating: they dispel the kind of settled tradespeople with whom you could gossip or leave messages for a neighbor or show off pictures of your children or cash a check, and they bring in the kind of transient trade that leads to increased littering and loitering, higher crime rates both on avenues and side streets, and disruptive behavior on the streets and sidewalks. I know that in my own part of New York City, sixteen fast-food places have opened up in the last seven years just in the six blocks along Sixth Avenue that I call my neighborhood, and the jingle-jangle they have produced has measurably harmed a once peaceful and cohesive community.

In economic terms, by destroying the economic texture of an area, the impact of the fast-food chains may be even greater. As bad money drives out good, they drive out not only the local self-owned restaurants ("When we move into a neighborhood," says Marriott Corporation executive Frederick Rufe with apparent cold-hearted delight, "local independents lose business") but a wide range of smaller stores that can't pay the new higher rents or can't survive with the new type of clientele in the area: specialty shops, neighborhood businesses like hardware stores and shoe-repair shops, and local small-scale manufacturers, all of whom are likely to have had deep financial roots in the community. And, economically, the invading chains contribute almost nothing to the area they settle in. A study by the Institute for Local Self-Reliance of Washington, D.C., showed that in a typical big-city McDonald's outlet, *74 percent* of the store's expenditures are sent out of the community where it is based: 41.81 percent goes for food and paper purchased (by contract) from the corporation; 20 percent for rent (paid to the corporate real-estate branch that owns the building), national advertising, headquarters accountants and lawyers, and corporate debt service; 6.53 percent for taxes to state and Federal governments; and 5.62 percent for corporate management salaries. Of the remainder, some 9.07 percent is unclear as to its eventual destination, and only 16.97

percent clearly remains in the community—15.04 percent for local labor and 1.93 percent for local taxes. That means that a hamburger place doing $750,000 worth of business a year would return only about $127,000 of that to the community that made the purchases and the neighborhood that gave it a home. The effect of this on any given locality is not unlike that of a leech on a bloodstream.

NOT EVERYTHING THAT is wrong with modern American goods can be laid to mass production, but the idea of producing identical items in enormous quantities with endless assembly lines in giant factories is certainly the root of much that is shoddy, unsafe, useless, wasteful, and expensive in our country. Sir Patrick Geddes, the Scottish biologist and pioneer city planner, called it making "more and more of worse and worse."

The quintessential mass-production commodity is the American automobile. It is far too big and too complex, it is wasteful of resources in the manufacture and the operation, and worst of all it is unsafe, exactly the quality that is *not* desirable when you put 100,000 of them speeding toward each other on a nation's streets and highways. In the first ten years of recalls mandated by the National Traffic and Motor Vehicle Safety Act of 1966, 51.7 million domestic cars had to be taken back by their manufacturers to repair defects potentially dangerous to life and limb—which is no less than *60 percent of all the cars sold in the U.S.* in that period. Admittedly not all of the defects were particularly serious or certain to cause malfunctions if uncorrected, but it would not be unreasonable to assume that this sizable number of unsafe cars has contributed in some measure to the 50,000 deaths and 5 million injuries we have on the highways every year. Mass production is responsible not only for the faulty workmanship in and inadequate inspection of the automobiles—those who have seen the gruelling and mindless assembly lines in Detroit need not ask why—but particularly for the swift and vast multiplication of any single error, which affects not merely one or two automobiles but hundreds, thousands, even millions—as in the case of the 6.7 million Chevrolets that were produced with dangerous engine mounts and had to be recalled in 1972 and 1973.

Mass production also accounts for the shoddy goods that flood our shelves: light bulbs that burn out within weeks (though individually produced bulbs are capable of lasting for years and, in the classic instance of that famous bulb in California that has been burning steadily since the turn of the century, decades); Teflon pots that chip and peel and put little bits of black plastic into your food (though we know that carefully processed Teflon works just fine, since that's what shielded the astronauts' re-entry vehicles on their return to the earth's atmosphere); wristwatches that don't keep proper time and stop working entirely in a matter of months (though carefully handcrafted watches have been known to keep running, accurately, for centuries). Not to mention paperback books that fall apart during the first reading, home hairdryers whose plastic knobs and hooks break off, window shades that never go all the way back up, plastic "sponges" that disintegrate in a week, rotary can-openers that rip and slip, kitchen towel racks that snap off after a month, ballpoint pens that run dry the second time you use them, charcoal lighting fluid that flickers away in moments, flashlight batteries that never work, dresses with buttons that fall off after the first wearing, costly towels that shred after a few washings, and toy trains that are in pieces by the day after Christmas.

Mass production also accounts for the shockingly unsafe products that maim an estimated 20 million people and kill some 30,000 each year; for the color TV sets that emit sterilizing radiation; football helmets that are never tested for impact strength; power mowers that throw rocks; home appliances that injure 25,000 people annually; toys that injure 700,000 children a year; food products that contain rodent excrement, insect eggs, and human hairs; and chemical food additives (none of which serve any function other than encouraging mass sales) that cause cancer, birth malformations, hyperactivity, and other diseases.

Mass production accounts for the wasteful overproduction of various commodities and the induced "throwaway society" of contemporary America—if you can make a lot of an item, you don't have to tend and fix any one specimen of it. That is why bikes that can last for twenty-five years are actually used in the U.S. for an average of only two years, industrial equipment that could go for twenty years is actually used for an average of twelve years, soda bottles that can have a useful life of five years (a typical deposit bottle is returned twenty times) actually last for about ninety-three days, and cars that can last ten years (in Africa I have seen some kept alive for thirty and forty years) are in fact replaced every 2.2 years.

Mass production accounts to a large degree for the incessant and needless turnover in products, made not to be better or more useful but only to be marginally new and infinitesimally different. Drug and grocery stores put 6,000 new products on their shelves each year in the 1960s, closer to 12,000 in the 1970s—that's 100 new products introduced *every month* for a decade—not because there was any great need or demand for them but because mass production could turn them out and mass marketing could put them in. Alvin Toffler has notedthat 55 percent of all the items found in supermarkets in 1970 *did not even exist* in 1960, and 42 percent of the items you could have bought in 1960 were no longer available.

And mass production accounts for the unnecessary and useless variation of goods, with each variant being turned out in great quantities by separate manufacturers—allowing Americans to choose, for example, among what I calculate to be 651 models of cars in 1977 (no fewer than 97 different kinds of station wagon), or 21,336 different kinds of chairs, or 83 different brands of aspirin all of which contain precisely the same ingredients.

Given all that mass production contributes to our society, it is not surprising that one of every four purchases, according to a Ralph Nader survey, results in consumer dissatisfaction; I suspect it would be higher if consumers were not by now inured to accepting as natural or inevitable most of the defects in the commodities they buy. Of course one should not overlook the other conditions besides mass production that produce faulty goods—there's greed, for starters, or the sway of fashion, or monopolistic power, or indifference—but the central fault is in the scale on which the goods are produced, which inevitably tends toward inept design, cheap materials, clumsy and inattentive labor, hasty inspection, careless packaging, and defective shipping. When production is not geared to human need, to the particularities of the eventual user, or to the capabilities of the individual producer, the end product will invariably have something wrong with it.

Geddes-izing, I'd call it: more and more of worse and worse.

INTERNATIONAL AID and cooperation, which have been the bywords of the world for at least the last thirty-five years, have proven to be impossible dreams because they depend on systems and organizations so vast and complex that they can't get the job done.

International conferences of every kind and under every sort of sponsorship have established a record of palpable failure, no matter what the time or place, no matter what the issue or cause, because the participants either do not agree on how to solve their differences or, more commonly, do not agree on what it is that they are actually negotiating. One weary veteran diplomat was quoted after the failure of the "North-South" conference in Paris in 1977, a fairly typical international meeting of the postwar era: "What it all proves is that international conferences just don't work any more."

International aid systems, no matter by what nation undertaken, have succeeded in helping the populations of the Third World not one perceptible iota. In the words of Peter G. Peterson, Secretary of Commerce in 1972–73 and subsequently chairman of Lehman Brothers, the international investors: "The vast international aid undertakings of the last decade have, by any reckoning, failed; more than 25 percent of the world's people still lead desperate, hopeless and deteriorating lives."

But the most glaring failure of all is the United Nations itself, the behemoth that lumbers about its business at the cost of some $700 million a year, or $2 million a day, without providing more than the most minimal benefit. It has not narrowed the economic gap between the rich and poor nations despite its repeated attempts—in fact the gap has widened and continues to widen every year. It has not been a forum for "harmonizing the actions of nations," as its charter promised, but rather a place of acrimonious conflict and discord where confrontation and belligerence are the common diplomatic modes. It has not been an effective mediator in any international war in its entire existence, with particularly notable failures in Korea, Vietnam, Africa, and the Middle East. It is unable properly to care for or resettle the refugees around the world that have been its special charges. It has not been able to forge significant international agreements on any contentious subject, and even in such a limited and non-bellicose area as the law of the sea it has sponsored two major world conferences resulting in precisely no accords whatsoever. Even its specialized agencies, those most adapted to applying solutions to day-to-day problems, have tended to become top-heavy bureaucracies and to be used chiefly for pettifogging propaganda forums; the two most successful, the World Health Organization and the Food and Agricultural Organization, though responsible for a few worthy achievements, have been particularly ineffectual in stamping out malaria and tropical diseases, in the one instance, or stemming worldwide famine, in the other.

THERE IS MUCH to be said about big cities, and it is an issue to which we shall return in the next section. Suffice for the moment to notice some of the known harmful effects of urban agglomerations. In general, it has been established that as cities get bigger they have increases in crime rates, traffic congestion, commuting time, pollution, income inequality, physical illnesses, personal expenses, municipal costs, taxes, crowding, stress, social disorder, and personal hostility. Moreover, as cities get bigger they generally have less park space, poorer school systems, worse recreation facilities, proportionately fewer hospital beds, and less money per capita for garbage removal, libraries, and fire protection. And in general, the bigger the city the worse the performance, with the very biggest cities almost uniformly ranking at the bottom.

But perhaps the most interesting evidence about big cities comes from a pair of sociologists at Temple University, Bruce H. Mayhew and Roger L. Levinger, who have made a careful and intelligent correlation between the size of cities and what they call the "density of interaction"—that is, the number of "contacts" with other people. They show that as the city's population increases, there is an increasing number of contacts and a decreasing amount of time to spend on them. That seems natural enough. In a big city, just walking down the street you have "contacts" with hundreds of people, and even if you can ignore all of *them,* your contacts with co-workers, business associates, friends, waiters, bartenders, tradespeople, and family are still numerous enough.

One tragic result of this, Mayhew and Levinger point out, is that in big cities it is not uncommon to encounter the "Kitty Genovese" phenomenon. Kitty Genovese, you will remember, was the woman who was murdered in 1965 in Kew Gardens, Queens, screaming for help over a period of thirty-five minutes, with at least thirty-eight people confessing later to having heard her screams and watched the killer return to attack her three separate times but deciding for reasons of their own not to intervene. The two sociologists suggest that this is a fairly inevitable—not to say rational—reaction to the number of incoming stimuli that any New Yorker has during the course of a day, so great that "the average amount of time a New Yorker would be expected to devote to a signal (contact) selected at random is considerably less than it takes to phone the police." Add to that the facts that there is a greater danger in a big city in "getting involved" in any tense situation, that the chances of the attack being genuine and the screams meaningful are quite uncertain, and that the unlikelihood that a call to the police would produce an adequate response in a short enough period to do any good, and you can begin to understand why no one reacted that night. And why the police say there are so many "Kitty Genovese" stories in big cities all across the land every year.

OLIGOPOLY AND CONCENTRATION are generally pernicious in any industry, but just taking the industry I know best—book publishing—their effects can be seen as particularly detrimental. There have been more than three hundred mergers in the publishing industry during the past twenty years, according to the Authors Guild, shrinking the number of major hardcover publishers to fewer than forty and allowing a few big conglomerates to occupy a "significant position in the marketplace," thus eliminating "an 'undue' number of competing enterprises" and placing "the surviving independents at a serious competitive disadvantage."

When book companies get too big they always forget the products they are supposed to be concerned with: books, and by extension ideas, literature, and culture. Here's the way it works. Conglomerate executives who have gained great experience in rental cars and laundry machines tell the president in charge of their publishing division that a certain amount of money is expected to be generated every quarter, since that, after all, is what the conglomerate is in the business of making; as a former editor-in-chief of one major house once admitted, conglomerates "are interested only in the bottom line—overall profit and loss figures." The president then asks his editors to come up with some variant of a five-year plan showing ever-increasing profits—the ruling principle, according to one editor, is that "next year must be bigger and better than this"—and he lets them know that the most important thing is "big books," books that, regardless of merit, can sell a lot of copies. The editors then go looking for books that will catch the current fashions (since no one on the corporate side is particularly interested in works that might be of lasting worth, or even sell very well many years hence), and they are prepared to pay large sums for those few while inevitably slighting or ignoring the majority of writers. In this search the editors will be guided primarily by two non-editorial departments: the sales department, which places the books in the bookstores and knows best which kinds of things seem to be selling that season, and the subsidiary rights department, which sells the hardcover rights to paperback houses and book clubs and which thus generates most of the large sums, the "cash flow" that the conglomerate is looking for. This in turn means considerable actual editorial control is exerted by two businesses quite outside the publisher itself: the paperback companies, which enjoy the only "modern" system of marketing in the book business—the "dumping" of books in drugstores, supermarkets, airline terminals, and the like—but whose editorial judgments are naturally based on what sorts of books appeal to the mass audience that flows into such places; and the bookstore chains, which similarly demand books appealing to a shopping-mall audience.

The deleterious effect of all this would seem to be obvious, a significant change from the days of the smaller, independent publishing houses. This is not to say, of course, that no more meritorious books get published by the conglomerates—only that the chances of that happening are considerably lessened and that it is doubtful that many artists will be nurtured for very long for their talent alone. Nor is it to say that in the past all the small-scale independent publishers were selfless saints without a hint of huckstering or skullduggery—only that when Alfred Knopf wanted to publish a book, he didn't have to base his decision on how many copies it would sell in paperback. The Authors Guild sums it up:

> Conglomerate-owned publishing firms and publishing complexes which have expanded by acquisition appear to be basing their publishing decisions more and more on "the bottom line," rather than on the professional standards that guided publishers when the industry contained many more independently-owned firms. There are indications that books of merit are being rejected because estimated sales, while sufficient to earn profits, are not high enough to produce the return required by "the bottom line."

The overall effect of bigness in the publishing industry is thus to strike at the very core of our culture and the means by which we transmit and preserve it. Because it reduces the number of publishers, it automatically reduces the chances of diversity, the opportunities for new authors and new themes, and the "uninhibited marketplace of ideas" that the Supreme Court has said is the reason publishers enjoy their First Amendment rights to begin with. Because it places a premium on a few big titles, it limits the number and kinds of books published (fewer new books have been published each year since 1974) and it discourages the production of controversial, experimental, innovative, and high-risk works, thus helping to stultify a culture that, God knows, has all the stultification it can handle. Because it encourages anonymity and buck-passing within the successive bureaucracies of the conglomerate houses, it diminishes accountability, moral responsibility, and integrity. Because it reduces the role and even the worthiness of the editors on whom the business has traditionally been based, it does not tend to nurture talent and creativity or to foster change. Because it encourages quantity over quality and fashion over value, it cheapens literature and stifles political expression. And because it produces throughout the publishing industry, from the agent to the bookseller, the attitude that publishing is a business rather than a profession or an art, it does not seek to advance the world of ideas or a civilization based upon them.

Something quite palpable is being stolen, and the thief is bigness.

THANKS TO MODERN construction methods, giantism in contemporary buildings seems to have no limits. ("This new size is frightening," as *New York Times* architecture critic Ada Louise Huxtable says, "and it raises all sorts of questions about density, concentration and sheer, inhuman scale.") Yet in general it is true that the bigger the skyscrapers become, the more costly, less efficient, less adaptive, less safe, and usually more hideous they are.

Sometimes they are gruesome jokes, like the John Hancock Life Insurance building in Boston, a needlessly huge and garish all-glass skyscraper, extravagant and wasteful to maintain, and so big that it would sway in the wind and pop out its mirror-glass sheathing onto unsuspecting pedestrians below. That last defect took some six years and at least $20 million to correct, and even then the company had a 24-hour crew of guards peering through binoculars up and down the surface to see where the next window was about to come loose from its moorings.

Sometimes they are foolish dehumanizers, like the Pruitt-Igoe housing complex in St. Louis, a project consisting of identically dull high-rise buildings so huge that the parents could never keep track of their children, so sprawling that some apartments were located two miles away from the nearest store, and so insipid and dispiriting to live in that crime and vandalism and littering were prevalent from the day it was opened. Ultimately it proved to be so disastrous that it enjoys the claim of being the only housing project that has had to be blown up and torn down even before all the buildings had been fully occupied.

But of all the oddities of architectural gargantuanism, none can quite rank with the NASA hangar at Cape Canaveral, Florida. It seems that a few years ago NASA decided to build a shelter for its huge rockets there, to keep them protected from the frequent rains that are an everyday part of the Florida weather. It brought in all its best scientists and highest-paid think-tank experts, its full range of technofixers and engineers, and after some considerable time and taxpayer expenditure they designed an enormous building, a single hangar-like affair bigger than several city blocks, said to be the largest building *ever built* in the world. The hangar was so gigantic that, as it turned out, it generated its own weather right *inside,* with hot and cold air streams fluctuating within it and pressure systems developing in the cavernous space, creating a steady climate of clouds and rainstorms so that the rockets in the end got rained on, inside their own shelter.

A FEW YEARS ago an angry middle-aged man in Atlanta, Pat Watters, wrote a book about angry middle-aged men. He examined himself, he looked around at his friends, he went out and interviewed hundreds of others in all kinds of jobs, and he recorded their seasoned views of American life in mid-century:

> They are supposed to be among the most powerful elements in the society, but they discover themselves virtually powerless. They have done all the things they were told would make them happy and secure, and after half a lifetime of bedeviled striving, they are neither. . . .
>
> What angry middle-aged men in America have to tell about their careers during the past twenty to thirty years forms a chronicle of the deterioration of [their] ideals in business and the professions, decline in quality and loss of such virtues as loyalty:
>
> "I was in my family's automobile agency, and at first I enjoyed it, selling people good cars, giving them good service. Then the automobile business went wild—wilder and wilder. The factories were bent on nothing but volume, ethics and service and caring for the customer just forgotten.
>
> "The founder of the company died, and that's when things began to slide. Then the company was sold to a big chain, and that's when loyalty to the employee went out the window."
>
> So they blame bigness—corporations, big business, big government—for much that has gone wrong for them and the country. They speak wistfully, but with little hope, of returning to smallness, restoring humanism to a computerized existence.

They do not know it, but they are truly disciples of the Beanstalk Principle—and victims of its violations.

5

Size and Shape

LEOPOLD KOHR TELLS the story of the *Treuga Dei,* the Truce of God, which was first propounded in A.D. 1041 and in successive decades slowly became the dominant code in European warfare. Originally simply a measure to limit the costliness of the battles that were fought from time to time among the cities, principalities, and duchies of central Europe, it held that all battles had to end on Saturday noon and could not resume until Monday morning, thus assuring that the sabbath was undisturbed as a day of peace. Gradually more and more territories adopted the code and gradually too it extended its reach: battles should not be fought in churches or in working grainfields, should not involve women and children, should not continue after sundown or begin before dawn. In time the truce was extended even further, with Friday being declared a day of peace, in honor of the day of the Crucifixion, and Thursday, too, in honor of the day of the Last Supper, and all day Monday as well, in honor of the day of the Resurrection. For nearly four centuries the *Treuga Dei* was observed throughout much of central Europe, particularly where the influence of the Catholic Church was strongest, not by every local ruler and state but by a goodly number, not in every contestation but in a surprising amount. For four centuries, the warfare of Europe was significantly moderated by a principle that seemed to work to the benefit of the many disparate states and cities. There were wars, to be sure, and some bloody battles, even among the adherent polities, but they were usually on a modest scale and not terribly destructive at their worst; and the severest and bloodiest campaigns were waged by those not aligned to the truce (most notably the nascent states of France, England, and Russia) or those who chose to forsake it.

But finally, with the emergence of more centralized states in the sixteenth century, the piecemeal application of the Truce of God and its apparent sanction of midweek warfare came under fire. Maximilian I, who took the title of emperor of the Holy Roman Empire in 1508, began the process of making the truce permanent and all-encompassing, under his royal auspices. If no warfare for five days a week, he reasoned, why not make it seven? If no warfare in churches and fields, why not include streets and towns and forests as well? If no killing of women and children, why not add all innocents, including men, into the bargain? Thus, he argued, instead of intermittent warfare there would be total and perpetual peace—and he set his empire and his armies in service of this noble goal. His success was limited, but after him came others, within the Holy Roman Empire and the budding nation-states of central Europe, who used their growing, centralized powers to enforce the idea of making the partial into the whole, the limited into the absolute, the small into the grand. The result, as Kohr notes, was a period of unprecedented slaughter and brutality leading to the catastrophic Thirty Years War and beyond, a period of all-out warfare fought every single day of the week, Sundays included, of battles on every piece of ground including churches and fields, of the killing not only of men but of women and children as well. And so it has gone down to the present, with increasing brutality and mounting casualties as big nations have continued to proclaim their devotion to everlasting peace, until the twentieth century ushered in warfare fought right around the clock, with the regular and indiscriminate killing of all human beings, and the experience of war extended to every part of the land and sea and in the air besides.

Something had become total and perpetual, but it was not peace.

IN THIS LITTLE TALE we may see the Beanstalk Principle at work in the political realm. As we might expect—and as these next chapters shall explore—when governments become centralized and enlarged beyond a certain limited range, they not only cease to *solve* problems, they actually begin to *create* them. It matters not the nationality of the government, whether it be Russian or American, Swedish or Brazilian, Cambodian or Ugandan; it matters not the ideology, whether it be capitalist or communist, liberal or conservative, Christian or pagan; it matters not the form, whether it be parliamentary or presidential, bicameral or tribal, dictatorial or republican. Regardless of any other attribute, beyond a modest size a government cannot be expected to perform optimally, and the larger it gets the more likely it is that it will be increasingly inefficient, autocratic, wasteful, corrupt, and harmful.

Let us examine for the next several chapters the example of the United States government, to test this proposition. Surely if it holds true here, in what may be properly regarded as a relatively beneficent nation, in what has been held up as a model of representative democracy, within a modern and technically advanced society and a settled and stable state, then we may suppose that it can happen in *any* nation when it approaches similar problems of size, according to its own territorial and population limits. If the United States, historically among the most ingenious, flexible, balanced, and resilient of societies, with immense riches to sustain its profligacy and wide spaces to contain its mistakes, with creative and hardworking peoples settling it generation after generation—if the government of that society shows deficiencies of size, then we may expect that others less fortunate will do so too.

It is appropriate to begin such an examination, as you would with a pumpkin or a car, with size and shape.

FIRST, IT IS necessary to apprehend the extraordinary size of the American Federal government, since despite the alarms about big government it is usually assumed that a nation such as this, without big-brother loudspeakers on the streetcorners and soldiers in the streets and national bureaucrats in the schools, must have a relatively limited governmental structure. Not so.

The U.S. Federal apparatus is charged with, or has taken onto itself, the job of overseeing the lives of some 220 million people, which is more than any government before the twentieth century had ever thought of trying to administer to; with guiding the destiny of the mightiest industrial economy ever fashioned, within an imperial arrangement that affects at least half of the 4 billion people on the globe; and with operating public systems of unprecedented scope and detail that stretch for nearly 5,000 miles and cover 3.6 million square miles of territory. For this task the U.S. employs more people than any other non-communist nation (China and Russia employ more, but of course in their systems far more people count as government employees), some 2.9 million civilian workers, one for each seventy-five of us, plus another 2.3 million people in the armed forces, and another 3 million who work exclusively for the Federal government though in private businesses (such as defense)—8.2 million citizens in all, 10 percent of the workforce. To this should then be added an additional 12 million people who work for state and local governments and may be regarded as adjuncts to the national apparatus as they are in most other countries, and 4 million or so who work exclusively for those governments in nominally private occupations. Taken all together, this comes to a total government sector of about 24.2 million people, *one quarter* of all the working citizens in the nation.

Moreover, this Federal apparatus is abetted by a system of "special district" governments that has grown up to govern various regions and control specific activities within the nation. As of 1972, according to the Federal Advisory Commission on Intergovernmental Relations, which looked into such matters, there were no fewer than 23,000 separate and distinct special district governments operating in the U.S., and the number was calculated to be growing at approximately 12 percent a year. As an example of this complex pattern, the commission pointed to the town of Whitehall, Pennsylvania, a perfectly ordinary settlement of some 16,500 citizens. It has a total of *seventeen* different layers of government influencing its daily affairs, starting with the United States government and the commonwealth of Pennsylvania, and moving on down through the Federal Air Quality Control Region, the Southwestern Pennsylvania Regional Planning Commission, the Western Pennsylvania Water Company, the Allegheny County government, the Allegheny County Port Authority, the Allegheny County Criminal Justice Commission, and so on and on. And that does not include the town's own government and its own independent agencies for sewage or public health or whatever, or the quasi-governmental layers such as the electric company and the post office, or such international bodies as the UN, NATO, or the Universal Postal Union, which may claim some authority.

It seems to be in the nature of all governments to expand, as Parkinson's Law demonstrated so well, but in the American government this has been accelerated by the unprecedented number of tasks that it is called on to perform. There are already so many Federal laws and regulations on the books that the *Federal Register* that records all of them was more than 60,000 pages long as of 1975 and covered a shelf fifteen feet long. Congress each year creates about two hundred new laws, and Federal agencies add to them some seven thousand regulations a year that have the force of law. (All told, legislative bodies throughout the U.S. create about six hundred new laws every single day, or about ten million in the course of an average person's lifetime, a somewhat sobering thought.) Naturally, to administer all these rules the government must create a variety of departments, boards, agencies, and commissions, and it must expand its employees with each new set of regulations. One single new Federal law—the Toxic Substances Control Act of 1975, which attempts to regulate 30,000 different chemicals used by various businesses—required an estimated 7,000 people for its initial enforcement.

It is hardly any wonder that, starting out as a negligible body only two centuries ago, the Federal government has today become the largest single entity of any kind in the country. Indeed, ever since the great period of expansion in the 1930s, every single President has appointed committees or boards to see if something couldn't be done to stop this growth. Since 1960, when the problem became acute enough to cause serious alarm in Washington, strenuous efforts have been made by all branches of government to cut down on the number of agencies and their personnel. The result? Between 1960 and 1977 there were established three new cabinet-level departments (a fourth was added in 1979), three new congressional committees (after a major reorganization), ten new White House units, fourteen new independent agencies, fifty new regulatory bodies, and more than two hundred new advisory boards. Total civilian employment (excluding those in espionage agencies, whose numbers we do not know) increased by 395,000, the executive branch by a mere 15 percent (though the President's Office has gone up 93 percent), the legislative by 43 percent, and the judicial by 72 percent.

But this government expansion has been not only in numbers, vertically in effect, but in reach and scope, horizontally as it were. Particularly since the Rooseveltian ascendency in the 1930s, the Federal apparatus has been charged with the task of overseeing the lives of its citizens and their organizations to an unprecedented degree. Federal laws and regulations now affect almost every aspect of our lives, starting with those that govern the hospitals where our mothers are sent for prenatal checkups and ending with those that determine the places in which our remains may be disposed. Washington's power now extends into virtually every nook of the society; where it does not control, it influences, where it does not dictate by virtue of law, it persuades by reason of wealth. As Washington veteran Richard Goodwin, certainly no cranky conservative, has noted:

> The central government has added more to its reach over the last few decades than during the previous century and a half of our history. In this respect, we are more distant from Herbert Hoover than he was from George Washington. Sovereignty withdrawn from states has gone to augment the immense, continuing, unimpeded accumulation of authority within the central government, especially its executive branch. The result is an Executive vested with almost exclusive power to take substantial initiatives; with a virtually unappealable veto over the actions of states and Congress, particularly in economic matters; and with the almost sole management of government bureaucracies that infringe upon the economic process at every strategic point.

As the case of Rudd, Iowa, testifies. In October 1977, the Federal government, through the Department of Health, Education, and Welfare, told that little hamlet of 429 people that its library would lose the money it gets from Federal revenue-sharing—almost all of its meager $3,500 budget—unless it built ramps up to the building for people with wheelchairs. There are no people with wheelchairs among the citizenry of Rudd, and should there ever be any, the librarian has promised to deliver personally any books they might desire. But the HEW people are adamant: no ramps, no funds. But, the people of Rudd reply, we haven't got the money for that sort of thing even if we wanted to build them—it was all we could do to scrape up the $8,000 for the library in the first place, and that was ten years ago when times were better. HEW is unmoved: rules are rules. The *New York Times* reporter sums it up, probably without even knowing the scope of what he sees:

> What Rudd is enmeshed in is a bureaucratic conflict not unfamiliar to thousands of cities and towns across the nation that have suddenly found themselves confronted with laws and regulations that have

filtered down from Washington through regional offices and state capitals to the local level for implementation. Still, the people who dwell in this quiet village dominated by towering grain elevators amid miles of corn and soybean fields do not often find themselves on a collision course with far-off authority. The whole matter is almost more than they can comprehend.

SECOND, IT IS necessary to realize the nature of the American bureaucracy through which this vast government expresses itself, for the shape of big government is ultimately as important as its size.

All bureaucracies have considerable deficiencies, but governmental ones, which are the largest, have the most—particularly those, like the American, that enjoy civil-service protection, a large public purse, and weak legislative review. This is not news. Ever since the days of Napoleon, when formal bureaucratic structures came to be solidified in Western governments, people have been pointing to the flaws—apparently inherent—in all agencies of government service. (As a matter of fact, Napoleon himself was a victim of certain of these flaws. Desperate for some means of transporting his troops across the Channel for an invasion of England, Napoleon considered many schemes but discarded them all because he couldn't muster a fleet that could outmaneuver the British ships. It seems, however, that a certain Robert Fulton, of the United States, had written to him telling of his successful experiments with ships that could be run by *steam* and proposing the construction of a whole fleet of steamships that could make the Channel crossing quickly and safely, easily outrunning the British sail-rigged fleet. Only, the letter had been dismissed by some governmental functionary as a preposterous boast, and Napoleon never saw it.) But with the growth of the modern welfare state, governmental bureaucracies have grown ever larger and increasingly more deficient.

We needn't linger long over these deficiencies—I assume that there is no citizen much above the age of ten who has not encountered them in one form or another—but among the most obvious may be listed the following. Bureaucracy is characteristically *inflexible,* operating by rules that cannot be altered "because it's always been done that way"; this is one reason, for example, that Jimmy Carter began issuing deadlines to HEW in the summer of 1977 for welfare programs he had no intention of actually introducing before the spring of 1978 ("At the White House," the *New York Times* reported, "officials said that Mr. Carter felt the deadlines were essential to overcome bureaucratic inertia"). It is *uncreative,* failing to reward new ideas, different styles, unconventional practices, or individuality in any respect—so that even the State Department, which every year picks the cream of the college crop, ends up universally recognized as the most unoriginal and hidebound department in Washington, scarcely able to act on an initiative when some non-career diplomat produces one. It is *unproductive,* in part because it is mired in regulations and overwhelmed with paperwork, but in part because there is simply no known way to measure whether it is being productive in providing public goods—since you can't quantify the amount of national safety the Pentagon is providing in return for its massive budget, you can't hold it accountable for greater or lesser productivity. It is *self-protective,* with an instinct for sheer preservation far greater than its instinct for public service—which is chiefly why the FBI paid to keep the Communist Party in existence for so long, and why the Pentagon comes up with new Soviet weapons just at budget time—and it therefore rewards loyalty and subservience rather than whistle-blowing and boat-rocking, as a goodly number of former civil servants can bear witness. And it is downright *inefficient,* an inevitable result of the hierarchic system that tends to distort information on the way up to the decision-makers and orders on the way down to the carriers-out; as Jimmy Carter learned when he signed regulations to "provide immediate relief" for drought-stricken farmers on the West Coast in May 1977 and discovered four months later, after the drought had gotten worse, that not one penny had reached a single farmer and probably would not until well after the emergency was over.

Those failings are bad enough, and go quite a way to explain why big governments fail to perform the tasks they take on. But there are two more characteristics that are particularly troubling about governmental bureaucracies, certainly in a country like America: its *irresponsibility* and its *authoritarianism.*

As to the ways in which bureaucracies breed a lack of responsibility—the "agent mentality" in which all connections between decision and deed are diminished, the familiar Eichmann malady—there are, alas, hundreds of examples. Nazi Germany was replete with them, and imperial Rome, and Napoleonic France, and "socialist" Eastern Europe: things happen, and bureaucrats order them and cause them, but everyone is just passing the orders through, and no one knows who is responsible for them and no one *feels* responsible for them. Gordon Liddy was a typical, if especially notorious, example in this country, a man willing to do whatever the hierarchy told him because it *was* the hierarchy and who took no responsibility for any of his actions, though indeed they led to the greatest rupture of the republic in a century. The one who sticks in my mind, though, is Robert McNamara, because it is he who personifies the amoral Washington bureaucrat at its height—or nadir. For years his waking and conscious self could go on prosecuting the cruelest and most pointless war of modern history, meshing neatly with bureaucratic gears both above and below, sending "personnel" to Vietnam, authorizing "strike forces" against "retaliatory targets," acting for all the world as if it were events and not he who had command. The problems of how to acquire and ship additional men and hardware became the bureaucratic objective, and what they would actually *do* when they got to the battlefield disappeared into some dim background of the psyche. Except of course at night, when, as McNamara's wife later reported, he used to grind his teeth in his sleep, every single night, while tossing and turning until dawn.

As for the anti-democratic nature of bureaucracies, the scholarly evidence is bountiful and remarkably uniform. Robert Michels, one of the greatest investigators, concluded simply, "Bureaucracy is the sworn enemy of individual liberty"; Max Weber held that "the great question" for our time was "what can we oppose to this machinery in order to keep a portion of mankind free from this parceling-out of the soul, from this supreme mastery of the bureaucratic way of life"; sociologist Robert Nisbet has determined that bureaucracy means "the transfer of government from the people, as organized in their natural communities in the social order, as equipped with the tastes, desires, and aspirations which are the natural elements of their nurture, to a class of professional technicians whose principal job is that of substituting *their* organizations, *their* tastes, desires, and aspirations, for those of the people."

But perhaps the most telling evidence of bureaucratic authoritarianism comes from the day-in-day-out experience of the American presidency over the last twenty years. The President, elected by popular vote, is, if anyone, the representative of the people's voice, the embodiment of the national will, and he is supposed to be the chief of the executive branch. In fact there are only about 2,200 out of 2.9 million jobs that he can directly control through appointment, hardly enough to impinge upon the settled bureaucrats and their chiseled ways, hardly enough even to get his orders transmitted properly to the army of civil servants, much less carried out. Every recent President has suffered from the impossibility of this task. John Kennedy tried, for example, to take over the Department of Health, Education, and Welfare bureaucracy with a handful of appointees: "A dozen of us showed up one day to take over 75,000 of them at HEW," a veteran of those days has recalled, "and we never got control. HEW has a heart and lungs of its own." George Reedy, an aide to Lyndon Johnson, has reported that it would take days for the President just to see if something he ordered done had even gone through the White House's own bureaucracy and weeks to find out if it had made its way through the department that was supposed to carry it out. Richard Nixon was so obsessed with being frustrated by the "Ivy League" bureaucrats that he tried to circumvent them entirely by operating the whole executive branch out of the White House and creating his own extralegal organizations outside of civil-service (or congressional) control. And Jimmy Carter has found that what he has called "bureaucratic inertia" is so strong that most of the domestic changes he wants to make can't be done—"It takes time," he once said in apologizing to Afro-Americans for not having done enough on their behalf, "to change the trends of history and to reverse the bureaucratic mechanism to one of support and compassion and concern and enthusiasm," and even that faint optimism was no doubt a result of his still being in his first year in office. As Robert Nisbet has said, "When bureaucracy reaches a certain degree of mass and power, it becomes almost automatically resistant to any will, including the elected will of the people, that is not of its own making."

As a final testament to the bureaucratic shape of the American government, we should take a brief look at the Department of Defense. One would think that if the bureaucratic machinery were to work anywhere within a modern industrial state it would be there, in the defense establishment. That, after all, is charged with the one central task that any government pledges itself to fulfill before all others: protection of the lives and territory of its citizens. Yet it turns out that there is no single agency in all of the Federal government that has performed more ineptly over the years, consistently and excessively, than the Pentagon. In 1970, after twenty years of various presidential efforts to reduce and modernize the defense mechanism, a committee of leading American industrialists released a report of their year-long study of the Pentagon, concluding that, by ordinary business management standards, the place was a shambles. They found that there were at least 35,000 too many employees, most of whom did nothing but shuffle papers from one agency to another; that billions of dollars are wasted each year in trying to develop weapons that would never be operational and would be inadequate if they were; that the sloppy assigning and reviewing of contracts wasted additional billions of dollars a year and the whole procedure should be completely revised; and that because of excessive centralization and bureaucratic sluggishness, decisions on many matters are often simply not made. The committee's chairman concluded: "We have not found personnel problems, but problems of organization. The astonishing thing is that anything works at all." Needless to say, the response to this was quite negligible, and when Jimmy Carter took office seven years later, his key defense task force reported that nothing much had changed. It found that the Joint Chiefs of Staff were unable "to provide guidance, to review contingency plans, and to resolve differences between commands"; that the "decision-making" process was so elaborate and complicated that it frequently failed to work at all; that the growth of the bureaucracy had "blurred responsibility" and made for increased inefficiency; and that the "management structure," particularly around the Secretary of Defense, was overbloated and overlapping and unable to resolve conflicts "over matters of contingency planning and the conduct of operations." It concluded: "Serious questions persist about the effectiveness of the command structure for the conduct of war, for peacetime activities, and for crisis management."

Carter immediately ordered the Office of Management and Budget to perform a "searching organizational review" and demanded action on "command and management problems that have resisted change for many years." No one could be found in Washington who took this seriously, certainly not across the Potomac in the five-sided building. An official there boasted that "the Pentagon goes along in its own way despite the whiz kids and efficiency experts," and disclosed that the OMB examiners were being called "Carter's Little Pills." Carter, frustrated, then appointed Stanley Resor, a former Secretary of the Army, to a new post as Undersecretary for Policy in mid-1978 with full presidential backing for "a significant restructuring of the office of the Secretary of Defense." Nine months later Resor resigned, victim of what one senior defense official said was "the Pentagon's ability to block any modernizing or change, especially when bureaucratic power was involved."

THERE IS A story sometimes told in Washington by the civil servants themselves, usually at the end of one of those days of bureaucratic madness that seem to require alcoholic restoration. It seems that three men, a doctor, an architect, and a bureaucrat, were arguing among themselves one night as to which of them had the oldest profession.

"Why, mine's the oldest," claimed the doctor. "Didn't God operate on Adam and remove a rib from him to make Eve?"

"Yes," replied the architect, "but mine's older still. After all, right at the beginning God constructed the universe out of chaos."

The bureaucrat sat back and looked at them both. "And who do you think," he said, "made *that?*"

6

The Failure of the State

THESE TWIN CHARACTERISTICS of big government—its enormous size and its inevitable bureaucratic shape—are the reasons for its failure. One might suppose that with such an edifice great things were being done, that it achieved, if only imperfectly, an increase in the abundance of the nation or the serenity of its streets or the livability of its environment or the happiness of its citizens. But that is not the case at all. On the contrary, it is precisely during this period of rampant governmental growth that the country has suffered its acutest problems, and is *still* suffering them, as we have seen. And it is precisely during this period that the citizens in poll after poll have announced their increasing dissatisfaction with the operation of the Federal government in almost all its aspects.

It is not because the people who are running our government are evil, or personally inept, or naturally inefficient, that the government has performed so inadequately. In fact we are told, and may believe, that they are an extraordinarily talented and dedicated lot; we know that they have the finest and most modern equipment available and the most current and pertinent information; and—since Federal white-collar pay scales exceed private-sector salaries, on average, by a third—we may assume that they are the finest government employees that money can attract.

But still they fail. Still the government does not perform, does not satisfy. The reason is, of course, its scale.

AGAIN TAKING THE American Federal government for our examples—though it would be perfectly simple to find the same problems in almost any other national government in the world, and in most major provinces and cities as well—let us briefly examine the reasons for the failure of the state.

Big government is overloaded. The volume of paperwork within the Federal government in a single year comes to about ten billion sheets—more than in all the books in all the public libraries of Baltimore, Boston, Buffalo, Chicago, Cincinnati, Cleveland, Detroit, Philadelphia, and Milwaukee combined—and is enough, according to the Federal Commission on Paperwork, to fill fifty professional sports stadiums. ("Paper breeds," as Parkinson once noted.) There is no way for this mass of information to be coherently administered, no matter how large the staffs or how numerous the computers. One gets the sense that the government will eventually reach the point envisioned in Stanislav Lem's fantasy, "Memoirs Found in a Bathtub," in which the bureaucracy sends around its millions of documents in a totally random fashion, the idea being that eventually every paper will arrive at the right desk.

Nor is it possible to make decisions coherently with a paper load of such size. Rational decision-making depends on these billions of documents being accurate and complete in the first place, then going up smoothly and without distortion through the bureaucratic hierarchy, and finally reaching the desk of someone capable of reading and digesting them swiftly and making sensible decisions on them, whereupon the whole process begins in reverse as the decision is transmitted for enactment. That's hard enough, but of course at any time during this procedure the original information may be superseded or some new law may have been passed or a different department may have issued a new regulation or a court case may have provided a new interpretation. George Schultze, in and out of the government at high levels over the last twenty years, offers the example of the Environmental Protection Agency, which has had to sift through several hundred thousand Environmental Impact statements since 1971, each one hundreds of pages long, and issue 45,000 separate plant permits, at the rate of about thirty every working day. It is forced, Schultze notes, "to make thousands of decisions based on detailed considerations it cannot possibly know and, even less, keep up with over time."

Big government is distended. By passing so many laws and then requiring so much paperwork to go with them, the Federal government has created for itself a Sisyphean hell. The Environmental Protection Agency was ordered by Congress in 1972 to assess the safety of the 35,000 pesticides available in the U.S.—a task of considerable urgency, considering the great dangers—but after four years the agency asked for an extension, the following year it asked for another, and in late 1977 it admitted that it would need at least another ten years for the job, providing no new chemicals are produced in the meantime. The Equal Employment Opportunity Commission had a backlog of 100,000 complaints in 1974, just ten years after it began work, and 135,000 in 1978, making it quite incapable of solving any of the grievances it gets within the time necessary to do any good for the claimants. The Federal Drug Administration is supposed to be responsible for overseeing 2,500 food additives, 4,000 chemicals in cosmetics, and 320,000 marketed drugs—at least $200 billion worth of products every year—but new drugs are invented faster than the agency can test them, new plants are opened faster than the agency can inspect them, and new foods and cosmetics are marketed faster than the agency can approve them. The Occupational Safety and Health Administration in its first six years managed to put out regulations affecting only seventeen of the more than fifteen-hundred carcinogens in the nation's workplaces, despite the undenied urgency of the industrial-cancer problem, and the Government Accounting Office estimates that it will take OSHA more than a *century* to regulate just those already identified as hazardous. The Office of Workers Compensation in the Department of Labor is generally considered by Congress to be the single worst bureau in the government, largely because it is asked to look into 30,000–40,000 claims a year, each of which takes an average of 630 days, so that to give disabled workers any satisfaction within even a year it would need a staff of 94,000 investigators, nearly six times as many as the total number of employees at present in the whole Labor Department. The Federal judiciary system handles about 200,000 cases annually, burdening each judge with an average of two hundred cases during the course of a working year, and Chief Justice Warren Burger has warned that unless the caseload is drastically cut there may be "a complete breakdown" of the judicial machinery.

Big government is contradictory. Obviously the right hand will not know what the left hand is doing, but sometimes they are actually doing opposite things. While the EPA was trying to stamp out Mirex, a cancer-causing pesticide, the Department of Agriculture was trying to get farmers to use it to stamp out fire ants. While the Endangered Species Scientific Authority was trying to protect the bobcat by banning the export of its pelts, the Interior Department was sending out its agents to kill bobcats as part of its "predator control" program. While the Department of Health, Education, and Welfare mounts a $5-million campaign to persuade people not to smoke, the Department of Agriculture hands out subsidies to the tobacco industry of more than $65 million a year. While one set of agencies works to improve employment in the ailing central cities, another set actively encourages businesses to move out to the suburbs. And so on it goes.

Moreover, the laws themselves tend to create contradictions, not only because there are so many of them, but because each one has to be complex enough to cover all eventualities in a huge and diverse nation. The Employee Retirement Act of 1974, to pick one example among many, was intended to insure protection for workers in small businesses who were covered by private pension plans that Congress thought were inadequate. The act ended up 247 small-print pages long, it was administered by three separate Federal agencies, each of which added its own set of regulations and interpretations, and ultimately it was so complicated and required so much paperwork that as many as a third of the companies that previously had pension plans decided to drop them. Those were, in the main, the small businesses Congress had hoped to reach in the first place.

It might all be summarized by the comment of one congressional staff member who is an expert on Federal aid programs: "The whole problem is that we've got a Government that is so complex, with so many different objectives, that the consequences of the mishmash are such that we cannot have a central philosophy. It is almost impossible to say what is right or wrong."

Big government is wasteful. Senator William Proxmire in recent years has taken it upon himself to be the monitor of government waste, and every month or so he comes up with another appalling example of people who are not doing the jobs they are paid to do or people who are paid to do jobs that never should be done. An example of the former is bureaucrat Jubal Hale, executive secretary of the Federal Metal and Nonmetallic Mine Safety Board of Review, no less, who willingly admitted that he read novels and listened to Beethoven on his $35,000-a-year job because his agency had not had anything to review in at least four years. As for the latter, the one that Proxmire calls "the most speculative, impractical, and redundant study ever paid for by the Government" was the $225,000 study for the Department of Transportation in 1977 that determined that if there were another Ice Age in America "a very large number of people will be forced or attracted to move to the South and Southwest to escape an undesirable climate," and if urban guerrilla warfare were common in the twenty-first century, "cities would need more transit police, automobile use in afflicted regions would become risky, and damage-insurance rates would rise astronomically."

Obviously these are only minor examples of government waste, and there are hundreds of more serious ones to be found throughout the Federal system, starting with contracts let without bidding, government-sanctioned "cost overruns," and unnecessary or redundant agencies. But just as obviously they make the point that waste is rife in any system so big that no individual or group is able to control it—so big, in fact, that Jubal Hale himself wrote several times to Congress urging them to abolish his position, without finding anyone who could or would do the job, so big that Senator Proxmire is able to come up with his bureaucratic absurdities year after year without creating any movement toward reform in the government at any level.

And that doesn't even take into account the amount of waste that government causes *indirectly*. Because of its labyrinthian shape and its circuitous regulations, it requires of its citizens billions of hours and billions of dollars just to carry on its everyday business with them. In fact the Federal Paperwork Commission, established precisely to look into this problem, determined that the cost to the economy of having individuals, corporations, and other governments filling out various Federal forms comes to about $40 billion a year—a sum that is greater, for example, than the amount that all elementary and secondary schools across the country spend annually.

Big government is costly. Waste aside, big government is simply a very costly proposition—disproportionately more expensive than small government. In part this is because it provides more services, and costly ones like defense and space exploration; but in greatest measure it is because it always adds a layer or two, or three, of administration as it gets bigger, and that means more employees, more salaries, more overhead, more paperwork, more inefficiency—and more money spent. Indeed, it has been estimated that eight of every ten Federal dollars goes not to performance but to upkeep—as for example, public television, out of whose $103-million budget in 1977 only $13.3 million went for actual programs and development of programs, the rest going for bureaucratic care and feeding. Thus it is that over the last thirty years, according to the *Statistical Abstract,* the Federal government has *taken in* on average $50 more per person in taxes than it has paid out in benefits. In fact, Federal taxes are growing so far beyond any reasonable rate that it has professional economists quite perplexed; they are growing twice as fast as costs of food, housing, or transportation, more than twice as fast as the Gross National Product, and more than seven times as fast as the growth of the population this money is supposed to be going to serve. It almost seems as if the same mysterious forces in Washington that are breeding paper are eating dollars.

Big government is inequitable. The bigger a government becomes, and the more complex, the greater is the inevitable role of special interests. In the first place, national governments that seek to involve themselves in all but the tiniest local affairs quickly find out they don't have enough information to do the job properly, so they must either go to the experts or ask the experts to join them. On the scale of the U.S. government, this means a daily working arrangement of the regulators with the people they are supposed to be regulating, and a personnel round-robin by which the regulators join the private regulatees after a government tour and the regulatees take jobs as government regulators while moving up in their private careers. If you wanted to design a system to ensure governmental favoritism, you couldn't do better; as a congressional investigating committee reported in October 1976, this arrangement guarantees Federal partiality "to the special interests of regulated industry and lack of sufficient concern for the underrepresented interests" of the general public.

In the second place, when government legislators seek to make laws in complicated areas beyond their own limited expertise, they also must rely on experts within the industries they seek to control and on the industry's allies in the bureaucracy. This accounts for the very important role played by various industrial lobbies in Washington and thus for the fact that direct subsidies to these special interests, according to the Congress's own study of the matter, come out of Congress at the rate of some *$100 billion* a year. That is an extraordinary amount of money, in fact greater than the total yearly corporate profits of all American business. In other words, it is that direct subsidy that enables American businesses to register any profit at all. No wonder so many of them feel it is a sensible part of their budget to allocate money, legal and otherwise, to influence these legislators and see to their reelection.

Big government is corruptible. By its sheer size and complexity, a big government is largely removed from public scrutiny, so the conditions for corruption—given human fallibility—are ever-present, particularly when it wields far-reaching power. At the same time its codes of bureaucratic self-protection discourage anyone within from calling attention to the crimes, so the chances of adequate self-policing are remote; a seven-month study by Vermont Senator Patrick J. Leahy, in fact, determined that "the fear of reprisal is so prevalent throughout the bureaucracy that waste and illegality are too often allowed to go unchecked." From what we know of the tip of the iceberg, this seems a reasonable conclusion: the sorry record of the U.S. government in just the last decade has included tax fraud, falsification of records, bribery, influence-peddling, misappropriation of funds, forgery, perjury—in fact the whole gamut of white-collar crime—in such agencies as the Small Business Administration, the Federal Housing Administration, the Immigration Service, the Drug Enforcement Administration, the Civil Service Commission, the FBI, the CIA, and the White House itself, in addition to both houses of the Congress. The total bill each year for fraud and theft in the Federal government is estimated by the Justice Department to be as high as $25 billion. Indeed, there is considerable evidence that the government has even been penetrated by elements of organized crime—for example, the continuing grand-jury investigation of Mafia influence within the Interstate Commerce Commission, the twenty-three inside-job murders of underworld informers working for the FBI, and the myriad connections between organized crime and the Nixon presidency. The verdict is not yet in, but it should not come as a surprise that the savants of the huge underworld apparatus have appreciated the possibilities open to them in the huge Federal apparatus.

IT NEEDS NO special political expertise, then, to be able to appreciate the degree to which the U.S. government, because of the characteristics of its bigness, must be branded a failure. We must acknowledge its several achievements, to be sure—the stockpiling of weapons, for example, the exploration of the moon, the building of interstate highways, the overthrow of unwanted foreign governments. But at the same time there's no doubt that, overall, from the point of view of "cost-effectiveness," as the economists put it, the performance of the Federal government is far from adequate. All that money, all those agencies, and yet the results inevitably so meager.

It's not a point that needs belaboring. Consider just the elementary things we ask a government do do:

Housing: government housing programs for the last thirty years have been a scandalous washout, providing billions of dollars of public money for developers and almost nothing in the way of shelter for the poor and the people of the central cities, resulting in the active decay of almost every single urban center in the land.

Food: government agricultural policies have totally failed to supply nutritious food at cheap prices for American consumers, to maintain or increase the farm population, or to feed those most in need—though they have succeeded in keeping millions of acres of fertile cropland from production, preventing the growing of billions of bushels of wheat and millions of tons of grains, and driving innumerable families off the farms and into the cities.

Defense: government defense policies have eaten up something on the order of $1.5 *trillion* in the last three decades, while the country has managed to engage in two calamitous wars (one of them the costliest in history), neither of which it was able to win, while government armament policies have considerably heightened the chances of war and nuclear escalation around the world.

Safety: government drug-protection programs have not worked, the pesticide control program is (according to a congressional committee) in a perpetual "state of chaos," environmental-protection measures have not improved air and water quality in most locations, food inspection does not keep tainted and poisonous goods out of the markets, air and automobile safety policies have come to virtually nothing, and consumer product-safety codes have not reduced the number of related accidents and deaths.

Health: government medical programs, particularly since the start of Medicaid in 1965, have eaten up a staggering number of tax dollars and increased the individual medical bill by about 1,000 percent, but the U.S. has not had a statistically significant per capita decrease in deaths since 1960 from any single illness except tuberculosis, syphilis, and hypertension, while the death rates for cancer and certain other diseases have soared.

LOOK AT IT this way. Suppose a kind of Rip Van Winkle cost-effectiveness statistician were to wake up suddenly in the present and seek to measure the performance of the United States government over the last thirty years. He would first of all want to know how much money was paid out, and we could show him the account books—the annual Federal budgets that show an outlay of something on the order of $5 trillion, give or take a billion or two. He would then want to know just what all this money had accomplished.

He would want to see how many more people had better housing, how many more were eating three full meals a day, how many more were free of diseases. He would ask by how much the crime rates had gone down, and the automobile death rates, and the suicide rates. He would inquire as to how much the economy had escaped perilous ups and downs, by what degree government policies had prevented inflation and steady price increases, whether there was any unemployment left, and how far income redistribution had gone in reducing the gap between rich and poor. He would want to know if vast defense expenditures meant that there had been no significant wars, no great loss of American lives, no escalating of nuclear dangers. He would be curious as to how much the legal system had become more effective and just, its punishments more equitable, its prison system more expeditious and humane. He would ask to know to what degree the quality of the cities had improved, how much cleaner their air, how much more rational their transportation systems, and he would similarly ask how much the small towns and rural areas had been revitalized. He would want to find out by how much the standards of education and effectiveness of the schools had increased and by what extent the benefits of culture had been absorbed into the populace at large. He would look to see by how much the racial harmony among the citizens had improved and to what degree inequities of social standing and status were erased. And he would wonder if the citizens declared their affection for the government more profusely or voted in greater numbers or showed more signs of patriotism or supported it more enthusiastically in times of war or trusted its motives and operations more.

He would put all the figures into his computer and come up with the reluctant conclusion that by every measure of cost-effectiveness productivity the U.S. government had failed the test.

7

Prytaneogenesis

BUT IT IS even worse than that. It is not simply that big government—as we see in the example of the United States—is such a failure because it cannot *solve* the multitude of economic and social problems. There is also the dismal fact, as we saw in the story of the *Treuga Dei,* that it actually works in a great many ways to *create* them. Simply as a result of the inefficiencies and inequities of their size, and quite often without intending to, big governments tend to set in motion forces that they are unable to control, whose progress they are unable to predict, with consequences, many deleterious, they are unable to measure.

Thus government regulations attempting to prohibit the sale of alcohol and to remove the scourge of alcoholism served to create a vast network of organized crime that set such firm roots that it still thrives to this day. Government laws and crackdowns designed to control the drug traffic and wipe out addiction have driven the trade underground, fattened the coffers of criminal gangs, generated billions of dollars of burglary and street crime, created an aura of challenge and danger that attracts the young, and failed to stem an increase in the estimated addict population from 50,000 to 560,000 in the last thirty years. Government tax policies designed to stimulate economic growth in underdeveloped sections of the country helped to create the pell-mell chaos of the Sunbelt in the last three decades and drain industries and populations from the ailing older regions of the Northeast. Government regulations intended to protect consumers and save them money have ended up increasing prices on every single commodity where they pertain, by about $180 billion a year, according to one congressional study, "only" $103 billion a year according to the figures of Washington University's conservative economist, Murray L. Weidenbaum. The National Highway Traffic Safety Administration itself, for example, acknowledges that regulations have added more than $200 (Weidenbaum figures $430) to the price of a new car in the last five years, for a total national bill of more than $1.8 billion annually. And, perhaps most telling of all because most unintended and most reverberative, government programs to spur interstate commerce and national cohesion through the $70-billion superhighway system have resulted in—but let the government's own independent Advisory Committee on National Growth Policy tell this one:

> The problem is that the planning which preceded construction of the highway system was narrowly focused and largely ignored anything not directly connected with technical design and construction *per se*. It was known in the early 1950s that concentrations of automobiles can cause severe pollution problems, but this was ignored. Routes were laid through central cities, ultimately requiring the eviction of thousands of people at a time when good housing was in short supply, but this was brushed aside in 1956. Obviously, the system would have (and has had) enormous and detrimental effects on other modes of transportation, but these too were overlooked. By opening up the suburbs to uncontrolled growth, the system facilitated urban sprawl and accelerated the decline of central cities, but this was not taken into account either. Tens of billions of dollars were spent on a program which promised to—and has—shaped our nation in concrete. The structure of domestic commerce was violently altered without consideration of the "spillover" effects that such a system would have on important sections of the economy and the regions through which the roads were to be built.

In short, one may conclude that the bull set loose in the china shop, though it may be quite well-meaning and may indeed intend nothing but peace and happiness, by its sheer size will create disaster.

It is a process that seems to me to deserve a name. Following the model from medicine, in which the term *iatrogenesis* refers to illnesses actually generated by a doctor, we may derive another Greek coinage, *prytaneogenesis*—from the Greek word *prytaneo* for the seat of government—to refer to damage actually generated by the state. Wherever we locate a problem and find some governmental law, policy, regulation, statute, code, or decree at its root, there we witness the operation of prytaneogenesis.

LET ME ILLUSTRATE the process with a short look at the awful urban phenomenon of the slum, a scar that seems as inevitable and as natural as, say, a desert. But we now know that deserts are not in the main caused by natural shifts toward hotter and dryer climates or a sudden drying up of rains and rivers. Deserts everywhere around the world are caused primarily by human societies ignorant of the ecology they are a part of, which overgraze, overcut, overcrowd, and overfarm the earth, weakening the natural vegetation necessary for the protection of the soil. In the same way, ghettos and slums are not caused by natural shifts in the economy or a sudden change in human habits and behavior. Slums are caused by human governments, through ignorance or intention, that overregulate, overmeddle, overrule, and overpower the city neighborhoods, weakening the human culture necessary for the protection of the community.

The South Bronx, that 12-square-mile section of New York City that has become synonymous with urban devastation, is a good case in point.

Today it is a place of such ruin—buildings burned out for block after block, their black windows staring pitilessly on the rubbled streets—that almost every visitor compares it to the bombed-out cities of Europe in the wake of World War II. Many sections are home only to derelicts and a few desperate welfare families; in others you find clusters of sad and determined people still working to keep up their homes while garbage piles up along the sidewalk, rats skitter through the vacant lots, and nearby buildings are gutted by fire after fire; three-quarters of the occupied buildings do not meet New York's legal minimum standards for health and safety. Hopelessness and fear are present everywhere: the crime rate is the highest in the city; more than a hundred youth gangs roam their turfs; heroin addicts cluster in every neighborhood and are serviced openly; arson-related fires gut buildings almost every single night of the year. The area is poor—the poorest in all New York, worse than even Harlem or Brownsville, with a median family income of $5,200—and unhealthy—with a quarter of all the city's known cases of malnutrition and an infant mortality rate almost twice the New York State average. It has been called—and there are few even among the local residents who are inclined to disagree—"perhaps the most wretched slum in America" and "a social sinkhole in which civilization has all but vanished."

But it wasn't always that way. It wasn't that way even forty years ago. It became that way mostly because of specific actions by large and sadly uncomprehending governments.

Forty years ago the South Bronx was not exactly a Gold Coast or a Georgetown, but it was a solid, safe, and stable community, primarily blue-collar white, with Italians, Poles, Jews, some Irish and Germans, only a sprinkle of blue-collar blacks and a dot here and there of the well-to-do, some with roots going back to the earliest years of the country. It had its tenements, six or eight stories, brick, dull, with fire-escapes on the front, not always very sightly to be sure, though well-built, substantial enough, and usually safe; but on many streets there were clusters of two-story row houses, owner-occupied, neat and brightly painted, with tree-lined sidewalks and little gardens; and in a dozen different neighborhoods there were large, handsome apartment buildings, with ornate carved cornices and friezes done by the first-generation craftsmen, lobbies of Italian marble, high-ceilinged and spacious rooms, home for one or another tightly knit and upwardly mobile ethnic community. A good deal of the industrial plant in the area was old and some of it obsolescent, but there were plenty of factories and warehouses operating all over—food processors and a piano manufacturer, some garment factories and small furniture manufacturers—and the area had an economic vitality still. And there were always the Yankees, who played in the stadium on the western edge of the South Bronx, and in the 1930s and 1940s they were champions.

Then came the deluge.

After World War II—and then in every housing act and in most tax bills for the next thirty years—the Federal government intervened in the housing markets with schemes that continuously induced people to move out of older neighborhoods like the South Bronx and into newer developments in the suburbs. Low-interest mortgages, tax credits, and government guarantees made it advantageous to buy the single-family homes being put up in rings around the central city and uneconomical to try to repair and rehabilitate existing urban housing: a family that wanted to buy a new, jerry-built $40,000 house in the suburbs could get it for about $2,000 down after all the credits and tax breaks, but a family that wanted to buy an old, solidly built $40,000 house in the city would have to put down about $13,000, assume a high-interest mortgage, and pay full rates for repairs. Naturally, most of those who could afford it chose the suburban alternative, and the backbones of communities like the South Bronx began to crumble. As the Urban Institute put it in its recondite way somewhat later:

> Directly or indirectly, provisions of the federal tax code have tilted the terms of economic competition in favor of suburban development and development of new regions of the country, thereby accelerating the abandonment of central city housing, contributing to the deterioration of the existing urban capital infrastructure, and precipitating decline of the central city tax base.

Other similar Federal tax policies also encouraged businesses—particularly the kinds of small manufacturing and warehouse businesses that had been the mainstay of the South Bronx—to relocate to new premises in suburban "industrial parks" or the growing Sunbelt states. Slowly, the jobs, and the economic ripple effect they produce in the attendant service industries, began to vanish.

And then came the interstate highway scheme. For the South Bronx, the unintended effects were almost extirpatory: it pre-empted funds for mass transit, the key to the region's economy; it encouraged the flight to Westchester and Long Island of still more businesspeople and homeowners; and it mandated the destruction of several stable neighborhoods so that freeways like the Major Deegan and the Bruckner could extrude their concrete sepulchers on top of them.

The Federal hand did not stop there. As it was drawing the more affluent (and often white) citizens out of the cities, it was forcing the poorer (and often black) citizens into them. Federal tax policies and subsidies helped to establish the agribusiness industry in the postwar years, particularly in the Sunbelt states, one effect of which was to mechanize and expand larger holdings at the expense of the poorer and more marginal farmers and farmhands, who then generally came to Northern cities in droves to look for work. At the same time, Federal welfare money, funnelled largely through states and cities in the North, also attracted impoverished newcomers by giving them a way to live even if they couldn't find jobs, drawing millions of unskilled workers into those cities, like New York, that agreed to participate, and into those regions, like the South Bronx, that were already in fluctuation.

By the 1960s, as a result of all this, the South Bronx was teetering on disaster. Enlightened policies—or, better, none at all—might have saved it. But that was not to be. Another outsized government, in this case New York City's, had its own contributions to make. City rent regulations discouraged the rehabilitation of marginal buildings by wrapping the whole process in complicated red tape, administered by sinecured bureaucrats and approved by not-always-honest inspectors. City rent-control laws allowed landlords to raise their rents with each new tenant, in effect encouraging them toward high turnovers and minimal repairs. City taxes rose steadily as city services expanded beyond those of any comparable municipality in the Western world, placing an increasing burden on the property owners, especially those with marginal rental housing. City welfare programs, severely overburdened, distributed clients wherever housing was cheapest, slowly pouring them into the deteriorating tenements of the inner city, creating a whole inbred welfare ghetto where the dispirited fed on the despairing amid the decaying. And thus, gradually, whole blocks of buildings began to sink into disrepair, landlords choosing to give up rather than resuscitate, tenants choosing to desecrate rather than maintain, and within just a few years the domino effect of abandonment began to topple its way down the South Bronx streets, gathering momentum as it went: between 1970 and 1975 at least 45,000 buildings were left to rot or burn. It was not long before an arsonist's racket grew up, teenagers being paid to burn down still-occupied but decrepit buildings so that the landlords could abscond with the insurance money—tax-free, thanks to Federal policies.

The flurry of Federal money thrown into the South Bronx in the 1960s and 1970s paradoxically worked to destroy the area even further. Federal housing programs paid for more than 10,000 dwelling units there, all of them in depressing high-rise projects whose sterile and anti-human buildings had the look of East European penitentiaries and served as little more than crude detention centers for the poor and elderly. "Model cities" projects were pasted together without thought for controls or safeguards, inviting massive frauds and scalawaggeries every year. The vast array of "war on poverty" programs, though clearly well-intentioned and well-financed, were similarly misguided. The grants actually went to paper agencies controlled by machine politicians and their henchmen who were able to build up "anti-poverty empires" to siphon off funds for their own purposes and leave very little behind for the people who needed help. Federal agencies remote from the South Bronx streets and honeycombed with overlapping bureaucracies had little idea where their funds were really going and in general cared more about satisfying their regulations than satisfying their clients. It took HEW, for example, three years to find out that one local satrap in the South Bronx was bilking the agency out of more than a million dollars, so lax were its financial controls—and even then it couldn't find enough hard evidence to prosecute the man, because he had easily covered his tracks months before. At least $500 million, perhaps $1 billion, was wasted on the South Bronx in this way, as useless as if it had been burned in the gutters.

The ripple effect of this Federal squandering was even worse. When money went to machine satrapies, it was not available for genuine community groups and neighborhood associations, and they slowly lost the allegiance of the populace and began to wither. People with no popular base other than their personal retinues and friends at City Hall came to exercise enormous power, and it was hardly surprising that many of them went whole hog into corruption, extortion, thievery, and other crimes; one boss, to pick a particularly callous example, took $312,000 from the New York State Education Department to give out free summer lunches to malnourished children and simply dumped the lunches in vacant lots to save the cost of dispensing them. Much of the money pouring into the area, which neither Federal nor city agencies were able or willing to keep track of, found its way into crime, particularly the heroin trade that flourished in the South Bronx from 1967 on and the police bribery that made it possible, giving the region not only one of the highest crime rates in the nation but one of the most corrupt police forces as well—as Frank Serpico found out when he was stationed in that very precinct and discovered nearly everyone but he was "on the pad."

One might find dozens of other examples of government actions that hastened the slide of the South Bronx into its present perdition. New York City came up with a Yankee Stadium "renovation" project that theoretically was to create new jobs and rehabilitate streets and shops in the stadium area but that managed to usurp more than $100 million of city funds away from far more needy Bronx projects without adding one permanent job or rehabilitating a single street. New York State financed the huge Co-op City apartment complex, designing it specifically for "middle-income" people (mostly white) and siting it off in the northeastern corner of the Bronx, thus effectively drawing most of the last remaining middle-echelon families out of the South Bronx and making it darker and poorer at a single stroke. And the Federal government provided the 1966 Banking Act, which permitted savings-and-loan banks, heretofore restricted to making most loans in the neighborhoods where they were located, to send their investments to any part of the country, thereby instantly drawing money out of the South Bronx, where investments might be risky, to the booming real-estate regions of the suburbs and the Sunbelt, beginning a process of "redlining" that effectively dried up all loans and mortgages in the area within two years. Given all this, it is perhaps fortunate that the grand multi-million-dollar schemes started by the Federal and city governments after Jimmy Carter's much-publicized visit to the South Bronx in 1977 have apparently fallen into such a morass of red tape and political bickering—there is a New York City office of sixty people that is spending $4 million a year simply to get the programs organized—that only a third of the promised money is actually being administered.

But the specific governmental measures are less important than their relentless cumulating effect: to bleed the life, slowly and painfully, from the South Bronx and similar communities in older cities across the land. Yes, the South Bronx is a slum today—but no, it was not always that way, and it didn't have to become so. It was *made* that way. It does not lessen the tragedy to know that the legislatures that created the measures of destruction often intended to be benevolent, that the bureaucrats who carried them out often were able and well-meaning, or that the judiciary that upheld them often was sensitive and reasonable. The bare and unappealing fact remains that it was the hand of big government and not any accident of nature that wrought the tragedy.

Prytaneogenesis.

THE SAD STORY of the South Bronx points not only to the creation of problems by governments grown beyond their efficient size but to the essential reason why this should be so: namely, that big governments tend to weaken and ultimately destroy the very organizations that might be capable of solving the problems, the local agencies, block associations, and community groups who know most about them. It seems that political power is finite and inelastic: when one instrument absorbs it, it no longer exists elsewhere. When the Federal government takes effective power into its own hands, it simultaneously removes that power from local and intermediate governments, and the more power it absorbs, the weaker do the local agencies become.

This is a pattern that is familiar to any historian of the nation-state over the past five centuries. As the state has grown in power, it has done so at the expense of the villages and cities that once were sovereign, of the duchies and principalities that once were independent, until ultimately the only permissible governmental organization of power has become the state. The process took several hundred years, but it was quite ruthless and quite methodical. The powerful families that set themselves up as royal houses and sought to build a large-scale apparatus around them worked systematically to deny to the smaller units powers that they had had from the time of the Roman Empire. Gradually, over decades, these nascent states passed laws against town meetings and folkmote government, they warred against the medieval "free cities," they outlawed local guilds and municipal unions, they took away communal farmlands and common grazing grounds, they established central control over independent universities, they replaced local coinage with royal banks, they abolished local tolls in favor of "King's Highways," they superseded municipal taxation with nationwide tax systems, and they sent their own agents and operatives to direct local affairs following directives from the central capital. They were often resisted, sometimes ignored, and in a few instances thwarted—but eventually, over the long sweep of centuries, these royal families succeeded in enlarging their powers, established ready allies among a few they were pleased to designate as nobility, amassed armies and fortunes to ensure their dominance, fostered bureaucratic machines to minister to their wishes, and ultimately forged what we recognize today as nations over most of the map of Europe. It was in its way an impressive achievement, but as Lewis Mumford tells us:

> The consolidation of power in the political capital was accompanied by a loss of power and initiative in the smaller centers: national prestige meant the death of local municipal freedom. The national territory itself became the connecting link between diverse groups, corporations, cities: the nation was an all-embracing society one entered at birth. The new theorists of law . . . were driven to deny that local communities and corporate bodies had an existence of their own: the family was the sole group, outside the state, whose existence was looked upon as self-validated, the only group that did not need the gracious permission of the sovereign to exercise its natural functions.

What happens with this consolidation of national power is that the state *makes itself necessary* by destroying the other organizations that were supplying public services to the citizenry before it came along. When the guild system operated in medieval cities it was generally guided by the *conjuratio*, the oath of mutual help by which each guild brother and his family would pledge responsibility for the welfare of other guild members and rally to them in time of need. But when the guild organization was destroyed by royal decree as being contrary to the interests of national mercantilism, there was no one to take care of the sick workman but his own small family, and more often than not it proved inadequate. At that point the state stepped forward and said, look, this poor man is sick and friendless, we must build a state hospital for him, he will surely die if left to himself—and, look, does that not prove the necessity of the state?

It happened all over, and it went on for centuries. As the size of the state increased and its control over localities grew, the instruments of self-sufficiency and cooperation by which these communities had come to live began to wither. As they withered, the state took upon itself the task of intervening to supply those services left undone, thereby justifying its existence by remedying the problems that its existence caused. Simple, circular—and devastating.

WE HAVE SEEN this same prytaneogenic process in the growth of the United States, of course.

In the earliest days when there was no government other than a chartered colony, the settlers had to do most things for themselves and thought nothing of it; official governments were so unimportant that the colony of Pennsylvania in the seventeenth century went without a governor or legislative council for twenty years without anyone visibly worse for it. The entire Revolutionary War was fought with nothing more than a temporary and intermittent "Continental" Congress, whose powers were restricted in practice to the raising and support of an army, and much the same system worked for the first eight years of the new nation's life under the Articles of Confederation, during which time local governments of only modest potency were easily the most important influence in the citizens' lives.

Then, for a variety of complicated reasons, some of the powers of the land decided that a more centralized state would be more beneficial for such enterprises as carving out the wilderness, exploiting the natural riches, fostering manufactures, and protecting international trade. Thus arose the Constitution of 1787 and the Hamiltonian principles of Federal power, with the adoption of which the United States embarked on the same kind of path the earlier royal houses of Europe had traveled in consolidating the nation-state at the expense of local governments.

That process was a long and complex one, but two disparate examples may serve to illustrate it: government control of the postal service and government dominance of education. Today we generally assume both to fall within the province of the state. It was not always that way.

In the earliest settlements, mail service was a matter of simple cooperation: letter packets would be off-loaded from ships at some dockside tavern or shop, townspeople would come in to collect their mail, and the few letters going into the hinterland would naturally be entrusted to the occasional traveler. Colonial governors set up systems of their own from time to time—a monthly service was begun between New York and Boston in 1672, another between Philadelphia and Delaware in 1683—and army units typically had their own couriers, but private citizens generally relied on travelers, friends, and neighborhood boys. Many settlements also depended upon some sort of informal cooperative arrangement along the lines that the Virginia colony made official in 1657, by which each plantation was expected to deliver the mail packet to the next outlying one, and so on from coast to hinterland. Private mail-service businesses also developed early on, charging rates according to distance, and this was dependable enough so that one William Goddard, a Baltimore printer, was able to run a service among thirty colonial cities at a steady profit. Even when the British crown took over the operation of official posts in the eighteenth century, such private services continued to be tolerated, but the government soon took the expedient of buying them out when they became too successful and threatened the income from official mails.

With the creation of the new nation, one of the first acts was to create an official postal organization (1789), under the Constitutional provision that "the Congress shall have power to establish post-offices and post roads"; it ran, naturally, at a deficit. Over the next few decades, it was authorized by Congress to extend its control over more and more territory, gradually displacing more informal methods of delivery, but somehow it never seemed to be able to match the private systems in either service or profitability. By the 1840s these private systems were so numerous and so successful—at least 200 of them in the Boston area alone—that the government decided to launch a frontal attack. Congress protected the government service with the 1845 postal act that explicitly prohibited "any private express" between cities running "by regular trips or at stated periods," and the Justice Department vigorously prosecuted the competitors, such as Lysander Spooner's American Letter Mail Company operating along the East Coast.

Thereafter the government set about determinedly to protect its monopoly over the postal business—not because it was so good at mail service, one hastens to note, but precisely because it was so poor at it. Occasional private services were tolerated in the Western territories—the Pony Express, for example, in the 1860s—but in the states, Federal laws required compulsory prepayment in 1855, the use of official stamps in 1856, the use of Federal equipment on rail routes in 1864, the elimination of all private carriers within cities (and even buildings) in 1872, and the exclusive patronage of official couriers for rural delivery in 1896. By the twentieth century, the postal service was a clear and powerful monopoly, its reign absolute over all classes of mail (with the exception of some types of parcel and third-class). Today it is a crime for any private service to engage in dispatch of first-class mail or door-to-door delivery; home mailboxes are specifically designated as government property and are forbidden to anyone else, paperboy or neighbor or political candidate or advertiser; and letters delivered by any person anywhere, even from your daughter to her classmates inviting them to a birthday party, are legally required to have U.S. stamps on them.

And in Rochester, New York, in 1976, Patricia and Paul Brennan were sued by the Federal postal service to keep them from continuing their thriving private first-class mail system, which guaranteed same-day delivery at a cost three cents under the Federal rate. The courts, naturally, upheld the suit. The *New York Times* declared: "The Brennans may have been enterprising, but they were also behaving like vigilantes—benign vigilantes, to be sure, but still taking a government prerogative into their own hands. There are some enterprises that can't be free, that must be monopolized by government." How far has the country come.

So the state justifies its existence by arguing that it is after all necessary to deliver the mails. That this service has seriously deteriorated in recent decades, running up deficits of hundreds of millions of dollars every year but one since 1945 (1979 was a temporary exception, following rate increases of 15 to 25 percent), while the quantity and quality of services have shrunk, suggests that in fact the state may not be a very good instrument for performing this task. That private parcel services have increased their business since 1960 from a few million packages to more than a billion a year, making a tidy profit out of that section of the mails in which they are allowed to compete, *and* providing better service, suggests that indeed the original system may have made the most sense after all.

EDUCATION REPRESENTS another area where government intervention has worked to undermine effective local power, even if somewhat more slowly and subtly.

For more than a hundred years, schooling in the U.S. was a matter for the home, the local public school, and the church or private school, with state and Federal governments remaining discreetly in the background. From the first New England township schools in the eighteenth century and down through the public school movement of the late nineteenth century, education was in the hands of local school boards, usually elected by members of the school district and empowered to raise their money through local taxation. Then at the beginning of this century, first the state capitals and afterward Washington began to intrude increasingly into these local provinces. State governments initially took to themselves the power to establish standards and examinations for all classes of schools, then to organize teachers' colleges and certifications, then to amass and distribute larger funds, and finally in recent decades to mandate the textbooks, classroom equipment, and types of instruction that would be used; and as local boards right across the country have gradually had to depend more and more on state financing, gradually have they had to agree to state demands by that same degree. The Federal government similarly has increased its role, beginning with grants for agricultural and vocational education during World War I and climaxing with the heavy war-on-poverty outpourings of the 1960s and 1970s and the creation of a separate Department of Education in 1979. Today, Federal money plays an increasing part in setting local school budgets—either directly or through state grants—and thus Federal directives play an increasing part in settling such local school matters as racial mixture, school architecture, pupil selection, curriculum content, textbook purchases, and teacher qualification.

The result of all this is that now local boards, though still empowered to *raise* money at the local level, are practically powerless to *spend* it as they see fit: from 90 to 99 percent of all expenditures now are mandated by state and Federal requirements and by contracts with government-sanctioned unions. The sad complaint of a president of a suburban school board association on Long Island can be heard almost from coast to coast: "We no longer have much control over finances because of the mandates, and we no longer have control over whom we can dismiss because of the unions and the courts. There has been a severe, accelerated erosion of local control."

And what is the ultimate result of that erosion of local control? A prytaneogenic decline in the quality of education, stark and irreversible. It's no mystery. It happens everywhere, in every kind of city in this country, in every similar system in the rest of the world. Where the control of education is taken out of the hands of the family and the community, and schooling gets further and further away from the people who have a direct stake in it, the quality suffers. It is that which accounts, in largest part, for the deplorable state of American education today. Yes, the government now controls education . . . but is it worth controlling?

SINCE THE U.S. government assumed control of the Trust Territory of Micronesia in 1947, it has spent more than $1 billion to bring the benefits of the modern welfare state to the 100,000 inhabitants of the islands. It has built schools and hospitals and centers for the elderly, it has established unemployment and income-maintenance and housing-subsidy programs, it has provided free school breakfasts and lunches, free medical care, and free food programs, and it has created a government bureaucracy that employs fully 70 percent of the workforce at salaries twice those of the private sector. One might therefore reasonably suppose that Micronesia today would be very close to a Pacific paradise.

Alas, no. In just three decades—a miraculous compression of the process that elsewhere had taken centuries—the American government managed to destroy and displace most of the stable traditions by which the Micronesians had guided their lives; a *New York Times* reporter noted succinctly, "American spending has distorted Micronesia's economy and undermined the beliefs and institutions that for centuries kept the islands in delicate balance with their oceanic environment."

Once, in Micronesia, concepts of individual possession were quite foreign and crime was virtually unknown; then American values were transplanted and encouraged by the territorial regime, the values of mutualism were largely discarded, and today crime, particularly theft, has become a serious and growing problem. Once, families and communities naturally accepted responsibility for the welfare of the young, the infirm, and the aged, in a system of human reciprocity that was held inviolate; then a dozen government welfare programs were introduced, those in need were encouraged to forsake the old ways, and eventually so many people came to depend on government handouts that the outlay amounted to 80 percent of the entire Micronesian budget. Once, local craftsmen carved sailing canoes and built thatch houses, the traditional island habitats; then the government started vocational programs and opened an Occupation Center that trains carpenters to use power saws and drills—of marginal usefulness in a land where almost none of the villages have electricity—and the old-fashioned craftsman has become a dying breed. Once, the country was self-sufficient in food, with abundant breadfruit trees, large fish populations, and fertile soil for vegetables; then the government inaugurated food-import programs (200 cases of prunes, in one instance) and tried to modernize the local agriculture, with the result that today breadfruit is allowed to fall and rot on the ground, there is very little farming done, tuna fish is imported from Japan in cans, food prices are 80 percent higher than in the U.S., and malnutrition has become endemic. In Micronesia today, unemployment among the young stands at 30 percent, alcoholism has become a major social problem, slums are growing within all the larger cities, poverty is spreading, and suicide has become the number-one cause of death among the young.

John MacInnis, an American who is an "economic development officer" for the islands, is somewhat bitter at what he has seen of the American impact on Micronesia. "If I were the District Attorney," he says, "I think I would take the case to the grand jury that all this was deliberately designed to induce dependency."

8

The Law of Government Size

IT IS AN interesting fact that when the peoples of Germany were divided into dozens of little principalities and duchies and kingdoms and sovereign cities—from about the twelfth century to the nineteenth—they engaged in fewer wars than any other peoples of Europe. During this period, according to the estimates in Quincy Wright's massive *A Study of War,* the Germans participated in only thirteen wars—whereas Denmark participated in fifteen, Sweden in twenty-four, Austria and Russia in forty, France in forty-two, Britain in forty-four, and Spain in forty-eight. From 1600 on, when modern warfare developed in Europe, the German states (even with the rise of Prussia) spent fewer years at war than any other nations except Denmark and Sweden, engaged in fewer battles, and suffered fewer casualties. Generally, it seems, the various small polities saw no particular advantage in trying to assemble unified forces for warfare, so attacks *by* them were few; and the outside powers saw very little advantage in trying to conquer and govern a bunch of disparate entities, so attacks *upon* them were few. Not that there was total peace, nothing so otherworldly as that. But there were long stretches without war, and those (mostly internecine) wars that did erupt tended not to be so intense or so lasting as those on the rest of the continent.

All that changed, of course, with the unification of Germany and the establishment of a government of 25 million people and 70,000 square miles. Then, within only a few decades, the German state embarked on major wars against Denmark, Austria, and France, conquered territories in Africa and the Pacific, and ultimately instigated two devastating world wars within the space of thirty years.

I find that quite suggestive. Nothing conclusive, of course, but it stands for me as a symbol of what large nations and governments so very often do, because of their size and power, that smaller ones usually do not.

And that same phenomenon is one that I find recurrent in history since Sargon's first empire in the second millennium before Christ. Governments, whether meaning to or not, always seem to create more havoc as they grow larger, and the largest of them historically have tended to be the most disruptive and bellicose. Some almost tangible quantity of power seems to accumulate in large-scale governments, boiling and seething until it has to find an outlet, sometimes emerging as domestic upheaval, often enough of the economic kind, sometimes as international warfare—and frequently as both. Indeed, so regularly does one encounter this phenomenon in the reading of history that I am emboldened to advance this as a full-blown maxim, what we may call the Law of Government Size:

Economic and social misery increases in direct proportion to the size and power of the central government of a nation or state.

Let us examine that through a brief overview of the historical record.

AS TO THE proof of this law in the ages before the rise of modern Europe, I can think of nothing better than the conclusions of Arnold Toynbee, whose masterful study of human civilizations is replete with evidence that any student may subject to microscopic examination. Time after time he shows that civilizations begin to decay shortly after they are unified and centralized under a single large-scale government, and he posits that the next-to-last stage of *any* society, leading directly to its final stage of collapse, is "its forcible political unification in a universal [by which he means united and centralized] state":

> For a Western student the classic example is the Roman Empire into which the Hellenic Society was forcibly gathered up in the penultimate chapter of its history. If we now glance at each of the living civilisations, other than our own, we notice that the main body of Orthodox Christendom has already been through a universal state in the Ottoman, . . . that the Hindu Civilisation has had its universal state in the Mughal Empire and its successor, the British Raj; the main body of the Far Eastern Civilisation in the Mongol Empire and its resuscitation at the hands of the Manchus; and the Japanese offshoot of the Far Eastern Civilisation in the shape of the Tokugawa Shogunate.

There is, Toynbee concludes, "the slow and steady fire of a universal state where we shall in due course be reduced to dust and ashes."

Lewis Mumford, from a different perspective, has reached these same kinds of conclusions. Throughout history, he has shown, the consolidation of nations, the rise of governments, has gone hand in hand with the development of slavery, the creation of empires, the division of citizens into classes, the recurrence of civil protests and disorders, the erection of useless monuments, the despoliation of the land, and the waging of larger and ever-larger wars.

Of early cities, Mumford writes:

> Once the city came into existence, with its collective increase in power in every department, [its] whole situation underwent a change. Instead of raids and sallies for single victims, mass extermination and mass destruction came to prevail. What had once been a magic sacrifice to insure fertility and abundant crops . . . was turned into the exhibition of the power of one community, under its wrathful god and priest-king, to control, subdue, or totally wipe out another community.

Such consolidated cities, the nuclei of growing states, succeeded by undermining "the positive symbiosis of the neolithic village community . . . by a negative symbiosis resting on war, exploitation, enslavement, parasitism. . . . The very means of achieving this growth oriented the community to sacrifice, constriction of life, and premature destruction and death." Whether in Sumer, Babylonia, Egypt, Assyria, or Persia, for all their differences these imperial metropolises all fell into the same pattern when they experienced successful consolidation: "Specialization, division, compulsion, and depersonalization produced an inner tension within the city. This resulted throughout history in an undercurrent of covert resentment and outright rebellion."

Later on, Mumford tells us, when the once-independent city-states of Greece were joined together into the successive Hellenistic empires, the same misuse of power, the same economic and political dislocations arose:

> These new states squandered human vitality and economic wealth on the arts of war, [and] would often crown their success in commanding slave power and garnering tribute by lavishing money on costly public works of every kind. Democracies are often too stingy in spending money for public purposes, for its citizens feel that the money is theirs. Monarchies and tyrannies can be generous, because they dip their hands freely into other people's pockets.

And still later, in the empire that was to pride itself on central power and imperial control, "all the magnitudes [were] stretched . . . not least the magnitude of debasement and evil." The Roman Empire, Mumford writes, suffered from a failure

> to make either the towns or the provinces more democratically self-governing and more self-sufficient: for too much of their surplus was destined to flow back to the center, through the very leaky channels of tax-gatherers and military governors. The cities were often given some

> degree of independence within this scheme; but what was needed was a method of encouraging their interdependence and of giving their regions effective representation at the center. This possibility seems to have been beyond the Roman imagination.

The result of such extreme centralization was "a parasitic economy and a predatory political system," both based in the Roman capital and both "built on a savage exploitation and suppression." Incessant warfare, the extremes of poverty, torture, and extermination, the devastation of the land—this was the life of Rome as it developed, in Mumfordian terms, from "parasitopolis" to "pathopolis," thence to "tyrannopolis" and finally "necropolis":

> Rome remains a significant lesson of what to avoid: its history presents a series of classic danger signals to warn one when life is moving in the wrong direction. Wherever crowds gather in suffocating numbers, wherever rents rise steeply and housing conditions deteriorate, wherever one-sided exploitation of distant territories removes the pressure to achieve balance and harmony near at hand, there the precedents of Roman building almost automatically revive, as they have come back today.

The glories that were Rome? Such as they were, they must have been produced at an awful price.

SO MUCH—at least in overview—for the earlier examples of the Law of Government Size. For it to be truly valid, however, it must be shown to operate particularly during those periods that saw the emergence and consolidation of the most powerful governments yet known in the world, the nation-states of Europe. And, since we deal here with more documentable history, it must make some direct and specific correlations between those periods of growth and the economic and political upheavals that are known to have taken place in Europe.

Dealing with amorphous categories like these, of course, one must treat with caution any attempt at too great a precision. Historical fluctuations are never exact, and movements do not usually fit nicely into those precise periods historians proffer us. Nonetheless, it is possible to ascertain certain broad trends over long historical reaches, certain valleys and peaks, and one can make some reasonable generalizations about patterns and correspondences. Or else what's a history for?

In the post-Roman era of Europe that saw the establishment of the modern nation-state—a process that altogether went on for eight hundred years or more—there have been, by common agreement, only four periods of marked solidification of governmental power: the dynastic era of the twelfth century, the absolutist era of the sixteenth century, the Napoleonic era at the turn of the eighteenth century, and the totalitarian era of the twentieth century. And thanks to the extraordinary work of some contemporary scholars, we are able with some accuracy to compare those periods with the times of most extreme economic and political turmoil over the same centuries.

The most precise economic evaluation comes from two Oxford University scholars, E. H. Phelps Brown and Sheila V. Hopkins, supplemented by material from the American business magazine, *Forbes*. Together they have constructed a price index for all the years since A.D. 1000, based on the probable amount of money an English working family (and presumably a family on the continent as well) would have spent for basic necessities as they were reckoned at any given time. This index shows with some clarity the periods in European history when prices were rising the fastest and the resulting inflation and economic dislocations were the greatest, as well as the periods of relative stability and tranquillity in between. For political evidence, the most accurate measure of dislocation comes from statistics on the nature and extent of warfare, much of which has been painstakingly compiled by Pitirim Sorokin, the Harvard historian whose masterful *Social and Cultural Dynamics* is a most detailed examination of the full sweep of post-medieval European history. Supplemented by the work of other military scholars, particularly Quincy Wright, these findings give a similar picture of the periods when European powers were most often convulsed by civil and international strife, and when they enjoyed times of relative peacefulness.

Remarkably, the economic and political indicators tend to coincide with great fidelity. More remarkably still, they coincide with the eras of governmental consolidation with almost uncanny historical precision. It is worth a somewhat more extended look.

THE FIRST PERIOD, from about 1150 to 1300, was the time in which royal dynasties were established and royal cities began to emerge in those parts of Europe where the medieval municipal traditions were weakest—particularly England, Aquitaine, Sicily, Aragon, and Castile. Sometimes under pressure from outside invaders such as the Mongols and the Turks, sometimes with the connivance of the Catholic Church looking for temporal power, baronies and principalities chose to unite under a single ruling family, which, though still with only limited power, worked to assert its control over wider and wider areas. Some of these dynasties were able to cause the creation of permanent royal centers now for the first time, with permanent archives, offices, courts, and bureaucracies. It is now that simple towns like London, Paris, Moscow, and Vienna begin to grow into true royal capitals, now that systems of delegated central authority and hierarchical administration begin to replace the loose pattern of mobile supervision characteristic of the earlier baronies.

This centralizing process is associated with the reigns of Philip Augustus (1180–1223) in France, Frederick II (1198–1250) in Sicily, and Henry II (1154–89) in England, and the rise of the house of Hapsburg (from 1273) in Austria, but the English example typifies them all. Henry succeeded ably in subduing the rival barons and establishing his control over much of the countryside from his royal town of Westminster, and his fight with Thomas à Becket—which Henry won, you may remember—was precisely a struggle about whether the state or the church would have control over ecclesiastical law. As one British historian has noted, "By Henry II's reign the English king had centralized so much authority under his immediate jurisdiction that all men of substance had frequent occasion to seek justice or to request favors at court." Though this centralizing process was partially curtailed by the greatly overrated Magna Carta, it was quickly re-established by Henry III (1216–72) and his doctrines of absolutism and then extended by Edward I (1272–1307), victor over the nobility in the Barons' War, and his all-embracing legal system, the Statutes of Westminster. After this Edwardian consolidation, England went through a long period of internal dissension, on-again-off-again warfare, baronial revolt, and finally the chaotic War of the Roses before royal power was reasserted some two centuries later.

This centralizing coincided with a period of rampant inflation unlike anything Europe had ever known during the early medieval ages—or perhaps ever before. This explosive epoch—economic historians call it the Commercial Revolution—disrupted the long-stable practices of medieval commerce and gradually eroded the self-sufficient free cities that had been the backbone of the earlier economy. Now, with newly powerful dynasties in the lead, the new economic order emphasized inter-city trade and court-protected trading routes and introduced cash money and "ghost money" (credit) backed by the emerging royal houses and their favored banking houses and entrepôts. The result was that for almost all of this century and a half, prices soared decade by decade, ultimately increasing by nearly 400 percent before leveling off in the early fourteenth century, a dizzying and devastating economic climb.

Over these same years, the political turmoil of accelerated warfare also increased, both internally as the royal consolidations met with baronial resistance and externally as the dynasties tried to extend their territories, trade routes, and sources of wealth wherever they could. The period was particularly marked by those chaotic campaigns that go under the name of the Crusades (1095–1291), whose effect was to create far more serious dislocations in the European countryside than in the Palestinian: the First Crusade, for example, involved the massacre of Jews in the Rhineland and civil wars in Bulgaria and Hungary, the Fourth was devoted entirely to carrying out Venice's plan for the sack of the Christian city of Zara in Dalmatia, the Children's Crusade degenerated into the selling of tens of thousands of children into slavery—and none of the bloody campaigns ever came close to succeeding in freeing the Holy Lands from the Moslems. Not until the end of the Crusades, and shortly thereafter with the decimation of the Black Plague, did European warfare enter a period of relative quietude that was to last for the next two centuries.

THE SECOND PERIOD of marked governmental growth, from about 1525 to 1650, saw the introduction of standing armies, forced taxation, centralized bureaucracies, national tariffs, royal customs collections, and extended territorial control. During this century the established orders that remained to challenge the royal houses—the church and the nobility—were pretty much vanquished, peasant uprisings in opposition to national control were quite brutally crushed, and sovereign cities that till then had resisted state dominance were brought to heel. The doctrines of absolutism became widespread, the notion of the "divine right" of kings was advanced and codified (particularly in France and England), and at the end of the period there appeared one of the most forceful of all justifications for the total sovereignty of the state. Thomas Hobbes's *Leviathan*—according to which the citizens "confer all their power and strength upon one man, or upon one assembly of men, that may reduce all their wills, by plurality of voices, unto one will." In England this is the time of the powerful Tudor rulers, the establishment of the state's own sanctioned form of religion (Church of England), the reunification of the country and its rise to international prominence under Elizabeth I, and the extension of state power into matters of currency, manufacturing, mail, labor conditions, agriculture, and the like, which had theretofore been private or communal affairs. In France, likewise, this is the time of the emergence of the House of Bourbon (from 1589) and the solidification of its administrative dominance under Richelieu and Mazarin (1624–61), and the ascendance of the man who came to embody absolute monarchy, Louis XIV, the "Sun King." "Throughout Europe," concludes Gerald Nash, a leading historian of government administration, "the absolute rulers created new institutions, such as efficient bureaucracies, to extend their powers into many aspects of the economic and social life of the nation."

In such a period, providing for the mere trappings of governmental power was bound to create economic strains and sustaining the substance beneath them would have to cause unprecedented financial turmoil. It did. Starting about 1525, after two centuries of relative economic stability, prices shot up steadily in an almost unending spiral, increasing by an unprecedented 700 percent in all before leveling off in the middle of the seventeenth century. Accompanying this unchecked spiral was the economic transformation known as the Capitalist Revolution, during which the primary mechanisms of capitalism came to be adopted throughout Europe, fueled by the new-found riches in, and colonization of, the New World. Then came, too, a whole new stratum of mercantile adventurers, fusing into what Karl Marx was later to call the bourgeoisie and marking the end of what remained of the communitarian economic arrangements of the medieval period: "good goods at a fair price" gave way to "what the market will bear," *fides publica* to *caveat emptor*.

At about this same time, Europe was also wracked on an unprecedented scale by the civil wars that the attempts to consolidate government power made inevitable, particularly the bloody Wars of Religion in France (1562–98), the Puritan Revolution in England (1642–49), and the ongoing internecine struggles of the city-states in Italy. More important, it suffered the single most destructive epoch of warfare between the ninth and the eighteenth centuries, the calamitous Thirty Years War (1618–48). That war, the first truly all-European war, fought from the Pyrenees to the Danube, from the Baltic to the Mediterranean, resulted in something on the order of 2 million casualties—twice as many as had been caused in the entire previous two centuries—and a "war intensity" that Sorokin marks as *seven times* as great as anything experienced in Europe before. Something there was that was absolute during this era, but it was not peace.

THE THIRD PERIOD of state consolidation, from about 1775 to 1815, comes with the adoption of what we would now think of as the modern system of government (and its rationale) over most of Europe. First through the American Revolution and then more drastically through the French Revolution, the existence of the nation-state was established as being even more absolute, more compelling, than that of the crown, or the mother country, or any subcategory of citizens within it: "The State alone has the duty to watch over the interests of all citizens," maintained the French Convention in 1793 to some workers on strike. "By striking, you are forming a coalition, you are creating a State within the State. So—death!" True national chauvinism was born in this era (indeed, it was a Napoleonic soldier, Nicolas Chauvin, who gave his name to it), with the accompanying fanaticism, zealousness, and flag-waving on behalf of the state that becomes part of modern patriotism.

The Napoleonic despotism that followed—not alone in France but in much of Europe—gave organizational form to such doctrines. Ancient provinces were officially abolished and replaced with new administrative units tied directly to the capital; newly empowered bureaucracies of increasing size took provenance over almost every human function; a Bank of France was created; a national police force was established (under Joseph Fouché, often called the father of the modern police state) for the first time in European history; national armies were drawn by conscription for the first time since the Pharoahs; and an all-encompassing *Code Napoleon* was promulgated, the first national code embracing all activities within the state since the Roman Empire. Napoleon's was a monumental creation of governmental power that, as Tocqueville himself remarked well after Napoleon had departed, was bound to outlive the Emperor:

> Centralization was built up anew, and in the process all that had once kept it within bounds was carefully eliminated. . . . Napoleon fell, but the more solid parts of his achievement lasted on; his government died, but his administration survived, and every time that an attempt is made to do away with absolutism the most that could be done has been to graft the head of Liberty onto a servile body.

Accompanying this achievement in almost every corner of Europe were the final blows to what, so persistently, remained of the communal villages and independent cities. France abolished the commune meeting, confiscated the remaining communal lands, and attempted to assign village mayors from Paris; England passed a flurry of Enclosure Acts from 1760 on, more than at any time before, effectively driving the villagers to the urban factories; and in Belgium, Prussia, Italy, Spain, Austria, and Russia, every royal house with sufficient power confiscated communal grazing and farming territories for its royal holdings or for division among the favored aristocracy.

The economic dislocations that accompanied all of this were of course severe. The inflationary curve shot upward during these four decades by more than 250 percent, breaking out of an equilibrium that had lasted for a full century, and at its peak in 1815 it stood higher than at any point in the past—indeed at any point in the future until the wild inflationary decade of the 1920s. This inflation and upheaval signalled the advent of the Industrial Revolution, particularly in England, during which the remaining agricultural and commercial economies were subsumed under the factory and wage-labor economies, with all their attendant maladics. The Industrial Revolution was to continue well beyond this period, of course, but it was during these first decades that the most intense dislocations took place, especially in Western Europe, and the patterns of rural ruination and urban overcrowding were established.

It was during these decades, too, that political turmoil and warfare reverberated throughout Europe, and through much of the Americas as well. The American and French revolutions were easily the most violent and disruptive of any revolutions until the Russian in the twentieth century, the French being a particularly bloody and terrifying cataclysm that resonated in Belgium and Holland, in Ireland and England, in Spain and Italy, and as far east as Austria and Prussia. These were followed immediately by the French Revolutionary and Napoleonic Wars

(1792–1815), which engulfed the entire European continent in total warfare for the first time since the Thirty Years War and involved more nations in more battles (an estimated 713) than in any previous thirty-year period in European history, more battles indeed than took place during the awful carnage of World War I; there were twice as many casualties for England and France alone as those countries had experienced at any comparable time since 1630.

THE FINAL PERIOD, the one most familiar to us, encompasses the preceding decades of the twentieth century, roughly from 1910 to 1975. This has been an era of unprecedented governmental growth that in almost every society, even the most liberal and benign, has meant large bureaucracies, universal conscription, compulsory taxation, state police forces, restrictions on individual freedom, and executive powers essentially remote from popular control. All European states have greatly extended the role of governments through wide-reaching social legislation and attendant taxation, and most of them have nationalized a great number of the industries and services of the land—some of them, *all.* The phenomenon of total power in the hands of the state, not only in Europe but today in remote American, Asian, and African lands as well, has even come to have a twentieth-century name: totalitarianism.

One indication of the extent of this progression is its effect on even such a nominally decentralized and traditionally anti-authoritarian country as the United States. In these last seven decades, the U.S. government has gathered powers to it that the Founding Fathers would have found unthinkable—gathered legitimately, as it were, because sanctioned for one branch by the other two—until it has become one of the strongest national governments in history. The process was slow and piecemeal—with the establishment of a national bank (1913), compulsory income taxes (1913), regulation of trade (1914), general conscription (1917), regulation of private power companies (1920), a national police force (1925), regulation of wages and working conditions (1935–38), an espionage organization (1942), and since World War II with a dizzying array of provisions covering most aspects of individual and communitarian life—but by the Bicentennial year there could be no mistaking the overall transformation that the American government had undergone in the direction of consolidated state power.

During these decades, the world has also experienced economic turmoil of a kind, or at least to a degree, unknown before. With the exception of an international depression in the 1930s, prices have risen inexorably every decade, up by an extraordinary 1,400 percent from 1910 to 1975, creating a rampant and apparently unabatable worldwide inflation that erodes people's earnings faster than they can compile them. Again, these economic dislocations have been directly associated with the growth of government, for the two most important underlying causes of twentieth-century inflation, economists are generally agreed, are, first, the considerable increase in the amount of credit in society created by national banks and Keynesian-minded governments, and, second, the significant expansion of "the public sector," people working not in the production of goods but for the newly enlarged governments themselves.

As to political turmoil, it is probably sufficient to point to the two most ruinous wars in all of human history within the space of just thirty of these years. But one might also add the innumerable second-rank wars that were far more destructive, even at that rank, than most of those that had preceded them—particularly the Russian Civil War, the Spanish Civil War, the Korean War, and the Vietnam War. By every measure—the number of troops engaged, firepower, duration of battles, military casualties, civilian deaths, destruction of property, and devastation of the countryside—this last has been the most violent era in human history.

IF ONE WERE to attempt the enormously difficult task of quantifying all this—the long process of history over more than eight centuries—the figures for price inflations and war casualties would provide, at the very least, a suggestive basis. These may not be precise—though they are compilations of diligent and thorough researchers—but they offer a general picture as few other statistics can. And when plotted on a graph, as on the next page, they give telling confirmation to the Law of Government Size. Again, I must caution that what we see in this graph is only an approximation, and it is based on statistics that must be considered open to modification and refinement. Nonetheless, the pattern that emerges does seem quite unmistakable and the correlations do seem to hold true with remarkable consistency: as government increases in size, it generates economic and political dislocations, and the bigger the governments, the bigger those dislocations.

I recognize that the *coincidence* of these three phenomena does not necessarily prove *causation:* one might make a claim that war needs big governments and drives up prices, or that inflation necessitates the diversion of war and increased government control. But obviously the initial causative element in the process has to be the government. It alone can set in motion the forces that lead to conflict, it alone declares and wages war, it alone sets the conditions for economic expansion or contraction, it alone determines the national protections and international opportunities within which the economy operates. Much happens without military reckoning, much without business considerations, but very little, particularly when the states accrue increasing powers with each passing century, happens without national governments.

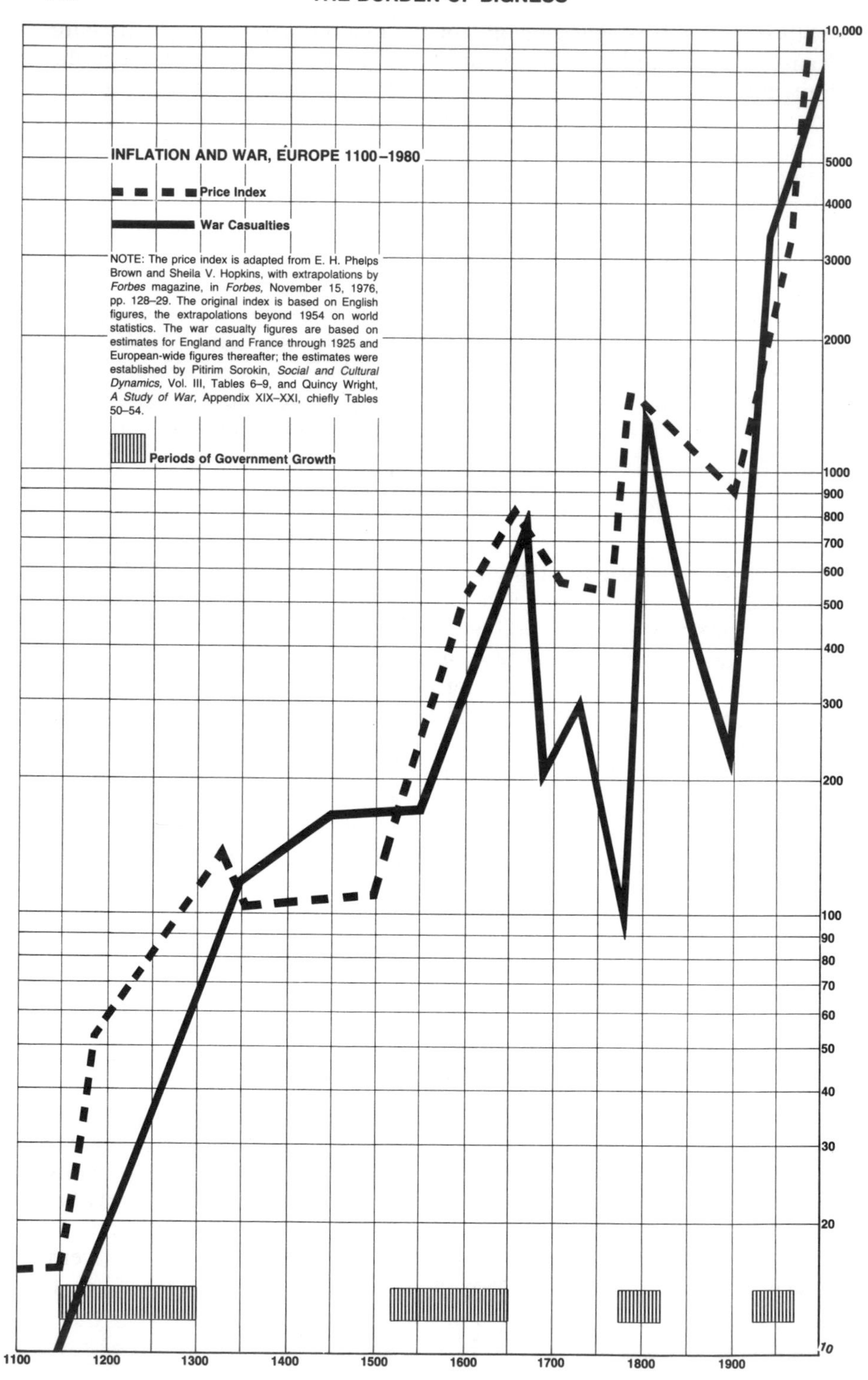
INFLATION AND WAR, EUROPE 1100–1980
Price Index
War Casualties
NOTE: The price index is adapted from E. H. Phelps Brown and Sheila V. Hopkins, with extrapolations by *Forbes* magazine, in *Forbes*, November 15, 1976, pp. 128–29. The original index is based on English figures, the extrapolations beyond 1954 on world statistics. The war casualty figures are based on estimates for England and France through 1925 and European-wide figures thereafter; the estimates were established by Pitirim Sorokin, *Social and Cultural Dynamics*, Vol. III, Tables 6–9, and Quincy Wright, *A Study of War*, Appendix XIX–XXI, chiefly Tables 50–54.
Periods of Government Growth
10,000
5000
4000
3000
2000
1000
900
800
700
600
500
400
300
200
100
90
80
70
60
50
40
30
20
10
1100
1200
1300
1400
1500
1600
1700
1800
1900

The processes are not mysterious. As a government grows, it must expand both its civilian and its military might, generally by extending the bureaucracy's influence in domestic affairs and the armed services' influence in external ones. New money must be found to pay for this expansion, and the greater the expansion the greater is the need for money. One way to raise money is through taxation—and that inevitably translates into inflation because businesses will always pass on their taxes to consumers in the form of higher prices, and higher prices mean inflation; result—as Mr. Micawber might say—social misery. Another way to raise money is by increasing the money supply, for example through central bank loans, measures of credit, guaranteed royal companies, or government-sponsored trade arrangements, and more money means inflation; result—again social misery.

With the need for money to pay for government expansion so important, a state will also naturally seek to increase its spheres of influence, and to enlarge its take of spoils, through warfare. And one government's expansion naturally runs up against another government's expansion more and more often, with more and more conflicts of greater and greater intensity, and greater and greater consequences to the social harmony. At the same time, war proves invaluable for fostering the spirit of nationalism that justifies the government psychologically to its people, and thereby for inducing them further to wage wars on its behalf. With warfare, too, goes additional government expansion and solidification: "War," as Tocqueville points out, "must invariably and immeasurably increase the powers of civil government; it must almost compulsorily concentrate the direction of all men and the management of all things in the hands of the administration." Result—more social misery, and oftentimes stark social tragedy.

The process is circular, neat, and powerful. If it were not, it would not have lasted so long. It is not by accident that it has produced the most powerful governments known to human history. And, for nearly a thousand years, the most severe economic and political turmoil.

AND SO WE HAVE COME, not by accident, back to the Beanstalk Principle: for everything—even a nation—there is an optimal size beyond which it ought not to grow, for at that point the critical mass of power accumulates to the point that demands an outlet, and thus is everything in that nation affected adversely. Or as Aristotle put it, anticipating the Law of Government Size two thousand years ago: "To the size of states there is a limit, as there is to other things, plants, animals, implements, for none of these retain their natural power when they are too large or too small."

There must be an alternative to nations grown too big, to governments grown too dangerous, to bureaucracies grown too inflated, to systems grown too complex, to enterprises grown too unwieldy, to corporations grown too immense, to production grown too massive, to cities grown too crowded, to buildings grown too vast, to tools grown too complicated, to relations grown too distended.

There is: human scale.

PART THREE

SOCIETY ON A HUMAN SCALE

I think that it will ultimately be proved that *scale* is a key factor in planning towns, neighborhoods, and housing developments. While very little is known about something as abstract as scale, I am convinced that it represents a facet of the human requirement that man is ultimately going to have to understand.

EDWARD T. HALL
The Hidden Dimension, 1966

Recovery of human scale is at the root of every enlightened concept of the city of our age.

ARTHUR B. GALLION AND SIMON EISNER
The Urban Pattern, 1965

If humanity is to use the principles needed to manage an ecosystem, the basic communal unit of social life must itself become an ecosystem—an ecocommunity. It too must become diversified, balanced and well-rounded. By no means is this concept of community motivated exclusively by the need for a lasting balance between man and the natural world; it also accords with the utopian ideal of the rounded man, the individual whose sensibilities, range of experience and lifestyle are nourished by a wide range of stimuli, by a diversity of activities, and by a social scale that always remains within the comprehension of a single human being.

MURRAY BOOKCHIN
Post-Scarcity Anarchism, 1971

1

Ecological Harmony, Ecological Hubris

A FEW YEARS AGO the World Health Organization, following its long-standing mandate, began to try to eliminate malaria in the Malay Peninsula by spreading DDT. Within a year or so there was a marked decline in the instances of malaria, but other problems cropped up, particularly an alarming increase in the rat population. It seems that the DDT, while killing some of the malarial mosquitoes, also poisoned roaches and other insects, which poisoned the gecko lizards who fed on the insects, which poisoned the cats who fed on the geckoes, which ultimately reduced the number of predators for the rats and allowed them to increase in prodigious amounts. The rat boom was a particular problem because the rats seemed to be carrying lice, and the lice in turn were carrying plague. The various governments of the area were about to embark on the natural and approved technofix solution—spraying rat poison over the country—when a couple of WHO ecologists rushed forward to intervene. Kill the rats, they explained, and you will merely force the lice and their diseases to find *another* carrier, most likely humans, and the plague that you want to avoid will come for sure. Their own proffered solution was decidedly non-technofix: bring in more cats, get the cat population up to previous levels, and they will eat the rats, lice and plague and all. This was naturally regarded with some disdain and condescension by the governmental authorities of the region, but there being no appropriately expensive, large-scale, and high-technology alternative, they eventually agreed. WHO was permitted to parachute hordes of cats over the countryside, and within a few months they had carried out their instinctive feline duties, decimated the rat population, and averted the threat of plague.

And the malaria? That was not so easily solved. As in most other countries around the world where the WHO DDT campaign was waged, the incidence of malaria showed a dramatic decrease for several years and then suddenly began to creep up again as the malaria mosquitoes developed resistance to the insecticide. Eventually the continued spraying of DDT was shown to have no measurable effect on the mosquitoes, and malarial rates were back to what they had been before the campaign, although the number of human miscarriages and malformations seemed to be on the increase. For several years now, the various health departments in the region have been asking for other strong insecticides, such as propoxur, but since the experience of Central America indicates that mosquitoes quickly develop an immunity to most of these as well, perhaps it will not be long before the conclusion is arrived at that non-chemical methods may prove the only answer. As a matter of fact, it has been known for some years—a 1972 article in *Environment* magazine presented the data—that the only permanent and effective cure for malaria comes not from some chemical technofix but simply from an improvement in the diet and general health of the affected population. The reason there is no malaria in the U.S., for example, is not that there are no anopheles mosquitoes in this country but rather that there are no malarial protozoa in our blood for the mosquitoes to pass along, and the reason for that is that the protozoa simply didn't find a sufficient number of hospitable—i.e., weakly—hosts; to take California as a case in point, it has been shown that malaria was eliminated there as a serious disease by 1930, long before the use of DDT, simply through improvements in the social and economic conditions of the majority of the inhabitants.

So much for the vaunted modern society. Here is a case once again of human technology, applied in ignorance of the natural ecosystems, not only failing to improve, but in serious ways threatening to damage, human ecology.

MUCH IS MADE of that concept "ecology," and so little understood about it. Yet it is quite simple. Ecology is that grand system of the interaction of all elements of the natural world—plant, mineral, animal, human—and a healthy ecology is one in which all of these are in such harmony that none dominates or destroys the others. The challenge for humankind today lies precisely in finding the scale at which it can live in that sort of harmony with nature—not eternally buffeted by it, defenseless and cowering, as the earliest peoples must have been; nor yet continually perverting it, altering and deforming it, as the industrialized nations so often do today. In fact it is just this challenge—establishing the balance between the human mode and the natural, in which human comforts are secured without ecological disruption—that should stand as one of the central occupations of any rational society.

You can figure it this way. On any given acre of land—let's take some part of the forest of North Carolina, since it so happens we have figures for it—the human animal is outnumbered by other animals by about 125 million to one. According to estimates of the biologists who calculate these things, there figure to be about 50,000 vertebrates making their homes on such a one-acre parcel—toads, turtles, snakes, and other reptiles; rodents, coatis, possums, raccoons, and other mammals; robins, sparrows, woodpeckers, crows, and other birds; plus fish and frogs and assorted amphibians. In addition there are some 124 million other animals—invertebrates—sharing that same turf, including 89 million mites, 28 million collembolans, 662,000 ants, 388,000 millipedes, 372,000 spiders, 90,000 earthworms, 45,000 termites, 19,000 snails, and a large number of various other crawling things, not including protozoa. And that says nothing about the vegetable life that may be found within that acre, from the pine trees and kudzu vines to the bacteria, molds, and actinomycetes, altogether amounting to some 5,000 pounds worth of plant life, roughly thirty-seven times the weight of the average human individual.

If all that suggests the need for some enterprise and ingenuity on the part of the lone human animal, it also suggests caution and a certain outnumbered modesty as well. Which is not to say that humans do not have their rightful place within that acre and are not as entitled to carve out their existence there—fishing with the earthworms, trapping the rodents, swatting the insects—as the rest of the creatures; one must not, after all, confuse the ecological ideal of living *within* nature with the somewhat more Eastern notion, recently popular here among the hairshirt wing of the back-to-nature people, of living *under* it. It is only to say that they do not have the right to create major imbalances there and wreak destruction upon either natural or animal configurations, for they will imperil themselves as well as the other species if they do so. And they certainly do not have the right to exterminate their fellow creatures, as human beings in fact have done, at the average of one species a year for the last two centuries.

A society guided by the human scale must begin right there, with an ecological consciousness that reminds us that we are only one animal among many, and one species among millions, that continually seeks the balance between the human animal and human arrangements, on the one hand, and the natural world and natural arrangements, on the other. Hans Selye describes a process he calls "intercellular altruism," the natural interdependence of the parts of the body so as to create a healthy organism; that is the obvious model for the wider world as well, an "interspeciate altruism" that interconnects all parts of the firmament so as to create a healthy ecology.

THE HISTORICAL RECORD is replete with evidence of the consequences of human societies disregarding ecological harmony, always to their ultimate ruination. Two examples may suffice.

The Roman Empire from the middle of the first century B.C. grew so extensively that it demanded an ever-larger army to secure its borders and an ever-larger bureaucracy to see to its imperial affairs; at the same time it permitted the accumulation of ever-larger numbers of people in its major cities, particularly Rome, which grew from perhaps 50,000 in the first century B.C. to more than 300,000 by the first century A.D. The feeding of these large non-agricultural—essentially non-productive—populations over the decades made severe demands upon the farmlands of Italy and eventually upon the conquered agricultural areas of Europe and North Africa as well. First Latium, then the Appenines, Campania, Sicily, and Sardinia, then Spain, parts of Gaul, Illyria, and finally Hispania, Macedonia, and all of northern Africa were turned into granaries for the insatiable populations of the empire, with little or no regard to the effect that this process might be having upon the natural systems. Overgrazing of animals depleted the grass covers and compacted the soils, with the result that the ground was able to hold less and less water and more and more of the topsoil was carried away by runoffs during the rains; overharvesting of crops led to an exhaustion of the soil's fertility and thus to a relentless deforestation of hills and mountains in the quest for new farmlands, with the result that whole regions were very quickly depleted and laid bare.

Ultimately this human rapaciousness led to the desertification of millions of acres around the Mediterranean, particularly in North Africa where the Sahara, once only a minor wasteland, soon became a desert from the interior to the sea. (It is easy to forget how recently this now-barren region must have been a reasonably thriving forest—this is, after all, where Hannibal got his elephants for the crossing of the Alps in 218 B.C.) Ultimately, too, it led to the creation of swamplands in impoverished and abandoned valleys throughout the Italian peninsula and around the Adriatic, and the mosquitoes that inevitably bred there soon carried malaria throughout the empire. And though the immediate causes of the downfall of the Roman empire were myriad, one need look no farther for the underlying causes than right here: what strength remained after the widespread starvation that marked the third and fourth centuries—and the consequent unrest and rebellion brought about by the increasing deficiencies of food—was surely sapped by the widespread ravages of malaria and other mosquito-borne diseases.

Across the world and several centuries later, a similar ecological vengeance took place in the Valley of Teotihuacán, on the central plateau of Mexico not far from the present-day Mexico City. There an autocratic priesthood and an ostentatious aristocracy rose to power sometime around the sixth century A.D.—it has been called the New World's earliest "superpower"—and determined that their temporal glory could be satisfied only with massive accumulations of cement blocks in the form of temples and citadels and monuments. They thus created the sacred city of Teotihuacán, which at its height covered more than eight square miles, with thousands of religious buildings from one edge to the other, including a citadel with a cement courtyard of 1.7 million square feet and the gigantic Pyramid of the Sun, 750 feet long and 216 feet high, the equivalent of a twenty-story building spread over two city blocks.

However, in the process of achieving all this monumentality they slowly depleted the surrounding countryside of all its forest cover. Not only did most of the buildings require beams and rafters made of wood, but the cement for the building-blocks themselves was made from quicklime in a process that required continuous fires and huge quantities of wood, about ten times as much as the resultant lime. At the same time, the capital city became swollen with people, perhaps as many as 150,000, and to feed this population more and more farmland had to be wrested from the surrounding hills, usually by the ruinous slash-and-burn method of cutting down forest areas and burning them clear just before planting. As the forests gradually disappeared to satisfy the continuing demands of empire, the unchecked rains began to erode croplands, unimpeded winds swept away fertile topsoils, underground springs lost their normal water supply, and the low-lying rivers that once irrigated the fields began to fill up with silt. Deforestation on this sort of scale, moreover, eventually affects the climate itself, diminishing the amount of water in both the ground and air and leading to a sharp decline in rainfall over quite large areas.

The resultant failure of the crops, attested to by archaeological evidence in the valley, and the ultimate starvation deaths of tens of thousands of people were undoubtedly the primary reason for the collapse of Teotihuacán sometime at the end of the eighth century and its later conquest by the rising Toltec kingdom. The bald hills around the ruins of Teotihuacán even today bear testimony to the fate of its imperial arrogance.

Obviously what we have here are striking violations, in ecological terms, of the Beanstalk Principle: for every state there is a size beyond which it cannot grow without disrupting its natural surroundings and eventually producing its own destruction. In fact it seems that throughout history it has always been the largest states that have been the most extreme in their ruthlessness toward, in effect their denial of, nature, and always eventually to their own detriment. From the Babylonians through the Byzantines, from the Bourbons to the British, we confront time and again in history the imperial passion for monumental construction, the maltreatment of the elements of nature, the dissipation of irretrievable resources, the subjugation of unwilling populations, the exhaustion of fertile lands, and the dislocation of ecological harmony. It is as if the natural world itself were held in some careful equilibrium, the violation of which by states grown overlarge and overconfident inevitably ends in their self-immolation.

A CERTAIN SATISFACTION may be had from these examples, perhaps, and we of a more sophisticated era may be inclined to an indulgent chuckle or deprecatory smile. Yet what is one to make of the brutal disregard for the ecosystem that characterizes our own period and our own nation? It is not simply the pollution that we scatter into the ionosphere, or the chemicals we spread upon the soils, or the nuclear wastes that we leave to radiate for the next 250,000 years, or the poisons that we scatter into our rivers and oceans, or the millions of acres of forests that we eliminate every year, or the animal species we obliterate in increasing numbers, though those are serious assaults indeed upon our ecology. It goes deeper than that. It is interwoven into all parts of our daily lives, into the institutions, the settlements, the habits and attitudes and religions, the laws and the history and the myths.

We are assured by Genesis—the opening lines of the West's single most important book—that we shall have "dominion over" all the earth: "Be fruitful, and multiply, and replenish the earth, and subdue it: and have dominion over the fish of the sea, and over the fowl of the air, and over every living thing that moveth upon the earth."

We are confirmed by the anthropocentric ideology of the Renaissance that the natural world is meant to be in service to the human: "It may be master'd, managed, and used in the Services of humane Life," as Joseph Glanvill put it in the seventeenth century.

We are persuaded by the evident triumph of capitalism that the resources of the earth are less to be contemplated and preserved than extracted and consumed for the betterment of our particular species, following the dictum of Adam Smith that "consumption is the sole end and purpose of production."

We are taught by the special progress of American history that a ruthless battle against the natural world is not only necessary for survival but in the long run beneficial in producing agricultural and mineral wealth for those who conquer the frontiers: "America is promises to / Us / To take them / Brutally," writes Archibald MacLeish, "With love but / Take them. / Oh believe this!"

We are convinced by the undoubted accomplishments of modern industrial technology that it is manifestly right to extend human control over the non-human world, believing, as Chauncey Starr of the Electric Power Research Institute has put it, that "science and technology are powerful and unlimited resources for bettering man's condition [and] have continually relieved the limitations on man's ability to live in the circumstances provided by nature."

We are encouraged by the specific laws of our nation, and by the example of the Army Corps of Engineers for more than a century, in the belief that we may divert any stream, develop any seashore, fill any swampland, farm any hillside, dam any river, and build on any surface with impunity, following the Corps of Engineers' homey motto, "If the good Lord didn't want us to build dams, he wouldn't have made cement."

We are even authorized by the deliberate provisions of Federal statutes to disregard any untoward consequences of the maltreatment of nature, since the Federal Disaster Act of 1950 and its successors provide full compensation to individuals whose building follies bring about natural disasters: "Our response to hazard, from colonial days to the present," investigator Wesley Marx has noted, "has made an about-face from an early policy of controlling hazardous development to one of controlling the hazard itself."

The Greeks had a word for it: *hubris*. It was the sin that most often brought down the figures in the Greek tragedies and wrought troubles upon their states. In its modern form it continues to do the same.

IT IS ECOLOGICAL hubris that has been responsible, for example, for the overdevelopment of the New Jersey shoreline in recent decades and the subsequent havoc visited by coastal storms. Ian McHarg, the Scottish ecologist who was called in to study the problem after the 1964 storm, pointed out that in an unspoiled setting, nature builds up vegetation on the seaward side of sand dunes, thus effectively anchoring them and permitting them to resist high winds and heavy waves. But when you build houses on the seaward side of the dunes, right up against the ocean, there is no way for that vegetation to take hold, and when the storms come along—as they will do, inevitably—they will sweep right into the sand and wash away millions of dollars of property every few years. McHarg's admonitions went without heed, and erosion continues to be a major problem along the Jersey shore—and, as well, along most other shorelines that are similarly overbuilt; the Corps of Engineers reported in mid-1978 that nearly half of the continental U.S. coastline was suffering from severe erosion, with 2,600 of the 12,383 miles classified as "critical."

It is ecological hubris that accounts for the persistence of the developers in Los Angeles who continue to buy up and build on the precarious hillsides of that region—in spite of the fact that in every major rainstorm hundreds of thousands of dollars worth of homes are swept away in mudslides and dozens of people die. (The toll from the savage rains of March 1978 was thirty-eight dead, several hundred injured, 50,000 homeless, and $1 billion in property damage.) There is no mystery as to why this happens: repeated studies have proved that careless overbuilding in the area has deprived the land of its natural absorbent cover and destroyed its ability to handle unusual amounts of water. The only mystery is why Californians somehow believe that nature proposes and man disposes: real-estate developers continue to build without regard for the conditions of the land; homeowners continue to buy without consideration of the inevitable consequences.

It is ecological hubris that explains the simple fact that many buildings throughout the Southwest are designed with flat roofs, without any drainage systems, as if there would never be any rains in the region or the rains would never accumulate in flat-bottomed areas. And when in May 1978 the undrained roof of a church in Garland, Texas, collapsed after a heavy rainfall, killing a nine-year-old girl and injuring fifty-seven others, a city official could only say that that kind of flat roof was "standard in this area" and "fully met the building-code specifications."

It is ecological hubris that leads to the attitude noticed among many New Yorkers during the 1977 blackout of the city when, as *Newsweek* columnist George Will put it at the time, "their strongest reaction to the blackout was indignation: Why was *mere nature* allowed to disrupt technology."

And it is ecological hubris, brought to what must be its highest point, that pervades this extraordinary letter, printed in the featured place at the top of the letters column of the *New York Times* in 1978:

> To the Editor:
>
> According to all the Judeo-Christian religions, as expressed in the biblical word which they all accept as divinely authoritative, God gave man, made in His image and likeness, dominion over all the earth and its hosts. Then why must we tolerate being subjected to the wild caprices of unstable air masses, including the ultimate obscenity called the tornado, in which a wayward, rapidly rotating air funnel deals death and destruction with impunity to whatever happens to lie in its path?
>
> Why don't the world's leading scientific minds and its political leaders, starting here in this country and possibly operating through the United Nations, give top priority to subjugating the earth's atmosphere, making it totally and abjectly responsive to the well-being and security not of mankind alone but of all life on earth, as well as the solid structures and tools of man's devising?
>
> My blood boils every time I read that a high wind or the pressure gradient in an air mass constituting a tornado kills people, wrecks buildings, tosses cars around like playthings, etc., often as not without the slightest warning. Let us have "weather by the consent of the weathered" and force the elements to respect our wishes! . . .

* * *

Two months later, the Federal Weather Modification Advisory Board, established under the National Weather Modification Act of 1976, reported that within twenty years, with proper funding, the U.S. could control rainfall in the Midwest, the amount of snow in the Rockies, the velocity of hurricanes in the Plains, and make "the science and practice of weather modification" a reality all over the nation.

ALL IN ALL, given attitudes such as these, it seems likely that the accumulated ecological disasters of the modern industrial nations will easily surpass those of Rome and Teotihuacán. An increasing number of scientific voices have gone so far as to echo the sober thoughts of G. Evelyn Hutchinson, who wrote in *Scientific American* a few years ago that things have come to the point where it is possible "that the length of life of the biosphere as an inhabitable region is to be measured in a matter of decades rather than thousands of millions of years," and then added: "It would not seem unlikely that we are approaching a crisis that is comparable to the one that occurred when free oxygen began to accumulate in the atmosphere."

It begins to appear that what separates the human animal from the lower orders is not a sense of humor or the ability to blush (although it is clear enough why it is that we had especial need to develop both of those), but rather a fully developed conceit, the conceit that declares us to be the rulers and shapers of the world and all its workings. Of course we could continue to indulge that conceit in a continuing and relentless battle with nature, for indeed we are armed with a considerable array of most effective technology and may be able to destroy many of the most basic elements of nature. But, as Fritz Schumacher used to say, if we win that battle we will be on the losing side.

It is perhaps wiser to contemplate a contrary course, the course of establishing a society based upon ecological harmony. Such a course is uncharted, to be sure, and as far as I know no one has attempted to determine just what might be the limits of an ecologically balanced world. But guided by the principle of the human scale and the matrix of the human form, I think it is possible to draw the rough outlines of what it must look like. Specifically, we might be able to determine the sort of technology that an ecological society would require; the limits at which an ecological society would construct its buildings, its communities, its cities; the kinds of social functions—energy, transportation, education, and the like—an ecological society would require and the scale at which they would be best supplied. Such is the enterprise upon which this section embarks.

IN 1946, FLUSH with its success in winning a war, Britain decided it could solve at least part of its domestic problems with the same kind of technological development and administrative concentration that had defeated the Germans. With guidance from the multi-million-dollar United Africa Company and upon the advice of its best technical minds, it established the British Overseas Food Corporation and proceeded on a massive plan for growing groundnuts (peanuts) in the East African territory of Tanganyika. And thus began one of the greatest ecological follies of the twentieth century.

Postwar Britain suffered from a shortage of cooking oil—essential for fish-and-chips, among other things—and it seemed natural enough to envision planting millions of acres of some tropical colony with the kind of crop that could supply such oil. The managing director of the United Africa Company, flying over the bare red-dirt scrubland of Tanganyika one day in 1946, figured that he had found the perfect spot, and he persuaded the British government to send out a team of experts to survey the area. The Wakefield Mission spent nine weeks in the field, testing the soil, checking rainfall charts, talking to local British administrators, and eventually it issued an enthusiastic report recommending the clearing of 3.2 million acres, proposing the planting of groundnuts within two years, and forecasting a harvest of at least 600,000 tons by 1951. The British government welcomed the scheme, allocated an initial $60 million, and dispatched 2,000 technicians to oversee 30,000 Africans on the job. No one doubted that a nation that could build the "Pluto" pipeline across the English Channel under fire, or checkmate the Japanese army throughout Asia, could tame the soils of Africa for the tables of Britain.

Nature is not so easily trifled with.

Hundreds of tons of earthmoving equipment were shipped into Tanganyika, and the huge operation of clearing the brush began. But the scrub roots and baobab trees were so tough and tenacious that they frequently broke the plows and blades used to clear them; that, coupled with the punishing climate, put at least three-quarters of the equipment out of order in the first three years. Then after the roots were removed, the soil, without anything to bind it, began to fall apart and turn to dust, rolling up in great clouds of dirt as the bulldozers moved through and dissipating the thin layers of fertile topsoil over the countryside. In some places the heavy treads of the earthmovers created tracks and gulleys in the fields that filled with water when the rains came, hastening the erosion further; in other places the equipment packed down the earth so solidly that it could not be worked or planted.

And after the scrub cover was cleared, the soil simply baked and withered in the African sun, its nutrients slowly drying up. One common weed that was eradicated early in the operation was later found to have been crucial for mobilizing the soil phosphates, without which the earth could not be healthy. The rainfall did not come as neatly as the charts had predicted: during the rainy season the soil, which turned out to have an unexpectedly high clay content, became sticky and unworkable, and during the dry season it cracked and shrank, imprisoning and choking off the roots of the groundnuts that did manage to get planted.

It was about that time that someone in the British Food Corporation pointed out that it was actually *West* Africa, a far different climatic area, that had been successful for many decades in groundnut production.

Shortly after the first crops were planted, on an inevitably reduced scale, late in 1948, the planners of the groundnut scheme began to worry about the overdependence on a single crop, the creation of an unstable "monoculture" unhealthy to any agricultural area. So they tried American soybeans, tobacco, cotton, even sorghum: none of it took. Finally they hit upon sunflowers, which grow practically anywhere and also produce a valuable oil. It was after planting acres of them, however, that they realized something was wrong: almost all of the bees necessary to pollinate the flowers had been driven away when the brush was cleared two years earlier. So at great expense they shipped out appropriate sunflower bees from Britain and Italy, only to discover that many of them could not survive in the harsh new ecosystem of East Africa. At least a third of the original sunflower crop had to be scrapped.

In 1954 the Great Groundnut Scheme was at last abandoned, a sorry tale of human hubris come full circle. Some 300,000 acres had been cleared by then, less than a tenth of what had been envisioned. Production amounted to about 9,000 tons a year, compared to the 600,000 tons that had been planned for. Total costs came to at least $80 million, about $1.6 million for every ton produced—roughly a dollar a peanut. It was as clear as any proof could be of the faultiness of the assumption, in the words of the great British Africanist Lord Hailey, "that mechanization would overcome the climatic and other defects of the notoriously unsuitable area." Or, as British scientist Ritchie Calder was later to remark, it was "a warning of the things you cannot do to nature and get away with." And he added, in a most telling comment: "On a small scale it would have been a matter for adjustment; on a large scale it was a chronic crisis."

2

Human-Scale Technology

THERE IS NO such thing as a society without technology. Even in Samuel Butler's utopian Erewhon, where the people had decided to abolish machines as the logical way of escaping enslavement, there existed such non-mechanical technologies as the cultivation of plants and the harnessing of animals, the two greatest technological developments before the industrial. The question is not of eliminating technology—it is no more possible for a society to live without technology than for a person to live without muscles—but of deciding what kind of technology should prevail.

On the one hand, it is perfectly possible for Western nations to go on with their present large-scale, centralized technology until they reach the point, in the name of efficiency and rational allocation of resources, that they will absorb all enterprises under their control—much as the French philosopher Jacques Ellul predicted some years ago:

> With the final integration of the instinctive and the spiritual by means of these [centralized] techniques, the edifice of the technical society will be completed. It will not be a universal concentration camp, for it will be guilty of no atrocity. It will not seem insane, for everything will be ordered, and the strains of human passion will be lost among the chromium gleam. We shall have nothing more to lose, and nothing to win. Our deepest instincts and our most secret passions will be analyzed, published, and exploited. We shall be rewarded with everything our hearts ever desired. And the supreme luxury of the society of technical necessity will be to grant the bonus of useless revolt and of an acquiescent smile.

Gloomy indeed, and by no means improbable; but not inevitable. For on the other hand, it is also possible for the developed nations to redirect themselves gradually toward an alternative technology, still complex and sophisticated but limited in scale and restricted in scope, and above all ecological. As the perils of large-scale technology become more apparent, it may be that the small-scale alternatives will come to seem not only more necessary for survival but more desirable and even in the long run more practical. Certainly the extraordinary growth of what has come to be called the "alternative-technology" movement over the past half-dozen years suggests that some people, at least, and among them what appear to be the brightest and most committed, have come to that conclusion.

Any alternative technology would logically have to be based upon the human scale, in the sense both of being designed for and controlled by the individual and of being harmonious with the individual's role in the ecosphere. What exact forms such a technology might take are too early to determine as yet, but two general provisions would seem to be essential.

First, a human-scale technology would need to be designed according to human needs, human capabilities, and human forms. As some of the academics who are at work in this area put it, it would have to be *prosthetic*—that is, created with human ends in mind, as with a cane or a pacemaker—rather than *cybernetic*—that is, created largely with mechanical or technological ends in mind, as with an assembly line or a spacecraft. (Or, as David Dickson puts it in his comprehensive *Politics of Appropriate Technology:* "Technology should be designed to meet human needs and resources—and not the other way around.") It would be built for actual human bodies in actual human circumstances—and hence it would not admit typewriters with keyboards demanding the greatest work from the weakest fingers of the left hand, or bathtubs with hard edges all around, or lawn mowers that are designed to spew out hard objects at great velocities. It would take form at a scale sufficiently small so that an individual could control it, sufficiently simple so that an individual could comprehend it, and sufficiently approachable so that an individual could fix it—as for example a plow or a shovel (even a steam shovel), rather than such objects as the "Big Muskie" earthmovers now used on construction projects, thirty-two stories high, as long as a city block, with 170 separate electric motors and several thousand different gears. It would be designed to promote decentralized operations—like the carrel-type workplaces of some of the Volvo plants in Sweden, rather than the impersonal assembly lines of American automakers—and thereby to allow more small-group decision-making and individual flexibility than most high-technology operations permit. And it would enhance the human users rather than alienate them, make them feel good rather than exploited, satisfy rather than frustrate the innate human desire for accomplishment and achievement—the difference between working a piece of rich-smelling leather into a shoe on a last and pulling a handle to stamp out uppers on a large and noisy machine.

Second, a human-scale technology would need to be guided by the principles of human accommodation to, and limitation within, the environment. It would of course not pollute the air or water or earth, or disrupt the natural cycles of either climates or animals. It would be very gingerly in its use of non-renewable resources in both the manufacturing and operation of its machines—hence a premium upon designing things to last rather than be thrown away—and it would be particularly careful about the use of fossil fuels, which we now understand to be so precious. It would seek solutions that are natural rather than chemical—a simple hose spray to knock beetles off vegetable leaves, rather than a crop-duster—and harmonious rather than disruptive—a system of little canals to contain water flows, rather than one big dam. And it would attempt to adapt itself to the immediate local surroundings, using local materials and energy sources, matching itself to local climates, meshing with local customs and cultures—following the general notion once laid down by Horace Walpole after a journey to France that "the French will never succeed in having trees, parks, and lawns as beautiful as ours until they have as rotten a climate."

(All this would be so obvious—in fact as I write it out it seems so elementary as to be hardly worth the stating—were it not that our present technology has somehow blinded us to these simple truths. We of the industrialized West live in an almost totally *un*-natural—not to say *anti*-natural—world. Concern for nature, despite all of the recent talk and legislation, is *not* an integrated part of our technology today. We design and develop our technology for three reasons: to achieve basic economic ends, such as profits and jobs; to allow greater institutional control and efficiency, as with automated supermarket checkouts and neutron bombs; and to make certain aspects of life easier for certain individuals, as with air-conditioners and golf carts. Nowhere do considerations of ecology necessarily enter the process, unless some catch-up regulation comes down from the government specifying that such-and-such a disaster is not to be permitted any longer. Take car manufacturing as an example. We know that Detroit still designs its cars not to fit harmlessly into the ecosystem and make optimum use of non-renewable resources, despite whatever regulations may have been forced upon them, but rather to appeal to the tastes and pocketbooks of a manipulatable market and to make a profit; and so far are we from a true respect for the natural world that our laws and our economic beliefs and our customs and our conditioned reflexes even declare that practice to be right.)

ALTHOUGH WE ARE obviously a long way from a world of human-scale technology, its attainment is by no means fanciful or impractical. Indeed, from a long-range perspective, the time has never been more apt.

In the first place, virtually all of the scientific understanding necessary for such a world is available, indeed common, right now. There is general agreement in the scientific community that, for the most part, we have all the ideas and theories that we need to construct any kind of technology we want and that what now remains is the task of applying them. As an editor of the *Smithsonian* magazine put it as recently as 1977: "It is hard to see that scientists are doing much more than adding details to the existing grand perceptions of nature, rather than creating new perceptions." Put another way, we seem to be coming to the end of the Period of Invention that has dominated Western technology for the past two hundred years or so—starting, say, with the perfection of the steam engine in 1769—and entering into the Period of Design, where we can refine and retool the inventions already on hand and make them more adaptable to the social purposes we deem most important. This is not to say that no new scientific ideas may come along in future decades, but only that we already have a sufficient body sufficiently understood to enable us to achieve almost any end we desire—as the existence of the thermonuclear bomb, on the one hand, or the pacemaker, on the other, would testify.

In the second place, much of the actual hardware of a human-scale society has already been invented, tested, and proven. It is not the stuff of fantasy: it is the stuff of backyard workshops and little enterprising laboratories and experimental entrepreneurs. In fact the people who have developed the great bulk of this hardware make up what is unquestionably one of the most active and successful movements that this country has seen for many years. Called variously the "appropriate," "intermediate," "radical," "soft," or "alternative" technology movement (I prefer the last since it makes its oppositional nature clearest), it is made up of quite a remarkably ingenious and inventive lot. Some of their technology brings back archaic and forgotten but still practical ways of performing tasks (cold frames, wood-burning stoves, hand plows, dirigibles). Some of it mixes old materials and methods with new systems (ordinary bicycle mechanisms to drive complex kitchen machines and lawn mowers, 55-gallon drums for composting toilets) or new materials and methods with old systems (plastic barrels for growing fish, windmills with aluminum or plexiglass blades). And some of it involves new inventions and techniques borrowed from large-scale industrial technology and developed for small-scale use and control (self-contained microcomputers no bigger than a typewriter, electric cars, photovoltaic cells, videotape systems, cold type).

As a self-conscious aggregation, the alternative-technology movement did not begin much before 1975 or so (though its sense and sensibilities are rooted in the New Left of a decade earlier), but its development has been swift and the end is nowhere in sight. It has already spawned several scores of magazines *(RAIN: Journal of*

Appropriate Technology, Co-Evolution Quarterly, Mother Earth News, Appropriate Technology, People and Energy, Undercurrents, Self-Reliance, Alternative Sources of Energy, Wood-Burning Quarterly, Appropriate Technology Quarterly, Solar Energy, Alternatives, etc.) with a combined circulation of what I calculate to be at least a million. It has formed into more than two thousand different organizations, some of them experimental (the New Alchemists in Cape Cod, Farallones Institute in California, Institute for Local Self-Reliance in Washington, Intermediate Technology Development Group in London, Volunteers in Technical Assistance in Maryland), some official (California's Office of Appropriate Technology, the Rural Development Department of the Agency for International Development, agencies in Tanzania, Zambia, Mexico, Nicaragua, Ecuador, and elsewhere), some profit-making (there were by 1979 more than a thousand corporations selling solar technology, at least a dozen making windmills, more than twenty offering woodstoves, and several hundred more providing other alternative hardware). It has even—for worse probably than better—gotten Federal government support, through the establishment in 1977 of the $3-million National Center for Appropriate Technology in Butte, Montana, which intends to finance certain small groups and inventors and hopes to set up regional small-technology centers throughout the country. And it has inspired literally hundreds of books and pamphlets, meaning that it is now possible to find instruction on—for starters—how to build underground houses and aquaculture greenhouses, how to design windmills and solar-powered bicycles, how to grow food by organic, hydroponic, or French-intensive methods, how to establish community-radio, videotape, and urban-homesteading projects, how to set up land trusts, food co-ops, and self-examination clinics, and how to construct practically anything you want out of earth, adobe, canvas, wood, stone, skins, logs, bamboo, or pneumatic balloons.

Of course some of that sounds a bit arcane, and there's no doubt that there is a good deal of crankiness and oddity—not to mention occasional outbursts of outright lunacy—within this movement. But there is no denying its achievements, no gainsaying the proof that it has offered of the practicality, the reality, of a human-scale approach to technology.

And that makes the present age unique. We now know that it is possible to achieve a technology that can provide for a full range of human comforts and still be kept within human dimensions and control and not do violence to the planet's limited resources or its various ecosystems. Before this era, a human-scale society with a technology sufficient to provide the standards of productivity and living that we of the West are accustomed to was simply not possible, for the technology was not advanced enough and the scientific knowledge was not developed enough. Now, however, we can see that we are on the brink of a different period and can enter it if we but choose to. We must not underestimate this achievement, this moment. Quite simply, it is now possible, probably for the first time since the human occupation of the earth, to evolve a technology that can allow us to avoid drudgery, escape poverty, provide protection, supply nutriments, and enjoy comforts, to expand personal freedom and power, and to live in ecological harmony.

ONE FURTHER POINT. It should be obvious that there is no necessary contradiction between sophisticated technology and human-scale technology. Rational technologies of the future would not *discard* everything about contemporary systems but rather *evolve* from them, leaving aside the dangerous and destructive aspects, absorbing the humanistic and communitarian ones. Obviously there is much in current high technology that is anti-human and brutalizing, but there is also part of it that, however it has managed to slip in, is potentially liberating. In fact in the last twenty years or so there has been a strong trend in the direction of smaller and more decentralized operations: miniaturization has brought about the silicon chip and the proliferation of sophisticated machines available to any home or office; the development of new materials like plastic and fiberglass has enabled manufacturing to take place in diverse locations far from iron and steel centers; the creation of machines that perform a multiplicity of functions, allowing a wide range of products to be built in a single plant, has opened the way for communities to have an increasing number of goods manufactured locally; and the development of solar energy has pointed the way to the time, not far off, when we can have a completely localized power source no longer dependent on centralized plants.

It is important to realize that technologies develop as they do because of the nature of science and invention. Political and economic systems select out of the range of current technology those artifacts that will best satisfy their particular ends, with very little regard to whether those artifacts are the most efficient or sophisticated in terms of pure technology. Technology is not neutral: that is a myth. The particular technological variation that becomes developed is always the one that goes to support the various keepers of power. Hence in an age of high authoritarianism and bureaucratic control in both governmental and corporate realms, the dominant technology tends to reinforce those characteristics—ours is not an age of the assembly line and the nuclear plant by accident. Nonetheless, it must be recognized that there are always many other technological variations of roughly equal sophistication that are created but *not* developed, that lie ignored at the patent office or unfinished in the backyard because there are no special reasons for the dominant system to pick them up.

For example. Sometime before the birth of Christ, Hero of

Alexandria designed (and probably built) a steam engine: a fire created boiling water in a cauldron and the steam from it was sent along a tube into a hollow metal ball; two other tubes on opposite sides of the ball expelled the steam, forcing the ball to turn steadily and creating motion that could then be harnessed. The trouble was that neither the rulers of Alexandria nor any other powers in the Mediterranean world had any particular need for such a device, since the muscle power of slaves seemed perfectly adequate and the economic advantages of such a machine were quite unappreciated. It was not until the eighteenth century, in an England of entrepreneurial capitalism, where slavery was outlawed and cheap labor unreliable, that the virtues of steam power were sufficiently appreciated to enlist whole ranks of inventors and investors, many of whom set about unknowingly reinventing Hero's machine.

Or again. By the late eighteenth century there were two kinds of machines capable of sophisticated textile production in England. One was a cottage-based, one-person machine built around the spinning jenny, perfected as early as the 1760s; the other was a factory-based, steam-driven machine based upon the Watts engine and the Arkwright frame, introduced in the 1770s. The choice of which was to survive and proliferate was made not upon the merits of the machines themselves nor upon any technological grounds at all but upon the wishes of the dominant political and economic sectors of English society at the time. The cottage-centered machines, ingenious though they were, did not permit textile merchants the same kind of control over the workforce nor the same regularity of production as did the factory-based machines. Gradually, therefore, they were eliminated, their manufacturers squeezed by being denied raw materials and financing, their operators suppressed by laws that, on various pretexts, made home-production illegal. It is interesting that it was against this technological double-standard that the early "Luddites" actually acted: they were not engaged in the destruction of all machines, as they are usually blamed for, but only those factory-centered machines that threatened to destroy their cottage-based—and therefore more individualistic—textile industry.

In other words, each politico-economic system selects out of the available range of artifacts those that fit in best with its own particular ends. In our own time, we have seen the great development of machinery that displaces labor (and hence does away with labor problems), but there is a vast array of machinery, as the alternative technologists have proved, that is of equal sophistication and effectiveness but is labor-intensive. A human-scale system would select and develop the latter kinds of machinery, at no especial sacrifice in efficiency but with considerable enhancement of individual worth and ecological well-being.

One day in 1973, Ridgeway Banks, then 39, a backyard inventor and amateur musician employed as a technician at the Lawrence Laboratories in Berkeley, came across a metal called nitinol that he discovered to have almost magical properties. A blend of nickel and titanium, nitinol, he found, was capable of changing its shape rapidly and contracting under heat and then springing quickly back to its original shape again when cooled—and then reverting to exactly the same contracted shape when reheated. This, Banks figured out, was exactly what he needed to make a practical steam engine run by solar power.

Working with a machinist colleague, Banks constructed a simple solar collector on a rooftop and used the hot water that it produced to fill up half of a small circular cylinder—about the size of a cookie tin—while the other half was filled with cold water. He then constructed a flywheel holding a series of nitinol-wire loops and set it into the cylinder. Because the nitinol loops contracted powerfully and straightened out as they went through the hot water—with a force somewhere around 67,000 pounds per square inch—they were able to turn the flywheel, and then because they quickly assumed their original shape in the cold-water bath they were able to exert their force again when they hit the hot-water side. The wheel thus could run continually, providing energy of up to about 70 RPM and capable of generating about a half a watt of electricity.

The first test run was on August 8, 1973. Five years and tens of millions of cycles later the machine was still running, without adjustment or hitch, with no sign of fatigue or failure—and even, for some unknown reason, running 50 percent faster than it had originally. It was by any measure a success: simple, sophisticated, effective, safe, cheap, nonpolluting, small-scale, easy to make, operated by renewable energy, and built of simple materials (both nickel and titanium, though nonrenewable, are fairly abundant). Human-scale technology at its best.

And the reason you've never heard of the Banks machine is that nobody is interested in developing it. The National Science Foundation put $113,000 into trying to develop a huge version of the machine and then gave up when the prototype didn't work properly; no one else, neither corporate nor military nor governmental, has expressed any interest at all.

One of Banks's friends, Bob Trupin, also a nitinol enthusiast, was quoted several years ago in the *Co-Evolution Quarterly* summing up, with a touch of bitterness, the fate of the Banks engine:

> Isn't it amazing? There's no money in this country, despite all the talk about the energy crisis, for a project that could end our dependency on oil. They don't intend to scrap any of their existing technology. And there's no way for the big utility companies to cash in on nitinol. It lends itself to a small rooftop unit where the sun can heat up a pan of water.

> You look at this wire and you see an engine that's *small.* I guess small is what they really hate. Only big is good.

But the engine, like hundreds and hundreds of other kinds of alternative hardware, is *there,* and probably still spinning as of this moment. Should we choose to want it.

3

We Shape Our Buildings

"We SHAPE our buildings," Winston Churchill is often quoted as having said, "and thereafter our buildings shape us." He was talking about the rebuilding of the House of Commons, which had been damaged after the war, and he was pointing out the way that the special oblong character of the Commons room had naturally led to a system of two distinct and oppositional political parties, both directed by a row of leaders rather than a single head, and—he might have added—whose debates and electoral styles were characterized by confrontation and challenge rather than cooperation and harmony. (A similar point might be made about the chambers of the U.S. Congress, though in that case the semi-circular rooms and radiating aisles led to blurred lines of distinction between parties, a lack of face-to-face debate, a sense of spectatorship as in a theater, and individualistic rather than communal political styles.) But his remark applies with equal force to all buildings, everywhere, indeed to all built environments of any kind: one way or another they affect us, the good ones with pleasure and serenity, the bad ones with stress and disquiet.

Studies to prove the point are numerous.

- One group of researchers examined a graduate students' apartment project at MIT—a series of large U-shaped courts, the open end facing the street—and found that the people who lived at the upper points of the U, with their front doors facing the street, had on average half as many friends as those living within the court itself—not, obviously, because those living within the court were more gregarious and charming but simply because they lived in a more communal setting. (As sociologist William Whyte once wrote, "The court, like the double bed, enforces intimacy.")
- A team in St. Louis spent three years interviewing residents of the infamous Pruitt-Igoe housing project and concluded that the chief reason

for the obvious social collapse there—crime, vandalism, litter, property destruction, hostility among residents, isolation, anomie—was that the buildings were designed without any semi-public places either inside or out where small groups of neighbors could gather informally and accidentally during the course of their daily activities. Never getting together with others, residents retreated into their own apartments and did not evolve the kinds of small networks that develop in normal neighborhoods, even slum neighborhoods, to provide social cohesion and exert social control.

▪ Another study showed that men living in large open Army barracks were able to identify, and usually selected for friendship, a far wider group of men than those living in walled-off, cubicle-style barracks; and though the cubicle soldiers tended to develop more intense and cohesive groups, they also were more likely to have disruptive fights with their companions.

Daily experience provides other examples. Airport waiting rooms, designed always with seats in long rows and with wide aisles between them, effectively prohibit any conversations except those between people sitting right next to one another—and it's none too easy for them since the seats don't move for face-to-face contact. Office buildings with enormous doors and archways, leading into tabernacular lobbies, succeed, perhaps not accidentally, in creating a sense of insignificance and powerlessness among the people herded there. Board rooms that are long and rectangular, with elongated tables and space for a single individual at the head, discourage discussion and open decision-making, while those ad-agency "creativity" rooms with circular tables and muted furnishings do in fact encourage give-and-take (for whatever pernicious ends) and collective judgments. Apartment blocks built in small clusters, two to three stories high and enclosing comfortable courtyards, create what Oscar Newman has called "defensible space," in which it has been shown that residents naturally identify with, supervise, and protect their own areas and neighbors. Government buildings in Washington, massive and hulking ponderosities, dwarf those who work within them: "They are designed," as columnist Russell Baker has pointed out, "to make man feel negligible, to intimidate him, to overwhelm him with evidence that he is a cipher, a trivial nuisance in the great institutional scheme of things," perhaps not too healthy a condition for those who theoretically guide our lives.

Now if we were to operate on the Churchillian principle, it would seem logical to construct our buildings so as to discourage certain kinds of behavior, stimulate other kinds. One may not be able to change human nature, exactly, but one might well be able to bring out its better side; as C. M. Deasey has put it in his provocative *Design for Human Affairs,* "The chances of redesigning a checkout line to make line-jumping impossible are much better than the chance of reforming pushy people in the world." (As a matter of fact, a few supermarkets have had success operating with a single feeder-line system such as that used in larger banks.) What it would take is simply an appreciation of the effect of buildings on people—what *their* lifetimes do to *our* lifetimes—and the creation of architecture that would enhance the activities and ennoble the sensibilities of the people who actually would be using them. It would take the conscious design of buildings to the human scale.

THE IDEA OF "human scale" is an old one in architecture, perhaps primordially old. We know that most pre-literate tribes build homes wittingly with the size of the human form and the activities of the human family in mind, and some tribes, like the Dogon of West Africa, according to anthropologist Daryll Forde, go so far as to build their huts in deliberate imitation of the human body. In the towns and cities of most pre-Hellenic civilizations—as we know from the archaeological remains of such cities as Çatal Hüyük in Turkey and Mohenjo-Daro in India—buildings generally were quite small and houses were divided according to family and social functions; even the few larger structures, such as the huge community center and public bath in Mohenjo-Daro, were well within the limits of what we would feel to be a human *public* scale. (There were of course exceptions—the Pyramids, the Tower of Babel, the temple at Eridu in Mesopotamia—but they were rare and always occurred in those empires that purposefully shunned the human scale and sought to create larger-than-life monuments.) With the Greeks particularly, the human body was taken as the central module by which all the built environment was measured—as we saw in the Parthenon, as one can see today in the buildings at Delphi, the temple at Bassae, the theater of Dionysus in Athens, indeed, one way or another, in practically all the known buildings of that civilization: "Man," as Protagoras often said, "is the measure of all things." And if the Romans for the most part ignored concepts of human scale in their major works—not to mention in the eight- and ten-story apartment buildings they put up in downtown Rome during its period of "grandeur"—it was not for want of being instructed: the first century-B.C. architect Vitruvius showed time and again how both individual buildings and public squares ought to be designed within the specific limits of normal human sight and mobility. Medieval European cities grew up with conscious human limits—houses of no more than three or four stories, regular public spaces throughout for markets and celebrations, the whole kept by walls and fortifications within an area whose boundaries could be easily walked in half a day; and a surviving print of the layout of the city of Bern, for example, shows what must have been a conscious imitation of the human form, with a main cathedral and fortified bridge at the head, radiating arm-like stretches of parks, the market and public square in the genital region, and buttressed fortifications splayed out like feet at the farthest end. And then with the Renaissance, when the human body was studied and celebrated as perhaps never before in history, architects built consistently in imitation of the figure, following such maxims as Luca Pacioli's, in his *De Divina Proportione,* "From the human body derive all measures and their denominations and in it is to be found all and every ratio and proportion by which God reveals the innermost secrets of nature"; one may still see it in the countless Christian churches built across Europe from the fourteenth century on, laid out deliberately in imitation of the body of Christ, sometimes with details in the apse to mimic mouth and eyes.

But the idea of human scale began to be lost somewhere in the early industrial era—around the time, not coincidentally, of the beginnings of the nation-state. With the rise of government-sanctioned science and the solidification of architecture into royal academies in the eighteenth-century "Enlightenment," the idea of objective "scientific" rules came to supersede subjective "humanistic" ones; this is when we get long, abstract building shapes like Wren's Naval College at Greenwich and absurdly rectangular formal gardens like those of Le Nôtre's Tuileries in Paris. Then with the rise of industrialism and the triumph of engineering in the nineteenth century, mechanical considerations came to outweigh visual and aesthetic ones and the "utility" of a building became its prime consideration; hence the look of those red-brick factory towns in industrial New England and of many squat and squalid prison and military installations still standing today. Ultimately with the rise of technological prowess and modern architecture in the twentieth century, the building took on a form of its own quite unrelated to human or even social perceptions, where the objects are to let the materials simply perform to their utmost—thus the triumph of steel in the Seagram Building, of glass in Lever House, of cement in the Guggenheim Museum, and so on—and to let technology provide, God help us, "a machine to live in."[1]

What happened in this long process over nearly three centuries was that gradually the *user* of the buildings was forgotten: the human scale was discarded. Yes, there was continual talk about it—hardly any great architects of this century have failed to declare their fealty to it—but usually they paid homage the way Mies van der Rohe did when he designed the Illinois Institute of Technology: he claimed he was building to the human scale because his windows were of human height, but in fact he put those windows in unappetizing rows along long, abstract boxes that look most like machine-made toys for giants, a perfect celebration not of the human form but the geometric. As Italian architect Bruno Zevi has noted, that is *not* what human scale means: "Scale means dimension with respect to man's visual apprehension, scale

1. The words are Le Corbusier's, as were many of the deeds: such anti-human machines may be seen in his Unité d'Habitation in Marseilles and in his plan for the ideal city, the Ville Radieuse.

means dimension with respect to man's physical size." And he adds, "In the last hundred years, however, enormous crimes have been perpetrated against it."

In fact the standard modern building, particularly the downtown office block, is *scaleless*—literally without any sense of human dimensions or perceptions either in its size, volume, proportions, or relations. Form follows function—but it is *material,* not human, function. Several years ago *Progressive Architecture* magazine devoted a whole issue to the problem of scalelessness as "the New Scale of Today" and the danger that "scaleless space seems to be on the increase":

> What does today's scale say about our age? Will future historians reveal that the 20th century, through its approach to scale, merely expressed the depressing inconsequentiality of the individual vis-à-vis the collective activity of society—merely the "scalelessness" of man himself within the vast megalopolitan complexes and the even vaster outer space?

The answer is almost certainly: yes. Those who are shaping our buildings seem to care little for the humans who inhabit them.

BUT WHAT WOULD human scale actually mean in contemporary building, anyway? Is there any consistent measure by which a society could plan its buildings not only so as to eliminate the terrible dehumanizing effect of scalelessness but actually to enrich and humanize those who experience them?

I suggest there is and, moreover, that modern science gives us the way to determine it.

In the last twenty years or so there has grown up a group of schools devoted to measuring the human body and its actions, working in conscious opposition to the dominant industrial aesthetic. They take different names, depending more or less on the disciplines their adherents come from—ergonomics, anthropometrics, sociometrics, human engineering, biomechanics, and the like—but what they have in common is the goal of scientifically discovering the way the human body actually functions and then designing tools, machines, appliances, furniture, living spaces, and workplaces to match.

Take the work of Alexander Kira, for example, a professor at Cornell University and one of the pioneers in ergonomic research. In 1966 he published a study called, forthrightly, *The Bathroom,* which was an exhaustive examination of the way people actually perform their human functions and the way most bathroom equipment is woefully misdesigned to serve them. For urination, for instance, he studied the role and workings of the kidneys, the distinctive features of the female and male anatomies, and the color, smell, and qualities of urine, and he then made empirical investigations of such things as stance, splash, noise, velocity of flow, target, and the like. He eventually concluded—as we all would expect—that the present Western toilet is totally inadequate, particularly for the male, in fact an outstanding example of an instrument in everyday use around the world designed entirely without sensible consideration of how people would actually use it. Kira then went on to design and build an improvement, a simple fold-down unit much like a rest-room urinal, exactly 24 inches above the floor, with an 8-inch opening and an elongated funnel that fits into the regular toilet bowl below and can be flipped back out of sight when not in use; it allows a man to relieve himself comfortably, easily, without embarrassing noise or unsanitary splash. The solution wasn't difficult, once the problem had been properly posed, but it had just never been studied that way before, with the human function, the human form, coming first.

Another important ergonomic designer, Niels Diffrient, has limned the problem precisely: "There is not a comparable data bank for people in their man-made environment," he says, "as there is on the performance of mechanisms and their related systems." And yet, "without comparable data on people's feelings, physical and emotional, as there is for machines to meet quantitative performance, the result is bound to be a continuously degrading quality of life." Much of the problem, he feels, is that giantism has tended to eclipse humanism: "The macro-scale tendency in technology has seriously affected design. It has spawned large-scale planning efforts far beyond anyone's comprehension." And in buildings this has meant "a large segment of design activity in which architecture is not so much the creation of humane spaces as it is the production of architectonic sculpture."

It is extraordinary that such voices remain so few and so often ignored. After more than a decade now of solid work in these new schools and their absolutely incontrovertible findings about human functions, most of their research is shamefully neglected or at best made use of by a few industrial designers seeking better knobs for windshield wipers. Yet it is here that we can find invaluable scientific evidence of the kind of characteristics a successful human-scale structure might possess and the kind of architecture a rational society might promote.

WE MAY BEGIN with the human eye. The visual effect of a building is, after all, the first, and always the most indelible, whatever other sensory impressions may later accrue; the eye, if not precisely the window of the soul, may still be regarded as the gateway of the brain and the organ from which our most basic mental and emotional comprehension derives. Various scientific studies of the eye have been undertaken over the years, though remarkably the first complete ergonomic studies were not carried out until Henry Dreyfuss, a New York industrial designer, set his studio to the task in the early 1960s. Their studies, and later refinements by the Dreyfuss team that have been published as a series of detailed charts under the apt title of *Humanscale,* have determined with some precision how the eye really functions:

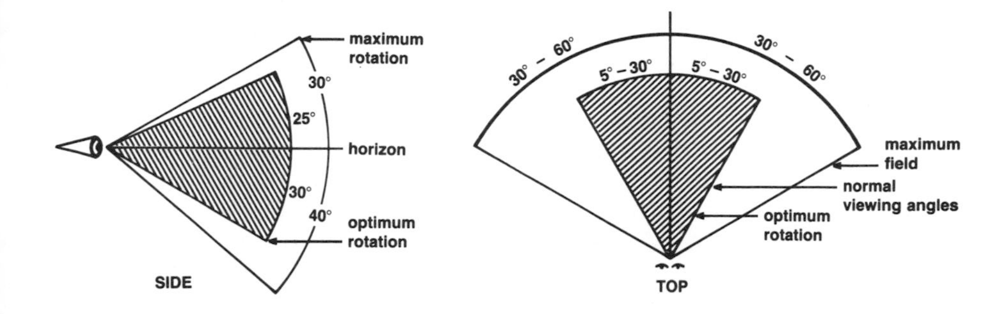

Now let us apply this to architecture. The best way to view a building is in its entirety, of course: as Aristotle among others has noted, it is always most satisfactory to see any object whole, at a single glance, so that its unity can be understood. If the maximum eye rotation above the horizon is 30 degrees, and the optimum is 25 degrees, as Dreyfuss's studies indicate, then the ideal vertical viewing angle should be something around 27 degrees. Now by a neat turn of trigonometry, this is exactly the angle established when you stand twice as far away from an object as it is tall, and the experience of centuries has confirmed that this is in fact a successful ratio for viewing both the details and the totality of any object. Thus to have a complete view of an ordinary two-story house at a single glance without moving your head, you would have to stand at a distance twice its height. Assuming 13 feet for each story and half of that for the pitch of the roof, it would then be necessary to stand some 52 feet away, thusly:

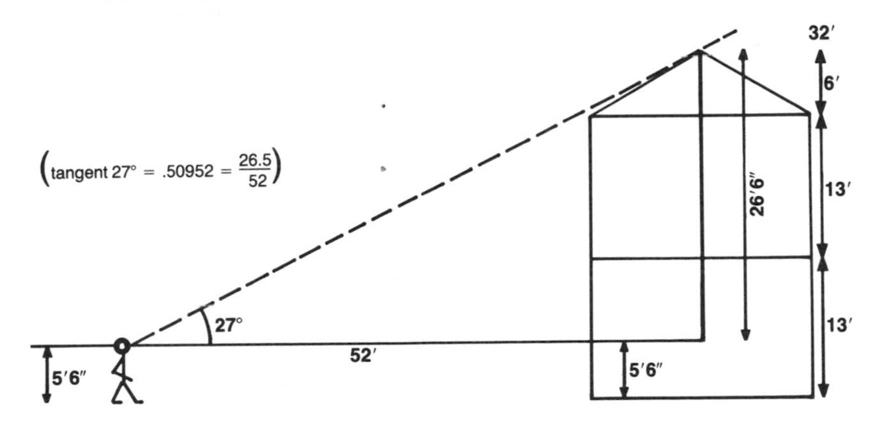

And, though horizontal viewing has been shown to be of far less importance in perceiving buildings—probably because we tend to look over buildings the way we do people, up and down—a similar measurement may be made for horizontal viewing angles. According to the Dreyfuss figures, the maximum horizontal field ranges up to 120 degrees and the normal viewing angle ranges from 10 to 60 degrees, so the average angle would be about 60 degrees, which is also the optimum angle of eye rotation, both eyes taken together. Assuming a house with a width no more than twice the height—the common proportions for the traditional single-family house—and thus about 60 feet long, to get a complete horizontal view at a single glance without turning the head it would be necessary—again—to stand some 52 feet away, thusly:

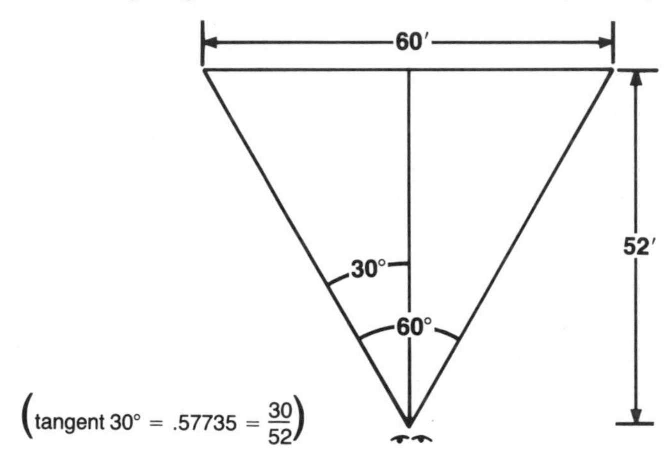

And that is why the most successful residential streets, those that satisfy in some indefinite way as you walk or drive along them, are those with houses set back from the street about 50 feet or so; farther than that and the buildings tend to get lost in the background, nearer than that and they give a sense of crowding, of looming.

And thus the human-scale module for any individual building.

In applying this scale to the design of cities, several further calculations based on the human form are necessary. Here we may be guided by a most extraordinary book that is fully as scientific as the modern ergonometric studies but in fact was written a century ago, by a German architect we know only as H. Maertens. It says something about our mechanistic age that although his work has been influential for a number of later city planners—particularly Werner Hegemann and Elbert Peets in their 1922 study, *The American Vitruvius,* and Hans Blumenfeld in his brilliant *The Modern Metropolis*—the original remains virtually unknown and, as far as I can tell, seldom read and never translated.

Maertens began with the clever assumption that the maximum

desirable vista for most urban life is that at which one person can recognize another. He then determined that the nasal bone was the smallest feature of the human face from which a person could be recognized, and he calculated the average nasal bone width to be about a quarter of an inch. From physiological optics he discovered that an object cannot be seen at a distance greater than 3,450 times its size—which is when the angle of vision is less than a minute—and he therefore concluded that a nasal bone could not be seen at any distance beyond $3{,}450 \times \frac{1}{4}$ inch $= 862.5$ inches $= 71.875$ feet. The basic module, then, for recognition and thus for any human-scale city would be roughly 72 feet; most particularly, that would be the maximum street width because that would be the maximum at which one person could recognize another across the street. In addition, Maertens determined that the maximum distance at which a facial expression could be easily read is about 48 feet (try it yourself; you'll see that it's about right) and therefore that would be the module for the ideal street width in those parts of a city, residential districts for example, where that kind of intimacy and community cohesion would be desirable.

Using the 27-degree formula along with Maertens's standards, we can then calculate the optimum heights for urban buildings. A street width of 72 feet provides a maximum building height of about 42 feet (tan. $27° \times 72' = 36'8\frac{1}{4}'' + 5'6'' = 42.2'$), or somewhere around three stories; a street 48 feet wide would allow building heights of about 30 feet (tan. $27° \times 48' = 24'5'' + 5'6'' = 30'$), or somewhere between two and three stories tall. Interestingly, these are the heights that even now we recognize as being the "normal" size for urban dwellings and shops except in the most built-up parts of our city centers, the size that feels most natural and comfortable; just a couple of miles away from the enormous spires of Manhattan, this is in fact the size of most sections of the so-called outer boroughs. Interestingly, too, these are the heights that allow us to observe the tops of trees over the roofs, providing that touch of nature that always seems so satisfying in the middle of a city, so necessary even to the most cement-hardened urbanite.

Of course buildings of such heights seem unusually small when we compare them to the soaring achievements of contemporary skyscrapers. And yet, when we reflect on the genuinely most satisfactory parts of the cities we have seen, they tend to be those observing something like that scale. It would seem to be no accident that the traditionally most durable urban neighborhoods in virtually every city have been those with buildings at this general height: most of Greenwich Village and Brooklyn Heights in New York, the French Quarter in New Orleans, Society Hill in Philadelphia, Telegraph Hill in San Francisco, Georgetown in Washington, Chelsea in London, Montmartre in Paris, Trastevere in Rome, Gamla Stan in Stockholm. At this height, it seems, residents and visitors alike recognize something special about an area, they tend to feel it is containable, knowable, somehow embraceable, and they feel themselves an organic part of it. And in almost all of the successful medieval cities, the crucibles in which our very concepts of urbanity were formed, both business and residential buildings tended to be three or four stories tall—partly, of course, because they were built before the invention of the elevator, but also because they apparently were felt to work most harmoniously at this scale. That scale is observable still today in a number of European cities: Tübingen, whose market buildings date from the fifteenth century and are generally four, occasionally five, stories; Oxford, where both college and city buildings from the fourteenth century onward tend to be two or three stories, interspersed with a few frilly six- and eight-story Gothic towers; Amsterdam, where certain older parts of the city, like the sixteenth-century Beguines section, show an intimate neighborhood scale, with all buildings at two or three stories; most of those traditional English villages, like the Cotswold villages whose streets are still lined with two- and three-story homes and shops; and the wonderfully preserved Dutch city of Naarden, still today very much as it was in the late medieval period, which has almost no structures over three stories except the central cathedral and where you can walk down the streets and easily view the handsome, solid buildings and still see the leafy green tops of the trees beyond, the whole thing giving a feeling of such comeliness and serenity that it is astonishing to believe that you are in the twentieth century.

ONE FURTHER DIMENSION of the human-scale city can be obtained by extending Maertens's ideas and considering the distance at which it is possible to distinguish the general outlines, clothes, sex, age, and gait of a person, which experience shows to be somewhere around 450 feet; this is the distance, too, ergonomic studies show, at which the basic colors can be easily distinguished. Using that module, one might then derive the optimum lengths for city blocks, for open spaces, for parks and malls and plazas.

Again, it comes as no surprise to find that this measure has indeed been used throughout much of history for many of the most successful public spaces: the maximum visual distance of the Acropolis, measured as the eye takes it in from the Propylaea to the farthest corner of the Parthenon, is about 465 feet and the maximum width is about 430 feet; the marvelous oval in front of St. Peter's in Rome is 430 feet wide; the Place Vendôme in Paris and Amalienborg Square in Copenhagen both have dimensions of 450 feet; the axis of the Imperial City in Peking, though it actually stretches for more than 8,300 feet, has no unbroken vista longer than 500 feet; and the maximum length of the Piazza San Marco, perhaps the most delightful plaza in a nation of extraordinary plazas, is 425 feet. New York City offers another telling example of public spaces: the blocks on the West Side of Manhattan, which as Jane

Jacobs points out do not work well either aesthetically or socially, are about 800 feet long, while those on the East Side, which are far more effective and inviting, range from 400 to 420 feet in length. And one always gets a sense of vague discomfort in those urban spaces that exceed this optimum distance by much—as with the mall in Washington, which stretches for more than 10,000 feet with only the slim Washington Monument to break it up, so that the Capitol at the end of it, which is supposed to be a grand and important monument at its imposing height of 307 feet, tends to get lost in the surrounding background. Such similar excesses of space also destroy the vistas of the Arc de Triomphe in Paris, the Assembly building in Le Corbusier's Chandigarh, the main plaza of Brasilia, Independence Hall in Philadelphia, the Civic Center in San Francisco, almost all suburban shopping centers, and many of the newer downtown shopping malls. Hans Blumenfeld, who is both an architect and a city planner and has designed buildings all over the world, has concluded: "There appears to be a definite upper limit to the size of a plaza, as to the length of a street, which can convey a strong spatial experience," beyond which "they can no longer be related to the human being: their scale is no longer 'superhuman' but colossal and inhuman." And he adds, "The measure of approximately 450 feet appears rather frequently in the most successful plazas."

Using this 450-foot module, by the way, suggests the optimum limit for the kinds of buildings that would ring such public spaces. To be taken in whole from the farthest side of a plaza of that length at an angle of 27 degrees, a building would have to be no higher than about 225 feet, or fifteen to twenty stories, a common and agreeable height for works meant as public monuments such as churches or civic centers; actually, following precepts common in public spaces since the Renaissance that the tallest buildings should be about a third of the longest dimension of a plaza, the optimum height might be closer to 150 feet, or ten to fifteen stories. In either case, one would avoid such urban design disasters as the Empire State Building, the World Trade Center, the Sears Tower in Chicago, the U.S. Steel Building in Pittsburgh, the Transamerica pyramid in San Francisco—indeed, one could say almost all contemporary skyscrapers—which are built so far in excess of the spaces they are set in that, incredible as it may seem, it is actually impossible to find *any* point from which they can be viewed whole. (To view the World Trade Center in its entirety, for example, it would be necessary to stand at a point roughly 2,700 feet—more than half a mile—away, and at that distance, unless you stood somewhere in the middle of the Hudson River, most of your visual field would be taken up with dozens of other intervening buildings.)

THERE IS, HOWEVER, something more to the human mode in architecture than just the height of buildings, as crucial—as *fundamental*—as that is. Those buildings that in some way or other relate to the human observer, that allow the person to feel a part of the building and to participate in its forms and spaces, also have other design elements.

It is extraordinary that, despite the years of talk in architectural circles about human scale, there is so very little written about these design elements and how they might operate, and only slightly less extraordinary that there is so little actual construction of them in the real world. Architecture still tends to be an abstract art, or at most a sculptural one, divorced in its inception as in its drafting from the people it will house. Professors Kent C. Bloomer and Charles W. Moore, in their cogent and most perceptive *Body, Memory, and Architecture,* put it this way: "The human body, which is our most fundamental three-dimensional possession, has not itself become a central concern in the understanding of architectural form." Yet obviously it is those elements that mirror or mimic or measure the human form that have been incorporated into all kinds of structures from the time that people started conscious building: doors and windows crafted to actual human dimensions, stairs reflecting the human foot and pace,[1] columns imitating the standing figure, domes and arches curved like the top of a head, doorknobs or latches formed to match the human hand, ceilings and walls easily measured at a glance, and so on. "These forms have been important to humankind," Bloomer and Moore argue, "because they accommodated the initial human act of constructing a dwelling, the first tangible boundary beyond the body, they accommodated the act of inhabiting, and they called attention to the sources of human energy and to our place between heaven and earth."

Such elements are not accidental and frivolous, gingerbread decorations out of some fevered Victorian mind. They are the vital parts of a building by which we know what *it* is and where, what *we* are and where. The whole business of seeing, after all, the very way in which we distinguish between two images in the eyeball, is the ordering of scale—that is, comparing known dimensions with those in our vision—so that although two trees appear on the retina at the same height we perceive that *that* tree back there is so big and *this* tree up here is only so big. "Such a scale is psychological," James J. Gibson has written in his pioneering *The Perception of the Visual World,* "it is something we carry around with us and it is implicit in the very process of perception." Where there is nothing to order this scale, we are always disoriented and confused—as in those rooms in psychology labs with chairs twice as big as normal and windows exaggeratedly small, or with those optical illusions where the perspective lines are distorted and the bigger object appears smaller. Gibson, among others, believes in fact that this sense of

---

1. It is from the Italian word for stair—*scala*—that we derive the very word "scale."

scale, of what he calls the "basic-orienting system," is one of the five basic senses, along with sight, sound, taste, smell, and tactile ("haptic") perception. Current psychological research would seem to support him, particularly that which suggests that a relatively clear *psycho-physical* orientation, a sense of being "centered" in the world, is vital in creating a healthy sense of oneself, and hence of one's relations to others; the modern "body-image" school even suggests, with considerable anthropological as well as psychological evidence, that there may be an innate, unconscious human desire to locate oneself in three-dimensional space within defined and knowable boundaries.

If so, then contemporary buildings, at least the agglomerations of them in larger American cities, fail abysmally. "Our cities are stacked up in layers," Bloomer and Moore write, "which bear testimony to the skills of the surveyor and the engineer in manipulating precise Cartesian coordinates, but they exhibit no connection with the body-centered, value-charged sense of space we started with [from childhood]." And their colleague, West Coast architect Robert Yudall, adds:

> Why are we not *moved* by our neighborhood shopping mall or city center office tower? Take, for example, a typical curtain-wall skyscraper. Its potential for pulling us into a realm of a movement or sound game is almost nil. We can neither measure ourselves against it nor imagine a bodily participation. Our bodily response is reduced to little more than a craned head, wide eyes, or perhaps an open jaw in appreciation of some magnificent height.

There are no human-scale elements in such buildings, no sense whatsoever of how many of *us* it would take to make up one of *them,* no statement to us as to how we are to see, enter, or use it, nothing to convey a sense of intimacy, or invitation, or sociability, or humanity. The UN Secretariat—sad to say, given the humanitarian purpose it is supposed to symbolize—is a perfect example of such a scaleless structure: 505 geometrically abstract feet, cold and forbidding no matter what the time of day or year, windows and floors that might have begun to suggest human dimensions hidden behind a bland and uncalibrated glass wall on the long sides and unbroken slabs of marble on the ends. And looking at two such structures together, or whole canyons of them, makes one almost catatonic.

Try standing on New York's Sixth Avenue in the 50s or Wall Street off Broadway, or the new Peachtree Street in Atlanta, or in Embarcadero Center in San Francisco, or Century City in Los Angeles, or any other of those urban compression chambers across the land, and you can immediately experience a peculiar sense of dislocation. The eye is continually confused by changes in scale, and, what's worse, by the lack of scale altogether, so that it soon is forced to shun the towering facades and seek relief in whatever street-level doorways and arcades there may be to break up the geometry; and yet there persists a feeling of looming mass, an unshakable awareness of hovering grey colossi, gathered above like the overcast of some cloudy day. That way, it is not hard to feel, madness lies.

ODDLY ENOUGH, it is in the much-maligned—and usually quite undistinguished—single-family house that we normally have to look today for architectural homage to the human scale. As Hans Blumenfeld remarks, "The field of the 'normal human' and especially of the 'intimate human' scale has been the field of the home builder rather than that of the architect." Here, in the ordinary home, are the elements that are generally missing from the urban mass. The natural world is present: there are lawns and trees and gardens, a certain amount of open space and windows on every front, a prominent fireplace and a chimney visible from the outside, and in the most favored settings a pond or a stream or a fountain; in other words, echoic images of the four primal elements, earth, air, fire, and water. The human body is incorporated: in stairs and railings with normal scales, doorknobs that fit actual hands at comfortable heights for an extended arm, lightswitches at near-shoulder height, doorways built for people and not some race of giants, windows positioned at the level of normal visual fields, moldings or arches or newels or recesses or railings or vaults or columns imitative of bodily shapes. And the social dimension is recognized: by the central room with a fireplace and hearth, which is a slightly magic area for us as it was for our distant ancestors, around which we still place our icons and treasures, our best furniture and favorite paintings and cherished artifacts, and too by the usual spacious kitchen/eating area like those in which people have gathered for food preparation and consumption from the time of the earliest hunting bands.

All of these things provide physical and psychological "centeredness," order, scale, and some degree of stability and serenity. But we simply do not experience them during the course of most of our lives. As Bloomer and Moore conclude:

> Offices, apartments, and stores are piled together in ways which owe more to filing-cabinet systems or the price of land than to a concern for human existence or experience. In this tangle the American single-family house maintains a curious power over us in spite of its well-publicized inefficiencies of land use and energy consumption. Its power, surely, comes from its being the one piece of the world around us which still speaks directly of our bodies as the center and the measure of that world.

We shape our buildings and thereafter our buildings shape us.

4

The Search for Community

ANTHROPOLOGICAL EVIDENCE is somewhat sketchy, but it seems clear that the oldest human institution is not the family, as is popularly thought, but the community—the tribe, clan, troop, or village, the larger setting. The family is very old, of course, but as a permanent and conscious social unit it seems to have emerged slowly over tens of thousands of years, well after the early hominids had formed into tribal communities. Bernard Campbell of Cambridge University suggests that the family may have originated sometime in the period of the *Australopithecus* when small bands of hominids, perhaps a male and one or more females and their children, broke off from a larger community during periods when adverse weather caused food supplies to dwindle. Such small groups were probably able to travel easier and farther in search of food and were able to exist more easily on limited resources, a pattern such as is found even today among certain baboon troops and among the so-called "Bushmen" (actually the Basarwa) of the Kalahari. In time, or with successive hardship periods, such small groups might have eventually developed into deliberate families, lasting a generation and then for successive generations, though probably always within the confines of a larger social grouping.

In fact it is reasonable to assume, as most anthropologists do nowadays, that it was some form of community living that made humans "human." In other words, it was not until some group of hominids meeting around some water hole began grunting and jabbering at each other in regular ways that speech developed, and those groups in which speech was most successful—where, for example, cries of warning or hunting directions were least ambiguous—were the ones that were more likely to survive. It was not until sufficient bands of people gathered for regular periods that fire, found in some treetrunk hit by lightning or dry grass sparked by a volcano, could be tended day in and day out, a task that was essential in those hundreds of thousands of years before our human ancestors had learned to create fire at will. It was not until individual hominids began operating as groups that they were capable of taking on the larger game animals and covering the kinds of geographical distances that would have been necessary for a regular supply of meat. It was not until groups began the regular sharing of food—a trait found in no other primate—that there were created the social ties and obligations, and eventually the kinship and mating systems, that distinguish humans from all other animals. It was not until early humans lived in settled groups that they developed art—such as the Lascaux cave paintings—and religion—symbolized by the Neanderthal grave sites in Iraq's Shanidar Cave—and manners and customs and traditions and codes and all that goes to make up a culture, however scanty and primitive. The very existence of the community it was that caused the development of so many of those things that over countless millennia shaped the animal that we know as "Modern Man."

Moreover, it can be said that the whole process of human success has depended precisely upon the ability of our species not only to live in small groups of twenty-five or so—for that is found in many other primates as well—but to create large communal institutions numbering into the several hundreds and to keep such gatherings intact by mutual aid and cooperation over years and decades and centuries. Amos H. Hawley, one of the pioneers in the study of human ecology, has argued that it has been this human capacity for living in community that has been the essential "adaptive mechanism" throughout human development. Human beings cannot adapt to their physical environment alone as individuals, he says; only by creating a "human aggregate," and thus practicing communal and cooperative efforts, are they able not only to survive but in many respects to prevail. One might even justifiably claim therefore that the instinct for community is as old and as vital and as powerful as the instinct for sex, since the former appears to be as necessary for survival as the latter is for procreation.

Indeed there is every reason to believe that the human need and capacity for communality is genetically encoded in modern *Homo sapiens*. René Dubos, the eminent microbiologist at Rockefeller University, argues that "the social organization based on the hunter-gatherer way of life lasted so long—several hundred thousand years—that it has certainly left an indelible stamp on human behavior." He points out that 100 billion human beings have lived since the late Paleolithic period, when the Cro-Magnons appeared, and "the immense majority of them have spent their entire life as members of very small groups . . . rarely . . . of more than a few hundred persons. The genetic determinants of behavior, and especially of social relationships, have thus evolved in small groups during several thousand generations." He concludes: "Modern man still has a biological need to be part of a group. [This] cannot possibly be altered in the foreseeable future, even if the world were to be completely urbanized and industrialized."

It seems safe to say, at the very least, that the organization of community is not simply *one* way of ordering human affairs but a *universal* way, found in all times and places, among all kinds of peoples. George Murdoch, one of the giants of American anthropology, undertook an exhaustive ten-year "cross-cultural survey" for the Institute of Human Relations at Yale some years ago, during which he and his colleagues studied some 250 different human societies all over the earth, of varying periods, geographical settings, and stages of development. He concluded:

> The community and the nuclear family are the only social groups that are genuinely universal. They occur in every known human society, and both are also found in germinal form on a subhuman level.

As fundamental as the family is, however:

> Nowhere on earth do people live regularly in isolated families. Everywhere territorial propinquity, supported by divers other bonds, unites at least a few neighboring families into a larger social group all of whose members maintain face-to-face relationships with one another.

Thus, though humankind has not shown itself to be adept at very many social relationships, it can be said that during the long eons of evolution it has probably had more experience with the small community than any other form and has learned to live in groups of that size more successfully than any other.

Given that apparently incontrovertible fact, it would seem sensible for any rational society to attempt to protect and promote the institution of the community. And yet, for all the tossing about of words like "community center" and "community action," we obviously are not doing so at present in this country or in very many parts of the industrialized world. Small towns are everywhere threatened, and rural populations dwindling; the newly burgeoning suburbs have shown themselves to be particularly weak in creating community cohesion and mutuality; and it is the rare section of the infrequent city in which any strong sense of communality is to be found any longer. It may be too strong to say, with Christopher Alexander, an urban planner at the University of California, that "Western industrial society is the first society in human history where man is being forced to live without" the intimate contacts of community, meaning that "the very roots of our society are threatened." Yet the increasing loss of communal life is undoubtedly at the heart of the malaise of modern urban culture and its disappearance clearly cannot bode well for the future.

But is it not possible to envision the criteria for an optimum community in the modern world? Just as the human measure can guide us in the design of technology and the creation of buildings, can it not provide a guide in the development of communities?

The cardinal task here, it seems to me, is to discover the limits of a human-scale community, the size beyond which—as the Beanstalk Principle would suggest—it ought not to grow. And here, thanks to various anthropological and sociological records, we have a considerable body of interesting evidence.

During most of its prehistorical eons, humankind did not live in settlements much above a thousand people and generally preferred places closer to half of that. John Pfeiffer, the anthropological writer who has perhaps as encyclopedic a vision as anyone in the field, goes so far as to call 500 a "magic number" because it recurs so often in human evolution as the limit of a community:

> The phenomenon becomes clear and meaningful only after taking census figures for a large number of tribes. Such studies reveal a central tendency to cluster at the 500 level, and this tendency is widespread. It holds for the Shoshoni Indians of the Great Plains, the Andaman Islanders in the Bay of Bengal, and other peoples as well as the Australian aborigines.

This number, Pfeiffer suggests, may have been determined by the nature of human communication and culture-sharing mechanisms: "There seems to be a basic limit to the number of persons who can know one another well enough to maintain a tribal identity at the hunter-gatherer level, who communicate by direct confrontation and who live under a diffuse and informal influence, perhaps a council of elders, rather than an active centralized political authority." That number may also have been determined by the optimum number of people for mate-selection. Martin Wobst of the University of Massachusetts has made computer calculations of prehistoric societies and determined that for a male to have an adequate pool of mates in his age group in any society in which incest taboos prevailed, he would need a total population of approximately 475 people.

Supportive evidence comes from a number of sources. Accounts of those primates that have developed herds and clans in addition to small-group units often indicate a range from under 100 to more than 700, with an average around 400 or 500: the gelada baboons of Ethiopia, for example, have herds up to 400, while langur monkeys of India tend to split into several herds after more than 550 or so. Joseph Birdsell, an anthropologist at UCLA who has spent considerable time among the Australian aborigines, has reported:

> The Australian data show an amazing constancy of numbers for the dialectical tribe [i.e., speaking one dialect], statistically approximating 500 persons. This tendency is independent of regional density. Since the data cover mean annual rainfall variations from 4 inches to more than 160 inches, the size of the dialectical tribal unit is insensitive to regional variations in climatic . . . factors.

Robert Carneiro of the American Museum of Natural History has observed that villages among the Amazonian people he studied in Brazil tended to split up at populations between 100 and 200, and among Peruvian villages most split before they reached 300. Pierre Clastres, a French scholar, estimates that the traditional Tupi-Guarani villages of South America had "a mean population of six hundred to a thousand persons," based on various sixteenth-century documents, and an actual census in the seventeenth century showed about 450 per village among the Tupis. Various anthropologists in the U.S. have estimated that the "long houses" of the tribes of the Iroquois Confederacy—the meeting places for village gatherings—typically held no more than about 500 people. And archaeological research at sites throughout the Middle East—for example, the work in Mesopotamia of Robert Adams of the University of Chicago and his colleagues—provides evidence that the general size of the earliest village settlements, established in the two millennia after 5000 B.C., varied from about fifty people to perhaps several thousand, but most often clustered somewhere around the 400–500 range.

We may leave it to René Dubos, once again, to sum it up: "The biology and psychology of modern Man has certainly been influenced by the fact that, during the past 10,000 years, most people have lived in villages of some 500 inhabitants."

THERE IS ANOTHER WAY of coming at the question of the human limits of a community. Hans Blumenfeld, the urban planner, suggests starting with the idea of the size at which "every person knows every other person by face, by voice, and by name" and adds, "I would say that it begins to fade out in villages with much more than 500 or 600 population." Constantine Doxiadis, after reducing thousands of data from various centuries, came to the conclusion that what he called the "small neighborhood" would hold approximately 250 people, a large neighborhood some 1,500, with an average around 800–900. Gordon Rattray Taylor, the British science writer, has estimated that there is a "natural social unit" for humans, defined by "the largest group in which every individual can form some personal estimate of the significance of a majority of the other individuals in the group, in relation to himself," and he holds that the maximum size of such a group, depending on geography and ease of contact, is about 1,200 people; he adds that business firms historically begin to face organizational problems at about this level. Terence Lee, a British sociologist who did a thorough survey of attitudes in Cambridge, England, reported in *Human Relations Journal* in 1968 that the people themselves thought of their "social acquaintance area" as containing from 0 to 400 houses (some people obviously were not too neighborly), or from 0 to 1,200 people, and the average was under 1,000. And examinations of successful communes over the last century indicate that for the most part they have tended toward an optimum size of 500—as Charles Erasmus, a University of California anthropology professor, has summarized the data, "Successful commune movements . . . have invariably been divided into communities averaging less than five hundred inhabitants." (He cites the upper limits of 600 for the Shakers, 500 for Amana groups, 300 for Oneida, 800 for Harmony, 500 for Zoar, and 150 for the Hutterites.)

Not many scientific studies have been done on what characteristics of the human brain may set these rough limitations on the size of successful face-to-face communities, but George Murdoch has provided some evidence here. He asked his students and friends to list the people they regularly associated with and knew on a first-name basis; the results showed such a surprising unanimity, ranging from around 800 to around 1,200, that Murdoch concluded that this represents the general "index of familiarity" for human groups. (Try it yourself sometime in a dull moment.) John Pfeiffer also notes that "there is an architect's rule of thumb to the effect that the capacity of an elementary school should not exceed 500 pupils if the principal expects to know all of them by name" (the average number of pupils in U.S. elementary schools in 1975 was in fact 409). He goes on to suggest that "the memory capacity of the human brain probably plays a fundamental role of some sort since that influences the number of persons one can know on sight."

It is possible, too, that the human visual capacity plays a role. Just as the optimal distance for determining the rough outlines of an individual has been determined to be around 450 feet, as we saw in the last chapter, the optimal distance for determining whether a distant object is human at all seems to be around 1,000 feet; this, too, is taken to be the normal optimum for the line of sight of an adult human. Now if we begin by taking this as the maximum space that an intimate community would occupy—the area in which the human form could be perceived from one end of the settlement to the other—the total space would then be 1 million square feet, or about 23 acres (1 acre = 43,560 square feet). Allowing a population density range of from, say, 15 to 40 people per acre (New York City, for comparison, averages 41 people per acre), that provides a community population of between 345 and 920 people. The eye, happily, seems to agree with the brain, again settling on approximately the same range around the "magic" 500 figure. Could that be a remarkable coincidence—or perhaps a reflection of the genetic coding that has effectively determined the limits of human social functions?

THERE IS ANOTHER NUMBER, or rather range of numbers, similarly "magic" perhaps, that recurs in the examination of community, suggesting another desirable, though somewhat larger, size for human groupings. Again, there is an interesting general agreement on the figures from a wide variety of sources.

Although it was the face-to-face village that was the primary communal unit, many societies, and particularly the more successful, often formed larger bands, or tribes, uniting these villages into a common culture. Anthropological evidence suggests that throughout prehistory the upper limit for such tribes—defined as those who speak the same dialect or language or who unite into an association of villages—was everywhere about 5,000 or 6,000. William Sanders and Paul Baker of Penn State University point out that tribal units sharing common customs, common language, and common territory might have grown as large as this limit but at that point almost always split into new tribes or else imposed a limit on further growth by establishing a central authority capable of governing population. The University of Arizona's William Rathje, after using general systems theory to interpret Mayan culture, has suggested that this 5,000 number may represent a level at which the social system, at least in early societies, had to limit itself to keep from overloading itself with complexities and burning itself out.

This was about the limit of the earliest cities, too, as near as we can reconstruct the sites—not only in the Middle East, an area that has been particularly well surveyed, but in India, China, North Africa, and Central America. A few might have grown to 20,000 and even 50,000 in their latest stages, shortly before collapse, but as a general rule the urban centers of the millennia before Christ seem to have stayed at between 5,000 and 10,000 people; in the words of Gideon Sjoberg, the urban sociologist, "It seems unlikely that, at least in the earlier periods, even the larger of these cities contained more than 5,000 to 10,000 people, including part-time farmers on the cities' outskirts." Constantine Doxiadis has calculated that a population of about 5,000 and an area of about two square kilometers were typical of almost all early cities: "Very few of the cities known to have existed during these thousands of years did not have these characteristics." His explanation for this was that the human animal might have a "kinetic field" set by a ten-minute radius, thus establishing the limits of its normal daily activities, and thus the limits of the area the city could cover and the population it could hold.

This same figure emerges again when we come to the medieval

period. Without doubt most places, at least in Europe, had fewer than 2,000 people, but the few trading centers that began to form after the twelfth and thirteenth centuries most commonly held up to 10,000 souls—as Chartres in the twelfth century (10,000), Ypres in 1412 (10,376), Basel in the fourteenth century (8,000). And when we look at the full-fledged cities, particularly those noted for their economic and cultural achievements, it is remarkable how often 5,000 is cited as the upper limit of the precincts, or quarters, out of which these new urban centers were created and into which they were divided. In fact "quarter" originally meant literally a fourth part of a city (or thereabouts), and since we know from Lewis Mumford that the medieval town did not grow beyond about 40,000 people—that was the population of London in the fifteenth century—this would confirm an upper limit of approximately 10,000 people per precinct. There is a print showing the layout of the city of Aachen, in 1649, that might serve as a neat example: it shows a circular city bounded by heavily fortified walls, with a large cathedral at the very center and, at the points of the compass, four smaller churches—the church, of course, representing the heart, the "community center," of the medieval neighborhood. The population at the time was somewhere around 20,000, and the quarters then probably held about 5,000 each.

Coming down to the present, a range of 5,000 to 10,000 shows up with surprising frequency in the recommendations of architects and city planners for the preferred size of a community. Clarence Perry, the grandfather of contemporary planning and the man who redirected attention back to the idea of small-scale communities in the 1930s when they were first threatened by the onrush of the twentieth-century metropolis, is typical. After years of study he hit upon an ideal "neighborhood unit" of from 3,000 to 9,000 people, with an optimum size at 5,000. His theory was that a neighborhood had to be small enough so that everything important—schools, playgrounds, shops, public buildings—was within easy walking distance, and large enough to support an elementary school and a variety of local stores and services. Both conditions could be satisfied, he determined, with a population of about 1,000 to 1,500 families, or an average of about 5,000 people, distributed at roughly 15–20 people per acre, with the total unit then occupying about a half mile square. Since Perry's formulation, a wide variety of other city planners, of different decades, philosophies, styles, and interests, has also arrived at about the same figure.[1]

1. *Architectural Forum,* in proposing the redesign of American cities after World War II, urged community units of 2,000–5,000. Planners Hermann and Erna Hervey and Constantin Pertzoff have recommended 500–2,000 families, or 2,000–8,000 people, as the "desirable unit size" based on "social activities and functions," and have drawn up a model unit of 5,000 people. Planner N. L. Englehart's model city has a "desirable" community unit of 1,700 families, or roughly 5,100–6,800 people. The American Institute of Architects

Perry's "walking distance" principle in particular has become standard among almost all urban planners who give any thought at all to community. Walter Gropius, the architect, has explained the rationale this way:

> The size of the townships should be limited by the pedestrian range to keep them within a human scale. . . . The human being himself, so much neglected during the early machine age, must become the focus of all reconstruction to come. Our stride determines and measures our space- and time-conception and pegs out our local living space. Organic planning has to reckon with the human scale, the "foot," when shaping any physical structure. Violation of the human scale will cause further degeneration of life in cities.

Like Perry, Gropius observed that the maximum distance a person would walk comfortably for ordinary community affairs was about half a mile, and thus he too came up with an optimum "township" size of roughly a half a mile square. Now if we assume that half the space within that area would be given over to public buildings, shops, pathways, and parkland, and if we assume for the remaining 160 acres residential densities somewhere around the models of Gropius and the "garden city" planners, we come up with a range of population—no surprise—around 5,000–8,000 people.

One final piece of evidence on community size comes from Leopold Kohr, who has done more thinking about this from the perspective of the social sciences than anyone to date. Kohr argues that it would take about eighty or a hundred adults to provide the *convivial* society, that is to say, the number to "fulfill the companionship function to the fullest" and "to ensure both variety of contacts and constancy of relationships"; but, he says, it would take more than that for an effective *economic* society. In a society with basic specialization—a shoemaker, say, and a baker and a builder and so on—there need to be enough people to consume the goods and services during the course of a year, and eighty or a hundred adults is too small a pool; "economic optimum social size," he estimates, requires "a full membership of 4,000 to 5,000 inhabitants." At this level, he argues, "society seems capable of furnishing its members not only with most of the commodities we associate with a high standard of living, but also of surrounding each person with the margin of leisure without

---

in 1972 recommended neighborhoods of 1,700–10,000 people. Paul and Percival Goodman in *Communitas* speak of basic urban units of 3,000–4,000. José Sert, in an article entitled "Human Scale in City Planning," has recommended neighborhoods of 5,000–10,000. And according to Barbara Ward, 6,000–10,000 is traditionally taken as the base community population in Britain because that is the number that is needed to sustain an elementary school; this was generally the scale followed there in the extensive New Town developments of the 1950s, including Milton Keynes, the largest of them and theoretically the showplace, which was specifically designed with a grid system of 5,000 people per unit.

which it could not properly perform its original convivial function." And for the optimum *political* society only a few thousand people need be added—"a full population of between 7,000 and 12,000"—to provide a sufficient number who can be spared from basic economic routines to perform legislative, legal, political, and security tasks.[2] This is the size, actually, of various real-world states that survive quite nicely, including the independent state of Nauru in the South Pacific, and such self-administered dependencies as Anguilla, the Cayman Islands, Montserrat, Falkland Islands, Saint Helena, and Tuvalu.

WHAT WE HAVE HERE, then, with our two ranges of "magic" numbers, is a rough measure of the two basic kinds of community that humans have apparently found the most useful and successful over their many millennia as social creatures. One is the face-to-face community, or association group, with somewhere between 400 and 1,000 people and an optimum of perhaps 500—what we might call the common *neighborhood;* the other is the extended association, a wider alliance of some 5,000 to 10,000 people, usually hovering around the 5,000 figure—what we would think of as the standard *community.* (I realize that there is a good deal of confusion in these terms, even among city planners and social scientists who should know better. But common usage suggests that "neighborhood" is the smaller unit—one's neighbors are typically those who live on the same or adjoining blocks—and "community" is the larger element—one's community generally includes nearby stores and parks and bars and schools, covering a wider geographical area and usually comprising several neighborhoods.)[3] These numbers are not meant to be hard and fast, of course. They can be only proximal, suggestive. But if humankind in all races and cultures has chosen to live in these aggregates throughout its evolution, right down to the current era, and if to some degree at least it has solved the problems of collective living at these levels, then they deserve our consideration in a most essential way when we contemplate the design and reconstruction of our social settings.

A community of less than 10,000 is, in modern terms, small, of

2. Some additional support to the idea of this range as politically optimum is provided by a study in Sweden some years ago showing that population units of about 8,000 people showed higher levels of political participation and effectiveness than any other sized units. It is a subject that we will return to in Part Five.

3. Doxiadis, in his elaborate and careful taxonomy, gives "neighborhood" a standard population of 800, and the next larger grouping, which he calls the "small polis," 10,000. Barbara Ward in *The Home of Man* speaks of the "concept of small neighborhoods as the basic building block of human communities." And Robert Dickinson, professor of geography at Leeds University, has attempted to establish British usage this way: "In urban areas [it is the] larger group, with 5,000 to 10,000, that is usually but erroneously referred to as a neighborhood. The term community area is more appropriate."

course—some might say hopelessly small. But, as Gropius says, "It is particularly the small size of the township with its human scale which would favorably influence the growth of distinct characteristics of the community," including regular associations between people, easy access to public officers, mutual aid among neighbors, and open and trusting social relations. Smallness is simply essential to preserve the values of community as they have been historically observed—intimacy, trust, honesty, mutuality, cooperation, democracy, congeniality. The record on this point is both ample and consistent and does not need to be rehearsed here; the school of historical scholarship beginning with Sir Henry Maine and Georg von Maurer (among many others) in the nineteenth century and coming down to Lewis Mumford and Murray Bookchin (among many others) at the present time has provided an abundant chronicle.

Of course we are a long way from the human-scale community today. But the sense that there is something ineluctably valuable in community has not yet disappeared. It is there among city planners, still using it as a model both in the design of new cities and in the rehabilitation of old ones; it is there among countless political organizers, particularly of the Left, who are seeking in cities across the country to draw people together around local grievances and issues; and it is there among ordinary men and women, in all kinds of social settings, searching in myriad ways—neighborhood organizations, block associations, community centers, groups of every kind and description—for the feelings that come from association, communion, and companionship. Perhaps that is the inescapable biological imprint of community at work.

It is, I would agree, somewhat fanciful to imagine a United States divided up into thousands of small communities. (Though it is not impossible, as is sometimes thought. You could divide the present population into some 45,000 separate communities of roughly 5,000 each, give each one as much as a mile to spread out on, plus a greenbelt of 15 square miles to separate it from its neighboring towns, and use up only about 720,000 square miles—less than a fifth of the nation's present area, less than half the amount currently given over to cropland.) But it is not fanciful to think of modern society seeking to achieve some such reordering of its social forms so as to recapture some of those communitarian values that have existed through the ages. If the Cro-Magnon caves and the Mesopotamian cities and the medieval towns and the English villages with their meager resources and understandings and technologies were able to know those values, then how much easier it ought to be for us today. And how much more necessary.

In 1898 a British army officer named H. Fielding Hall, who had spent a decade in Burma after the British conquest of that country, wrote a book, *The Soul of a People,* describing in some detail the workings of Burmese society, from all accounts an accurate picture of the land of that time. Hall was particularly struck by the way in which Burma was a land of villages and villagers, without large cities or even many sizeable trading towns and, with the exception of a king and tiny royal family, without hierarchies or a fixed nobility or ecclesiastic powers or any notable division into classes or castes. He wrote:

> Each village was to a very great extent a self-governing community composed of men free in every way. The whole country was divided into villages, sometimes containing one or two hamlets at a little distance from each other—offshoots from the parent stem. The towns, too, were divided into quarters, and each quarter had its headman. These men held their appointment-orders from the king as a matter of form, but they were chosen by their fellow-villagers as a matter of fact. Partly this headship was hereditary, not from father to son, but it might be from brother to brother, and so on. It was not usually a very coveted appointment, for the responsibility and trouble was considerable, and the pay small. It was 10 per cent on the tax collections. And with this official as their head, the villagers managed nearly all their affairs.
>
> Their taxes, for instance, they assessed and collected themselves. The governor merely informed the headman that he was to produce ten rupees per house from his village. The villagers then appointed assessors from among themselves, and decided how much each household should pay. Thus a coolie might pay but four rupees, and a rice-merchant as much as fifty or sixty. The assessment was levied according to the means of the villagers. So well was this done, that complaints against the decisions of the assessors were almost unknown—I might, I think, safely say were absolutely unknown. The assessment was made publicly, and each man was heard in his own defence before being assessed. Then the money was collected. If by any chance, such as death, any family could not pay, the deficiency was made good by the other villagers in proportion. When the money was got in it was paid to the governor.
>
> Crime such as gang-robbery, murder, and so on, had to be reported to the governor. All lesser crime was dealt with in the village itself, not only dealt with when it occurred but to a great extent prevented from occurring. You see, in a village anyone knows everyone, and detection is usually easy. If a man became a nuisance to a village, he was expelled. . . .
>
> All villages were not alike, of course, in their enforcement of good manners and good morals, but, still, in every village they were enforced more or less. The opinion of the people was very decided, and made itself felt. . . . So each village managed its own affairs, untroubled by squire or priest, very little troubled by the state. That within their little

> means they did it well, no one can doubt. They taxed themselves without friction, they built their own monastery schools by voluntary effort, they maintained a very high, a very simple, code of morals, entirely of their own initiative.

And then Hall adds a particularly interesting comment on the Burman:

> And so I do not think his will ever make what we call a great nation. He will never try to be a conqueror of other peoples, either with the sword, with trade, or with religion. He will never care to have a great voice in the management of the world. He does not care to interfere with other people: he never believes interference can do other than harm to both sides.
>
> He will never be very rich, very powerful, very advanced in science, perhaps not even in art, though I am not sure about that. It may be he will be very great in literature and art. But, however that may be, in his own idea his will be always the greatest nation in the world, because it is the happiest.

5

The Optimum City

IT IS QUITE remarkable that in the several thousand years people have been given to musing about that odd arrangement of society known to us as the city, they have continually come up with approximately the same idea of how big it ought to be. In general, it was agreed, a city had to be bigger than a village, even a small town, because a certain number of people were necessary to obtain what were regarded as the advantages of urban life: anonymity, diversity, complexity, opportunity, tolerance, specialization, innovation, self-expression, stimulation. And, it was also agreed, it had to be smaller than an entire nation, tribe, or empire, because with too many people these advantages became submerged under accumulating disadvantages: crime, filth, pollution, congestion, overcrowding, inhumanity, loneliness, political ineffectuality, social disintegration. A community of 5,000, or even 10,000, successful as it might be in providing cordiality, economic self-sufficiency, and democratic government, would still be too small, at least in post-classical experience, to provide the urban values: one virtue of the community, after all, is that it does *not* encourage anonymity. But similarly an agglomeration of 10 or 20 million, though presumably capable of offering considerable variety and opportunity, would probably have more of those than any one person could tolerate and so many other assorted ills that living there wouldn't be worth the effort.

Plato, one of the first people in the West to confront the subject of city sizes, was quite specific: a city, he said, should contain 5,040 citizens—i.e., male heads of households—which along with families, slaves, and metics, in the usual Greek fashion, would make a total population of perhaps 35,000 to 40,000 people. Aristotle, though not so precise, cautioned that "a great city is not to be confounded with a populous one" and suggested a limit of "the largest number which suffices for the purposes of life, and can be taken in at a single view," which scholars have generally agreed to be in the area of 30,000 to 50,000 people. Leonardo da Vinci posited an ideal city of 30,000. Thomas More's Utopia had cities of 6,000 families, or about 20,000 people. Montesquieu offered no numbers but favored small city-states on the order of his own Bordeaux—certainly under 60,000—and asserted: "It is in the nature of a republic that it should have a small territory." Rousseau similarly talked in terms of an ideal city comparable to his own town of Geneva, then with about 25,000 citizens, and at one point suggested 10,000 as a democratic optimum. Ebenezer Howard envisioned "garden cities" of roughly 30,000, surrounded by agricultural belts containing another 2,000 or so.

The few modern scholars who have dared to dip their toes into these waters have been inclined to push the ranges upward just a bit—probably because modern technology is capable of increasing communication and efficiency somewhat—but still favor cities of about this size. The Ruth Commission in Great Britain, which was charged with studying these matters in the 1950s, concluded that cities of about 30,000 to 50,000 would be ideal, while the later Royal Commission on Local Government in the 1960s raised the limit to 100,000–250,000. The most enterprising American investigator, Otis Dudley Duncan, once showed that cities between 50,000 and 150,000 had all the urban facilities any city would need, and a more recent counterpart, Werner Z. Hirsch, has suggested that the most democratic and efficient cities were "medium-sized communities of 50,000 to 100,000 residents." Gordon Rattray Taylor has suggested an upper limit of 100,000, and his countryman E. F. Schumacher—though as we know biased in favor of smaller forms—argued that there should be "ideally no major town more than a couple of hundred thousand or thereabouts." Ralph Borsodi, among his exhaustive researches into the human condition, has written: "It is very probable that every need of a high culture with a high standard of living could be provided with cities of around 25,000 population and without any cities of over 100,000." And Robert Dahl of Yale, one of the leading American political scientists, has argued that for the most workable kind of modern city there might be "an optimum size in the broad range from about 50,000 to about 200,000."

Scholarly opinion aside, this small city is also the model that the practical planners of the world, both East and West, have tended to favor in recent decades. The ones that have been built, I hasten to add, have not all been unalloyed successes, but their deficiencies have nowhere been blamed on their sizes and the fact that they have all been designed around a specific population range is a revealing indication of an apparently cross-cultural perception of urban optimality. The New Towns that were planned in America in the late 1960s ranged from 30,000 to 120,000, with an average at about 73,000; as of 1975 the two most successful were Columbia, Maryland, with 43,000, and Reston, Virginia, with 35,000. The British New Town program, the most extensive in the world, created cities with populations ranging between 35,000 and 135,000, with an average of about 60,000. The Chinese have designed regional centers in the 50,000–80,000 range; the Russians have built model cities of 50,000–200,000 (with the "ideal communist city" said to be 100,000); the Dutch new towns have targets of 100,000; Sweden's "primary centers" are planned for 50,000–100,000; and Israel's new cities are to be built for about 40,000.

As for the reason for such a striking degree of harmony on such a difficult matter, I think it is probably to be found in one central fact: as near as we can tell, for most of human history, no matter what the continent or climate, regardless of political or economic conditions, cities only very seldom actually grew beyond the 50,000–100,000 range.

During most of its celebrated life, Athens as a city seemed to have hovered around 50,000 people, though at periods of particular power the surrounding Attic state may have grown to perhaps 150,000 or 200,000. Italian cities that began and nurtured the Renaissance, as we have seen, did not grow to more than 80,000, and most of them had closer to 50,000—the Rome of Michelangelo with perhaps 55,000, the Florence of Leonardo 40,000, and Venice, Padua, and Bologna at their height probably 50,000–80,000. New York and Philadelphia at the time of the American Revolution had fewer than 30,000 people, Boston—the cradle—no more than 15,000. In fact, it seems that only on very rare occasions did pre-industrial centers ever go much beyond 100,000, and then only temporarily, when serving as the bloated capitals of empires, as with Babylon, Syracuse, Rome, Alexandria, Constantinople, Edo, Hangchow, and Peking; even then, Rome at its height probably had no more than 300,000 or 400,000, Babylon 110,000, and Alexandria and Hangchow perhaps half a million. It has only been in the last two centuries, as we have seen, that giant conurbations have emerged and lasted—though even today there are still fewer than a hundred cities over a million and the great majority of the world's population, something like 80 percent, lives in places of fewer than 20,000 people.

The conclusion of the protean Constantine Doxiadis, after a lifetime of categorizing such things, seems quite on the mark: "If we look back into history . . . we find that, throughout the long evolution of human settlements, people in all parts of the world have tended to create urban settlements which reached an optimum size of 50,000 people." Indeed, he argues, the fact that the few larger cities that were tried did not survive for long suggests that humanity has solved the problems of living in cities of up to 50,000 or so, but obviously not in units much above that size.

Thus we seem to have arrived at another set of "magic" numbers, similarly inexact and somewhat elastic but similarly suggestive, which may provide some indication of the nature and the extent of the human-scale city. Obviously we must think of the desirable city as a congeries of neighborhoods and communities, for these have to be the building blocks out of which any larger entity is built; they must continue to supply the rootedness, even as the wider society supplies its diversity. But within that context it is still possible to find enough evidence—sociological, economic, political, and demographic—to permit some reasonable conclusions as to the validity of these magic numbers and the limits of the optimal city.

SOCIOLOGICAL

THE SOCIOLOGICAL VERDICT is generally that the quality of life in the larger American cities is—I can think of no better summary—solitary, poor, nasty, brutish, and short. There are no doubt compensations for it all, in the minds of many, but the studies suggest that, as a rule, as cities get bigger they increase in density, fragmentation, deviance, criminality, social stress, anomie, loneliness, selfishness, alcoholism, mental illness, and racial and ethnic segregation. These ills begin to gather, it seems, somewhere around the 100,000 level, and without doubt the biggest cities are the worst.

There is considerable debate just now about whether it is crowding or higher densities or sheer numbers or the amount of interaction that produces the individual and collective pathologies of big cities. Biological studies show almost uniformly that crowding and high densities in animal populations produce such things as hypertension, stress, aggression, violence, exhaustion, sadism, mental disorder, infertility, disease, suicide, and death. For the human animal, rather more adaptable to its surroundings, the evidence is not quite so unequivocal, but ever since the work of Georges Simmel at the turn of the century it has been fairly well recognized that accumulations of people create accumulations of social problems, and the more people the more problems.

Simple enough, really. With more people there are bound to be more contacts among them in any given day—the New York Regional Plan Association, alarmingly, has determined that there are 11,000 people within a ten-minute radius in Nassau County, 20,000 in Newark, and 220,000 in Manhattan—and so to prevent a psycho-social "overload" a city dweller must either have fewer contacts, spend less time on any single contact, or be less intense and involved during any contact. That's what produces the familiar patterns of big cities. People create fewer contacts by developing ethnic ties and consequent separation from "outside" groups, by having unlisted phone numbers, by not making eye contact with people on the street, and by ignoring life-long neighbors. They spend less time per contact by cutting out traditional courtesies such as "sorry" and "please," by treating supermarket checkers and short-order waitresses as faceless and history-less, by refusing "to get involved" when some unwanted contact intrudes. And they reduce the intensity of each contact by limiting their "span of sympathy" (in sociological parlance) to the immediate family, by supporting governmental agencies that take over their personal commitments (welfare, for example), and by emphasizing their privacy through personal-space "cocoons" in an elevator or at a lunch-counter.

A goodly number of scholarly studies have compared the civility of people in large cities with those in smaller places, and with rare uniformity they show big-city dwellers to be, as the stereotypes would indicate, far more selfish, hostile to strangers, unhelpful, and unfriendly. In one study a group of researchers dialed phone numbers at random in various-sized cities and, concocting a plausible story about an emergency, asked the people who answered to look up and call a number for them and leave a vital message; people in the smaller cities were overwhelmingly more responsive. In another study, stamped envelopes were casually left around at restaurant tables, countertops, and the like in different-sized cities, and the ones from the smaller places were mailed back far more often. Students at the City University of New York compared the responses of people in Manhattan with those in small cities in upstate Rockland County when asked by strangers if they could come in and use the phone; an average of 27 percent of the New Yorkers allowed people in, compared to 72 percent of the others. And one researcher placed abandoned cars in middle-class neighborhoods, first near New York University in a city of 7 million and then near Stanford University in a city of nearly 20,000, and discovered that the car in New York was quite rapidly stripped and gutted while the one in Stanford was actually guarded by passersby against attempts to damage it. What is interesting in all these cases is that the very qualities of civility that the urban setting has traditionally been supposed to foster—the word "civility," after all, comes from the Latin for city—are least in evidence there.

Aside from sheer civility and friendliness, smaller cities outperform larger ones in any number of social variables, and it is those in the 50,000–100,000 range that most often perform the best.

Crime is a particularly good index, not because the individual statistics are always so reliable but because they are plentiful and, year after year, for example, show a very sharp jump when a city population goes past 100,000: cities of 25,000 to 50,000 reported 343 violent crimes per 100,000 people a year, those of 50,000 to 100,000 people reported 451, but cities of more than 250,000 reported 1,159, and cities of more than a million averaged 1,175. Similarly, murder rates per 100,000 for that year were 5.7 for the 25,000–50,000 cities, 7.2 for the 50,000–100,000 cities, but they jumped by nearly 300 percent to 21.4 in the cities over 250,000 and then to 29.2 in cities over a million.

Health statistics point to a similar, if somewhat less dramatic, pattern. Though big cities show a slightly better record than any others in terms of infant mortality, after that the data consistently favor small cities: there are fewer pollution- and stress-related diseases, lower death rates for cancer, heart disease, and diabetes, markedly lower incidences of bronchitis, ulcers, high blood pressure, alcoholism, and drug addiction. In general, there are proportionately fewer serious illnesses and fewer people laid up during the course of a year in small-city metropolitan areas than in either rural sections or bigger cities.

As to *mental health,* the common-sense suspicion that big-city living increases nervousness and tension, with resulting high levels of irritation, overfatigue, and social friction, is borne out by the majority of studies. Schizophrenia is significantly more common in bigger cities, suicide rates are somewhat higher, and the percentage of admissions to mental hospitals is greater in more crowded areas. As put succinctly in a comprehensive report to the 1971 UN Conference on Human Environment, "the incidence of neuroses and personality disorders is considerably higher" in large urban settings, due to the "terrifically harsh, intensively individualist, highly competitive, extremely crude, and often violently brutal" social life of the big city. At the other end of the spectrum, some recent evidence indicates that those in remote rural areas, where families tend to live in isolation and community bonds can become weak, have rates of mental illness as high as, and sometimes higher than, city dwellers.[1]

Recreation facilities in small cities have been shown to be usually better than those in large cities and easier to get to. Cities in the 50,000–250,000 range have more park space per capita than cities of any other size, and their parks serve a clientele that is more mixed economically. Small cities in general spend just as much per capita to build and maintain their parks, and they have a higher percentage of people employed in recreation and entertainment than cities of any other size.

Finally, *education* statistics confirm the social desirability of the small city. One pioneering study using data from the U.S. and three other industrial countries determined that there were usually more people employed in education, for each million dollars of regional income, in cities in the 40,000–50,000 category than in either smaller or larger cities. Another recent survey found that employment in education and health is proportionately higher in metropolitan areas from 50,000 to 200,000 than in either rural areas or larger cities. And the Committee for Economic Development recently determined that, in economic terms, once a school system has more than 2,000 pupils, "advantages continue

---

1. It should be borne in mind that there is no absolute evidence that large cities *cause* mental morbidity. It may be that they simply attract a great number of disturbed people or that they have more poor people and the poor are more disturbed or that the healthier residents have moved elsewhere. But the fact that mental illness relating to stress and tension is commonest in big cities is suggestive, while in rural areas diseases relating to loneliness and persecution are commonest. And ultimately, of course, it is not so much the *cause* as the *condition* that is important.

to accrue until a school system reaches perhaps 25,000 students"—which in this country means cities with a maximum of 70,000 to 90,000 people.

The one area of social activity in which the small city has been most often faulted is that of high culture: symphony orchestras, opera companies, ballet and theater troupes, art museums. It is true that, particularly in the U.S.—though, interestingly, not in such countries as Italy, Germany, and Sweden—high culture has tended to be a product of the very biggest cities, principally those of the Northeast. But it does not follow from that that big cities are *necessary* for these cultural institutions. Small cities between 50,000 and 150,000 in this country as a rule sustain both libraries and museums of considerable quality, and attendance figures show that at this level they are used more per capita in smaller cities than in larger ones, suggesting that this might be the right population size for peak efficiency and usefulness. Small cities can and do sustain symphony orchestras—of the 166 premier orchestras listed by the American Symphony Orchestra League, 51 percent are in cities under 150,000. Small cities have as a rule a far higher rate of *participation* in cultural matters, far greater contributions from all age, race, education, and economic sectors. Small cities have equal access to what is today the primary source of cultural dispersion—records, radio, and public television—and the quality there is generally superior to all but the best live performances. And small cities offer their cultural amenities without the prices, dangers, and distractions of big cities: the citizen of Albany who goes to hear the local orchestra on a moment's notice, with easy traveling, free parking, modest prices, cheap babysitters, and no fear of street crime, can hardly be said to be worse off than the New Yorker who goes to Lincoln Center for a $20-a-seat concert punctuated by subway rumbles and is mugged on the way home.

Moreover, the old system by which culture was siphoned off into one or two major cities is unquestionably changing, as those cities change. On the one hand, local and regional cultural institutions have blossomed in recent years—there were 150 major regional theaters in the U.S. in 1977, for example, and 70,000 ongoing theaters in all—and university towns, most quite small, now provide libraries, museums, symphonies, and theaters (as well as resident creative and performing artists) not significantly inferior to those of the former cultural "meccas." On the other hand, as the drama critic Robert Brustein, among others, has noted, there is today "a crisis in the performing arts" brought about largely by the "inflationary increase in expenses" attached to doing business in the bigger cities, coupled with a decline in both the private and public money those struggling cities can raise. There will be, he argues, a "profound change" in the arts in the next few years, including a "serious depletion" in the number of companies at work and a dispersal of talent as culture moves outward where the people and money are going.

ECONOMIC

IT HAS BEEN traditional for economists to justify large cities with the doctrine of what they rather inelegantly call "agglomeration effects." This holds that big cities are most efficient and are necessary for a healthy economy because they lump everything together in one place: a large and varied labor pool, a large and easy-to-reach market, a wide variety of specialized goods and services, and so on. In the words of William Alonso, a professor of regional planning at the University of California, "Bigger cities are more efficient engines of production."

The only trouble with this doctrine is that, in at least three crucial respects, it is no longer true.

In the first place, in big cities today the effects of agglomeration are most likely to be *dis*advantageous for businesses. There are higher transportation and distribution costs because of traffic congestion; higher business costs because of a decrease in the number of hours worked per worker; higher maintenance and cleaning costs because of air pollution; higher energy costs because of the "heat island" effect over cities in the summer and the shading of dense buildings in the winter; higher security costs and higher property-loss rates because of crime and vandalism; higher costs in training new workers because of inferior schools; higher land costs and greater building-construction expenses; and higher insurance rates, higher wages, higher costs of living, and higher taxes. Many of these costs can be pushed onto the cities themselves—traditionally cities have been economic for businesses *only* because the citizenry at large paid for the streets, police and fire protection, water, garbage collection, snow removal, and the like—but at some point these begin to push taxes higher for everyone, including the businesses.[2]

Second, manufacturing in general is no longer located in big cities—the classic Detroit pattern no longer holds, not even for the auto

2. One economic disadvantage of the big city was forcefully borne in on me some years ago. I was working for the *New York Times* as part of a team planning a new afternoon paper that was going to make a fortune filling the gap left by the recently defunct *World Journal Tribune*. We worked diligently for weeks, writing clever copy and designing snappy layouts, but finally one day we were told to abandon the project—not because our material was no good but because the *Times* executives had figured out there was no possible way of getting it out to the customers. The executives had decided that any afternoon paper attractive to the *Times*'s kind of classy audience, particularly the audience they hoped to pick up in the suburbs, would have to include the closing stock prices and therefore couldn't go to press until shortly after three o'clock. But that would leave only a little over an hour to get it to the railroad stations in time to reach the homeward-bound commuters, and it turned out that just wasn't enough time to get the delivery trucks through the horrendous traffic jams of a normal mid-Manhattan afternoon. There was talk of using motorcycles, even helicopters, but in the end the idea was abandoned and the executives resigned themselves to giving thanks that the *Times* itself was a morning newspaper and could be delivered through the Manhattan streets at night.

manufacturers (GM's new plant is in Lordstown, Ohio, Chrysler's in Belleville, Illinois, and Volkswagen's in New Stanton, Pennsylvania). Ever since the 1920s, in fact, and particularly since the 1950s, larger businesses have been moving *out* of cities, to smaller places where the costs are cheaper and social control over the workforce easier, where there is more space for assembly lines, where their executives prefer to live, and where an increasing percentage of its market is also moving. Transportation has changed, favoring road systems in the countryside over decaying rail and water systems, and manufacturing itself has changed, with a trend to miniaturization and high technology and to finished goods rather than assembly operations. In the 1950s and 1960s the business-district vacuum was filled by corporate headquarters and financial and administrative offices—hence business's support in the 1960s for urban renewal and cultural projects—but even they have begun to move out now to the suburbs where their executives live anyway (hence the growth of a place like Stamford, Connecticut) and where improved communications can keep them in touch easily with the necessary downtown contacts.

Finally, there is the simple fact that big cities have not had any better economic record than smaller cities, as you would expect them to if agglomeration worked as the economists say. Studies show that they do not have greater economic stability; they are not better able to weather economic crises; they do not have any better record on unemployment rates; and they do not have a higher percentage of their population employed. In fact, they do not have any measurable superiority in economic diversity, in the broad categories of employment that they offer. A study by two urban economists from Columbia University has shown that small metropolitan areas, with major cities likely to be between 25,000 and 100,000 in size, have just as great a range of economic functions as the giant cities of a million or more—*and* a considerably higher proportion of people in agriculture and construction, appreciably higher in retail trade, domestic work, medicine, and education, and slightly higher in utilities. That size city is particularly advantageous for retail trade—50,000 people has been found to be the minimum for the development of all the 65 types of retail outlets classified in the U.S. census data, and according to an expert in the quarterly *Land Economics,* the "optimum size from the standpoint of availability of specialized retail facilities is in the 50–100,000 range."

Moreover, no one who has been reading the newspapers needs to be told that most of the nation's larger cities have been particularly vulnerable to the economic crisis of inflation in the 1970s. New York's case has been much publicized, and it is grim indeed, but other major centers have also been hard hit: Boston has cut the number of on-duty policemen, Philadelphia has closed down its only public hospital, Atlanta has cut its trash collection by half, New Orleans has reduced both police and sanitation departments, Chicago is going without new traffic signals or street repairs, Seattle has cut its park-maintenance crews, San Francisco has cut back on school and library expenditures. In the largest sixteen U.S. cities, a survey by the *New York Times* determined in early 1977, "reduced resources and higher prices" characteristic of big cities have brought about a "decline in the quality of life that urban Americans have come to expect." This decline, according to a follow-up report from the Department of Housing and Urban Development, is substantial and will continue.

And that brings up another crucial economic fact: smaller cities, it seems, are more efficient in supplying municipal services than any other form of government. Urban economists are agreed that there is, in their terms, a "U-curve" for city efficiency that works as follows. From 2,000 people to about 30,000, cities are required to spend a lot of money to supply a large number of urban services, but at that level there are not so many economies of scale, start-up costs for each new service are considerable, and the tax base is still too limited to recompense the municipality sufficiently. Over that limit, and on to about 100,000, cities have a large enough tax base to fund new services, can buy supplies and build plants with economies of scale, and are still limited enough geographically so that they can supply services efficiently. Above 100,000 or so, all costs tend to rise slowly but inexorably as the city gets bigger, while its ability to get the services to the citizens just as slowly and inexorably declines. The optimum point in this U-curve comes somewhere between 50,000 and 100,000 for most city services, with some studies pointing to 200,000 or 300,000 for a few municipal obligations, but most falling in a range that I have calculated averages out to 63,000. Percival Goodman, the architect and city planner, has put it succinctly: "It requires no stretch of credulity to believe that from the viewpoint of conservation economics the future belongs to compact cities in the 50,000 to 150,000 population range."

It is because of this U-curve efficiency that per capita expenses of cities for all services begin to soar past a certain fairly low limit. The Urban Institute a few years ago calculated annual expenses per individual this way—

in a city of 50,000–100,000	$229
in a city of 100,000–200,000	$280
in a city of 500,000–1 million	$426
in a city of 1 million plus	$681

—or in other words almost a 300 percent increase from the smaller city to the largest. And no one suggests that this increase comes about because of higher quality service—far from it. Service levels in the bigger cities actually *decline* as population increases. One recent study of police service in eighty metropolitan areas, for example, showed that every time a big city adds five police officers to the force it actually ends up with eight *fewer* officers out on the 10 P.M. shift (apparently because of increased bureaucratic complexity); and smaller departments with five to ten officers have almost twice as high an officer-to-citizen ratio as big departments with 150 or more officers.

Werner Hirsch, a professor of economics and director of the Institute of Government at UCLA, has been probably the most diligent examiner of municipal performance in recent years. A few geographically diffuse services, he says, such as public health and electricity and water supply, may best be handled "on a district or countywide basis," but most municipal services can be operated best by small cities: "Local urban governments, particularly if they serve 50,000–100,000 citizens, can effectively provide education, library service, public housing, public welfare services, fire and police protection, refuse collection, parks and recreation, urban renewal, and street maintenance programs."

And if that academic opinion seems too esoteric, there is the practical opinion from the city manager of Boca Raton, Florida, who spent several years studying city governments around the country while fighting through the courts for the right to limit the size of his city. "We estimate," he told the *New York Times,* "that the cost of providing municipal services increases sharply over 100,000. That's the breaking point when everything starts to cost a lot more."

POLITICAL

FOR SEVERAL THOUSANDS of years there has been considerable agreement among students of political theory that a truly democratic community, a *polis* in the Greek sense, must have a relatively small scale. British historian G.D.F. Kitto tells a nice story to this point. Imagine, he says, a citizen of Periclean Athens transported to modern London, talking to a member of the posh Athenian club and being chided for the petty quality of political life in his ancient city compared to the marvellous grandeur of London:

> The Greek replies, "How many clubs are there in London?" The member, at a guess, says about five hundred. The Greek then says, "Now if all these combined, what a splendid premises they would build. They could have a clubhouse as big as Hyde Park." "But," says the member, "that would no longer be a club." "Precisely," says the Greek, "and a polis as big as yours is no longer a polis."

Yet it is precisely in this no-longer-a-*polis* world that most of us live today. Our major cities cannot lay any serious claims to being governed democratically. Citizen participation is limited almost entirely to quadrennial elections for mayor and city council, the candidates offered having been selected usually by one or another form of clubhouse politics. And even those elected officials are the first to admit that they are mostly in the grip of unelected forces—the bankers, the business elite, the city bureaucracy, and the municipal unions—thus making the citizens' connection to the actual governing of their city very tenuous indeed. It is this that accounts for the common findings among residents of large American cities of political apathy, cynicism, alienation, lethargy, and non-participation, all of which apparently increase as city size increases.

Indeed, Douglas Yates, a political scientist at the School of Organization and Management at Yale who has been studying urban politics for more than a decade now, has summarized the state of affairs in the title of his book, *The Ungovernable City*. He is obviously not happy about it, but his analysis leads him to the grim conclusion that "the city problem is a problem of government, that the large American city is increasingly ungovernable," and that this cannot be cured by weeding out corruption or improving the lot of the poor or any other palliative, since it is "a product of the city's basic political and social organization." His verdict on the big city is uncompromising: "Given its present political organization and decision-making processes, the city is fundamentally ungovernable . . . incapable of producing coherent decisions, developing effective policies, or implementing state or federal programs."

And here again, the smaller city apparently has a decisive edge. Political scientist Robert Dahl after considerable study formulated it into a kind of law: "The larger the place, the less likely is the citizen to be involved as an active participant in local political life." And the corollary: "The smaller the unit, the greater the opportunity for citizens to participate in the decisions of their government." Such studies as have been made on political participation generally support this commonsense view. One found that people with a high school education or better felt "more influential" and "significantly more confident" in cities under 100,000, and it concluded, "Persons living in small towns were significantly more likely than those in large cities to feel that they could influence political decisions." Another determined that "democratic participation" was enhanced in cities from 50,000 to 100,000, particularly in influencing decisions about city services. Still another determined that "the larger a city generally the less involvement in and 'attachment' to community affairs."

Of course there is no great mystery about this. To begin with, the sheer scale of events and organizations in the smaller city invites participation and creates the feeling that individuals have, or at least can have, some control over the events that affect their lives. Smaller cities also can be more efficient and responsive in meeting citizen needs, since there is likely to be more two-way communication, more and better message-sending to the people in charge, and easier access to their offices. And smaller governing systems are far more adaptable to any crisis, have far better information to rely on, and can depend on greater cooperation from the citizens. One measure of this effectiveness is found by one survey that shows the kinds of cities that are able to be governed by city managers—that is, those that are most amenable to efficient management or can escape the pull of factional divisiveness, or both:

City Populations	Percent with Manager System
10,000–25,000	40.0
25,000–50,000	52.8
50,000–100,000	50.5
500,000–1 million	26.7
1 million +	0.0

Could it be that, given human frailty, a modest limit on political effectiveness is inevitable?

Dahl certainly seems to have come to this conclusion. After worrying it over in several books and a number of articles, spending more time on this than any of his academic colleagues have done, he finally decided: "I think that the optimal unit is, or rather could be, the city of intermediate size, bigger than neighborhood, smaller than megalopolis. . . . The appropriate size looks to me to be a city between about fifty thousand and several hundred thousand inhabitants."

DEMOGRAPHIC

PERHAPS THE MOST persuasive, or at least the most interesting, fact about small cities as a place to live is that people seem to like them best and in recent years have begun an unmistakable movement toward them.

Polls in population preferences over the last thirty years have shown time and again that many people who live in large cities on the whole don't like it there—at least 36 percent of those in big cities, according to the latest survey, would move out if given the chance—and that people in all kinds of places think they would rather live in small towns and cities than anywhere else. A compilation of a wide variety of polls taken in the U.S. by various organizations between 1948 and 1972 provides these average figures:

17 percent prefer big cities
25 percent prefer suburbs
32 percent prefer small towns and cities
26 percent prefer rural areas

Another survey, perhaps the most careful one done to date, reported in the journal *Demography* in 1975, was even more precise:

9 percent prefer cities over 500,000
16 percent prefer cities of 50,000–500,000
17 percent prefer cities of 10,000–50,000
39 percent prefer cities under 10,000 within 30 miles of a city of 50,000
19 percent prefer rural areas not within 30 miles of a city of 50,000

In other words, only a very small fragment of the population likes big-city living, and around two-thirds would prefer to live in or around small cities in the 10,000–100,000 range. It seems odd that such a unanimity of opinion has received so little consideration in the course of national debates on the organization of our cities and the shape of our political structures.

Not everyone can back up these preferences by "voting with their feet," but population movements over the last forty years show that a great many of those who can, do.

The first great migration out of the cities was during the suburban boom years of the 1940s and 1950s, when city dwellers tried to find the best of both worlds: small cities with pretentions to community, some part of nature, and better schools, from which they could dart into the big cities for Saturday-night amenities. The second wave began some time around 1960 and continues down to this day, a movement beyond the suburbs to the exurban rings, to country areas, and to small towns and cities quite far from the metropolis. Together they make up what RAND demographer Peter A. Morrison has called "one of the most significant turnabouts in migration in the nation's history," since for the first time probably since this country was founded people are not moving *into* the cities but *out of* them. (Big cities, as a matter of fact, cannot sustain themselves except by immigration from the countryside; Warren Thompson of the Scripps Foundation for Research in Population Problems has written that "in the Western world as now organized, all the evidence indicates that no urban population of 100,000 or more, and probably even in cities of over 25,000, will long continue to reproduce itself.") In the years since 1970, central cities, particularly in the Northeast quadrant but in many Sunbelt areas as well, have been losing population at an accelerating rate, while the smaller cities and smaller metropolitan areas have been gaining at an unprecedented rate. Juan de Torres, an economist with the Conference Board, a business-research firm, has noted: "People now want lots of space, low population densities. They're seeking out the smaller metropolitan regions where the squeeze for space is less." Some are going to rural areas, but most are heading for modest-sized urban places: "It's not back to the farm," de Torres says, "but back to the factory in the smaller city."[3]

City size	1950	1970	Change
1 million +	11.5	9.2	−2.3
500,000–1 million	6.1	6.4	+0.3
250,000–500,000	5.4	5.1	−0.3
100,000–250,000	6.4	7.0	+0.6
50,000–100,000	5.9	8.2	+2.3
25.000–50,000	5.8	8.8	+3.0
10,000–25,000	7.8	10.5	+2.7
5,000–10,000	5.4	6.4	+1.0
2,500–5,000	4.3	4.0	−0.3
Rural areas	36.0	26.5	−9.5

"Back to the factory" points to one reason why the new movement is taking place: as businesses have moved out of the city, they have quite naturally taken the jobs with them, and people tend to move to where the jobs are. But it is not only that. As a 1978 RAND study discovered, jobs also move to where the people are, particularly the skilled workers. And it is by and large the younger, better educated, and more affluent, the professional and upper-echelon workers, who have led the migrations out of the big cities, who have the confidence and money to move to areas they find more congenial and then go about seeking jobs. The fact that they tend to head to the small and medium-sized cities suggests that possibly there are many others—poor, old, and unskilled—who would be doing the same thing if they had the chance; in other words, that a great many people are in big cities today not because they love them so much, or even find much virtue in their anonymity, diversity, complexity, and so on, but because for one reason or another they are stuck there. A movement toward a human-scale city, therefore, might not be quite as difficult and utopian an accomplishment as it might first appear. In the words of James L. Sundquist, a Brookings Institution economist and the author of the authoritative *Dispersing Population:* "A population dispersal policy and the programs to execute it can be conceived simply enough. They are likely to be popular. They are not unduly costly. And they work."

Nor is it a difficult matter, come to that, to achieve this spreading-

---

3. Census figures over the last thirty years show that small cities have consistently grown while big cities and rural areas have shrunk. Figures on the percentage of the total population in various places show this pattern:

out in sheer geographical terms. The entire American population could be divided into cities of 50,000 people and if each city took even ten times as much space as city planners normally allot for that size population—allowing room for both agricultural and green belts around them—the whole territory would add up to only about 310,000 square miles (4,400 cities with 70 square miles each), which is not more than the present metropolitan-area acreage today and represents only a fifth of the most fertile areas of the nation at present under cultivation.

CERTAINLY THERE ARE very few who can any longer champion the big cities as they are, as they have been, increasingly, for several decades. Some may defend them because they themselves are still able to enjoy the benefits and cushion themselves from the ills—they can afford the tickets to the show, the dinner at the exclusive restaurant, and the taxis to and from work, and have air-conditioning, summer homes, and vacations in the Bahamas. But they are a lucky, and a dwindling, few. Some may also believe that, in time, the big cities will arrive at miraculous solutions—Jane Jacobs, for example, who loves cities and apparently loves them at any size and density no matter what anyone says, argues that attempts to limit their size are "profoundly reactionary" because cities can always solve their problems by "progress" and "new technology." But this seems unduly optimistic—on the record, the cities' successes at solving such problems as traffic congestion, pollution, crime, and social decay after several centuries of trying does not inspire a lot of confidence in future technofix solutions.

The trend of the future, rather, seems clearly entropic. The big city, it is not too much to say, has probably outlived whatever usefulness it may have had (as, for example, a center of political control or of industrial technology). The traditional reasons for the existence of the big city—defense of the citizen, abundant employment, easy commerce, class and ethnic mixture, creative innovation—clearly no longer pertain generally. Indeed, it is no longer a city at all, in the traditional sense, a coherent, visible, determinable place—Los Angeles, for example, could not be called anything more than a political fiction, certainly not a city, a place of defined spirit and cohesion, a place with a center, an agora or a market or a square, a place with *city-ness.* Murray Bookchin, whose critique of the city is one of the most penetrating we have, says:

> If the word "city" traditionally conveyed a clearly definable urban entity, New York, Chicago, Los Angeles—or Paris, London, Rome—are cities in name only. In reality they are immense urban agglomerations that are steadily losing any distinctive form and quality. Indeed, what groups these cities together under a common rubric is no longer the cultural and social amenities that once distinguished the city from

> the countryside, but the common problems that betoken their cultural dissolution and social breakdown.

And that is why demographers have been forced to give new names to these things—metropolis, conurbation, megalopolis, necropolis. For they are, alas, no longer cities.

6

Energy: The Sun King

It is the perception of Sean Wellesley-Miller, a British-born architect now working with a solar engineering firm in California:

> Of all elements of the built environment, the home is distinguished as being a microcosm of the macrocosm it is set in. It reflects, on the level of the biological unit, the same needs that society as a whole exists to provide.

And, he proposes:

> We can apply much the same criteria to the home as suggested for society as a whole. The object becomes to develop a home that is heated and cooled by natural energy, grows 70 percent of its food requirements and recycles most of its wastes. This implies a shift in domestic architecture from the Le Corbusian paradigm of the modern home as a "machine-for-living"—a direct extension of fossil-fuel powered technical society—to the concept of architecture as an extension of the natural environment fueled by the same forces that drive the rest of the biosphere.

At one place, at least, the shift has already been made.

At first sight, from the long driveway, it looks like a normal-enough modern home, though it does seem to be all roof and no windows, and it is snuggled rather deeply into the hillside as if to protect itself from the northern winds. Closer up, though, and especially from the other side, there is no way you could mistake it for a conventional house. Stretching along the whole top story of the south-facing wall is a series of what appear to be small oblong windows—solar panels, actually—and below that is a huge roof of translucent plastic slanting down for nearly two stories into a wall of small double-paned windows. Inside that roof you can see a greenhouse full of neat tables with plants and flowers and just beyond that two rows of 500-gallon plastic barrels filled with greenish water. Outside, off to the left of the structure, is a row of pine trees providing further protection from the wind, and off to the right, set apart in a field, is a metal tower with a strange-looking device like a tiny airplane with a diving board stuck through its tail, in fact a modernistic windmill.

Inside, it is even more unusual. Though there are elements that are quite conventional—a kitchen, living/dining room, a study, three bedrooms, and bathrooms, spacious enough to suit five adults—none of it operates much like a normal house. The kitchen is supplied with its own greenhouse, where flowers and a year-round supply of fresh food—herbs, vegetables, fruits—are grown in raised benches. The living room, half-panelled with rich wood from the surrounding countryside, is centered around a large Jötul wood-burning stove (a tertiary heat source if the other solar fixtures need a back-up) and overlooks the main greenhouse, with its fresh smells wafting up from below. The bathrooms, with hot water supplied by the solar panels, have a Swedish-designed Clivus Multrum, a composting toilet that very effectively uses aerobic decomposition to turn toilet and kitchen wastes into useful fertilizer. The greenhouse itself, the house's "life-support system," is full of a variety of plants, some grown hydroponically, some in deep chemical-free beds, productive enough to supply not only the house itself but also some of the surrounding neighborhood with fruits and vegetables. And the 500-gallon drums inside it are actually full of tropical fish, a hardy, fast-growing, vegetarian species called *tilapia* that provides a regular source of protein for the inhabitants and feeds on algae created from the house's vegetable wastes. The whole building is heated by a combination of solar devices: the solar panels collect heat that is sent to a large hot-water storage tank in the basement and is blown through the house by a small fan; direct sunlight flows through the greenhouse roof and walls; and the fish drums store heat in their water at temperatures up to 95 degrees—so successfully in fact that the house can survive in winter for at least three days comfortably without any other heat source whatsoever.

This is the "Ark," a name chosen with conscious allusion by its creators, a group of young scientists and social experimenters who call themselves the New Alchemists. It is perched a stone's throw from the ocean on Spry Point, Prince Edward Island, a rather desolate province off the eastern coast of Canada. And it is one of the few examples in the world of a building that tries, for the most part with success, to apply the principles of ecological balance and alternative technology to a human-scale existence.

The Prince Edward Ark was established in 1975, under a grant from the Canadian government, as a sort of laboratory to test the principles of self-sufficiency that the New Alchemists had derived after a half-dozen years of wrestling with the problem of small-scale ecological living and a couple of years of initial experiments at a "Mini-Ark" at their headquarters in Wood's Hole, Massachusetts. The extraordinary thing about the Ark is how it is able to be so completely self-contained on the one hand and yet so completely natural on the other. (Fond of science jargon—"aquaculture," "homeostasis," "interphasing"—the New Alchemists call it a "bioshelter.") It is able to produce all its own heat, through the solar panels, heat-retentive barrels, and wood stove. It generates much of its own electricity from the windmill ("wind-driven power station") and is hooked into the island's electricity grid, able to take extra power when the winds are light and to return power to the network when it has an overload. It can produce most of its own food, both animal and vegetable—the *tilapia* are said to be a gourmet-quality fish—with the exception of certain grains and dairy products (though, should they wish to expand, those would be easy enough to add on); and it supplies most of its own liquids (so far none alcoholic, but that too would be simple enough to do). It can recycle almost all its own waste—human, kitchen, and garden—through biological composting, creating up to 6,000 pounds of fertilizer a year out of it. Not only that, but the systems interact; the fish are eaten by the humans, human wastes go to fertilize the soil beds, fish wastes and water are used to irrigate and fertilize the vegetables, and weeds and scraps of the plant life go to feed the fish.

The Ark does all this in the soundest ecological fashion. Depending for the most part on the sun, it does not use polluting or non-renewable fuels, and because it is carefully built with heavy insulation and wind protection, its energy needs are small. It does not dump toxic wastes into the nearby soils or the water table, nor does it require an expensive and polluting sewage-treatment plant. It uses very little water (and that from its own well), since irrigation is done largely by recycling water from the fish tanks and since the composting toilets use none at all. It is built primarily of local and quite basic materials—wood, glass, cement—with the exception of the large greenhouse roof, made of acrylic glazing (it permits the passage of valuable germicidal ultraviolet waves that regular glass does not) and the solar panels. And its systems are directly derived from specific local conditions—high regular winds, ample sunlight, and abundant wood.

Now the Ark, obviously, is not yet a panacea. For one thing, it operates more on a household than a communitarian scale, with minimal connections to the surrounding populations—though more because of the undeveloped nature of social institutions on the island than any particular fault of its own. For another, it has had only a few years to test its systems and may not be able to live up to its potential over the long haul—though its performance during the hard early years certainly would suggest optimism. And above all, it has been an enormously expensive undertaking, depending on a total grant of $345,000 just to get started—though that, of course, was probably inevitable in a place that intended to be more a laboratory (it has a system of thirty-three sensors and print-out recorders) than a home, an experiment whose ways, once perfected, can then be adapted to other settings at far less cost.

Still, the importance of the Ark is beyond a doubt. As John Todd, one of the founders and leading lights of the New Alchemists, has put it:

> The Ark is one of the first synthetically framed explorations of a new direction for human habitation. With its use of diverse biotic elements, energy sources impinging upon or generated at the site, and internal integration of human support components normally exterior to households, it begins to redefine how people might live. The Ark is not an end point, but an early investigation of a viable new direction.

IF ANY BUILDING comes close to being, in Wellesley-Miller's terms, a microcosm of a human-scale society, it would probably be the Prince Edward Ark. As such, then, it is a fitting link for us between the social *settings* that we have been discussing—human-scale buildings and communities and cities—and the social *functions* appropriate to them that we must now consider: energy supply, food production, waste disposal, transportation, education, and health facilities. Such social functions, it seems logical to think, might also have their appropriate limits, their optimal scales, at which they could satisfy all the needs of a comfortable modern society without the dangers and expense of the larger and more fragile systems we have with us now.

In examining these functions it is important to consider not only those we have traditionally asked governments to provide—education and waste disposal, for example—but also those we have given over to private businesses—such as energy and food supply. That they are all in truth *social* functions—activities that are literally vital to the maintenance of the society and with profound effects on the ongoing patterns of that society—is without doubt, despite some of them being in private hands. The social impact of the automobile is an oft-noted case in point: though the decisions as to its use have all along been made by a small handful of private individuals, operating by corporate charter and individual whim primarily in pursuit of increased profits, the effects on the society as a whole have been so enormous, from Los Angelization to highway slaughter, as to need no further retelling.

OF ALL THE vital functions of a community, perhaps none is more central than the provision of energy. Without energy, no wheels spin, no pistons pump, no cogs mesh, no mills grind, nothing moves. Energy is the bedrock social function, without which most of the others are just not possible. Thus a successful society must have a regular and abundant source of energy, preferably cheap, widespread, and harmless, and ideally ecologically benign while being gathered, shipped, used, and disposed of.

For most of human history, societies depended upon indirect solar energy—largely wind, wood, and water power—supplemented by human and animal power. Just a few generations ago, our American forebears depended almost exclusively on solar energy—direct sunlight plus the primary converters of sunlight, plants and animals—and probably at least a third of the world's population does so today. It was only comparatively recently that fossil fuels—first coal, then oil, and still later natural gas—were developed and then exploited widely—the first electric power plant began in 1882, less than one hundred years ago—and it seems likely that we will very shortly reach the end of the fossil-fuel era. Without going into the quarrels and quibbles about how many barrels of this and tons of that may remain in the earth's crust, and how much can actually be developed, it is incontrovertible that sometime not too long hence, whether ten years or twenty or even longer, we are going to run out of economically recoverable fossil fuels. And then we will be able to see that era for what it was: an extraordinary exception, an aberration, a bump on the long line of energy use in which solar has held pride of place since humans domesticated fire and which is now coming back to the fore as current technologies enable us to improve upon the cruder methods of the past.

Even if we were not running out of fossil fuels, even if the most optimistic estimates of supplies were to prove correct, it is obvious that developed societies cannot go on using energy sources that are destructive, wasteful, unreliable, inflationary, and expensive. We simply cannot afford them.

Large energy systems use fuels that create environmental damage in their recovery and transportation, deaths and accidents in their production and storage, air and water pollution when they are burned and expelled, and inevitable inflation as their supplies dwindle and use increases.

Large energy systems, among the most capital-intensive enterprises in the world, may not be able to find the necessary investment money to sustain themselves in the immediate future—it will take more than $1 trillion to finance the energy industry before 1985, and that represents about three-quarters of *all* the net private domestic investment for all causes during that period—and even if they did, they would severely cripple all the other social systems needing that money.

Large energy systems are vulnerable to a variety of natural and human hindrances—strikes, slowdowns, equipment malfunctions, foreign blockades and boycotts, accidents, summer lightning, harsh weather—and have regularly for the past few winters produced coal and natural gas shortages, job layoffs and plant shutdowns, school closings, freezing homes, economic dislocation, and incalculable numbers of deaths and illnesses. (W. S. Gilbert once remarked that "Saturday afternoon, although occurring at regular and well-foreseen intervals, always takes this railway by surprise," and one is tempted to say the same thing about winter and the American energy establishment.)

Large energy systems depend upon large and intricate bureaucracies—with all the consequences of *that*—and upon complex and often inefficient political support mechanisms at city, state, and Federal levels, the whole producing a network of such complexity that it is in fact beyond the comprehension of any single person, or even any single body of people. "No human designers," says social planner Rufus Miles, himself a veteran of the Federal bureaucracy, "can know or comprehend all the factors that need to be taken into account, and their interrelation, sufficiently to make the current set of systems work well."

Large energy systems depend upon large and intricate sub-systems extending all over the world, from the North Sea to the Trucial States, from Zaire to Houston, involving ever-larger projects of exploration, exploitation, allocation, transportation, storage, construction, distribution, and security, any part of which—the oil tanker or the offshore rig, the mine shaft or the nuclear reactor—is liable to failure at any given moment, with reverberations up and down the network.

Lastly, large energy systems are inevitably the most costly. Their components generally cannot be mass-produced and are therefore expensive; their extensive spreads entail elaborate distribution mechanisms, which are expensive to build and maintain and cause the loss of vast amounts of energy to boot; and their facilities take decades to plan, build, and make operational, exposing them to a wide range of increased interest, labor, and material costs.

But why go on? No one with an open mind who has lived through the accumulated crises of recent years, large and small, should need convincing. Remember the gasoline shortages of 1973–74; the collapse of the Teton Dam in 1977; the $2.8-million Browns Ferry nuclear plant accident in 1975; the coal strike of 1978; the oil spills off the coast of Brittany in 1977 and the coast of Texas in 1979; the half-a-million-gallon leakage of stored fuels from the Hanford, Washington, storage site from 1971 to 1976; the New York City blackout of 1977; the refinery explosion that rained gasoline on Staten Island in 1977; the temperature-inversion smog that cost Los Angeles businesses $5 million a day in July 1978; the regular failure of the Turkey Point reactor in Miami from 1974 to 1976; the radiation leakage at the Platterville, Colorado, nuclear plant in 1978; the LNG explosion in the storage tank on Staten Island in 1977; and

Three Mile Island in 1979. . . . The systems are clearly fragile and dangerous and unreliable, they are inadequate and expensive and ecologically hazardous.

Some other way, it is agreed by almost all students of the energy problem, must be found.

For a while it was fashionable to put forth a variety of large-scale nuclear solutions, but since they all *magnify* the drawbacks of the present systems and add a host of their own, that future is regarded as increasingly unrealistic, even by former advocates. The number of nuclear reactors on order fell from twenty-seven in 1974 to zero in 1978, while eighteen other plants have been cancelled and more than a hundred "deferred." Then-Secretary of Energy James Schlesinger acknowledged in Senate testimony in 1978, "Nuclear power plants simply are no longer a viable option for the majority of the nation's utilities."

Nowadays several other kinds of technofix solutions are being offered—gasification of coal, for example, and ocean thermal conversion—and many of these are favored by aerospace and electronics companies looking for new work. Some even get funding from a Federal government that is hoping for a few large-system miracles, following the operation of Plunkett's Law (as stated by Dr. Jerry Plunkett, head of a small energy firm in Denver, to the Senate Small Business Committee in 1975): "The Federal government has what I believe is an almost incurable habit of undertaking large-scale projects. Given two equally valid technical responses to a national problem . . . the technology that is larger in scale will invariably be preferred to the smaller more decentralized technology." Yet so far none of the technofix schemes has been able to overcome the inevitable big-system problems, and all of them face the difficulty that it takes so much energy to manufacture and maintain them that, when you add it all up, the net energy gain to the society may be quite minimal, if it exists at all. In short, so far the grand technofixes seem to be future technologies whose time has passed.

So increasingly, it seems, as the whole political direction of the 1970s has been telling us, there is really only one solution to the energy problem, one obvious renewable, non-polluting, free, and benign energy source: the sun.

NOT ONLY IS solar energy the solution to the energy problem, however, it is an explicitly *human-scale* solution.

▪ Solar energy is ecological. It does not pollute; it is silent, odorless, and safe; it does not use irreplaceable materials; and it is completely natural.

▪ Solar energy is conservational. Not only does it save fossil fuels for other restricted uses, but it forces people to make daily connections between the switch on the wall and the resources behind it and to consider the long-term consequences of their use of nature.

▪ Solar energy is inherently democratic. The sun's rays fall in roughly equal proportion, given polar and equatorial variations, on every man, woman, and child around the globe and are available to any person, any family, any neighborhood, any community. Sunshine falls at an average of 17 thermal watts per square foot all over the U.S.—varying only by a factor of two between sunny Arizona and cloudy Washington—and enough of it descends in just half a day to supply, were it fully captured, the entire energy needs of the nation for an entire year.

▪ Solar energy is localized, and adaptable to different geographies and settlements. In the U.S., it would take different optimum forms in different regions: direct sunlight in the Southwest and Southeast, wind power on ocean and lake coastlines and in higher altitudes, water and wood power particularly in the Northeast and Northwest, methane production in rural areas of the South and the Plains States. In the rest of the world, it is particularly beneficial in direct form for the poorer nations, which tend to be in the hotter climates.

▪ Solar energy is communitarian. It is uniquely suited to any kind of small-scale operation, and those energy tasks that cannot be done optimally in a single building can be done very neatly on a community-wide scale.

▪ Solar energy is decentralized. Since the whole operation is controlled by those who use it—who thereby become active producers of energy instead of passive consumers of it—it can be specifically adjusted to what engineers call "end-use needs," i.e., the requirements of the individual family or community. It cannot be dominated or monopolized by large corporations or central governments, cannot be bought and sold, cannot be jacked up in price from year to year. (It is conceivable that a company or government could try to monopolize the *equipment* used for collecting and storing solar power, and there's every indication that this is in the minds of certain energy companies right now. But most of the hardware is so easy to produce from common materials, in any location, and there are so many different kinds of systems that can be built, that this attempt does seem problematical. No doubt one reason that there were no fewer than a thousand separate firms in the solar business as of 1979.)

▪ Solar energy is flexible, adaptable to a great range of social and political conditions, foreseen and unforeseen, unlike present more rigid systems. An entire solar unit can be constructed, even for a fairly large building, within a matter of days—a nuclear plant, by contrast, takes decades—and it can be enlarged or modified any time, as needs suggest.

▪ Solar energy is efficient. Very little of the original energy is lost through *conversion*—as when it changes from heat to hot water—and even less through *transmission,* since any solar unit is relatively self-contained. This is in contrast to a normal electric utility, which loses *50 to 65 percent* of its primary energy in conversion and transmission.

▪ Solar energy is free for all, and though the means of collecting and dispersing it of course are not, most of that hardware is inexpensive or competitive with existing systems; once in place, solar gear requires only the most minimal of maintenance costs and of course no fuel costs at all. What equipment expenses there are can be decentralized, absorbed by the respective buildings or communities, and therefore there need to be no large, ongoing financial burdens on the nation as a whole, as at present.

▪ Solar energy is economical in its scale. It does not require large prediction and planning studies, complex delivery systems, enormous storage capacities, elaborate back-up mechanisms, expensive bureaucracies, and such overhead costs as advertising budgets, security systems, and labor.

▪ Solar energy is especially adaptable to alternative technologies. It is simple to understand—anyone who has ever gotten into a closed car on a hot day knows how it works—and simple to build, maintain, and repair, using standard common skills known to any do-it-yourselfer. Much of the technology—for windmills and stoves, for example—is very old and well tested, and most of the newer elements are simple, cheap, and amenable to backyard innovations.

▪ Solar energy is safe. It has, true enough, been used before as a weapon—Archimedes is said to have set fire to the Roman fleet using a mirror to reflect the sun's rays in 212 B.C.—and there is the story of that woman in California who surrounded herself with aluminum on a sundeck and got baked to death, like a potato. But those seem to be the limits of solar danger, and probably containable.

▪ Solar energy is adaptable to any sort of thermodynamic job—uniquely so among energy sources, because it operates from low temperatures to higher instead of the other way around. It is gathered on the surface of the earth at about 70 degrees (the temperature needed for space heating) and then if necessary can be transformed by cells and concentrators up to 150 degrees (for hot water), 200 degrees (for heat-driven air-conditioning), 1,000 degrees (for electricity generation), or even 2,500 degrees (for steel and metal manufacturing). Oil, by contrast, has to be burned at about 500 degrees in a furnace before it can be used at all, uranium becomes useful only when it reaches 2,400 degrees, and then all that heat just has to be drained away to make it useful in heating a house to 70 degrees, a process of unimaginable inefficiency: it is, as the ecologists have been saying for some years, "the thermodynamic equivalent of cutting butter with a chainsaw."

BUT OF COURSE questions persist. Solar energy may be as good as all that, *but*—is it practical, is it affordable, and is it sufficient?

Let's take each, briefly.

Is solar energy practical—now? Emphatically yes. *All of the technology necessary for the processing of solar energy into both heat and electricity is available now, has been in use on a small scale for many years, and is competitive in price with even the cheapest of alternative fuels.* There is nothing arcane here, nothing still to be invented, nothing awaiting a technofix solution. There will be refinements and improvements, of course, new materials and more efficient systems, innovations and developments in all parts of the hardware—with an alternative technology like solar, that is to be expected. But as to the practicality of the existing systems there is no question. In the words of Friends of the Earth's expert physicist, Amory Lovins, "Recent research suggests that a largely or wholly solar economy can be constructed in the United States with straightforward soft technologies that are now demonstrated and now economic or nearly economic."

Space heating (and cooling) and water heating by solar power, once an oddity, is now a commonplace all across this country: at least 200,000 buildings were fitted with collectors as of 1980, and the Solar Energy Industries Association, the trade group of manufacturers whose business this is, is confidently predicting 11 million by 1985. In Japan there are said to be 2 million buildings with solar collectors, In Israel 250,000, and many other countries—France, Austria, Denmark—have had solar homes for decades. Passive solar systems, too—using greenhouses, trombe walls, heavy insulation, south-facing windows, protective berms, steel drums, bead walls, water-tank storage, and other simple, ingenious devices—are increasingly common in the U.S., probably now built into more than 20,000 homes. Solar heating systems are known, tried and tested, efficient, and affordable—and since it is heating that takes up 58 percent of our energy use today, this is a technology that single-handedly goes a long way to solving the energy crisis.

Solar-generated electricity is a bit more complex, but there's no question about the workability of the devices. Though costs have so far kept down their widespread use, photovoltaic cells, which convert sunlight into electricity, have been used and proven for a decade now. Such cells were used to provide internal power for most of the NASA space satellites; Arizona and New Mexico state police have installed solar cells for a series of radio-relay stations; cell-generated irrigation systems have been used by a number of Sunbelt farms since 1975; the U.S. Forest Service uses cells for the power supplies of remote watchtowers; the Coast Guard has been using cells in its buoy-signal system off the coast of Florida since 1973; and the Ivory Coast has installed small cells to power some 4,000 television-relay stations throughout its back country. Moreover, various experimental household installations have shown that, as the National Science Foundation – NASA Solar Energy panel reported back in 1972, "about three times the present household consumption of electric power can be collected from average-sized family residences, even in the northeastern U.S."

Wind, water, and wood systems are centuries old, of course, but there are many alternative-technology refinements to them nowadays that have made them even more efficient. The most important work has been done in wind conversion, where both small and large turbines have been shown to work at unusually high efficiency and with unusually low material costs. Over 6 million windmills have been built in the U.S. since the middle of the last century, *Science* magazine estimates, and at least 150,000 of them are still going. Their use for electrical generation, water-pumping, and both heating and cooling is amply proven—many parts of rural America, after all, depended upon windmills for decades as their only source of electricity. At least ten windmill types were on the market in the late 1970s, and more are coming every year—two-blade, three-blade, paddle-blade, prop-blade, silo-type, and various kinds of back-yard designs in between. Water power is a simpler matter, for here all of the technology has been developed and tested for generations, and the applications of modern materials only makes the operations more efficient. Most of the nation's water potential is unused, but enough unused back-country dams exist in the U.S. *right now,* according to the Federal Power Commission, to supply the entire annual electrical needs for a population of 40 million people—more than the Rocky Mountain and Pacific regions combined—if only they were equipped with generators. Sites for at least 23,000 additional dams exist now on the Atlantic seaboard watershed; there are another several thousand in the Great Lakes area; and at least ten thousand more are usable in the Northwest and Rocky Mountain regions. And wood power is even simpler still, since fireplaces and wood stoves have been in use in millions of homes for years—the 1970 U.S. census reported that a million households listed wood as their primary heat source, and sales of wood stoves have been averaging a little under a million a year since then—and modern devices developed in the last few years have made them safer and more efficient than ever before.

It is true enough that all of this solar hardware does generally not represent, as the engineering profession likes to say, a "hands-off technology." Solar machinery needs some kind of hands-on care in the best of conditions, and given the likelihood that the earliest models will exhibit uncertain workmanship, faulty materials, naive engineering, and probably the greediness of fly-by-night profiteers, there might be more than a little of such care for some homeowners. And yet none of it should tax the competence of the able around-the-house putterer, and all of it draws the energy user into a profitable relationship with the surrounding environment.

* * *

Yes, but is it economical? Again, emphatically yes. With the single exception of the photovoltaic cell,[1] all of the equipment associated with solar energy as of this writing is cost-competitive, most of it is somewhat cheaper, and so many improvements are coming along every year now that it is sure to be cheaper still. *Provided,* of course, that it is kept at a small scale. Large-scale solar projects, the kinds that have generally been funded by the Federal government over the last few years—"power towers," space stations, solar "farms," ocean-thermal processes, gigantic wind machines—have turned out to be either very costly half-successes or very costly failures. And no wonder: all attempts to take a basically decentralized form of energy, centralize it with some large machine, convert it into electricity, transmit it back to decentralized users, and convert it again to heat, will inevitably prove as sensible as killing a fly with a cannon; that such systems have been built at all testifies only to the operation of Plunkett's Law. Kept at a small scale, however, there is no question whatsoever about the economic benefits of solar power.

And the fuel is free and will never go up in price.

Furthermore, solar lends itself very neatly to various cost-saving operations. First, active solar technologies can be combined with passive solar techniques to reduce overall energy requirements considerably; there are literally thousands of houses that now are able to get through even the coldest of winters with a minimal use of solar-panel or wood-stove heat just because they are well insulated and designed to make maximum use of direct sunlight. Second, solar technologies are uniquely suited to simple combinations by which energy tasks can be integrated, as when electricity from a wind system is used to run a fan circulating circulating heat from a roof collector; a Vermont dairy farm uses the heat expelled from its solar milk cooler to heat up its anaerobic digester, which otherwise couldn't be used during the harsh winters, and the digester produces methane that is used for both heating and cooking over the entire farm. Finally, solar technologies lend themselves to communitywide generation, transmission, storage, and equipment pooling that can produce significant economies, particularly when several

---

1. Photovoltaic cells produced electricity at $200 a watt in 1959, but since then so many improvements have been made, with minimal government support, that the costs have plummeted, to $20 a watt in 1976 and $14 a watt in 1978. That's still nowhere near competitive with electricity from nuclear plants (about $.46 a watt in 1978) or coal-fired plants ($.30 a watt), but with present methods it is planned to have cells down to about $1 a watt soon after 1980 and the Federal Energy Department target is $.50 a watt by 1984. But one U.K. team has come up with a process that could produce electricity at $.28, a Massachusetts firm, Tyco, has projected $.25 for its process, and various other U.S. manufacturers believe they can get the price even below that within the next ten years. With new silene-gas treatment, amorphous silicon sheets for electricity has also become possible, enabling an average roof to generate from 3 kilowatts (cloudy) to 15 kilowatts (full sun) of power at current (1980) costs of less than $.60.

communitywide systems are used together. Cogeneration—the process of using excess steam from factory production to heat or electrify nearby buildings—is especially suited to a community scale.

There has been an intense debate over the precise costs of solar energy ever since it began to be taken seriously after the oil crisis of 1973, and nothing can be taken for certain in an area where everything is fluctuating so rapidly. But Amory Lovins has produced a set of detailed and exhaustive figures in his 1978 book, *Soft Energy Paths,* that leave no doubt, even with his conservative statistics, that solar technologies, "calculated traditionally as internal costs per unit of energy or power, are arguably attractive compared with those of competing hard technology systems." That would seem definitive enough—though initially challenged, his calculations have stood up against all criticisms—but it leaves out what are actually the most important economic considerations in judging energy sources: *hidden costs* and *social costs.*

The *hidden costs* are those that the consumer actually pays for but isn't aware of because they are not in the fuel bill every month. These costs come in three broad categories:

▪ Taxes paid to Federal and state governments, which are used to provide tax breaks and subsidies to energy companies, to provide research-and-development money for fossil-fuel companies, to create and run departments to regulate and supervise utilities, to maintain agencies for controlling pollution caused by conventional fuels, etc.

▪ Costs added on to the general run of consumer goods by all businesses passing on the expenses they must pay for increased energy bills, increased costs of energy-intensive materials, added expenses for government-mandated pollution devices, added expenses from productivity lost due to pollution sickness, etc.

▪ Direct consumer expenditures and losses from such things as declining real-estate values in smog-affected areas, increased medical bills for pollution-connected diseases, homeowner devices for preventing or cleaning pollution, etc.

It is hard to put an exact cash value on all of these, of course, but taking conservative figures from Federal and corporate sources, I would estimate that, in all, these hidden costs would have added about $1,080 to the household budget in 1975 (and even more in subsequent years because of the rising price of fuels), a very considerable addition to—in most cases a *doubling* of—the flat fuel bill.[2]

2. The hidden costs are estimated as follows, current dollars, 1975:

a. Federal subsidies to the oil industry	$10.3 billion
b. Federal subsidies to utility corporations, including uncollected taxes and the 20 percent tax subsidy on new plants	2.7

And that does not even take into account the hidden costs of *general* business subsidies and tax breaks, amounting to $18.4 billion in 1975, of which the energy and utility companies share some part; nor does it take account of the inflationary effects of these energy companies on the general economy, both when they raise their prices to pay for ever-higher fuel costs and when they bid up interest rates in borrowing the billions of dollars they need for their development.

The *social costs* of conventional power are, by contrast, totally incalculable in monetary terms. Some are regrettably obvious: the deaths and accidents, the pollution, the environmental destruction, the loss of irreplaceable resources. One has only to think of the terrible physical and human destruction of an area like Appalachia, a sad symbol of our society's energy priorities, turned into a pathetic social backwater by the relentless and careless extraction of coal over the past century. It is a place where death is common, children are left parentless, and no family is without "the pain in the chest that just won't never heal," as one old West Virginia coal miner puts it; where the accumulated perils of air and water pollution have touched every single person, from the shoeless child with nephritis to the man bedridden with incurable black-lung disease; where whole towns have been forced into the service of a single company and then abandoned when their soils are exhausted; where the countryside, once one of true beauty, has been left scarred and unsightly, with piles of slag heaps that become potential avalanches every time the rains sweep off the strip-mined hillsides and swell the foul rivers.

---

c. Federal subsidies to coal and nuclear industries	3.8
d. Federal expenditures on its own power projects	11.0
e. Federal agency expenditures fo utility regulation (by FPC, etc.), *excluding* pollution	1.0
f. Federal expenditures for energy agencies	3.5
g. Total Federal and state government expenditures for pollution control *(excl.* auto)	4.8
h. State government purchases of utility power	11.2
i. Business expenditure on pollution control	6.5
j. Business expenditures related *(excl.* auto)	3.0
k. Consumer expenditure *(excl.* auto)	1.0
l. General pollution costs (sickness, building deterioration, real-estate values, *excl.* auto)	18.0
TOTAL	$76.8 billion

The total costs divided by the approximate number of households in the U.S.—71.1 million—works out to an expenditure of $1,080 per household. (Sources: a., c., Congressional Joint Economic Committee, "The Economics of Federal Subsidy Programs," January, 11, 1973; b., *Progressive,* February 1977; d.–h., *Statistical Abstract of the U.S.,* 1975, 1976; i.–k., *N.Y. Times,* January 9, 1977, p. 39; l. *Vanishing Air,* Nader Report, 1970, pp. 19–20.)

But some social costs are more subtle. There is the social dislocation when an oil strike transforms a settled town, when people are forced off their lands by strippers and drillers, when smogs force people out of their neighborhoods and towns. There is the anguish and bitterness of countless citizens and groups who go up against the utilities and energy combines and are met with the unanswerable well-if-you-don't-like-our-energy-you-can-always-burn-candles attitude. There is the ever-expanding government apparatus that tries to control the increasingly powerful energy establishment with growingly complex laws and enlarged bureaucracies. There is the declining role of individuals and communities in the setting of energy policies and a consequent shrivelling of their political forms, ultimately leading to a perceptible decrease in the democratic nature of American society. There is the increasingly fragile position of the dollar on world currency markets, the increasingly vulnerable position of the U.S. in world affairs.

These effects are incalculable, true. But they are no less real. As the pioneer scholar of social costs, K. William Kapp, has said: "To dismiss these phenomena as 'noneconomic' because they occur outside the market complex is possible but neither ingenious nor tenable in view of the fact that, apart from their obviously human aspect, they have a price for both the individual and society." They unquestionably have to be factored in to any calculations of the costs of solar energy.

Can it supply all the energy our society needs? Once again, emphatically yes.

Of course that depends a bit on how you define "needs." Does America really need to consume the vast amounts of fuel it does—between 30 and 40 percent of the world's supply for 6 percent of the world's people? Does it need to use energy equivalent to the bodily labor of 500 slaves for every single man, woman, and child? Does it need to have electric toothbrushes using up to 800 million kilowatt-hours of energy, at an annual cost of $3.2 million? Does it need to lead a life of electric knives and aluminum cans and self-cleaning ovens and plastic packaging and 15-mile-a-gallon cars? Does it need to throw away, through carelessness and overindulgence and conversion wastes, some 34 quads of every 75 quads of energy it produces?

Of course not—that is greed and thoughtlessness, not need. As we have been often told recently, Sweden and Switzerland are able to live at even higher standards of living than the U.S. with a per capita energy consumption approximately 60 percent that of America's; even America itself lived quite comfortably with only 40 percent of its current energy use—*just twenty years ago.*

So if we assume a "need" for energy not 40 percent of what we use now but 50 percent, there is no doubt whatsoever that solar processes could be supplying the bulk of it, the essential core of it, within a dozen years. An early study by the Federal government's Solar Energy Task Force, in 1974, predicted that it would be possible to supply 39 quads of energy a year from solar processes by the year 2000—and that is a little more than half of the 75 quads we are now using. If we assume not only a less wasteful use of energy but a concerted policy of conservation, a continuation of the movement toward passive solarization (including heavy insulation in both new and old buildings), and a modest program of recycling such energy-intensive materials as aluminum and plastics, then it would be possible to have solar energy systems supplying virtually all of the U.S. energy needs by the first years of the next century.

One other "if." If the American population continued to follow its trend toward optimal redistribution, and if the giant metropolises were to shrink to smaller, energy-efficient cores, an additional immense amount of energy could be saved. Big cities, contrary to the myths of the metropollyannas, are actually terribly wasteful of energy, their surrounding suburban periphery almost as much so, and the business of going back and forth between them most wasteful of all. Large cities are so far removed from an ecological equilibrium that they must consume extraordinary amounts of energy just to do what nature itself can do in a small community. Waste disposal, for example—a small town can recycle almost all garbage and sewage or return it to the soil beneficially, but big cities have to use garbage trucks, impacters, sewage plants, secondary-treatment facilities, water purifiers and the like. As New Alchemist John Todd has said:

> The contemporary [energy] dilemma has been created by the establishment of high technology industrial and urban regions which have long overshot nature's healing capacity. Our attempts at correction and purification of these ecologically unsound areas will actually run down available high quality fuels at a more rapid rate. If we stick with our present system we are trapped, because we will need to use a disproportionate amount of energy to sustain a livable environment which in turn will leave less energy for primary work. For future societies to thrive, growth limits should be set by ecosystems rather than by economic dictates which span only a few years. It is unlikely that new forms of energy, even nuclear energy, will be able to bail us out if we don't restructure the human landscape of this country.

Add in the fact that cities have to build systems and produce energy to the most exacting specifications—to keep hospitals running at all times, and elevators, and traffic lights—even though 90 percent of the end-uses of the population require far less energy. And then add in the added energy use in traffic jams, elevators, high-intensity street lights, neon jungles, taxi fleets, high-rise buildings, office blocks that actually use air-conditioning to dispel the heat created by their lights, and traffic lights on every streetcorner for both cars and pedestrians, and one may fairly conclude with Rufus Miles: "Without huge amounts of energy, every modern city would come to a mechanical stop and turn to chaos."

IT WILL NOT have escaped notice that there is much about solar energy that favors communitarian operation.

There is no difficulty in imagining efficient solar energy systems on an individual-building level, but if those buildings are linked in a fairly small system then a whole variety of problems can be eased and definite efficiencies and economies obtained. Amory Lovins again:

> There are often good reasons to share even simple energy systems among, say 10^1–10^3 people. For example, neighborhood solar heating systems (for individual or cluster housing) can clearly offer substantial economies over single-house systems through freer collector siting and configuration, reduced craft-work, reduced surface-to-volume ratio in storage tanks, more favorable ratio of variable to fixed costs, and perhaps even a bit of user diversity (different people using, say, hot water at different times, though this would be a very small term).

Communal storage is particularly valuable in a solar operation. Since maximum storage systems, be they water or rock or salts, must be fairly large for long-term heat storage, it can be cheaper for a number of users to get together to create one large neighborhood storage chamber. This would be particularly desirable for annual (or "trans-seasonal") units that can store enough surplus heat in summer to provide complete heating needs all winter; according to the magazine *People and Energy*, "Annual storage systems are most economical for multi-family dwellings and small (in area) communities."

Electricity production, too, can usually be more efficient at this level. There are economies of scale in power plants up to the point at which there is more heat lost in transmission than actually gets through to the users, but that begins to happen after only a few square miles. A number of studies have shown that small plants (ideally photovoltaic and photothermal) serving about 5,000 people—the magic number, again—are most cost-efficient. A Spectrolab Corporation report to the Project Independence solar-energy team, for example, concluded: "Smaller plants, located close to an end-user, may compete with power worth up to three times the central power station costs due to the savings in distribution costs and the higher fuel costs of conventional 'intermediate load' plants"; it suggested an optimal plant of 4 to 10 megawatts of electricity, suitable to supply a community of 800 to 2,000 homes (3,000–6,000 people).[3]

---

3. An additional study by the Portola Institute of California, published in their *Energy Primer* (1978), indicates that economies of scale for wind, water, solar, and biomass energy

Communitywide systems, even more than individual systems, are especially suited to a mix of energy sources. Such "total-energy systems" typically capture the considerable quantities of "waste" heat (up to 70 percent of the original energy) that is expelled by a power plant and use it for space heating and air-conditioning of nearby buildings. A small total-energy unit of that kind could supply the full power needs of an area of more than a million square feet—a small urban neighborhood, a shopping center, an apartment complex—for 40 percent of the costs of separate units; in fact these are now coming increasingly into use in factories in certain high-energy industries like chemicals and paper production. Larger units operating up to two square miles are possible before transmission wastage occurs, and even larger units, such as the "district heating" systems that supply up to several hundred thousand people in parts of Sweden and West Germany, are possible before conventional power systems are seriously competitive.

There is something to be said, too, for the sheer communitarian values of a locally based power system, approachable, guidable, and controllable by local citizens. There seems to be extraordinary support for such a system, judging by the reactions of people all around the country when the Department of Energy in the summer of 1978 held ten public hearings on energy policy. As reported by the Institute for Local Self-Reliance, contracted to monitor the hearings:

> The dominant theme of every hearing was strong support for the decentralizing and self-reliant characteristics of solar energy. Decentralization was emphasized not only because of cost advantages, but because it leads to a different relationship between the consumer and the energy he or she uses.
>
> Many people saw solar as a way to build community. One speaker said, "Solar creates jobs in the local community and keeps the money there, rather than sending it to companies . . . and institutions miles away." Speakers repeatedly urged the government to decentralize its funding programs and emphasize small businesses, individuals, and small research organizations.

Community power control is not such a weird idea in this land, either. As of 1975 there were 1,497 power plants operating in this country owned by public entities such as cities and towns, as against 2,123 privately owned plants. They were almost all in low-population areas—the typical system served only 15,000 customers, as against an average of 250,000 for the private utilities—and operation at that scale provided them special efficiencies, particularly because of moderate

---

production begin to operate in a community with somewhere around fifty houses and continue fairly evenly at least up to a hundred.

service limits and citizen cooperation. This, combined with the fact that they didn't have to pay any dividends to private shareholders, and to a lesser extent their exemption from income taxes, meant they operated 30 percent more economically than private utilities.

Rufus Miles is one of those who has perceived the liberatory effects of such community-based energy systems, concluding that "individual and community control over energy use and policies . . . would in fact help us to preserve and enhance democratic values." If small-scale solar plants did nothing more than that, they would be invaluable. The fact that they can do that *and* conserve energy, save money, and protect the environment makes them perfect instruments of a human-scale society.

IN 1492, SPAIN was a successful but by no means prosperous nation, and its gamble in sending an aging Christopher Columbus out to discover trade routes and treasures was considerable. The gamble paid off, however, and handsomely. The amount of wealth that flowed back to the Spanish monarchy from Columbus's discovery of the new world and its golden hoards is quite incalculable, but probably no less than twice the value of all the recovered mineral wealth known to the European world at that time.

Spain thus enjoyed a period of unprecedented power and grandeur. On the basis of seemingly endless supplies of mineral wealth, it consolidated its hold over the entire Iberian peninsula, erected a monarchy absolute in its domain and powerful throughout the length and breadth of Europe, embarked on repeated military campaigns to enlarge the power of both its mercantilism and its Catholicism, and inaugurated a century of intellectual and cultural achievement—in Spanish it was called, of course, *siglo de oro*—that sustained such giants as Cervantes, El Greco, and Velasquez.

But in fact the supplies were not endless. Sometime around the end of the sixteenth century, imports of gold and silver from the new world sharply declined: the pot was not bottomless, and its depths had been scraped with all the cruelty that the Spanish could muster. Almost all the sites known to the Indians had been exhausted, and those few that remained could not be mined because Spanish rapacity and European diseases had so decimated the local population that there simply was not enough manpower.

Very quickly the Spanish society began to feel the effects. When Philip II died in 1598, after forty years of imperial glory, the royal treasury was in arrears. The economy was similarly in disarray: "The wealth of America gave the government and some of the upper classes an easy income," as historian R. R. Palmer says, "leaving them without incentive to encourage commerce or manufacturing and in any case ignorant of real economic questions." Thus, as foreign wealth dwindled, there were no ready sources of domestic wealth to take its place. The thriving farms and small industries that had once existed throughout the country had been ignored and allowed to wither during the era of gold; Seville, which in the 1520s had been humming with at least 16,000 looms, had only 400 in operation a hundred years later. Spain's military might, too, swiftly crumbled. Defeat of the Spanish Armada by England in 1588, followed by the secession of the Netherlands in 1609, the flight of the Moriscos in 1610, and eventually the secession of Catalonia and Portugal, signalled the end of imperial Spain, and it has not risen since.

Spain's period of golden glory, based on what it had assumed were limitless riches, lasted just about 120 years, from Columbus's initial voyage to the secession of the Netherlands in 1609, which symbolized Spanish powerlessness.

The high-energy era in the United States began with the first successful oil well of Colonel E. L. Drake, of Titusville, Pennsylvania, in 1859.

7

Food: The Broken Loop

IN A SIMPLE ecosystem, fertile fields produce grains and vegetables, which are then eaten by animals and humans, who then produce waste and manure, which is then placed in the fields to provide the nutrients for the soil, which then may continue to produce grains and vegetables. A closed loop, and for many thousands of years—roughly from the domestication of crops at many places around the world in the sixth millennium B.C. until this century—a successful one, in most places most of the time.

But now? Taking normal U.S. practice as an example, increasingly infertile fields produce grains and vegetables, which are then combined with a great many additives and processed so that they lose many of their nutrients, and they are then transported considerable distances to be eaten by animals and humans, who then produce waste and manure, which is then flushed away with water and burned into the air or poured into rivers and oceans. The fields are then treated with synthetic fertilizers, made mostly of petroleum products, and chemicals, and sprayed with insecticides and herbicides, also made from petroleum and chemicals, and doused with heavy amounts of water in an attempt to restore the moisture that the chemicals take from the soils, and they manage to produce additional grains and vegetables, though with declining productivity per acre as the soils become depleted. A broken, not to say energy-intensive, highly exhaustive, and somewhat poisoned, loop. But for the space of approximately thirty-five years—from the post-World-War-II agribusiness boom to the present—a successful, extremely successful, one, in most places most of the time.

BUT NOT WITHOUT its consequences:

- From 1940 to 1975, U.S. farm output increased by 90 percent. Fertilizer use increased about 900 percent.

▪ Because high-chemical, high-energy farming is most congenial to large industrialized farms, particularly those operated by large corporations, they have been best able to take advantage of the new pattern of American agriculture—which has come to be called, interestingly, agri-*business,* with the *culture* part of it discarded. Large farms have increased by 500 percent over the last thirty years, while more than 2.5 million one-family farms have gone out of business.

▪ In Hopewell, New Jersey, a family farm owned by the Howells is being turned into a farm museum, with tractors, carriages, tillers, cows, chickens, horses, and samples of the normal produce of the American farm of legend. Does it not say something about how far the nation has come when a *farm,* a perfectly ordinary farm, is regarded as something that must be delicately preserved, as if a parchment under glass or an endangered species like the snail-darter?

▪ The average American ate two pounds of chemical additives in food in 1960 and ten pounds in 1978, a fivefold increase in less than twenty years. Most of these additives were put there *not* to preserve shelf-life or retard spoilage, as is usually claimed; more than 90 percent of the additives (both by weight and by value) were there to deceive—that is, to make the agribusiness product look, taste, feel, and nourish more like the real thing.

▪ James Gordon, a pest control specialist for California citrus growers, quoted by Daniel Zwerdling in *New Times* magazine:

> It's a merry-go-round. First you spray for thrips. But when you spray for thrips you destroy the natural enemies of the red spider mite and scale, so you get mite and scale outbreaks. So you have to go in with mite and scale treatments, but the more you spray the more you wipe out their natural predators, and the more the mites bounce back and the more you *have* to spray. . . .

With the result that spider mites, once insignificant insects in California, have developed into one of the most troubling and costly pests in that state.

▪ In 1949, some 11,000 tons of fertilizer nitrogen were used in the U.S. per unit of crop production. In 1968, some 57,000 tons were used to produce the same yield, and in 1975, some 95,000 tons.

▪ Chemical and industrial farming, being extremely expensive, puts a premium on mechanization and the displacement of human labor. Much work has gone into the creation of a tomato that will have a thick enough skin so that it can be picked by a mechanical harvester. The MH-1, developed by the agricultural specialists of the University of Florida, has been bred to such toughness that it is able to withstand a drop of more than ten feet on a hard tiled floor, exceeding by more than 250 percent the minimum safety standards set by the Federal government for bumper safety in collisions of new-model cars. Thomas Whiteside, the reporter who discovered this fact, remarks: "This undoubtedly represents a great step forward in tomato safety. Yet such further advances cannot but leave one nagging doubt in my mind: Now that the food industry has succeeded admirably in breeding tomatoes superior in gasability and crash-worthiness, where is the flavor?"

THE STATISTICS ABOUT the changed nature of farming in the U.S. since World War II are copious, but basically repetitive. Distilled, they describe a simple picture: the decrease of the family farm and the increase of the industrial farm, blue serge taking over from blue denim.

The consequences for the *economy* have been unstable but increasing prices (particularly since about 1970), excessive use of high-priced energy, and the rising influence of big corporations, multinationals, and monopolies. The consequences for the *ecology* have been severe depletion of the soil, toxic leakage and runoff into streams and rivers, chemically laden air in some farm areas, the depletion of water supplies, and an increase in both pests and plant diseases in an increasingly unstable ecosystem. The consequences for the *society* have been the destruction or diminution of many thousands of rural communities, an exodus of millions of rural poor into cities that proved unable to absorb the burden, and greater control over rural life by a smaller number of larger institutions. The consequences for the *consumer* have been higher prices, lower quality food, an increased intake of chemicals, the ingestion of untested additives, and increased incidences of cancer, heart disease, and obesity. And the consequences for the *farmer*—well, let Wendell Berry, a writer who is also a farmer and one very much attached to the soil of his native Kentucky, describe that:

> The concentration of the farmland into larger and larger holdings and fewer and fewer hands—with the consequent increase of overhead, debt, and dependence on machines—is thus a matter of complex significance, and its agricultural significance cannot be disentangled from its cultural significance. It *forces* a profound revolution in the farmer's mind: once his investment in land and machines is large enough, he must forsake the values of husbandry and assume those of finance and technology. Thenceforth his thinking is not determined by agricultural responsibility, but by financial accountability and the capacities of his machines. Where his money comes from becomes less important to him than where it is going. He is caught up in the drift of energy and interest away from the land. Production begins to override maintenance. The economy of money has infiltrated and subverted the

> economies of nature, energy, and the human spirit. The man himself has become a consumptive machine.

It is of course no accident that this sad process has swept so much of the country in the last generation. It is an example, once again, of the influence of big institutions over the lives of little people, and particularly of the biggest institution of them all, the Federal government. American agriculture is another victim of prytaneogenesis, the process by which government actually creates ills in the body politic.

Ever since the New Deal, Federal policy has been to build up corporate over communal or individual agriculture and large, successful farms over smaller, more marginal ones. Despite the recurring rhetoric about the "family farm as the backbone of the nation" in every farm bill, the laws enacted have had the effect, as the General Accounting Office of Congress determined in a 1976 study, "of consolidating and reducing the number of family farms, and continuing the trend toward greater mechanization and more energy-intensive methods of production." Price supports, soil-bank arrangements, direct payments, export controls, research-and-development funds, disaster-assistance payments, marketing agreements, tax write-offs—*all* have been designed to work chiefly to the benefit of the largest, usually corporate, farmers, and have done so for more than forty years. Direct subsidies, for example, go disproportionately to the largest corporate farms: in 1975, 65 percent of the payments went to farms with annual sales of $20,000 or more, only 14 percent to those with sales under $10,000. Similarly, the Farmers Home Administration underwrites loans every year overwhelmingly for chemical-based, machine-intensive, monocultural, and large-scale farms, thus setting the pattern for local banks and credit institutions and also for equipment and chemical suppliers.

And because Federal funds have accounted one way or another for *between 20 and 40 percent* of all farm income since 1955—easily the largest single source—what the Federal government does is the single greatest element in determining the character of American agriculture. If the tune in the American farm country is "Get Big or Get Out," that is because a government that favors bigness is paying the piper.

Wendell Berry, for one, has understood what that has led to:

> Those who could not get big have got out—not just in my community, but in farm communities all over the country. But as a social or economic goal, bigness is totalitarian; it establishes an inevitable tendency toward the *one* that will be the biggest of all. Many who got big to stay in are now being driven out by those who got bigger. The aim of bigness implies not one aim that is not socially and culturally destructive.

BUT OF COURSE the big American farm is more efficient, more productive, more economic, the thing that keeps alive a hungry world—right?

Dead wrong.

The research is copious, and it points to one inescapable conclusion: the most technically efficient and optimum-sized farm—the human-scale farm—is a small unit of a few hundred acres run by a family or one or two people. The U.S. Department of Agriculture has issued a number of studies over the last fifteen years, all in general agreement with the position of the 1973 report:

> We are so conditioned to equate bigness with efficiency that nearly everyone assumes that large-scale undertakings are inherently more efficient than smaller ones. In fact, the claim of efficiency is commonly used to justify bigness. But when we examine the realities we find that most of the economies associated with size are achieved by the one-man fully mechanized farm. . . .
>
> The fully mechanized one-man farm, producing the maximum acreage of crops of which the man and his machines are capable, is generally a technically efficient farm. From the standpoint of costs per unit of production, this size farm captures most of the economies associated with size.

A 1967 study showed that the "modern and fully mechanized one-man and two-man operation," not the giant industrial farm, is optimal for economies of scale; a 1972 study showed that, in California, where the average corporate farm was 3,206 acres, the ideal size in terms of efficient production was actually only 440 acres; another indicated that although the optimum farm varies according to the crop being grown, the ideal vegetable farm would be only 200 acres and only a wheat-and-barley spread in Montana need be more than 800 acres. According to Michael Perelman, a professor of economics at California State, Chico, "small farms have higher yields than large farms" and "in any state the value of the crops grown on the average acre tends to be larger when the average farm is small." Finally, despite the beliefs of most bankers, it turns out that small farms are actually more stable and better credit risks than large farms; as a government official acknowledges, "We know from our studies in the Department of Agriculture that the rates of foreclosure and delinquency are greater on big-farm loans, for the large-scale farm units, than for smaller loans on family farms."

And just for good measure, evidence from international studies confirms the connection between output and size. Research in Britain, Ceylon, Thailand, the Philippines, Brazil, and Guatemala has found time and again that the smaller the farm generally the higher the yield per acre, in some cases 60 percent higher. It was on the basis of these studies, in fact, that the World Bank a few years ago began to support small-farm development rather than pour all its grants into Third World agribusinesses trying to look like America's. The British publication, *The Economist,* in commenting upon this shift, wrote tellingly:

> This is not the romanticism of seeking out the noble peasant. It is a hard-headed calculation that small farmers, working for goals and returns they understand, on land where they have security of tenure and with enough co-operative credit and services to enrich their labour, produce the world's highest returns per worker and often per acre. And basically it is upon this strategy of backing the small men . . . that the hopes of feeding most of mankind in the long term depend.

It stands to reason, actually. Large farms have administrative costs and bureaucratic inefficiencies that small farms don't, they depend upon hired labor and often on distant decision-makers, they are forced to be highly mechanized in order to operate such acreage (and a tomato-harvesting machine costs at least $60,000), they have significantly higher labor costs, including overseers and corporate officers, and they have rental costs, overhead costs, equipment costs, and transportation costs that the smaller operations do not have. Whatever savings they make in cheaper bank loans, bulk buying, or market domination do not make up for those diseconomies. And no one realizes this more than the giant agribusiness firms themselves: a Ralston Purina official once admitted, "The individual farmer or family corporation can meet, and many times surpass, the efficiency of the large corporations that operate with hired management," and an official Tenneco report in 1975 conceded, "From the standpoint of efficiency, there is no effective substitute for the small-to-medium-sized independent grower who lives on or near his farmlands, who is knowledgeable in the science of horticulture, and who has a deep personal involvement in the outcome of his efforts."

Even the Department of Agriculture's insistence on "mechanized" ought not to be taken too rigorously, for smaller farms are able to operate without most of the high-cost machinery that big spreads demand: you can go out and hand-spray an organic pesticide on an acre or two, and you don't *need* an expensive crop-dusting plane. Economist Perelman's work indicates that, "all other things being equal, mechanization tends to decrease yields." The Amish, for example, are a perfect case study in non-mechanized efficiency. They make use of contemporary scientific information about breeding and biological pest control, but they use no internal-combustion machinery whatsoever, relying wholly on animal and human labor and simple tools. Yet they consistently produce more per acre than their neighbors, various studies have shown, and of course, judged in terms of energy, harvest far more per unit of energy invested. The experience of organic farmers, too, over the last ten years—there were at least 1,500 successful organic farms in the U.S. as of 1979, most of them perfectly ordinary operations mixed in with chemical neighbors—has shown that it is possible not only to operate without chemicals and their attendant machines but to match crop yields at the same time, improve quality, and eliminate expenses. An oft-cited study from Washington University's Center for the Biology of National Systems that examined sixteen organic and sixteen conventional farms in the Corn Belt indicated that organic farms actually operate at about $2 an acre *better* than the conventional chemical farms—and use only about a third as much energy.

All of which is not to say that optimum farms have to go back to the horse-drawn plow—far from it. With appropriate use of alternative technologies it is possible for any small holding to reduce the back-breaking part of farm labor without the budget-breaking enslavement to a fossil-fuel economy. On a small scale—and on a small scale only—it is possible to run a tractor on methanol, an auger feeding and irrigation system on wind power, a milking machine on methane, a farmhouse and food-drying process on solar, and so on—in short, to find the fitting balance between the mechanical and the human.

SMALLER FARMS POSSESS other virtues, too. They are able to switch to and use solar energy more easily, and without much difficulty become energy self-sufficient if they wish to. A series of tests by the Small Farm Energy Project in Nebraska has shown that retro-fitting and insulation, wind and biogas systems, and solar heat and hot water collectors can be installed cheaply in single-family farms and that the savings they produce are almost immediate. At least a dozen farms in the U.S. get their full energy needs from methane systems powered by the manure of their own animals; no fewer than 150,000 depend for at least some of their power on windmills. Smaller farms can more easily switch to and maintain such methods as organic, hydroponic, and biodynamic farming; hydroponics, using liquid fertilizers instead of soils, and biodynamic methods, using a compact-bed system practiced for years in France, have both been shown to increase crop yields from two to eight times normal. Smaller farms, because they can be located closer to cities, can also supply food directly to local markets, eliminating the extremely costly and energy-wasting practice of refrigerating and shipping vegetables from coast to coast—and providing far fresher and better food as well.

Smaller farms are especially valuable because they are uniquely suited for communitywide cooperation—and indeed the history of agriculture in this country for the most part was exactly the history of rural cooperation. The largest farms have little interest in cooperation, since they normally have all their own equipment and can throw their weight around in the marketplace, and the corporate farms of agribusiness of course operate completely vertically, back and forth to the corporate hierarchy, rather than horizontally, with the other farms in the area. But small farms have always felt the need to band together, whether for a barn-raising in the eighteenth century or to establish marketing co-ops in the 1870s or to buy bulk fuel in the 1920s. Indeed, it is the family farm that remains the backbone of the considerable cooperative movement that exists today: 7 million people participating in some 879 rural electric co-ops (which own 42 percent of all distribution lines in this country), 3 million in some 5,000 marketing co-ops, 3 million also in the 2,700 supply co-ops, and 700,000 in the 239 telephone cooperatives. The co-op movement, in fact, according to *Business Week,* accounts for a third of the $165-billion agricultural market and is "the most important force in U.S. agribusiness."

And of course the cooperative tradition at the other end of the food pipeline, also with deep roots in this country, has enjoyed a new flowering in the past decade. No accurate number is possible, but one estimate has it that there are at least 20,000 food cooperatives that have operated for five years or more; some are solid, multi-million-dollar operations, like Berkeley's Consumers Cooperative (28 stores, $78 million sales volume in 1976), others are more modest operations of twenty or thirty families operating out of a church basement or an unused storefront. And in many places food co-ops have joined forces with other collectively run businesses up and down the whole food chain, forming food federations of farms, truck lines, grain mills, warehouses, bakeries, markets, and restaurants; the People's Warehouse in Tucson, for example, is a federation that serves almost all of the Southwest with a volume of more than $1 million a year, and the Intra-Community Cooperative in Madison, Wisconsin, has boosted its volume to more than $2 million a year.

Ultimately it is not hard to imagine this extraordinary communal tradition, providing it can survive the increasing hold of corporate agribusiness, evolving into a network of truly self-sufficient food communities. Indeed, there is not one region of the country that could not eventually become self-sufficient in food, with only the smallest dietary changes. Self-sufficiency is not as chimerical as agribusiness would have us think. We do not *need* to depend on complex intercontinental—and international—distribution systems. We import snails from France, for example, with great fanfare and considerable cost, but the fact is that the exact same animal, *Helix aspera,* is found naturally almost every place in this country; we get tomatoes from Florida even during the summer months because distributors and supermarkets are locked into marketing contracts, but the fact is that from May to September tomatoes can be grown anyplace in the entire country, and in greenhouses they can be grown locally the year around.

And if it be countered that our coast-to-coast contrivances now let us have plenty of good, farm-fresh foods no matter where we are, it should be noted that per capita consumption from 1950 to 1975 has *dropped* for eggs (389 to 278), milk and cream (348 pounds to 241), butter (10.7 pounds to 4.7), fresh fruit (108.8 pounds to 81.5), fresh vegetables (115.2 pounds to 99.3), and wheat flour (135 pounds to 106). In other words, now that we have built up an enormous system of transportation, refrigeration, and warehousing that allows us to ship any fresh food any distance at all, we actually eat *less* of it than we did thirty years ago.

Of course, when you stop to think of it, it's obvious that most areas of the country were self-sufficient for the greater part of our history, indeed until quite recently. A citizens-action group called Vermont Tomorrow, for example, has shown that Vermont was self-sufficient at the turn of this century, producing its own fruit, vegetables, meat, grain, and dairy products, and by today's dietary standards supplied more than adequate amounts of everything except citrus fruits, though it had plenty of vitamin C from tomatoes and berries. Today, however, it is dependent on the large agribusiness networks for almost all its foodstuffs—even for milk, though dairy farming is the leading agricultural enterprise in the state. In response to this, Vermont Tomorrow has launched a campaign to try to restore something closer to the earlier pattern by setting up a loose network of food and grower cooperatives, community gardens, canning centers, farmers markets, and restaurants to create what they call "LIFE": a Locally Integrated Food Economy.

Self-sufficiency at an even smaller level is also feasible. Although the standard figure is that it takes an acre to feed four people, the Department of Agriculture has printed a study indicating that an acre of land can provide full garden crops for approximately fifty-five people, and the Institute for Local Self-Reliance in Washington estimates an acre could feed between forty and seventy people, even more if certain intensive-agriculture methods are followed. (Livestock production, of course, would take somewhat more space, but almost all animals except cattle—especially fish, chickens, pigs, and rabbits—can be tended on very small plots.) That means that a community of 5,000 people would need only about 100 acres, or less than a sixth of a square mile, to grow its needed foodstuffs. Imagine part of that acreage covered with solar-powered greenhouses, and there's no reason to suppose that the self-sufficiency could not be obtained year-round.

Nor should cities be excluded from considerations of self-sufficiency. The experience of a dozen different organizations has proven that urban agriculture is not only possible but comparatively simple and successful, within certain obvious constraints. There are, surprisingly, plenty of sites: rooftops, providing only that the roofs themselves are reinforced; back gardens, particularly successful with greenhouses and hydroponic systems; and vacant lots, which are always more numerous than most people suspect (there were 29,000 empty lots in New York

City in 1977, for example) and which can be converted to use with topsoil and compost. Community gardens, too, are easy and successful, if municipalities make cleared spaces—in city parks, say, or along rivers—available for individuals to plant; Britain has had community gardens for more than 150 years, with at least 300,000 of them flourishing today, and by 1978 five states in the U.S. had sponsored community gardens, 160 of them in California, 236 in Massachusetts (with 56 in Boston itself). A city the size of New York could not be food self-sufficient—another deficiency of the overlarge city—but a city of 50,000 would need less than two square miles to grow all the garden crops it needs.

But—it is appropriate to interject here—what about the hungry world? Doesn't it need our bounty to keep its starving people alive? Self-sufficiency for America is one thing, but it can hardly be defended if it means starvation for the rest of the world.

Of course there is no reason that American communities could not find some way to transfer their unneeded agricultural surpluses to parts of the world suffering from genuine famine. But the fact is that, except in extreme cases, that's probably not really necessary: American self-sufficiency in itself would go a long way to *helping* the rest of the world rather than hurting it.

In the first place, it would allow Third World countries to use the land that is now given over to fancy exports for American tables to grow food for themselves instead. Mexico, for example, grows strawberries and tomatoes for export to the U.S. while millions of its own people go hungry; Senegal, in the heart of the famine-riddled Sahel, grows groundnuts and vegetables on some of its richest land, all for the European market; Brazil has increased its production of soybeans destined to be exported and fed to American and Japanese livestock by more than twentyfold over the last decade while its production of food crops has actually declined. It is the American (and to a lesser extent European) market, in other words, an incredible gobbler of goods, that is determining what most foreign nations will plant in their soil, and it is American corporate money that is dictating that it be largely cash crops for export rather than food crops for local consumption.

In the second place, American agricultural self-sufficiency would mean an end to the exporting of the so-called "Green Revolution"—high-chemical, high-fertilizer, high-technology agriculture—that has favored those largest and richest Third World farmers who could afford expensive First World technologies and driven off the smaller and marginal farmers. Such large spreads, in the hands of a wealthy elite, not only grow cash crops instead of food crops—like the Colombian entrepreneurs who switched from grain to carnations—but they force thousands of people off the land and into the cities, where of course the food crisis becomes doubly acute.

Lastly, it seems that American food exports that theoretically are keeping alive the world's poor actually do nothing of the sort. There has been no noticeable reduction in the number of hungry and starving since the American food programs were undertaken, not even in years when we have particularly good harvests, and on-site studies suggest that very little in the way of useful food actually gets put on the plates of the Third World needy. Indeed, such programs generally work against food production in the recipient countries by substituting systems of dependency (profitable for grain dealers, middlemen, warehousers, money lenders, and the like) for those of self-reliance; extraordinary as it seems, there is not a single country to which the U.S. exports grains that could not grow those grains itself, should its priorities take that direction.

The fact is that the problem of starvation stems not from a lack of food in the world but a lack of food in the bellies of the poor people. The world right now grows enough to feed everyone on earth adequately; C. Peter Timmer, a food economist at Cornell University, has shown that there is enough food to feed everyone in the world a nutritionally adequate diet of 3,000 calories, and 65 grams of protein, a day. And if all arable land were brought into production instead of being ignored or allowed to lie fallow, there could be still more—Africa, for example, cultivates only 22 percent of its arable land, South America just 11 percent, and even if much of that would be uneconomical to bring into production now, there are still millions of acres that could easily be providing foodstuffs. As Martin McLaughlin, a fellow of the Overseas Development Council think tank has said:

> It is the lack of adequate *demand* (or income), not of supply, that keeps people malnourished; they are hungry because they are poor. . . . Even in the crisis year of 1973–74, there was no shortage of food. Many people simply could not afford to pay for it. . . . The adequate supplies so often mentioned these days do not get to hungry people, because neither the international food system nor the national and international political structure permit them to grow or buy their food.

In short, the rest of the world has enough land to grow adequate food on (though at present it is badly used) and there is enough food in the world to feed the hungry (though at present it is badly distributed). One does not want to minimize the obstacles, but it is clear enough that a system of world self-sufficiency would be perfectly possible, even at present population levels—and the quickest way to get there would actually be American self-sufficiency instead of American domination.

ABOVE ALL, TO RETURN TO OUR THEME, smaller farms are simply better for the social texture of rural America, for people. It cannot be maintained that the developments of the last three decades have been *beneficial* for these areas, whatever one may try to say about the resultant productivity. Professor Earl O. Heady of Iowa State University, a strong defender of American agribusiness, nonetheless has acknowledged:

> The change in the very nature of farming, with its higher productivity and greater degree of mechanization, has severely affected rural communities in agricultural areas. With the decline in the farm population the demand for the goods and services of business in country towns has been eroded. Employment and income opportunities in typical rural communities have therefore declined markedly. As people migrated out of the rural communities, there were fewer people left to participate in the services of schools, medical facilities and other institutions. With the lessened demand such services retreated in quantity and quality and advanced in cost.
>
> Nonfarm groups in the rural communities took large capital losses as country businesses closed down and their operators moved elsewhere, in many cases leaving their dwellings to decay.

And one wonders about those individuals who "migrated out"—how many of them actually wanted to leave, how many had any say in the manner or timing of their leaving, how many found productive lives elsewhere, how many preferred their urban lives of subordination and dependence to their previous lives of comparative independence and responsibility?

Two unusually comprehensive studies done in the premier agribusiness state, California, offer interesting confirmation of what this rural change has meant to America.

The first was undertaken by a task force of the Small Farm Viability Project and presented to the state government in November 1977. It engaged a team of sociologists from the University of California at Davis to examine 130 towns in the San Joaquin Valley and compare the quality of life in those places with small-parcel holdings (under 160 acres) as against those with large cropping patterns (640 acres or more). Invariably, they found, the small-farm communities generated more local businesses, higher levels and diversity of employment, and better civic services and had more lawyers, dentists, doctors and medical specialists, elementary schools, movies, hospitals, pharmacies, farm-equipment stores, police, firefighters, and post office workers. They concluded:

> Many have argued that large businesses are a financial boon to an area. Whether this assumption is valid or not, the suggestion here is that

> large-scale farming may have a less positive effect on the local community than small-scale farming. Large-scale agriculture offers the local communities no substantial advantage. The smaller scale farming areas clearly tend to offer more to the local communities than their counterparts.

And they provided as additional evidence a finding by Berkeley economist George Goldman that converting a region from farms averaging 1,280 acres to ones averaging just 320 acres actually *improves* the economy of the area—generating 540 new jobs, increasing retail sales by $16 million a year, and raising personal income by $6.2 million—because more of the money stays in the locality and through its "multiplier effect" enriches the whole population.

The second study is a sociological classic that was originally conducted by Walter R. Goldschmidt of UCLA in 1945–46 and then updated and validated by two further investigations in 1970 and 1977. It compared two California towns of approximately the same size—Dinuba, with 7,404 people, and Arvin, with 6,236—but with totally different farm communities, one of the traditional small-holding style and the other changed by agribusiness. Dinuba had 635 farms, averaging 45 acres, most of them family-owned, and only a third of its population was hired labor (down to 14 percent in the 1970s); Arvin had only 137 farms, averaging 297 acres, and many large operations, some of them absentee-owned and almost all dependent on hired workers, who made up two-thirds of the population (down to 38 percent in the 1970s). The difference between them was like corn and cattle.

Dinuba had a significantly higher standard of living, with more than half the population in farming or in white-collar occupations compared to one-fifth in Arvin, and 15 percent in skilled trades and professions compared to 6 percent. On a "level-of-living" index (measuring the number of homes with electricity, appliances, automobiles, etc.), Dinuba had 38 percent of its residents in the top quarter, Arvin only 18 percent. Sales of home supplies and building materials were three times higher in the small-farm community, and it had almost twice as many small businesses and 61 percent higher retail sales (by 1976 that had grown to 70 percent). The Dinuba population had a higher percentage of high-school graduates—38 percent to 19 percent—and supported four elementary schools and one high school, against a single elementary school for Arvin; Dinuba had two newspapers, each larger than the single paper in Arvin. Dinuba had twice as many church, civic, and social organizations, with higher levels of participation. And it had "more institutions for democratic decision-making and much broader citizen participation in such activities," one result of which was that it had "far better" public facilities, including sidewalks, roads, sewerage, garbage service, playgrounds, and parks.

A tale of two cities: where people are close to their own land, where the small family farm is the basic social unit, the attachment to the local culture, it seems, is stronger, the texture of the local community is richer. A tale that suggests what has been lost to America in the last thirty-five years and what might be worth recovering.

8

Garbage: There Is No Away

OURS HAS OFTEN been called a "throwaway" society, and that is an accurate enough description of a nation that creates something like 3 billion tons of solid waste material every year. There's one fundamental thing wrong with it, however. There is no "away."

For five years, between 1947 and 1952, a chemical company in Niagara Falls, New York, dumped 21,800 tons of toxic wastes in heavy metal cans into a landfill site called the Love Canal, where they were covered over with garbage and dirt. Thirty years later, the chemicals began to leak out of the rusted containers and seep to the surface. Pools of foul poisons would appear after every rain. Children playing in empty lots would suffer chemical burns on their skin. The smell of pesticide hung in the air. Trees and grass suffered acid burns. Suddenly residents of the homes that had been built over the Love Canal began to report a large number of miscarriages and birth defects, mostly serious. Finally the air over the Love Canal was tested: it contained eighty-two different compounds, eleven of them carcinogens or suspected carcinogens. In August 1978 the New York State Health Commissioner called it "a great and imminent peril to the health of the general public." The state bought 239 homes there, moved out the families, and boarded up the windows, probably forever.

For more than fifty years New York City and coastal New Jersey municipalities have been dumping their sewage and garbage, treated and untreated, into a section of the Atlantic Ocean eleven miles offshore known as the New York Bight. There it has slowly polluted all of the surrounding area, killing ocean vegetation, marine animals, and fish. Every year during the 1970s extraordinary "fish kills" have been reported in or near the area, and every year since 1976 various beaches in New York City and Long Island have had to close at one time or another because of sewage contaminants.

Cattle feedlots in the U.S. handle 90 percent of the 42 million cattle slaughtered in the U.S. every year. Those cattle, it is no surprise to learn, defecate, and at the rate of approximately 100 pounds per animal per day: 2 million tons of bullshit a year. Very little of it is used for manure because the salt content of the feed used for beef cattle is high—unnecessarily and unhealthily high. One inventive company in Oklahoma is under contract to the city of Chicago to supply methane gas from such wastes, and the town of Lamar, Colorado, has begun the largest biogas plant in the U.S. to convert 350 tons from its feedlots into one million cubic feet of methane for its 10,000 residents. But for the most part these wastes are simply piled up near the feedlots into small hills, which become large hills, which become mountains, sitting there, rotting, growing at the rate of 2 million tons a year.

There is no away.

THE CHINESE IDEOGRAM for "crisis," it is said, is a combination of the signs for "problem" and "opportunity." It was never more appropriately applied than to the issue of waste disposal in this country: this is a problem that is just loaded with opportunity.

Let us take the matter of the broken food loop first, for that is one where food policies and waste policies overlap. Here, on the one hand, we have farmers using millions of tons of artificial fertilizers to accomplish the same sort of things that could be done more easily and cheaply by using organic waste, and without leaking pollutants into the ground and air. And on the other we have millions of tons of organic garbage and sewage piling up in the cities of the land, with no place to put it all and no way to dispose of it that is environmentally safe or economically sound. Obviously the simple answer is to use the *opportunity* of organic recycling to solve the *problem* of urban waste disposal.

Much of this waste (though waste it should not be) is in the form of sewage. The idea of water-borne sewage must have been a good one once, but it has clearly outlived its day. The process takes high quality and very valuable drinking water, pollutes it with human and industrial wastes, and deposits it directly in a nearby river or lake or at a city plant where no one quite knows what to do with it. We have tried drying and incinerating the stuff, but that creates pollution in the air and produces an ash residue that then has to be put somewhere else; we have tried letting it settle and then carrying it out to the oceans or dumping it in the rivers, but the billions of pathogens produced by huge populations linger on and carry their toxins into the waters; we have tried settling it and then treating it with chemicals to kill the poisons, but that proves to be extremely expensive and the environmental effects of the chemicals themselves are very uncertain; and we have tried pouring it into landfill sites, but there, too, toxins leach into surrounding groundwaters, and anyway the U.S. Conference of Mayors has warned that at least half of our cities will have run out of landfill space by 1980 and thereafter there will be no place to put it at all.

Obviously the thing to do with the millions of tons of sewage is to put it back where it can do the most good: on the land. There is nothing arcane about such an idea, however unpleasantly it may first strike the American sensibility. Recycled raw sewage holds no particular danger, providing that toxic chemical wastes are excluded: the soil uses such nutrients as nitrogen, phosphorus, potassium, calcium, sulfur, zinc, and copper; it can neutralize a normal load of human pathogens within a few weeks; and it can absorb low levels of such heavy metals as nickel and lead without toxic effect, particularly if the fields are left fallow for a year after being treated regularly. (Some big-city sewage has been found to contain high levels of such metals as cadmium and mercury and such poisons as PCBs, but it is simple enough, should a city wish, to locate the source of most such contaminants and isolate them from the sewage stream.)

Raw sewage has been used on farmlands in Europe and Asia for centuries, and still is in many places. Most large Chinese cities today have facilities to settle sewage into usable sludge and "honey trucks" that cart human and vegetable wastes from buildings that are not connected to the central sewage system; it is all deposited as a nutrient-rich effluent on farms in the surrounding areas. The city of Melbourne, Australia, at present dumps all of its raw sewage on a single 42-square-mile farm outside of town, where it is used to grow the forage for cattle and sheep whose sale brings in the equivalent of $675,000 a year. And in the U.S., perhaps a thousand cities and another two dozen or so industrial facilities deposit their wastewater on surrounding agricultural land; the city of Janesville, Wisconsin, a fairly typical one, carts 25,000 gallons a day in sludge trucks to fertilize nearby farms and the city's own parks and golf courses, and it has done so for the past fifteen years.

Of course Janesville is a town of 50,000, and many things are possible, as we know, at that size. Indeed, population size does have an important bearing on sewage. Large metropolitan sewage systems accumulate so many human pathogens and industrial poisons that if they want to eliminate them they have to use elaborate and expensive "tertiary treatment" centers, processing the waste with chemicals before it can become useful fertilizer. Only when population levels are kept to a modest, ecological level can the natural buffer lands be used effectively to recycle human wastes without serious effect. In fact, it has been calculated that with an ideal environmental balance, sewage systems would be quite unnecessary; according to S. Fred Singer, an economist at the University of Virginia, "It should be noted that if people were perfectly distributed, according to the available surface water, then the

United States could accommodate 250 million people without requiring any sewage treatment."

The question of water deserves some amplification. Polluting it as we do, 400 billion gallons every year, is clearly foolish, particularly when we are just forced to spend tens of billions to try to clean it up again—just so that it can once again be polluted. And the foolishness is compounded when the chemicals used to clean it up destroy certain of the "hard-water" nutrients valuable for bones and blood vessels and may even combine to create cancer-causing agents; the Environmental Protection Agency in 1975 identified sixty-six possible carcinogens in the water supplies of every single system it tested in eighty cities. Changes in household habits can eliminate some of the problem (for example, a composting or biological toilet that uses no water at all could save roughly 13,000 gallons of water per household per year) but for any serious revision of the water-use pattern, the changes will have to be made by agriculture (which uses 50 percent of the available water) and industry (40 percent). That, in turn, is not likely to come about until there is some sort of local control over those enterprises, so that the individuals who are directly affected by, say, industrial pollutants in the drinking water can decide as to whether the pollutants are to be used in the plant originally.

ASIDE FROM SEWAGE, the remaining portion of organic material in the American waste stream is organic garbage—and that, too, is all potentially useful.

There is a staggering amount of it. According to recent estimates, American agriculture alone is the source of some 590 million tons of dry organic wastes a year, at least 23 million of which would be easily collectable and reusable. Yet American farms use only about half a million tons of that as fertilizer; the rest is simply burned or carted away or left to pile up or fed to farm animals or passed into septic and riverine systems. (And then the farmers buy 40 million tons of chemical fertilizers to put on the fields.) In addition, American cities generate 141 million tons of organic refuse a year, 73 million tons of which are estimated to be readily collectable and recyclable; and American industries generate another 100 million tons or so, 10 percent of which is recyclable.

Some of this huge amount of organic waste might be useful to recycle just as is—reprocessing paper garbage into newsprint, for example—and some might be most profitably turned into a cheap source of fuel. But most of it, and particularly the agricultural wastes that are right there on the farms to begin with, could easily be returned to the land; if just the most minimal kinds of compost systems were installed, American agriculture could get more than half its fertilizer from organic sources. To that could easily be added the 22 million tons of food wastes, say, and the 24 million tons of yard wastes found in the cities, which together make up nearly 40 percent of all the solid municipal wastes in the nation. All that's needed to collect this is a modicum of individual responsibility, with every household separating organic from inorganic material, and a minimum of systematic collection, to keep the two types separate. And it can't be all that hard: systems of this sort are quite common in the U.S. even now, in scattered areas—mostly small towns and suburbs—where city officials have tumbled to the virtues of selling separated wastes to recycling companies. Organic collection and compost systems now operate in cities in California, Oregon, Utah, Texas, Nebraska, Georgia, South Carolina, Pennsylvania, and New York; Odessa, Texas, has been composting organic waste for half a dozen years now, using the product for the surrounding range land, and in the South Bronx a community group has begun a pilot program to compost 600 tons of organic garbage each week, half to be used locally and half sold for fertilizer elsewhere in the state.

The Chinese have a saying that, in the light of all this, seems indisputable: "All waste is treasure."

THIS PROCESS OF recycling waste has other dividends beyond the agricultural one of restoring the food cycle, and they are supremely important for a properly balanced ecosystem. Most of what this country throws away—or, rather, tries to—is easily reusable, either as fuel or compost or raw material. Municipal garbage, on average, contains approximately 40 percent recyclable paper, 10 percent glass, 10 percent ferrous metals, 3 percent rubber and leather, and 4 percent wood—a marvellous, and essentially untapped, storehouse. And whatever it would cost to put these materials back into productivity is many times less than trying to create them from scratch, and the energy savings are considerable as well.

Aluminum, for example—it makes up 2 percent of the nation's total garbage stream and at present is mostly thrown away. But it can be separated and recycled quite simply, at one-twentieth of the environmental (energy and pollution) costs of mining and smelting virgin ores. Or copper—fully a quarter of the nation's copper consumption, about 375,000 tons a year, could be obtained by extraction from municipal garbage, according to Professor Earl Cook in the quarterly *Technology Review;* what's more—incredible as it may seem—some municipal dumps have higher percentages of copper than some mines that are operated profitably right now in the Rockies. It is obviously very close to criminal to discard resources such as these in a world of growing scarcity.

The logic of recycling—not to mention the current profits therefrom—is so overpowering that virtually no one now disputes the virtues.

However, the general governmental response has been to apply technofix methods, particularly the huge "resource recovery" plants that have been built in a number of big cities in the last few years. They certainly have the right idea, but so far they have proven to be dismal failures—they are very capital intensive, they are wasteful of resources, they turn out to be inefficient and susceptible to recurrent breakdowns, and they even end up adding to pollution by creating dust during their shredding processes and air pollution during their incineration phase. Moreover, since they all need a tremendous volume of garbage to be able to produce sufficient amounts of recycled materials for resale, they place a premium not on conservation but on waste, and the more the better; the city of New Orleans, for example, has had to *pay* its recovery plant for every ton of garbage below the contracted level that it is unable to deliver. Not one of these plants has so far operated as expected, and several of them—in Baltimore, St. Louis, New Orleans, and Seattle—have gone sharply into debt or even been abandoned as hopelessly uneconomic. It will come as no surprise that the Federal government has spent about $100 million to perfect these high-technology plants, virtually nothing for other recycling systems.

A small-system, community-based approach, however, has worked out very well in a number of towns and cities across the country. According to Neil Seldman, the solid-waste expert at the Institute for Local Self-Reliance in Washington, there were between 150 and 200 cities with source-separation and recycling programs as of 1978, up from only 20 in 1975. (Most of them in smaller places; in the words of Bill Bree of the Oregon Office of Recycling Information, experience has shown that "easier communications, and more established patterns of social cohesion, make smaller towns easier to work with in setting up recycling operations.") The ORE Plan in Portland is among the most successful of them. There the residents of four neighborhoods pre-sort their garbage into four types—paper, glass, metal, and organic—and place it at the curb once a week. It is picked up by separate pick-up trucks, one for organic, another for inorganic, and hauled off to be composted or sold to bottlers and metal processors. The system has been found to be the most energy-efficient municipal garbage system in the country, and it even offers the household subscribers a lower collection rate than they would have to pay for ordinary private haulers.

Small-scale recycling operations in the future can be made even more efficient, by the application of both communitywide support and alternative technologies. Studies have shown that the optimum efficiency for a recycling program is reached when every household in a community of 5,000 participates—at that level, a once-a-week collection can produce a large enough quantity of various materials to make a 15–20 percent profit on resale, and it will still not cover so much territory that it loses money in collection and transportation costs, which account for

70–80 percent of the overhead of an ordinary system. Collection could be even more economically done with small electric carts or trailers—the ORE system has used golf carts profitably—or ultimately with a simple network of underground, solar-powered conveyor belts that, in small areas, could be operated as efficiently as such belts are in any large factory today. *That* would be the most efficient and equitable separation and recycling operation of all.

A small-scale system would be most economic, too, if the community were to reuse everything within its own borders as raw material for its own products. Small-scale aluminum and de-tinning facilities can now be constructed for less than $5,000 and, using ordinary cans, could turn out metal for local production of bicycles, wheelbarrows, machine and auto parts, and the like. (The Oregon Appropriate Technology group has estimated that in a city like Eugene, with a population of 80,000 or so, it would be possible to recover about 16 tons of aluminum, a ton of copper, and about 54 tons of high-grade iron and steel each year.) Paper-processing plants could easily convert wastes into newsprint or reusable paper or even into cellulose-fiber insulation. Bottling centers could either recap and reuse the bottles as they come in or crush them to use in road or building construction or, with currently available machinery, remold them into canning jars for neighborhood use. A somewhat more complex, but still small-scale, process can be used to convert plastic wastes into building materials, furniture, auto-body parts, or fish tanks. Human and food wastes could be fed to community livestock—pigs, as everyone knows, are marvellous waste-eaters and process even human excrement to a valuable manure[1]—or composted collectively for community gardens; "graywater" (household water used for washing or bathing) could be piped from each house and, after a simple filtering, used for irrigation.

There seems to be a scale, that is to say, at which even such a process as garbage recovery can make economic and ecological sense. It is small, and human.

THE LOVE CANAL site was not the only one in Niagara Falls where toxic wastes were buried over the years. It turns out that there are at least three other sites that pose equally serious health hazards: the Hyde Park landfill, a 16-acre stretch on the north side of town where at least 80,200 tons of hazardous wastes were buried, and the 102nd Street fill and the S-Area dump, both of them close to the water-treatment plant that is the source of the city's drinking-water supply. A Federal-State Interagency

1. One experiment a few years ago showed that when the excrement from one pig is fed to another, and then the second pig's is fed to a third, it is the third pig that is the fattest and the healthiest.

Task Force on Hazardous Wastes, formed after the Love Canal episode gained nationwide publicity in 1978, has also identified no fewer than eighty other companies in the Niagara area that dispose of potentially harmful industrial wastes.

The Hyde Park dumping area, used for at least five years by the Hooker Chemical and Plastics Company, a division of Occidental Petroleum, is known to be contaminated with an estimated 3,300 tons of the chemical trichlorophenol, a highly toxic defoliant. One of trichlorophenol's byproducts is the chemical dioxin, a poison so powerful that if 3 *ounces* of it were introduced into the drinking water of New York City it would be sufficient to wipe out the entire population. An estimated 2,000 *pounds* of dioxin are buried in Hyde Park.

Some of the chemicals from this site, including dioxin, have been found to be leaching into a stream, appropriately called the Bloody Run, that flows beneath the fill and out past a residential area and a university, and then into the Niagara River and Lake Ontario. Residents of the area along the stream have reported a high percentage of serious illnesses among their children, and it is said that none of the neighborhood pets ever lives beyond two or three years.

According to the Federal Environmental Protection Agency, there are no fewer than 51,000 sites in the U.S. containing hazardous wastes, piling up at the rate of 40 million tons a year, and in at least four hundred of them there has been measurable environmental damage since the agency first began keeping track, in 1972.

There is no away.

9

Transportation: From Radial to Cyclical

YOU MAY RECALL that fascinating scene in *Ragtime* where the father, Tateh, and the little girl take a series of trolley rides all the way from Hester Street in New York City to Boston, transferring from one urban system to another. The New York trolley line took them all the way to Mount Vernon, some ten miles north of Manhattan, where they transferred to a New Rochelle line, then on to the Greenwich, Connecticut, trolley, which joined with the Bridgeport system, and then successively on to the lines of New Haven, Springfield, Worcester, and finally Boston:

> The great wooden car swayed from side to side. The wind flew in their faces. They sped along the edges of open fields from which birds started and settled as they passed. . . . The car barreled along its tracks down the side of the road, and whenever it approached an intersection its air horn blew. Once it stopped and took on a load of produce. Riders crowded the aisle. The little girl could not wait for the speed to be up. Tateh realized she was happy. She loved the trip.

The whole journey took them two days, stopping overnight in New Haven, and cost $2.40 for the adult fare and just over a dollar for the child's.

> Tracks! Tracks! It seemed to the visionaries who wrote for the popular magazines that the future lay at the end of parallel rails. There were long-distance locomotive railroads and intcrurban electric railroads and street railways and elevated railroads, all laying their steel stripes on the land, crisscrossing like the texture of an indefatigable civilization.

At the time—the early 1900s—the U.S. had more than 20,000 miles of trolley tracks and nearly 5 billion trolley passengers yearly.

Beginning in the 1930s, General Motors, which had achieved a virtual monopoly of bus manufacturing in the U.S., decided that there would be a ready market for bus transportation if only American cities would dispense with their electric trolley systems. Consulting no one but its own board of directors, it formed a United Cities Motor Transit division whose sole function, according to later Senate Judiciary Committee testimony (Ninety-third Congress, Second Session, 1974), "was to acquire electric streetcar companies, convert them to GM motorbus operation, and then resell the properties to local concerns which agreed to purchase GM bus replacements." By 1949, GM was able to compel the abandonment of more than a hundred trolley systems—including important lines in New York, Philadelphia, Baltimore, St. Louis, Salt Lake, and (most tellingly) Los Angeles—by economic or political pressure that forced the scrapping of trolley cars, the dismantling of power lines, the paving over of tracks, and the installation of a system of buses. In 1936, the nation had 40,000 trolley cars in service; in 1949, 6,500, and that would drop by half in another ten years. Throughout the nation a cheap and efficient, essentially non-polluting, form of mass transportation was replaced by an expensive and clumsy, exceedingly dirty form, poisoning the urban air, driving up city budgets, congesting city streets, and eventually impelling millions of Americans into private automobiles (another development GM could hardly have minded), with ripple effects that even now touch every corner of society.

Ultimately GM was hauled into court to face the consequences. In 1949, along with Firestone and Standard Oil of California, its co-conspirators, it was convicted in U.S. Federal court in Chicago of criminal conspiracy and anti-trust monopolization. It was fined $5,000.

It is not often that we find real conspiracies in history. This is about as close as one can get.

And it suggests that there is more to the terrible bollix of America's transportation than sheer accident or some psychic desire for speed on the part of the average American. We didn't get into this mess all by ourselves; we had help. It is probable that we won't get out of the mess without finding some way to dispose of that help.

ABOUT THE EXISTENCE of the transportation mess there is no question.

It begins with the motor vehicle. Responsible for 24 million accidents a year, 50,000 deaths, and 5 million injuries, it has appropriately been called the most lethal device since the invention of the machine gun. It causes approximately 60 percent of all air pollution in the nation, 82 percent of all carbon monoxide poisoning, and much of the noise pollution, producing a significant if incalculable number of deaths and illnesses. It uses up one-fifth of all the energy consumed in this country, and a high percentage, perhaps as much as half, of all the petroleum, plus large amounts of other resources (74 percent of all rubber, 63 percent of all lead). It demands enormous amounts of space, both in the countryside, where it has so far caused 60,000 square miles of land to be paved over, and in the cities, where roughly half of all the land (in Los Angeles *62 percent)* is given over to its needs. Its purchase and upkeep cost American consumers approximately $200 billion every year (1975 figures)—for many of them it eats up a quarter of their incomes—in addition to the estimated costs to the economy of traffic jams ($5 billion), accidents ($35 billion), road maintenance ($28 billion), and pollution ($7 billion). It is an instrument of necessity rather than desire, in most instances, for it is used for simple fun only a fraction of the time—according to a 1970 American Automobile Association survey, only 2.5 percent of the total auto mileage is used for vacationing, 3.1 percent for local pleasure driving. And it sustains a *radial society,* with a matching radial psychology, that is ever more centrifugal in its effects, distending the social fabric, pulling families and generations apart, spreading out neighbor from neighbor, ruining urban neighborhoods, creating suburban wastes.

I am constantly made aware of the personal impact of our nation's autocentricity because I live for half the year in Manhattan, where I can walk to almost any place I need to get to, and half in rural upstate New York, where I am dependent for almost all my needs upon a car. Moving from a foot-world to a car-world does more than simply involve one in an expensive and often unreliable new means of transportation—it somehow *creates* distance. In the country everyone in my family needs a car to do all the things they did in the city on foot: shopping, visiting friends, going to the movies or the library or a restaurant. Suddenly the distances that we have had to travel are changed from blocks to miles, multiplied not merely arithmetically but exponentially, for everything has been built with the car in mind. And worst of all is the fact that, when we have this enormous instrument that goes up to eighty miles an hour and can zip us around the landscape from this point to that, everything takes *longer* to do than it does when we travel by foot: shopping becomes a matter of hours rather than minutes, getting to the movies becomes an afternoon's expedition. The car, in short, adds miles and time, rather than shrinking them as it is theoretically supposed to do.

All of this is quite irrational, of course—and if the American motorist had to pay the full cost of the arrangement, its irrationality would become instantly manifest. If drivers had to pay directly for the cost of building and maintaining American streets and highways, for snow removal, highway patrols, and traffic courts, for traffic lights, bridges, and rush-hour police, for the costs of congestion, accidents, and pollution, for the losses to a city of having half its land off the tax rolls—if all these costs were added on to the price of fuel at the gas pump instead of hidden and absorbed by the society at large, the average gallon of gasoline, it has been estimated, would cost $7.50 more. In New

York City alone, public subsidies of this kind to the car user, things that the general taxpayer supplies for the benefit solely of motor vehicles, amount to $189 million a year, according to the careful figures of the Citizens for Clean Air in that city. Worse still: that $189 million comes to four times as much as the public subsidy for the mass transit system—although in New York only 17 percent of the commuters use automobiles, 83 percent use mass transit.

And therein lies still another problem with America's transportation. Mass transit systems of all kinds have, in almost every case, failed. They seem to be totally unable to pay for themselves, no matter how much they get in subsidies, and they are inefficient, inadequate, and unreliable. Subway and rail systems repeatedly lose money every year, whether in New York or San Francisco, and the cost today of building new lines is so astronomical, even with massive Federal subsidies, that it seems sure that the small networks just completed in Washington and Atlanta will be the last ever in this country. Bus systems, aside from being unmatched polluters, are generally inadequate in most cities and used primarily by those who have no choice: the poor. Railroads have consistently mistreated and discomforted their commuter passengers, even while charging ever-higher rates, and are going bankrupt.

There is no serious question about the ability of mass transit systems to move the greatest number of people for the least expenditure of energy, but they simply cannot seem to do it economically—not even when supported by public subsidies, which in 1975 amounted to as much as $26.93 per rider nationwide. The experience of San Francisco's Bay Area Rapid Transit seems typical enough. The most advanced mass transit system in the world when it began in 1972, it cost $1.6 billion and runs at a deficit of $40 million a year. Its stations are placed so far apart (to permit high speeds on the line) that passengers find it inconvenient to get to, up to 40 percent of its (brand-new) equipment is in the repair shop at any given time; and not surprisingly it has attracted only half the number of riders it needs to break even. To cover its deficits—it must spend about $6.80 now to provide a *single* rush-hour ride—BART will eventually be forced, like all other rapid rail systems the nation has known, to raise its fares, thus decreasing the number of riders, thus increasing the deficits, and thus forcing an increase in fares.

Ultimately, it is the city, of course, that must pay the fullest penalty for the chaos of the U.S. transportation pattern. It is here that the congestions impinge the most, that drivers waste countless hours, many of them spent in mental frustration and physical lassitude, creating the urban thrombosis that slowly deprives the city of its lifeblood. It is here that the pollution lingers, eating as much into human lungs as into public monuments, and in some places—like Riverside, California, for example—forcing people to live the greatest part of their lives indoors. It is here that businesses fold because people can't reach them, that neighborhoods disintegrate as families and factories leave, that communities are bisected with one or another new concrete or iron attempt to alleviate a problem by creating more of it.

It is somehow symbolic of the chaotic state of American transportation that flowers and shrubbery beside the urban freeways turn grey and die from the vehicle exhausts. In many cities, it seems, transportation agencies have made the decision to replace this real vegetation with plastic bushes and plants. Not because it is cheaper to do so, for in fact it costs about as much to keep the plastic varieties clean as it does to spruce up, or even replace, the real ones. Rather because it is held to be more "efficient," more "modern," and even—how telling—more "natural."

IN ONE SENSE, America's automania is "reasonable." For, given the spatial arrangements of America created by the preponderant use of the car, the car is the most sensible instrument to use to get around them. Since the car has created suburbs and scattered-site housing and low-density cities, the car is just about the only way to travel in and between them. Houston and Los Angeles and Phoenix don't have a lot of cars just because people *like* them—such cities were built precisely because of cars, to and for cars, and people *need* them.

A human-scale solution to the problem of the car, then, begins not with transportation but with settlements. The first and most essential way to deal with urban traffic, in other words, is urban design.

Now, traditionally, traffic experts have operated with one objective: to move people into and around cities as rapidly and efficiently as possible. Even some of the clearest thinkers—Doxiadis, Fuller, Soleri, Goodman—have sought to create elaborate mechanisms to whisk people from point to point. Speed becomes uppermost, and the fact that it is never obtained, no matter what contrivances the engineers make, never seems to deter them in their pursuit of it. Seeking the swift and unimpeded movement of people and machines, they do away with plazas and squares, remove traffic lights and put in beltways, eliminate intersections and construct cloverleafs. It sometimes seems that they may not rest until the most monumental traffic obstacle of all—the city and its impedimental buildings and people—is removed.

But of course that is no solution—in fact it's all backwards. The thing to do is not to move people through cities but to stop them there. Isn't that why cities are built in the first place? Isn't that why cities are traditionally located where they are the endpoints of traffic, both land and water? New York is built where a harbor offers shelter from an Atlantic voyage, where river traffic down the Hudson comes to an end, where land journeys from the southern slopes of the Adirondacks and the Catskills come to an end. New Orleans is built at the midpoint of the gulf that is the southern gateway to North America, and at the end of the

Mississippi River and its enormous valley system. San Francisco is built upon the harbor that ends the voyage across the Pacific, at the terminus of both the Sacramento and San Joaquin valley systems, where movement from the Sierra Nevada's western slopes reaches its focus.

Cities are meant to stop traffic. That is their point. That is why they are there. That is why traders put outposts there, merchants put shops there, hostelers erect inns there. That is why factories locate there, why warehouses, assembly plants, and distribution centers are established there. That is why people settle and cultural institutions grow there. No one wants to operate in a place that people are just passing through; everyone wants to settle where people will stop, and rest, and look around, and talk, and buy, and share.

Cities, in short, should be an end, not a means. Rationally one wants to have traffic *stop* there, not go *through,* one wants movement within it to be *slow,* not *fast.* Therefore, in thinking about urban design in connection with transportation, one might cleave to such concepts as these:

▪ Cities should not try to move people to facilities but provide facilities where the people are. Or as Barbara Ward has put it succinctly, the goal is "access not mobility." In a city built of roughly self-sufficient communities, each one should be able to provide all the important daily services within walking distance, thus eliminating the need for any but the thinnest ribbon of streets and freeing most space for pedestrian (or bicycle) paths or parklands. For the elderly and infirm, golf carts (miniature electric cars, actually) and multi-passenger electric tricycles would be a simple expedient.

▪ Cities should be small enough so that inter-community trips, when necessary, could be managed either on foot, by bike, or with some simple subway or trolley system. Obviously some cities today are so spatially unmanageable that getting around them by bike would be folly: these demonstrate the dangers of concentrating too many functions in a single center. What is wanted, instead of one large city glutted with all the attractions, is a series of smaller cities each with its own glitter and charm. This is the carnival principle: you don't put all the cotton-candy stands in one place and have everybody jam up there; you spread them through the grounds so that people will be dispersed.

▪ Cities should attempt to slow down the flow of traffic, particularly with plenty of squares and plazas and parks, places where wheeled vehicles are forced to halt, endpoints that invite stopping and resting. It is not for nothing that medieval cities had squares wherever one would go and that the most charming cities of the world even now—Salzburg, Venice, London, San Juan, Savannah—are those with repeated interruptions of streets. Of course squares are not good for traffic, and that's why most cities have removed them (New York City even tried to do away with historic Washington Square Park a few years ago), but they are good for people and for trade. It is around the square that people will stop and shop and sit and drink and learn the flow and pace of their community.

▪ Cities should try to bring home and workplace back together—a point so vital that it needs a little amplification.

FOR MOST OF urban history, it was assumed that people would work near where they lived. Tradespeople lived above their stores, craftsmen and their apprentices lived in the same building as their workshops, innkeepers lived in their taverns, professionals hung shingles outside their residences. Not until the Industrial Revolution at the end of the eighteenth century did this pattern slowly begin to change, and even then, even when factories employing hundreds of people drew laborers from all over, the workers normally lived within a short distance of the mill; the urban pattern for most of the nineteenth century was to have the workers' residential district chock-a-block with the industrial district. It was around the end of the nineteenth century that the real distention began. Trolley lines pushed the city and its industries outward, and factories that began depending on road transportation moved away from the congested centers. At the same time workers' districts in the heart of town were taken over by retail trade or by houses for the well-to-do, and workers were forced to resettle in isolated pockets in undesirable spots around the city and in new towns around the periphery, now usually many miles from their workplaces. (Such resettlement also had the desirable effect, from the point of view of the capitalists in charge, of reducing the workers' opportunity for common action, lessening the chances that workers from the same factory would live near each other, and thus dampening the labor militancy that had been so virulent at the end of the nineteenth century.) The divisions in the city thus created were solidified by zoning legislation, created at about this time, which assured that the residential sections would be kept far from the industrial and commercial ones.

But now that the workers were over *here* and the work was over *there,* urban transportation became a serious problem, and so mass transportation and urban highway systems had to be built to get them together. But then it was not long before the availability of the private car encouraged people to move even farther from their places of work, out deeper and deeper into the suburbs, extending urban sprawl in all directions by the postwar period. In effect, America decided to turn each of its cities into *two* cities, one large one for daytime use when everyone would congregate, the other a web of small ones for nighttime use when everyone would disperse; each would be left more or less vacant while the other was used, each would have to create its own lighting, sewage, road, telephone, police, education, medical, and political systems. And that meant again that the workers who were now way out *there* had to get to the jobs that were either still in *here* (particularly service jobs in commercial centers) or else all the way over on the *other side* of the city—hence the ever-growing networks of freeways and beltways and superhighways. Hence the madness of American transportation.

This story leads, I think, to only one conclusion: no solution of the transportation puzzle is possible until work and home are put back together (it is, after all, the comfortable condition of the nation's President) or at least within walking distance of each other.

It is obviously not impossible. Many small towns operate that way in America today—in fact, according to the last census, 2.7 million Americans across the country work at home, 5.7 million walk to work, and another 1.9 million use non-motor means, mostly bicycle, and that comes to a million *more* than use all of public transit in the nation. A number of the new towns built in Europe over the last two decades are also designed to operate that way, their success depending largely on their distance from a major city. Most successful of all, it seems, are the British New Towns, built precisely with the idea of providing walking-distance, or at the most short-hop, jobs. Not all were able to attract enough businesses to provide employment for all the residents, but they have done far better than comparably sized *unplanned* towns in England—and far better than American New Towns such as Reston and Columbia, which were built without this principle uppermost and which have degenerated essentially into bedroom suburbs. Barbara Ward has written of the British towns:

> Since each has achieved a reasonable degree of local balance between employment and residence, their commuting to the center [London] is still less than that of people still living in the scattered, unorganized "slurb" areas. According to estimates made in the mid-Sixties, less than ten percent of the New Town workers were commuting into London and the percentage seems to be shrinking.

Assuming, then, that urban design provides the essential answer to urban traffic problems, it may well be that with the harmonization of urban residence and urban employment will come the answer to urban design.

ANOTHER ASPECT OF a human-scale solution has to do with the transportation itself.

Obviously one desirable means of transportation would be something efficient, non-polluting, simple to manufacture and repair, energy-conserving, cheap, and harmless. It just so happens that such a means is available, indeed widely available, and has been for nearly a century: the bicycle. S. S. Wilson of the Department of Engineering Science at Oxford puts the case neatly:

> The contrast between the bicycle and the motor car is a very good illustration of technology of human scale. The bicycle is a supreme example of ergonomics—the optimum adaptation of a machine to the human body, so that it uses this power efficiently. Hence the worldwide success of the bicycle and its derivations in meeting the real needs of the people in both rich and poor countries, with a minimum demand for energy and raw materials or ill effect on the environment. The motor car, on the other hand, is a machine of inhuman scale as regards its size, its weight, its power (from 100 to 1,000 times that of the driver himself) or its speed.

In terms of translating energy into transportation, there is *nothing,* neither animal nor mechanical, that is superior to a human being on a bicycle; as the chart on the following page makes clear. Pity the poor mouse watching us ride by on our bikes.

Bicycles, of course, have been used for many years, in many parts of the world, and for many purposes, so their abilities and limits are well known. Most urban transport in Asia is by bike, particularly in China, where the private car is forbidden. The North Vietnamese relied primarily on bikes to move equipment along the Ho Chi Minh trail in the 1960s. Most cities in Africa, many in Europe, have more bikes than cars; in the Netherlands there are 8 million bicycles, only 4 million cars. Even in this country, there were estimated to be 95 million bicycles in use in 1979, with annual sales running at about 10 million (down from the peak of 15.2 million in 1973, the oil-embargo year, but well up from the average sales of 4 million a year in the 1960s).

And the possible developments of the bike—long overlooked in this country until quite recently because the Model T came along only a few years after bicycles became popular and quickly displaced them—seem to be almost endless. Victor Papanek, the designer, once set a team of students in Sweden to work on the various potentials of the bike, and among the other models they came up with was an ingenious three-wheeled machine with a seat in the back that could be used for hauling heavy loads, had a gear system so that it could be pushed uphill with ease even when loaded, could carry stretchers or planks, and could be connected in tandem to make a short train. Three-wheeled bikes are commonly used in China for carrying loads—and increasingly in this country in retirement communities—but the British foundation OXFAM has developed an especially efficient model that has three speeds, a differential drive, and a cargo space with a payload of 336 pounds. Various alternative-technology experimenters in this country—the New Alchemists, for example, at their Cape Cod headquarters, and the

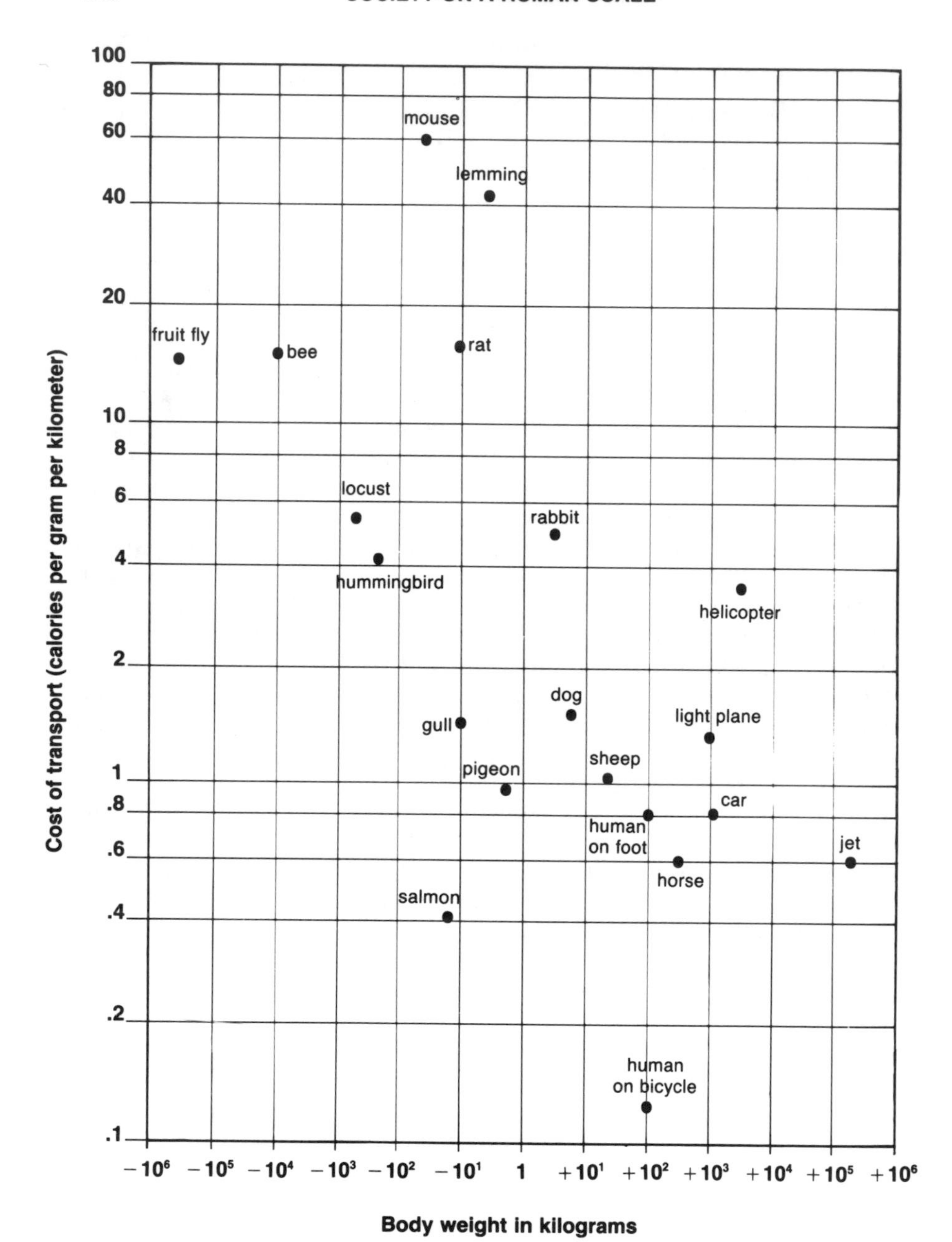

ENERGY EFFICIENCY OF DIFFERENT MEANS OF TRANSPORT

Rodale Press people at their farm in Pennsylvania—have even explored the uses of stationary pedal power, using bicycle pedals and drives to power grinding machines, corn mills, water pumps, washing machines, lathes, and similar machines; fixed bicycles, it turns out, can easily be used to generate electricity in short hauls, equivalent to a power output of 75 watts at a steady pace.

Moreover, bicycles can be readily equipped with little "booster motors" for traveling over difficult terrain or uphill, or for the elderly and infirm. The one- and two-horsepower gasoline engines that are now found on mopeds represent one kind, and they don't use enough gas or cause enough pollution to be much of an ecological threat. Even better, however, are electric bikes, which can get up to 40 miles an hour with no more than a twelve-volt battery attached, use no fossil fuels, and create no pollution. Such machines have been around for years—about 4,500 electric bikes are now manufactured each year around the world, 600 in the U.S.—but recently their designs have been improved and costs reduced. The Campbell bike, for example, invented by Dr. Peter Campbell of Cambridge, England, runs on a new disc-shaped electric motor that allows high torque even at low speeds and operates at a remarkable 75 percent efficiency, 20 percent better than most electric vehicles; the in-production cost is estimated to be less than $200.

Which leads us, of course, to the electric car. Nothing new about it, to be sure—it was common in the early automobile days, accounting for as many as one out of every three cars in 1904—but the refinements that have been made on it in just the last half-dozen years of accelerated interest show that it will very swiftly play a part in the American transportation scheme. Already there are more than a million electric vehicles in use, and though most of them are golf carts and forklifts, the number of on-the-road vehicles is estimated to be 10,000; 1977 production figures for the U.S., according to the estimates of *Electric Vehicle News,* showed 22,000 golf carts, 21,000 forklifts, 1,500 passenger cars, 640 trucks, 40 buses, and 7,000 other assorted carriers. There are electric buses now being used in Washington, Philadelphia, New York, and Long Beach, California; the Post Office has purchased more than 2,000 electrics for mail delivery and both AT&T and the Long Island Light and Electric Company are experimenting with electrics for bill delivery and meter-reading; the Electric Vehicle Council, the trade organization, calculates that there will be 100,000 electrics on the roads in the early 1980s.

The advantages are obvious. Electric vehicles do not pollute, they are quiet, simple (roughly twenty moving parts, compared to three hundred in the average internal-combustion engine), and comparatively cheap (they run at two cents a mile, half that of the normal car, and can be built for under $3,000). Although they don't go much over forty miles an hour and are just now limited to ranges of about fifty miles before recharging, they are ideally suited for urban travel—since in fact 60 percent of the car trips in urban areas are less than two and a half miles long. (The incredible backyard do-it-yourself enthusiasm that electrics have inspired, plus a recent Federal grant of $160 million, are also sure to improve the technology and performance of electric vehicles within a few years.) It is true, of course, that electrics do not necessarily cut the total amount of pollution, especially if they are charged from conventional currents, since that merely transfers the pollution from the streets to the central power stations. But it is certainly easier to contol pollution at that single source, even with available "scrubber" technology, and if the power station were solar-operated there would be no damage at all. Ultimately, the most efficient arrangement would be to have each car-owner or community recharge batteries with a windmill, or even for a windmill during its working phase to charge a series of batteries that could then be picked up and used as needed, not only for cars but for any other electric device.

It seems almost incredible, really, that the two most obvious mechanical solutions to urban transportation—the bicycle and the electric car—have already been given to us. They don't have to be dreamed up, or invented, or prototyped, or developed. They are here, they are well known, and they are widely available. They await only our intelligence to use them.

BICYCLING BURNS BETWEEN 250 and 600 calories an hour, depending on the speed and the weight of the cyclist. It is extremely valuable not only for weight loss but for muscle tone, particularly for legs, thighs, and stomach, and for heart-muscle exercise. What is one to make of a society that suffers from gross obesity and heart disease, that annually spends more than $150 million on dietary drugs and weight-loss gimmicks, that jogs and tennises and health-clubs with unending passion, and then turns its back on the bicycle as the principal means of daily transportation? Can a society that is willing to spend $419 on a "Vitamaster Motorized Exercizer" or $159.95 for a rowing machine, or, God forbid, $129.95 for a "Rotocycle" (nothing more than a bicycle with a stand in place of the back wheel so that it remains stationary) and still devotes a quarter of all its waking time to the care and feeding of the automobile—can such a society be regarded as completely rational?

ONE FINAL ELEMENT of human-scale transportation: inter-city locomotion.

At present, because we have created a system of multiple and interlocking dependencies, we use most of our transportation to ship things vast distances, at great cost, with enormous waste, to little point.

The average American, it has been calculated, depends each year on the movement of 7,000 ton-miles of freight. And why? So that (these are all actual examples) the wheat grown in Iowa can be shipped down to St. Louis to be made into bread that goes back to the supermarkets of Iowa; or so that the Dickinson jam company of Tigart, Oregon, can send its jams to customers in Roseland, New Jersey, while the Polaner company of Roseland is shipping its nearly identical jams to the customers of Oregon; or so that Florida tomatoes can be sent to Texas, and California oranges to Florida, and Washington apples to New York, and Idaho potatoes to Long Island, and California peaches to Georgia. Such strange hurry-scurry does provide a good deal of makework for a good many people—in 1975, for example, there were 6,871 inter-city truck haulers, most of them duplicating routes, facilities, and equipment—but looked at from the point of transportation sanity, energy economy, or resource efficiency, it does seem odd.

And the movement of people at present is almost as bizarre. Apart from commutation, most inter-city travel (in terms of passenger miles) is by airplane, about a tenth of it by bus, a fifth by railroad. The largest percentage of air travel, almost three-quarters, is for "business"—though it is difficult to know how much business cannot in fact be transmitted by letter, phone, or Telex: I have flown with an insurance executive who was flying from Indianapolis to New York merely to get a single piece of paper signed; with an oil-company engineer who was going back to Dallas after a meeting at the New York headquarters at which not a single piece of business transpired; with a Hollywood agent whose sole reason for going to New York was to see a play that might remotely have some bearing on his job; with an airlines executive who was taking a free ride to Chicago simply to check up on whether the pilots' on-time reports were accurate. No doubt there are some who find the speed and comfort of a plane important in their lives. But if they calculated it all, if they measured the time they really spend (an average of about one and a half hours of ground transport and waiting time for every hour of flying time on domestic flights), if they factored in the social costs (pollution, congestion, noise, accidents, deaths) of these elaborate air networks, they might well conclude that the frenetic game is not worth the kinetic candle.

In short, probably most inter-city traffic is basically unnecessary, or at least represents a luxury rather than a necessity. But we can take for granted that there are some instances in which the movement of people and objects over long distances would be valuable or even vital. What would be the most desirable instruments?

The trolley or electric train would probably be a sensible means—basically non-polluting, quiet, simple, and safe—providing that services weren't needlessly duplicated. Electric cars and trucks could supplement this service for the door-to-door pick-up and delivery that is important for things that are, for example, perishable or fragile. Water transport is another natural, ideal both for Europe and the U.S. since most major cities are already connected by rivers and canals (about 20 percent of American freight traffic now goes by inland waterways). Conventionally powered ships and barges are among the best forms of transport from an energy point of view, but sail is a practical alternative: what with modern sail designs using contemporary aerodynamic knowledge, hulls made out of the newer light-metal alloys, and such improvements as an electric winch system to raise and lower sails (reducing crew space and increasing cargo area), sail power could reduce fossil-fuel use to a bare minimum and still provide long-distance transport within a reasonable period of time. As to air transport, if necessary it might be wise to think in terms of dirigibles—LTAs, or lighter-than-air craft, as the aficionados call them—which use about a third of the energy of a jet airplane (and should be able to operate on solar, given their large flat surfaces), need very little space for takeoff and landing, are essentially silent, and are now regarded as safe, safer in fact than a conventional plane. Goodyear, which has had a lot of experience with dirigibles, has shown that they are cost-effective for carrying cargo almost any distance, and in Ghana a West German firm has been using them to transport both freight and people economically throughout the country for several years now.

But of course the variety of alternative transport is extensive—after all, there were a lot of machines produced between the second millennium B.C., when the first long-range sailing vessel was developed, and the contemporary fossil era. The particular alternatives are not so important as the principle behind them—of finding a reasonable alternative to the unworkable present. Given the impetus, there is no reason to doubt that it would be possible to create a sophisticated and efficient system of inter- and intra-city transportation that would not be dependent upon the car, the truck, and the airplane—and that would have a few inestimable side effects like clean air and community cohesion as well.

One British writer has reflected rather interestingly on what the effects of that might be:

> The key to alternative transport lies in the alternative *to* transport. As values change, mobility will be recognized as having been essentially an ingredient of the throwaway society, a temporary middle-class cult, which had some connection with an immature need for constant stimulation. And another connection will become apparent: that when people are able to buy mobility on the cheap, the difference between one place and another disappears: variety throughout the world is lost, and a Hiltonesque sameness descends over everywhere. Conversely, as travel and transport are made to bear their true social and energy costs, they shrink into their correct proportions. Each place then develops in

> its own way according to the differences inherent in its people, its geography, climate and other life forms.

And diversity, the scientists tell us, is the essential ingredient for the survival of a species.

It is Ivan Illich's contention that there have been four great revolutions in the history of transportation. The first was, of course, the invention of the wheel, sometime back in Neolithic times. The second was the simultaneous use of the stirrup, shoulder harness, and horseshoe during the Middle Ages to improve the efficiency of the plowhorse and permit greater crop rotation and more cultivation of European farmland. The third was the development of sturdy ocean-going vessels by the Portuguese in the fifteenth century, which opened up the globe for the extension of European capitalism.

The fourth was the invention in the latter part of the nineteenth century of the ball-bearing, the spoked wheel, and the pneumatic tire—and thus the creation of the bicycle.

10

Health: Heal Thyself

THE HORSESHOE CRAB happens to be a very sensitive animal, particularly in its reaction to various endotoxins of gram-negative bacteria that it may find in its ocean habitat. Its cells contain a coagulating agent that begins to work when it gets the message that an endotoxin is present, and that agent so clots the cells that the invading bacteria are immediately trapped and rendered harmless. However, something quite strange happens to this efficient mechanism when a horseshoe crab is injected with a small amount—no more than a microgram—of purified endotoxin in the laboratory. Then the animal, quite simply, overreacts, and instead of one cell or another coagulating to capture the endotoxin, all of them begin to do so, and the whole blood system quickly clogs up and slows down. Within a few hours, the crab is dead. There is no need, from a purely biological point of view, for the animal to react this way, for not all of its cells are attacked and it could easily take care of the invaders within the few cells affected; but somehow it gets the message that the endotoxin is everywhere and it responds throughout its whole body. Its benign defense mechanism, in other words, actually turns into a suicide mechanism. Dr. Lewis Thomas, one of those who has participated in this research, has commented: "The defense mechanism itself becomes the disease and the cause of death."

Just this sort of overreaction, it has occurred to me, characterizes the American health-service establishment. The parallel is by no means farfetched. Over the years American medicine has so blindly resisted new ideas and methods—everything from acupuncture to paramedics, refusing to undertake serious nationwide preventive medicine programs, rejecting community health-care systems (even health-maintenance organizations), performing like a petty sect of tribal priests—that it has become disastrously choked and coagulated through its whole body: no less than the President of the United States was moved to declare that doctors "have been the major obstacles to progress in our country in having a better health care system in years gone by." And a certain kind of suicide seems to be imminent. Unable to provide service to millions of people, unable to handle the growing number who do come forward for its services, unable to control the costs, the medical establishment is unable finally to do much in the way of making Americans healthier and more disease-free; even though something like $200 billion a year is now spent on medical services in the U.S.—more than $900 a person—it is estimated that one-third of the citizens are totally unserved by a doctor and another one-third inadequately served, and America continues to slip in the ranks of the world's nations in infant mortality (fifteenth in 1975), female life expectancy (sixteenth), and male life expectancy (thirty-fifth). The defense mechanism of the health-service system is, in a very real sense, the disease that is killing it.

There is even a sense in which it is true that American medical practices *cause* disease in this country. One does not have to agree with everything adduced by Ivan Illich to acknowledge the force of his argument about *iatrogenesis,* the process by which doctors actually generate illnesses. For examples:

- The American College of Surgeons and the American Surgical Association in a joint study admitted that about 30 percent of the surgical operations performed—that's about 4.5 million a year—are completely unnecessary, and an additional 50 percent perhaps beneficial but not essential to save or extend life; and that means, assuming only *half* the general operation-mortality rate, approximately 30,000 quite needless deaths a year.
- Hospitals are almost always overutilized—by a strange process known as Roemer's Law, which has shown that the *supply* of beds in a place increases the *use* of beds—and yet it has been shown that about 5 percent of all patients admitted to hospitals develop *additional* infections while staying there and 15,000 die from them every year.
- Every year at least 300,000 people suffer such severe reactions to drugs prescribed for them by doctors that they are forced to go for treatment to a hospital, 18,000 patients who are given drugs while in hospital die from the side effects, and some 10,000 people suffer life-threatening reactions from various prescribed drugs in the antibiotic category alone. Add to that the sicknesses caused by poorly run hospitals, poorly educated or overbusy doctors, outright malpractice, the malfunction of technical equipment, and the special drug-oriented blindness of American medicine, and it is not hard to imagine that there is a strong, if largely hidden, case to be made for current medical practice being as much a threat as a cure.

Indeed, all that helps to explain the otherwise quite astounding finding—by John and Sonja McKinlay of Boston University, in a monograph published in 1977—that, all other things being equal, the *fewer* doctors there are in a population the *lower* is the mortality rate. And the otherwise quite confounding evidence that whenever there is a doctors' strike—as there have been in recent years in the U.S., Canada, England, and Israel—the death rates in the affected areas actually *fall.*

BUT A CATALOGUE of the ills of modern medicine is hardly necessary—the criticisms are many and trenchant, they have been made often before, and in fact most of them are well known to the medical profession itself (e.g., a compilation called *Doing Better and Feeling Worse,* a medical self-examination that came out in 1977). Indeed, anyone who has sat in an emergency-ward waiting room of a major city hospital, or who has even had to pay a few non-covered medical bills recently, knows a good deal of what's wrong. We might, however, be aware of the habits of mind and culture that have come to make us dependent upon this clumsy behemoth.

The American medical system today is a product of three intertwined beliefs that have become ingrained in the American ethos:

1. that the human body goes around pretty much at the mercy of Darwin's nature-red-in-tooth-and-claw, particularly the little microbes and bacteria that you can't even see, constantly out to get us;

2. that it is given to human beings to conquer nature and its little pests (that anti-ecological attitude again); and

3. that some sort of technofix, either mechanical or chemical, can always be devised to accomplish just that.

None of these is true. The human system, after having lived with microorganisms for millions of years, has developed built-in defenses and responses to most of them, even the unexpected; as Dr. Thomas points out, "It is certainly not true that they are natural enemies" and in fact "it comes as a surprise to realize that such a tiny minority of the bacterial populations of the earth has any interest at all in us." Nor is it true that we have any real way of "conquering" the effects of these bacteria on any very large scale, even for the diseases we presume to understand, much less those that remain mysterious (like cancer) or emerge suddenly (like Legionnaire's Disease); moreover, all our recent experience has shown that efforts to do so will entail an enormous expenditure of time, money, and resources, are likely to prove fruitless, and will probably produce an unforeseen range of side- and after-effects. And although medicine has indeed come up with a few technofixes in the past—notably penicillin and smallpox vaccination on the chemical side, dialysis machines and heart pacemakers on the mechanical—there have been far fewer of these than is generally supposed and their overall effect on American health far less than is popularly portrayed.

For the fact is that it is *not* modern medicine that has improved the health of Americans significantly in the last eight years—it is rather, quite simply, improved nutrition and improved hygiene. In almost every case—particularly scarlet fever, tuberculosis, typhoid, measles, streptococcus, bronchitis, pneumonia, and pertussis (whooping cough)—the mortality rates had shown their major sharp declines well *before* any medical therapies or drugs came along; to take measles as an example, the mortality rate declined from 13.3 per 100,000 population in 1900 to 0.3 in 1955, prior to the licensing of a measles immunization vaccine. And the reason for the decline has in greatest measure nothing to do with medical "breakthroughs" or even preventive measures but rather with better sanitation and water facilities, in the case of air- and water-borne diseases, and with increasingly healthful diets, in the case of most others. In their important study, Sonja and John McKinlay found that between 1900 and 1973 there was a 69 percent decrease in the overall mortality rate in the U.S., almost all of it attributable to the declining death rates of infectious diseases; that in turn, they found, was accounted for by the introduction of specific medical measures in 3.5 percent of the total cases, while the general improvement in the standard of living during that period was responsible for the remaining 96.5 percent. The much-praised vaccines of the twentieth century? Except for tetanus and polio shots, the impact of immunization on the mortality rate has been virtually negligible, probably offering even briefer and less effective protection than going through the disease itself. And our vaunted antibiotics? Studies in Sweden, generally accepted here, indicate that death rates from the major bacterial diseases (with the exceptions of syphilis, certain types of meningitis, and septicemia) all decreased at exactly the same rates even after the introduction and widespread use of antibiotic drugs.

But what about life expectancy—isn't *that* at least a triumph of modern medicine? Well, no. Aside from America's dismal standing compared to other nations, it turns out that the only significant improvements in life expectancy have come about by the decreased mortality rates of childhood diseases noted above plus better obstetrical and neo-natal care for infants; life expectancy *past the age of 45* has not increased appreciably in this or any other country in the world in the last eighty years. Not only that, but the gains in expectancy-at-birth that have been made in this century—from 47 years in 1900 to 71.9 in 1975—have actually stayed roughly static over the last 25 years (the 1955 rate was 69.6 years) and seem likely to remain so; in the words of Dr. Robert P. Whalen, New York State's Commissioner of Health:

> There has been a relatively small gain in longevity over the last decade, when health spending more than doubled and government spending for health quintupled. Most experts agree that further gains in life span seem unlikely in the near future.

Moreover, it is a solemn fact that modern medicine is not even capable of having an effect in any but the smallest area in the field of health, paradoxical as that may sound. Research by Aaron Wildavsky, head of the Russell Sage Foundation, indicates that the ministrations of doctors account for less than 10 percent of an individual's well- or ill-being, no matter what the economic status—more than 90 percent is determined by factors over which doctors exert no real control whatsoever (they could, of course, but they don't), including personal eating habits, smoking, exercise, and stress; the healthfulness of the air, water, and food ingested; and the conditions of the workplace. Health, in short, is really more a personal and political than a medical matter. In the words of Herman Somers, a leading health specialist, "the greatest potential for improving the health of the American people is probably not to be found in increasing the number of physicians or hospital beds but rather in what people can be motivated to do for themselves." And not merely in their own bodies, but their own communities.

In sum, though no one wishes to deny its significant achievements, the American medical system must be kept in careful perspective, and its Ben Casey mythology gently amputated. It is really irrelevant to the basic problems of the health of the citizens, it is largely inadequate and unresponsive, and in some very real ways it positively threatens the well-being it so genuinely wants to safeguard. And it may even be moving backwards, as crabs do.

A HEALTH SYSTEM in a human-scale society, then, would ask that at least four associated processes be at work.

Nutrition. The very first priority would be to assure beneficial nutrition, since we know that much of the reduction in disease morbidity and mortality in our time is due to improved food habits. Imagine how much further our health could be improved if some sort of agricultural self-sufficiency were common and people had access to locally grown food that would not need to be processed or refrigerated and would arrive fresh on the dinner table at its nutritional peak. Or if our food were free of all the 2,000 unnecessary additives—some of them highly suspect as causes of disease, and the majority not even tested—now injected into it. Or if people had as equal and automatic access to nutritional as to junk foods, at equal prices, and with equal information. The staff of life itself may be symbolic here. At present most bread is made from white flour, from which 90 percent of the nutrients (including all the vitamin E, most B vitamins, most of the protein, and many minerals) have been processed away, and to which is added a great variety of substances (softeners, whiteners, preservatives) and enough vitamins to replace only about a tenth of the nutrient value that has been lost. (Dr. Roger J. Williams of the University of Texas not long ago kept one batch of experimental rats on a diet of white bread and another on white bread plus vitamins and minerals. All of the supplement-diet rats were alive and growing after ninety days; one-third of the bread-only rats were stunted, two-thirds of them had dropped dead.) It would be far staffier, and do more for life, were it simply baked locally from local whole-grain flour—a process not so difficult that our grandparents couldn't manage it.

Living patterns. The second step would be restyling a few of the basic patterns of life to prevent most of the general, non-microbial causes of disease, the 90 percent the Wildavsky study found were outside a doctor's purview. Probably no single change would be more beneficial, particularly against heart and respiratory disorders, than strenuous exercise, through the increased use of bikes, encouragement of hands-on labor, regular opportunities in the community gardens, and emphasis on participatory instead of spectator sports. Relocation into optimal-sized cities and into rural areas would significantly reduce the stress of life as given to us by the big cities, which creates by itself various heart and nervous-system ailments and which leads on to those debilitating companions of stress, alcohol and tobacco. One could also imagine real protections being taken against job-related sicknesses—black-lung disease in coal mining, brown lung in textile production, cancer in asbestos plants, and so on—if workers themselves were to set the policies of the workplace and communities had some say in economic priorities. And similar worker and community decision-making could lead to serious controls over industrial and municipal poisoning of air and water, which by itself would probably reduce the incidence of cancer in time by 70 to 80 percent.

Above all, smaller and less crowded settlements could simply *avoid* a good many diseases because, it turns out, most disease microorganisms need a large population in order to sustain themselves—as in the case of measles, which needs nearly half a million people in which it can operate before it can perpetuate itself. Interestingly, anthropological studies examined in some detail in William McNeill's *Plagues and Peoples* suggest that Paleolithic hunter-gatherers, operating in bands of fewer than a thousand, were apparently completely free from infectious diseases, or at any rate from those caused by microorganisms developed specifically to live off humans; it was not until large cities grew up that regular infectious chains were established.

Self-help. A third element to a rational system would be a reduction in the dependency on both doctors and drugs.

The extent to which quite ordinary people with quite limited training can tend to most everyday medical problems and often to even sophisticated and complex medical tasks must not be underestimated. The North Vietnamese, it is reliably reported, were able to teach lay people to perform complicated eye operations even in the middle of an unstabling war. The Chinese system of "barefoot doctors" seems to be agreed upon—whatever one may think about the Maoist superstructure above it—as highly successful, and the job these minimally trained people did in setting up community clinics, teaching preventive techniques, and guiding campaigns to rid the country of such diseases as schistosomiasis as something close to a human miracle. China today still depends upon a million such doctors, one for every 800 people or so, who divide their time between preventive measures in their community, visiting fields and factories to check on health conditions, and operating clinics for the needy. Even in this country, similar lay operatives have shown themselves to be medically adept: paramedics, for example, with less than a year's training, have been used with considerable success in all but eight states of the union to provide front-line medical care (there were some 266 paramedic training centers in operation as of late 1978); and nurses or "physicians' assistants" with no more than two years of medical teaching now operate what the *New York Times* estimated in 1978 to be "several hundred" primary care clinics across the country, with more to follow.

Beyond that, the value of the simple and most elemental medical treatment—self-help—has been proven incontrovertibly. The remarkable growth of the women's health movement in the last decade, for example—symbolized by the quarter-million sales of the do-it-yourself book, *Our Bodies, Ourselves*—has shown that women the country over are capable of taking care of many of their own health problems without professional intervention. The so-called "holistic" health movement similarly has shown that people who work at controlling the health of their own bodies—eating natural foods, in limited amounts, fasting from time to time, exercising rigorously, meditating daily or through times of stress, using vitamin therapies, and the like—are unquestionably freer of disease than the general population. As a matter of fact even the most quotidian kind of self-help that most people practice when they get sick—take two aspirin and don't call anybody in the morning—turns out to be probably the best treatment in the great majority of illnesses; studies in both Great Britain and Denmark have shown that the largest percentage of people begin to treat their ailments correctly even before they visit a doctor, and in most cases the doctors merely continue that treatment.

As to overdependence on drugs, there is not a single voice within medical circles or without that denies that far too many prescription drugs are being administered these days—340 million prescriptions in 1975, to perhaps 150 million people—and few that would disagree with the finding reported in *Scientific American* that 35 to 45 percent of these drugs are prescribed for conditions in which they are totally ineffective. There are many voices, too, that argue that most prescribed drugs could simply be eliminated from the pharmacopoeia and, where necessary, substituted for by natural, or "herbal," elements—since, after all, that's what most of the laboratory drugs were based on in the first place.[1] In addition, there is a great variety of *non*-drug therapies around, ranging from acupuncture and homeopathy, both proven for many decades but resisted by the medical establishment, to yoga and other Eastern psychosomatic techniques, whose results are sometimes impressive even if their workings still lie beyond the understanding of Western medicine.

Facilities. Lastly, after such steps as these, then and only then might one turn to consideration of medical facilities themselves. Obviously far fewer would be needed than now—as a matter of fact the U.S. has the equivalent of at least 700 more hospitals at the moment than it can properly use—but obviously some centers would probably be necessary. On the neighborhood level it might be advantageous to have one or two lay medical advisors with rudimentary training operating out of their homes to offer elementary advice and comfort, rather like self-starters for self-help. At a community level, a small clinic with a nurse or a paramedic would easily suffice—experience under the present system has shown that a population of 5,000–10,000 can support a small facility, with patient visits averaging between ten and twenty a day. Just such a primary-clinic system operates right now in several states, notably North Carolina, where twenty-one facilities are staffed by nurses and paraprofessionals serving small-town and rural populations; the clinics have proven to be far more convenient than big-city hospitals, they offer quicker service, they treat their patients on a first-name basis, and they have a full range of doctors all over the area for higher-level consultations if necessary. This is the level, too, at which the best kind of care can be given for both mental and geriatric patients, since they benefit particularly from intimate treatment. One English study reported in the *Journal of the American Geriatric Association* found that "size seemed to be the major factor in success or failure of a given unit," with twelve- and fifteen-place geriatric hospitals proving far more successful than twenty-eight-place and larger units; and another researcher, Brian

---

1. It is well enough known that digitalis, used nowadays to regulate heart rhythms, is derived from the purple foxglove, an ancient Welsh medicine, and that rauwolfia, used to control high blood pressure, comes from a kind of snake root used for centuries as a sedative in India. It is less well known that there are effective herbal contraceptives, like the Chinese concoction from pine-tree sprouts, good for three years, and the South American Indian mixture reported by Nicole Maxwell, in *Witch Doctor's Apprentice,* that lasts for seven years unless a contra-acting fertility potion is taken. You can understand why Western drug companies have shown little interest in such long-acting agents to replace daily-pill dependency.

Abel-Smith of the London School of Economics, has concluded that "there may be positive advantages in certain types of psychiatric hospitals being small and intimate to avoid the rush and authoritarianism which tend to be the characteristics of general hospitals throughout the world."

For larger populations larger facilities would naturally be required, but as might be expected there is a point here too where the Beanstalk Principle operates. Studies of optimal-sized hospitals are for some reason notoriously unreliable—one of the ablest investigators, Sylvester Berki, has simply thrown his hands up and declared the whole matter intractable—but there are a few established facts, and they tend to favor hospitals of 100 to 200 beds (considered to be the upper range of the "small" hospital). Economies of scale in purchasing and supplies begin to operate over 100 beds and decline somewhere after 600 beds; but hospitals of more than 300–400 beds begin to show the strains brought on by problems of staff administration, recruitment, and communications; they suffer from all of the bureaucratic deficiencies any large organization experiences; and because they inevitably demand greater travel time they are consequently used by a decreasing percentage of the population. Moreover, hospitals of 100–200 beds in the U.S. today would seem to be as close to the optimum as we can get in practice with current standards: they have more admissions than any other sized facility, smaller or larger, see more outpatients and attend more births than any but the giant 500-plus-bed medical centers, and yet they are more efficient in terms of expenditure per patient than all other larger units, twice as efficient as the largest centers.[2] And on the less quantifiable side of things, it appears that the smaller hospital has certain intangible advantages—in the words of the trade journal *Hospitals,* the smaller hospital has "this talent . . . to make its patients feel at home, cared for by friendly, attentive staff," and practically all reports agree that is of major significance in restoring most people to health.

And the population necessary to sustain a hospital of 100–200 beds? Both in Europe and in America such facilities have been shown to be most efficiently used and sustained in cities of 30,000–60,000 people.

2. Statistics may be found in any issue of *Hospitals* magazine; these are from September 1977:

Beds	Admissions	Outpatients	Births	Length of Stay	Expenditures
25–49	118,000	573,000	10,000	7.6 days	$102,000
50–99	326,000	1,622,000	26,000	9	372,000
1–200	618,000	3,490,000	55,000	9.7	804,000
2–300	538,000	3,046,000	53,000	10.9	796,000
3–400	405,000	2,851,000	42,000	11.6	650,000
4–500	308,000	1,953,000	33,000	11.7	536,000
500+	563,000	4,131,000	61,000	12.5	1,245,000

Extensive findings based on the comprehensive Chicago Regional Hospital Study in the 1960s indicate that, taking in a wide variety of variables—including economies of scale, travel time, degrees of care—the optimum patient population for any given center is somewhere between 20,000 and 40,000. And the Rutgers Community Health Plan has shown that health-maintenance centers—called HMOs in this country—seem to operate with an optimum enrollment of 32,000. Not to overdo magic numbers—but the small city comes through once again.

Above this optimum level for everyday facilities, it may be desirable to have certain highly specialized medical services, not large in themselves—in fact normally fewer than 50 beds—but which could serve large populations of 500,000 to 1.5 million or more through some sort of regional cooperative arrangement among a group of communities and cities. Crude forms of such an arrangement operate in a good many places today (New York City, not having a burn center, sends its burn patients to other Northeastern facilities, and northern Minnesota has a network system for kidney dialysis and transplant treatment) but in general each city tends to set up its own high-priced clinic with its own high-technology machinery even if it is consistently used way below its capacity. (A 1969 survey established that fully 30 percent of the 777 hospitals equipped for closed-heart surgery performed absolutely no heart operations during the year, and of the 360 hospitals with equipment for open-heart surgery only 4 percent used it at the optimally efficient level of four-to-six operations a week while 77 percent used it less than once a week and 41 percent less than once a month.) It would take not millions of dollars, nor millions of people gathered in a city, but only the simplest kind of cooperation and a very elementary form of ambulance transportation to create and maintain regional medical facilities that genuinely and rationally served a region.

Serious questions must intrude here, however, as to whether a society at one with its health would necessarily wish to invest huge resources in such things as burn centers and dialysis machines and CAT scanners and the rest, high-technology machines that artificially sustain life; or whether a society in ecological equilibrium could justify the construction and operation of such machines at great waste of resources and energy for the prolongation of a comparatively very few lives. Ivan Illich would answer—as he does in his *Medical Nemesis*—that such machinery is never morally justified and that an adjustment to the inevitable process of dying and death is one sign of a healthy society. Yet it might also be said that, insofar as such technology can be kept within certain community-established restraints—ecological, economic, ethical—there is no better purpose it could serve than the saving of human lives and limbs. Ultimately this would seem to be a decision that any given settlement or region might decide on its own terms and depending on its own priorities.

THOSE ARE THE obvious steps to a rational and human-scale approach to health, and though they may seem in some regards utopian, they are all really very easy and well within the bounds of our capabilities. Just as is the utopian vision offered by Ernest Callenbach in his enlightening little novel, *Ecotopia.*

Callenbach imagines a time in the not-too-distant future after a new nation carved out of the Pacific Northwest from Santa Barbara to Canada secedes from the United States and establishes a decentralized, humanistic, and ecologically minded government. William Weston, a New York journalist, is invited to Ecotopia to report on its strange customs, and among the dispatches he sends back is this one on health care:

> The greatest difference between Ecotopian hospitals and ours is in scale. Though the medical care I have received seems to be at the highest level of sophistication, from the atmosphere here I might be in a tiny country hospital. There are only about 30 patients all together, and we are practically outnumbered by the nursing staff (who, by the way, work much longer hours than ours, but in compensation spend as much time on vacation as they do on the job). . . .
>
> In one respect the Ecotopians have taken a profoundly different direction than our modern hospitals. They do not employ electronic observation to enable a central nurses' station to observe many patients at once. The theory, as I have gathered it, is that the personal presence and care of the nurse is what is essential; and the only electronic gadget used is a small radio call set that can retrieve your nurse from anywhere on the hospital premises without bothering anybody else. . . .
>
> The clinics and hospitals are responsible to the communities—normally to the minicity units of about 10,000 people. Thus the power of the physician to set his own fees has evaporated, though a doctor can always bargain between the salary offers of one community and another, and in fact doctors are reputed to have among the highest incomes despite the fact that they are much more numerous than with us. Doctors perform many duties that nurses or technicians perform in our more specialized system; on the other hand, nurses and technicians also perform a good many of the services that our doctors reserve for themselves.
>
> Intensive-care units are also not developed as highly as in our hospitals. This clearly involves a certain hard-heartedness toward terminally ill or very critically ill patients, who cannot be kept alive by the incredibly ingenious technology American hospitals have. This may be partly an economic necessity, but also Ecotopians have a curiously fatalistic attitude toward death . . . when they feel their time has come, they let it come, comforting themselves with their ecological religion: they too will now be recycled.

> On the other hand, the Ecotopian medical system has a strong emphasis on preventive care. The many neighborhood clinics provide regular check-ups for all citizens, and are within easy reach for minor problems that might develop into major ones. No Ecotopian avoids getting medical care because of the expense or the inaccessibility of health facilities.

On second thought, that's not so utopian at all.

11

Education: Big School, Small School

IT IS THE THEORY advanced by Roger B. Barker, professor of psychology at the University of Kansas and director of the Midwest Psychological Field Station, a research center there, that it is possible to gauge the effects of size in any given institution by examining what he somewhat gracelessly calls its "behavior settings." If there are a lot of people in any one setting, he says—a meeting, for example, or a classroom—then each person has less influence on it, less chance to participate in it, less sense of responsibility for what goes on within it. If there are only a few, the chances are that each person involved will participate more, influence the events more, and have more intense reactions to what goes on.

Barker and his colleagues have tested this theory in a great variety of situations over the years—in small towns, churches, factories, even jails—and it has been borne out with striking regularity. The setting with fewer people, they have found, because it is "undermanned," calls forth a much greater response in the participants. They work harder toward the desired goal, are less judgmental of their fellows and more cooperative with them, and end up with a greater sense of individual importance and self-worth. And although there usually have to be somewhat lower standards in settings with only a few people—a small high school orchestra, for instance—and hence lower levels of performance within them, the chances of the group's ultimately succeeding at what they are doing are greater. There is, in sum, what Barker calls "a negative relationship between institutional size and individual participation."

Probably the most striking work that Barker and his colleagues have done to date was their research on educational settings. In the late 1950s they spent three years studying thirteen high schools in eastern Kansas, some small, with only about forty students, some quite large, with two thousand and more pupils. They drew up lists of "behavior settings" for each institution—athletic contests, for example, classes, the school play, school elections—and then through careful observation and by subsequent surveys and interviews measured the amount of student participation in, and satisfaction with, each of them. Their findings, published in 1964 as *Big School, Small School*—by now a classic in American psychology, though largely ignored by American education—were consistent and unqualified:

> The large school has authority: its grand exterior dimensions, its long halls and myriad rooms, and its tides of students all carry an implication of power and rightness. The small school lacks such certainty: its modest building, its short halls and few rooms, and its students, who move more in trickles than in tides, give an impression of a casual or not quite decisive educational environment.
>
> These are outside views. They are illusions. Inside views reveal forces at work stimulating and compelling students to more active and responsible contributions to the enterprises of small than of large schools.

For school activities, particularly in music, drama, journalism, and student government, "participation reached a peak in high schools with enrollments between 61 and 150," and was anywhere from three to twenty times as great as in the largest school. Although the large schools offered slightly more extracurricular settings—chess club *and* yearbook, volleyball *and* tennis—"the small school students participated in the same number" of activities as the students of the large school, and they were more likely to hold "positions of importance and responsibility" in those activities. The small school students reported that they were more aware of the attractions and obligations of non-class activities, and also felt "more satisfactions relating to the development of competence, to being challenged, to engaging in important actions, to being involved in group activities, and to achieving moral and cultural values."

As to classroom work—though, interestingly, this was found to account for only about 20 percent of the school's behavior settings—the findings are similar. Although big schools, as one might expect, did offer more kinds of subjects, "the large school students participated in fewer classes and varieties of classes than the small school students." In the case of music classes, studied in detail, it was found that the big schools produced more specialists—since individual students didn't have to learn several genres and could concentrate on a single instrument—but that "musical education and experience were more widely distributed among the small school than among the large school students."

The size of the town in which the school was located, the Barker team discovered, also turned out to be educationally important, for in the smaller places the students participated more in non-school affairs and the community members participated more in school affairs. Comparing the settings of the smallest schools—communities of 1,000 to 2,000—with those of the largest—a city of 101,000—Barker found "a wider participation by the adolescents in the business, organizational, religious, and educational settings of the towns than of the city." Like the small schools themselves, the small communities "provided positions of functional importance for adolescents more frequently," the cities less frequently. Moreover, although the urban environments presumably offered museums and theaters and other resources that the small towns lacked, this did not seem to make up for the "relatively meager" use made of them by the big school students.

In sum, the Barker study gave significant support to the idea that, the myths of most modern educators to the contrary, the small school actually best nurtures the values that educators have long sought to foster. How small a school? Well, Barker's group, being academics, naturally shied away from naming any particular optimum numbers, but it was emphatic that "a school should be sufficiently small that all of its students are needed for all of its enterprises"—and that condition, its fieldwork shows, pertains generally at enrollments of roughly one or two hundred students.

Just about, in fact, what common sense would suggest.

AND YET the trend in American education over the past three decades has been toward bigger and bigger schools, larger and larger school districts. In 1950 there were 139,000 elementary schools in the U.S.—60,000 of them small, one-teacher affairs—serving an enrollment of 21 million children. By 1975 there were only 79,000 schools, barely a thousand of them one-teacher size, although the enrollment had risen to 32 million. In other words, the average school used to hold 153 children; now it holds 405. Similarly, though there is not the same sharp decline in the number of high schools, the average enrollment per school has gone up from 229 to 543, and in metropolitan regions to nearly twice that (and the New York City average for academic high schools is 3,344). In countless areas of the country "consolidated" school districts have grown up to displace local community-run districts, tearing down the smaller schools and busing the children to large central schools, with professional educators all the while saying that this was the "modern" way, the "most efficient" system.

Probably the most influential underpinning of this process was the Conant Report, issued in 1959 with all the authority of one of education's most prestigious figures, former Harvard University president, James B. Conant. It came down unequivocally on the side of large consolidated districts and large centralized high schools, determined that any high school under 750 pupils could not offer a sufficiently comprehensive curriculum, and recommended as its most important step the elimination of the small high school in America. That there was no supporting data in the report for such claims seems surprising, inasmuch as this was supposed to be an academic document; that contrary evidence available even then was ignored seems more surprising still; but that the education profession accepted these dicta without doubt or questioning is most surprising of all. Yet in district after district across the country the heavy weight of the Conant Report was used year after year to beat down those voters who clung to the idea of small schools and local control and to shore up those who favored the modernity of larger institutions.

And what has happened to the quality of education in all these years of bigger and bigger schools? As we all know, it has been declining precipitously, worst of all in those big cities where the school populations are largest and school sizes the greatest. The record is woefully clear:

- Schools are not teaching the most elementary skills. Although nearly 90 percent of our children have gone through twelve grades of schooling over the last twenty years, functional illiteracy in the population has not declined in the least and seems even to be growing. Surveys indicate that at least 27 million Americans over 16 cannot read even the simplest English phrases—newspaper want ads, job applications, cooking directions—necessary for normal functioning in contemporary society.
- Academic achievement is declining. Between 1966 and 1977, for example, the number of people scoring above 600 (800 is top) on the verbal part of the Scholastic Aptitude Tests has declined by 36 percent, on the math part by 11 percent. Textbooks are now routinely written for students at a level *two years* below the grade at which they are to be used. A survey of more than half the entire primary and secondary school population of the U.S. in the late 1970s found that "average achievement scores for all grades above third or fourth not only fell, but fell simultaneously, and each year the same children dropped farther and farther behind."
- Two Vs—violence and vandalism—have replaced the Three Rs. Schools, particularly in the big cities, have become places of mayhem and terror. Cases of aggravated assault—meaning real physical damage, to either students or teachers—now run at more than 150,000 a year, and some teachers carry handguns. Vandalism is said to cost school districts at least $600 million a year.

American public education is simply not working, even on its own terms. It is not preparing its students adequately for the jobs they are supposed to take over in the larger world, particularly since more and more of those jobs demand special skills. It is not providing the young with the understanding of politics and the language of citizenship upon which a representative democracy is supposed to depend. It is not passing on the great cultural traditions of the West to more than the barest handful—Greek and Latin have not been taught for years, grammar and rhetoric are all but forgotten, politics and history have been replaced by "social studies." It is not pacifying and channeling the young, socializing them to accept the world around them without complaint, and indeed the amount of violence and open rebellion suggests the opposite. And it is not even—to measure as capitalism and many current educators do—economically efficient, considering that some $820 billion has been spent on it in the last twenty years and its quality declines without abatement.

(It does, of course, make manifest the value of competition and compartmentalization, the virtues of obedience and submission, the importance of grades and tests, and the arts of self-promotion and apple-polishing, none of which are lost on very many of the young. Yet I would submit that an education system in which this is the primary achievement is probably not worth the space it occupies.)

Perhaps not all of this decline can be laid to bigness alone—no doubt other elements of the postwar world, from family disintegration to social anomie, have contributed. But coincidence certainly suggests cause, and the deficiencies of bigness are so glaring that they clearly offer themselves as a major if not exclusive reason for the deterioration of the schools. Moreover, that this deterioration has taken place while other key educational elements have gotten better—student-faculty ratios have actually improved over the last decade, and money spent on education has markedly increased—lends support to this explanation.

It does seem rather interesting that the big schools nowadays are all working rapidly, at great expense and with much hoopla, to implement such innovations as open classrooms, peer-tutoring, multi-grade classes, individualized instruction, and community participation—the very things that existed *inherently* in the small school in the small community, particularly the one-room schoolhouse, and that were so often lost in the pell-mell rush to Conant-sized institutions.

A HUMAN-SCALE EDUCATION, thus, would obviously begin with the small school.

In addition to Roger Barker's pioneering work, there have been a host of later studies underscoring the virtues of small size. Allan Wicker, for example, has reported that the "cognitive complexity" of students in smaller high schools—with junior classes of twenty to fifty members—was "significantly higher" than those where junior classes numbered four hundred or so. Leonard Baird, in an unusually wide sample of 21,371 high school students, found that "high school size has a considerable effect on achievement," with smaller sizes being particularly advantageous for both academic and non-academic success in writing and dramatics, of some advantage for music, and without much effect on science and art. And another pair of researchers was able to conclude simply: "Higher achievement results correlated with smaller schools at both elementary and senior high school levels."

Perhaps even more compelling is the unequivocal finding that emerged from the famous report by James Coleman and his colleagues in 1966 *(Equality of Educational Opportunity),* based on a U.S. Office of Education survey of 645,000 pupils, the most extensive educational study ever conducted. They concluded that the major determinants of classroom success had very little to do with the actual content of courses or the amount of educational equipment or even the competence of the teachers, but rather with "the attitudes of student interest in school, self-concept, and sense of environmental control"—in other words, with the student's sense of being at one with the school. Moreover, the factor "which appears to have a stronger relationship to achievement than do all the 'school' factors together is the extent to which an individual feels that he has some control over his own destiny." It is exactly that control, that sense of oneness, which the small school has always been able to foster, simply by virtue of its size in relation to the individual student; it is exactly that which has been lost in the urban blackboard jungle.

That sheer *size* should play such a significant part should not be all that surprising. We all know that a young child going into a new school almost always feels nervous, and we know that when the school is large and the staff distant and the surroundings impersonal the nervousness can develop into real terror and dread, feelings that some children never get over even when they learn to get through the day. There is probably a biologically set level to which a child can adapt, beyond which a feeling of stress begins to accumulate and mental stability falters, and the younger the child the smaller are those levels: first the family, then the playgroup, next the intimate nursery class, much later the large classroom and the larger school, and only with full development the community at large. The World Health Organization once observed:

> So many wise books and reports have appeared in recent years on the principles which should be followed in organizing institutions for children that little discussion is called for here. All are agreed that institutions should be small—certainly not greater than the 100 children suggested by the Curtis Report—in order to avoid the rules and regulations which cannot be avoided in large establishments. Informal and individual discipline based on personal relations, instead of impersonal rules, is possible only in these circumstances. . . .
>
> There is no difference of opinion regarding the size of the group [within the institution]; all agree that it must be kept small.[1]

1. Indeed there is evidence that smaller schools can actually be of help with students who are regarded as mentally "maladjusted" in larger ones. Arthur Morgan *(Community Comments,* Yellow Springs, Ohio, October 1970) has reported the case of a number of hostile and "retarded" children being transferred from a big-city school to a small,

The extent to which the small school interacts with its community is another crucial point in its favor. The large big-city school rarely establishes any contacts with its immediate neighborhood on any level above that of recurrent complaints; it will almost never make an effort to use the community elders as a resource; its teachers most probably live many miles away. The small school, by contrast, normally serves as a genuine community center, where townspeople gather for adult meetings and rallies as well as for the children's school plays and Christmas sings; the teachers are likely to know everybody in town and be able to call upon them to contribute their time and skills; and the children view their teachers not as some kind of foreign beings but as community fixtures, whom the parents have known for years and who are invited home for dinner at least once a year. The small school's board is most likely to be intimately tied to both community and school, knowing the bulk of the children and known to the bulk of the parents; a 1975 Gallup poll found that far more small-town people than big-city people felt that the school board represented their views, and 43 percent of them could actually name a recent action of the board compared to only 25 percent in the city.

It is true that such intimacy can sometimes breed insularity and a certain narrow-mindedness. (Not the least of which is having a teacher say, "Oh, you're one of the Johnson boys, always *have* been trouble-makers.") Yet the teachers in the small school inevitably have an understanding of the underlying culture of the town—no matter if they move to it late, for they soon learn—and of the kinds of ideals and expectations it wants instilled in its children; they *respect* the community, for it is theirs as well. Whatever prejudice they exhibit is most probably that shared by the townspeople and is far preferable to the kind of insidious superiority so often exhibited by those teachers at larger schools who do not understand the community culture. One West Virginia mother, recorded by psychiatrist Robert Coles, poignantly relates:

> We had a fine little school here. It was small, and the teachers knew how to get on with our kids. We had no trouble sending them off; everyone loved school. Then they 'consolidated.' It was supposed to be the best thing in the world. It was called 'progress.' Well, we were all for that! But what happened? What did we get? We ended up with our kids being lost in that building; and hearing how 'backward' we are here, and how 'forgotten,' and how 'ignorant.' It's no good, when you

---

ungraded primary school of sixty-five students and three teachers, where they all showed improvement: "The more personal 'human size' situation made it possible for the 'confused' child to feel accepted and at home in the smaller group, and to have a feeling of 'self-worth.' . . . Where a shunning of school attendance had been the result of retardation or other embarrassment [in the city] it became possible for some of the children to complete four grades in two years."

> have your kids coming home and telling you that the teacher is all the time looking down her nose at people like us.

To which Coles adds: "An old, sad story, alas."

One particularly thorny problem that emerges whenever the idea of genuinely community-based school arises is what to do about segregation, should it happen that preferred residential patterns result in there being a lily-white school in one community and a coal-black school in another. It has been an article of faith that denying community cohesion and mixing up the schools—usually by closing the neighborhood black schools and busing the children to consolidated ones—is the fairest and most egalitarian way to further education and redress the past. But in fact there is no real evidence to show that black children do significantly better in integrated schools after about the third or fourth grade, and the reason for improvement up until then does not have anything to do with classroom integration but with the quality of the white school and its teachers. And there is indeed some research that indicates African, West Indian, and Afro-American children actually tend to develop at a slightly faster pace than white children, have more advanced motor skills, demand a "more active" environment, and therefore throughout their schooling need more open and active classrooms—in other words, their own schools. As Harry Morgan, professor of education at Syracuse University, points out, when they are put in the "formal and didactic" classrooms that have been created for white school children, "black children fall behind":

> In the upper grades they work very hard at loosening up the classroom to make it more supportive of their style of learning. Too often, though, their release of energy and the school's reaction constitute a miniature battleground.

With the result that black children are given drugs or disciplined or suspended, at about four times the rate for white children—and more often than not the black children are the ones to drop out of school the soonest with the least education. Obviously nothing about the much-vaunted integration of schools is very valuable for *them,* and clearly having such children beside them cannot be helping the other children very much either. When this is compounded by the disintegrative effect on the community brought about by the absence of the school as an educational and social centerpiece, and still further by the remoteness parents begin to feel from the children's learning experience, it is hard to see that the children of any race are being well served by heedless integrative policies.

TO CONCLUDE the case for the small school it is pertinent to dispel a myth or two about the large school.

▪ Characteristically, the large, consolidated school is justified on the ground that it will produce economies of scale—the savings supposedly derived from bulk buying, joint administration, consolidated plant, and the like. Some economies of course do exist—it may be cheaper to have one large furnace than several smaller ones—but it turns out that they are more than offset by a number of diseconomies (not even reckoning in the *social* costs). There are much higher transportation costs in busing the children to and from the school (new buses, garages, equipment, drivers, mechanics); increased administrative and salary costs with the new consolidated bureaucracy (where once the teacher could put in an order for supplies, it is now necessary to have several full-time purchasers); increased inventory and storage costs; additional costs from the inevitable delays, inflexibility, overbuying, and sheer mistakes of the larger system (instead of a dozen wrong pencils, several gross of them). And of course construction of the new and larger facility is an enormous expense in itself.

Nor are larger school *districts* economically advantageous. Werner Z. Hirsch, the same economist who has analyzed the performance of cities, reported after an examination of school districts nationwide that he was "unable to find significant economies of scale" in larger school systems.

▪ As to the claim of greater efficiency, several studies indicate that larger schools actually spend *more* money per pupil (higher teacher salaries, more administrative overhead, higher maintenance costs) without significantly improving classroom quality. One comprehensive study of schools in Vermont suggests that there may even be greater *in*efficiency in big schools because of "a pervasive isolation of administrator from the community and the educational process" and the "impersonality and alienation associated with larger size."

▪ Finally, the argument that large schools are better because they can afford more and better equipment and more highly qualified teachers—which indeed they usually can—is not borne out by the mass of evidence, from Christopher Jencks on down, showing that these elements don't really have much effect on basic school achievement. Nor is it borne out by such figures as we have about the percentage of high-school children who go on to college: small schools do every bit as well as large ones, and their students seem to have the same rate of success in college. One analysis for Vermont showed that six of the top ten schools in percentage of graduates entering college were small (with less than sixty in the graduating class)—*and* they managed to produce these percentages with an operating cost per pupil of $225 *less* than the large schools.

OF COURSE A small school would be almost inevitable if society were reorganized into human-scale communities of 5,000–10,000 people.

Taking contemporary U.S. figures for average family size, a community in that range would have about 1,000 to 2,000 school-age children. Assuming, from the Barker studies, that a school of about 100–150 people is optimal—perhaps smaller in the elementary grades, on the larger side for high schools—that would mean around eight schools altogether in the smaller community of 5,000, twice that in the larger. (This accords well with models of classroom size—with 15–20 people per grade in an elementary school of eight grades, the pupil population would be between 120 and 160; in a high school with two or three separate classes of 15–20 people in each grade, the population range would be from 120 to 240.) That may seem like a lot of schools for such a small setting, but in fact it is only a fraction more than the ratio this country had in 1950, when there were 27 million school children and 166,000 schools nationwide. And obviously if that number put a real strain on a community's budget, one alternative might be to provide a variety of ways of offering educational experiences *without* schools—and indeed, even *with* the requisite schools these ways make sense.

The idea of education without schools is no longer such a heretical one, thanks to the work of a number of astute education critics—John Holt, Paul Goodman, George Dennison, and particularly Ivan Illich, whose *Deschooling Society* is a ringing denunciation of formal school systems in the West. The appeal of this notion is obvious: if schools aren't educating, we can hardly do worse by eliminating them and probably do better by establishing creative alternatives. After all, we long ago gave up the practice of making church attendance compulsory—why do we still make school attendance compulsory? Aren't there hundreds of other ways that children can learn, particularly since we now know that only a minimum of learning takes place in the school anyway? And aren't there some odious side effects of compulsory attendance and regimented activities and enforced obedience and rote learning that do not jibe well with our notions of a free society? If it didn't serve to keep most youths off the streets and out of the grown-ups' world; if it didn't keep them out of the labor force for an extra ten or fifteen years; if it didn't allow both parents to go off to work; if it didn't provide employment to so many otherwise economically marginal people and sustain such a significant part of the economy ($120 billion nowadays); if—in other words—it didn't play such a primary economic role in our nation, would we bother with school at all?

After all, where do people really learn? Most of us grasp the fundamentals—language, sociation, motor skills, and so on—in the home as children, pick up our everyday styles and worldly knowledge in the streets as kids, learn our craft or discipline on the job after graduating from school, and for the rest of our lives continue to gain such knowledge as we have through books, television, acquaintances, travel, and work.[2] Schools generally play only a minor part. Ask yourself where you have learned the most in the course of your life, and the chances are that it is *not* in a school or college, even less likely in a formal classroom. For my own part, I think I can say unequivocally that I have learned far more in the 42nd Street Library of New York City than I ever did in sixteen years of schooling.

Thus it is possible to imagine children in a human-scale society taking advantage of a wide variety of other educational settings in addition to the school—or mix-and-matching them as their temperaments and interests dictate until they choose one adult niche or another. Some might choose to spend their time in a library, serendipitously led from stack to stack and shelf to shelf. Some might choose to use the community itself as a resource, poking into this or that enterprise, investigating this or that office, traveling to various sections of the area to see how the place really runs. Some children might welcome a return to a system of apprenticeship in which they could train in jobs—say from the age of 13, the traditional point of adulthood, as the bar mitzvah reminds us—in which they could train in jobs consistent with their physical and mental ability, protection from abuse lying in the watchful eye of the community itself. Some older children might be encouraged to set up their own self-sufficient units, complete with farms and energy stations and handcrafts, and learn about life by living life. And some might simply choose to hang around a community center, horsing around, playing sports, watching films, reading magazines, talking with elders, doing odd jobs, waiting for the right moment in their lives to lead on to something else and learning unconsciously the while.

The possibilities are almost limitless, once we remember what it is that we really want of our children. As William Godwin put it long ago: "Let us not, in the eagerness of our haste to educate, forget all the ends of education."

WHAT IS TRUE of the schools is true as well of the institutions of higher education: they are by no means the only places to learn, and they are not necessarily suited for everybody, but if you're going to have them, they will perform best if they are small.

The original universities formed in the medieval era were very small, usually with somewhere between five hundred and a thousand participants, some with only a hundred or so; yet it was these small institutions that restored classical learning and nurtured Western civiliza-

2. Allen Tough of Canada's Adult's Learning Projects estimates that adults undertake a "major learning effort"—studying for high-school equivalency, learning to drive, etc.—a median of eight times a year, spending on average 700 hours, almost all of it without schools, 70 percent with self-teaching.

tion almost single-handedly for several hundreds of years. In general they were limited to just three faculties—law, theology, and medicine—and there was no impulse to grow large and ever larger: what mattered to these scholars was the *communality* they were creating, within which ideas could be developed and exchanged and transmitted, and they realized that a population of modest size was necessary for that. It is not an accident that they tended to restrict their size to the numbers that we have already seen characterize the face-to-face association, or neighborhood. And even when they began to grow bigger as new disciplines and faculties were added on in the eighteenth and nineteenth centuries, they kept the principle of communality by creating discrete self-governing colleges within a university.

It was not until the twentieth century that universities, and particularly those in this country sponsored by state governments, began to evolve student bodies in the several thousands. And it was not until just a few decades ago that we began to see the peculiar phenomenon of the giant "multiversity," campuses with 30,000 and 40,000 students. In 1930 the average university student body was just 781; in 1950 it was 1,460; in 1975, it was 3,100. As late as 1958 there were only ten institutions with more than 20,000 students, only one larger than 30,000; in 1970 there were sixty-five over 20,000 and twenty-six over 30,000; in 1978, despite some falling enrollments, there were seventy-eight over 20,000 and forty over 30,000, plus another twenty-six junior colleges over 20,000. Obviously, whatever it was that such places were seeking, it was not communality.

As I noted before, such enormous institutions are competent at turning out graduates in great numbers—today a million or more people get B.A. degrees each year—but whether anyone would want to call it education is another matter. Quite recently the *New York Times* offered us these sentences written by a college graduate—white, middle-class, native-born, and, we were told, typical—who was teaching in an elementary school and studying for a master's degree at a major state university:

> As a teacher, children whose mother and father are dead or divorced are often angry and are hurt and are expressing themselves. . . . If you know the problems, the children are difficult to evaluate with.

It is a step or two above functional illiteracy perhaps, but it is not the English language.

Again, the coincidence of declining standards and increasing size is too regular to be merely accidental: there is a specific correlation between the quality of education and the number of people who are to receive it. Alison R. Bernstein, a leading education expert, has concluded: "There is a limit to a person's ability to learn and grow in an environment that makes him or her disappear in a sea of faces." It is not that a giant university cannot turn out competent graduates, for we all know that places like Berkeley (30,000) and Texas (43,000) and Michigan (45,000) sometimes manage to do that; it is rather that as a rule these large institutions have grown so fast and far that they cannot maintain whatever quality they had, and in order to spread what remains over so many thousands of students they are forced to water it down. "It appears," writes Alexander Astin, professor of education at UCLA, "that the net result of the massive expansion of the public system and the relative demise of private higher education is that the total benefits to college students have been diluted."

The effect of college size is shown strikingly in the results of the largest nationwide study of college student development ever undertaken, a ten-year survey by the American Council on Education and UCLA, covering some 200,000 students and 300 institutions. Alexander Astin, who wrote the final report, *Four Critical Years,* summarized it thus:

> When it comes to student achievement and involvement, the results clearly favor smaller institutions: Students are more likely to participate in honors programs, to become involved with academic pursuits, to interact with faculty, to get involved in athletics, and to be verbally agressive [in the classroom] in small institutions. At the same time, they are more likely to achieve in areas of leadership, athletics, and journalism. Students in smaller institutions are more satisfied with their faculty-student relations and with classroom instruction. . . . Small institutions foster a greater degree of altruism and intellectual self-esteem.

Astin's survey, incidentally, also shows that large institutions have no significant economic advantages over small ones and that "economies of scale are mainly illusory": "Large institutions actually spend somewhat *more* per student for educational purposes than small institutions." In fact, he argues, a policy of creating small colleges rather than large state multiversities "would actually lead to real savings in terms of *cost per degree produced."*

The reasons for the educational deficiencies of large institutions were outlined rather well a few years ago by a pair of scientists at the University of Washington School of Medicine, Jonathan A. Gallant and John W. Prothero, writing in the magazine *Science:*

Lack of community. "If it is true that a community constitutes a good environment for scholarship, then university growth beyond a rather small size becomes progressively more dysfunctional as it eliminates, at one level after another, the possibility of community."

Overspecialization. "In a small college, the individual scholar's microenvironment can include the entire faculty: men of letters, artists, scientists," but in a multiversity there are so many large departments full of narrow specialists that "a biochemist's immediate community is two dozen other biochemists, rather than zoologists, chemists, and mathematicians, let alone humanists."

Bureaucracy. As universities grow they add units, and "as the number of units increases, the number of coordinations required increases disproportionately," and the result is bureaucratic inefficiency, garbling of information, a loss of creativity, and impersonalization.

Alienation. As "the community becomes a crowd, anonymity, impersonality, absence of community, and bureaucratic complexity combine to diminish the possibility of fruitful human interaction," creating a distinct loss of morale among the faculty and alienation among the student body.

Growthmania. The desire for academic or institutional status often results in "the perpetual expansion of individual university units," since "size is often taken as a mark of status and if a mediocre program cannot be good, it can at least be big."

Gallant and Prothero shy away from putting absolute numbers to their analysis, but they seem to feel that for purely educational purposes the small colleges—500 to 1,000 students, one gathers—is the most advantageous, though for economic purposes a larger university of several thousand may be beneficial. They find no justification at all for the multiversity: "It is difficult to see," they conclude, "what further advantages, other than the possibility of United Nations membership, can accrue to a university population above 10,000 souls." In other words, their analysis seems to support the idea that for scholarly pursuits a small group built upon face-to-face association would be best, and a small college could take on the intimacy of a *neighborhood,* much as in medieval times. And for certain grander tasks, particularly where special equipment or a comprehensive library was thought to be valuable, a gathering of several thousand might be more appropriate, such a research center taking on more the character of a *community,* with primary association confined to smaller groups but interaction with a larger body still easily possible and a certain cohesion still retained. Either would be well within the human scale.

THOMAS JEFFERSON KNEW something about universities and about their sizes. In planning the University of Virginia, toward the end of his life, he rejected at the start the notion that "one large and expensive building" could possibly provide the right kind of setting for a true education, since this would turn into "a large and common den of noise, of filth and of fetid air." Instead he proposed a setting with a variety of small buildings, a separate one for each department, linked by small houses for the professors and smaller dormitories for the students, all of them kept within walking distance and all surrounding a space "open and green"—"for many reasons, particularly on account of fire, health, economy, peace, and quiet," and to provide the true "academical village."

Jefferson knew well that, as our buildings shape us, they shape our perceptions, our education, our very sense of culture. Built on the Brobdingnagian scale—as most of the newest campuses are, stark collections of geometric immensities—a place of learning would find it hard not to teach powerlessness and pettifoggery and incapacity along with whatever virtues it might pass on; on a Jeffersonian scale, it can instill refinement and harmony and quiet rectitude. John Russell, an art critic for the *New York Times,* once observed that the increasingly out-of-scale universities of the world "are places at which things have a predisposition to go wrong," and added:

> If free minds survive three years in the university that glowers down from what were once called the Sparrow Hills above Moscow, it is by a miracle. But the buildings at Nanterre, outside Paris, are no less an affront to human dignity. Every country has its failures in this respect, and every country will pay for them.

There is a scale at which learning best takes place—a scale of classroom, of school, of campus, of community. Insofar as we cherish our schools, insofar as we believe them to be the means to transmit the better parts of the human tradition and not merely engines to empower our economy, we must resurrect that scale: the human scale.

PART FOUR

ECONOMY ON A HUMAN SCALE

The conditions of a right organization of industry are, therefore, permanent, unchanging, and elementary. . . . The first is that it should be subordinated to the community in such a way as to render the best service technically possible. . . . The second is that its directions and government should be in the hands of persons who are responsible to those who are directed and governed.

R. H. Tawney
The Acquisitive Society, 1948

Human institutions should not be allowed to grow beyond the human scale in size and complexity. Otherwise, the economic machine becomes too heavy a burden on the shoulders of the citizen, who must continually grind and re-grind himself to fit the imperatives of the overall system, and who becomes ever more vulnerable to the failure of other interdependent pieces that are beyond his control and even beyond his awareness. Lack of control by the individual over institutions and technologies that not only affect his life but determine his livelihood is hardly democratic and is, in fact, an excellent training in the acceptance of totalitarianism.

Herman E. Daly
Steady State Economics, 1977

Plainly people want jobs *and* beauty, they should not in a just and human society be forced to choose between the two, and in a decentralized society of small communities, where industries are small enough to be responsive to each community's needs there will be no reason for them to do so.

The Ecologist, A Blueprint for Survival, 1972

1

"Gloom Is Spreading"

ECONOMIST E. F. SCHUMACHER liked to tell a story which he felt explained not only what he had in mind with his term "appropriate" technology but the nature of the economic world in which that technology would operate. One day, he said, a philosopher out walking in the wood came face to face with a figure in a radiant beam of light, none other than God Himself. Awed only temporarily—he had spent a lifetime, after all, pondering His existence—the philosopher came directly to the point.

"You are the Lord, I presume."

"Yes," said God, "I am."

"Well then, my Lord, I wonder if you would be good enough to answer for me a few simple questions that have been troubling me for some time."

"Certainly, my son."

"Is it true, Almighty, that what is for us a million years here on earth is for you nothing but the merest moment?"

"Yes, my son, quite true."

"And is it also true,"the philosopher went on, "that a million dollars here on earth is for you nothing but a paltry penny?"

"Also quite true."

The philosopher paused only a moment. "Then I wonder," he said, anxiousness showing, "if it would be possible for you, if it is not too much trouble, to give me a penny?"

"Why certainly, my son," said God. "I'll be back in just a moment."

Different worlds inevitably operate at different scales, and it would seem to be best all around if the human—no matter how great the temptation—is not confused with the celestial. And yet it is something very like just that confusion, it takes only a moment's thought to see, which in fact has come to mark our present world economy, the economy of global reach, of multinational systems and organizations, of interlocking worldwide production, distribution, and consumption. For it is an economy, whatever else may be said for it, that is simply beyond the competence of any one person, or group, or nation, beyond the very real but very limited capacity of humans in any form, to plan, manage, control, or repair, and no amount of computerware technofixing, no scheme of trade agreements or common markets, can change that fact. Were it operating efficiently, providing healthy and productive lives for all the world's citizens without ecological or social disruption, its fragility and complexity might still be a cause for some concern. Operating as it does, however—as it has for decades and gives every sign of continuing to do—producing chaotic and devastating mixtures of inflation and stagnation, dividing the world into apparently unbridgeable camps of rich and poor, desecrating the resources of the Third World to indulge and pollute the rest, in which no nation can be said to be truly prosperous and most exist in abject poverty, its uncontrollable enormity, its very non-human scale, becomes a matter for justifiable alarm.

Our experience in all these decades past with operating a celestial economy has, for the greatest part, been a failure: we tried playing God, and it didn't work.

The question then becomes: is there a realistic alternative, some sort of economy built closer to the human measure, with institutions designed more for human control? For even if it is accepted that a reduction to the human scale might lead us to better health, more efficient transportation, better schools, nutritious food, recycled wastes, and all of that, is it realistic to think that such small units as these could work economically? Even if we were to agree that a human-scale *society* might be possible, with its reduced cities and ecological balance and alternative technologies, is there any reason to believe that a human-scale *economy* could actually survive? And if it did, could it provide the equal of our present standard of living without plunging us into the economic chaos of the dark and dismal past?

The answer is, in every case, yes.

What the human-scale economy means, as we shall see, is perfectly simple. It opposes to the present system of American industrialism and its oversized units—corporations that can be shown to be too complex for innovation or efficiency, factories that we can demonstrate are too large for optimum profitability or economy—a world of smaller, more manageable, more human enterprises that can provide us with all our needs and most of our wants without sacrificing anything substantive in our true material standards. It is founded on the all-important concept, beginning to make headway in economic circles in these last few years, of the "steady-state" economy, one of stability rather than growth, preservation rather than production, and one that, far from being implausible, can actually be seen to be taking hold right now in the bosom of industrial America as more and more people turn away from the standard consumer economy. It coincides, too, with the movement toward workplace democracy that has begun in many parts of the world, including this one, under the banners of "self-management," "worker control," "community ownership" and the like, and it takes from a number of the existing models of worker democracy that we shall explore the attributes of size and productivity and humanity that promise most. And finally, it points ultimately to a world of ordered self-sufficiency, of cities and communities that are able to achieve steady-state balance and workplace democracy because they function with modest and delimited economies able to combine the best virtues of modern technologies with the truest values of ancient artisanry.

And, as I trust will become evident, this human-scale economy is not only desirable for what it allows the individual as well as the community to become—it is, above all, possible.

IT IS WITHOUT ANY DOUBT *necessary*. The crisis in the worldwide—particularly capitalist—economy has reached a point where, even conservative economists now agree, something drastic has to be done. I pick at random from my file a clipping from the *New York Times*—it happens to be from November 1978, but it could have appeared at any time over the last several years:

GLOOM IS SPREADING AS PROBLEMS GROW IN WORLD ECONOMY

MANY NATIONS FACING TROUBLE

Sluggishness and Unemployment Afflict Industrial Countries—Poor Lands Heavily in Debt

PARIS, Oct. 31—In the business and financial centers of Europe an air of gloom is spreading over prospects for the world economy and its capacity to tackle the mounting problems of debt management, sluggish growth and rising unemployment.

The highly publicized difficulties of Britain and Italy have diverted attention from the fact that at least a third of the industrialized countries are in some serious financial trouble. As one analyst put it, "We're reaching the point of critical drag, without even considering the impact of new oil price increases."

France, Denmark, Ireland, Finland, Belgium, Australia and New Zealand are among the other countries in the Organization for Economic Cooperation and Development, an economic directorate of the industrial democracies, where yellow signals are flashing.

The poor nations of the developing world are in even worse shape, struggling under a mountain of $135 billion of debts, still unable in some cases to feed themselves. . . .

> "Everyone can see the dangers ahead," another expert said, "but nobody knows how to avoid them."

The United States has suffered the effects of this worldwide crisis most acutely, because it has had to absorb the loss of the world hegemony it once commanded, as symbolized by the drastically deteriorated value of the dollar and the rise of the OPEC challengers. Inflation has taken hold as a permanent feature in the U.S., driving prices inexorably higher every year, and still the high employment and increased productivity that are supposed to go along with it just haven't materialized, nor do they seem likely to.

In the meantime, the results of economic chaos continue to flourish:

▪ The unequal distribution of wealth, which hasn't changed significantly in the last thirty years: the richest fifth of American families had 41.8 percent of the national income in 1955, 41.1 percent in 1975, an estimated 41.3 percent in 1980; and only *half* of all families are estimated to have a net worth of even as much as $3,000.

▪ The rise of personal debt, putting American families in hock to the extent of $3.5 trillion, according to *Newsweek,* which enormous sum happens to be *twenty-six times* the total foreign debt under which *all* of the nations of the developing world are said to be struggling.

▪ The increasing shortage of just plain capital, the wherewithal of American business, which will amount to anywhere from $1.5 trillion (according to Chase Manhattan) to $4.5 trillion *(Business Week)* by 1985, meaning that business will inevitably be forced into retrenchment and decline.

▪ The decline in productivity of American corporations—a result of old plants, bad management, unhappy workers, and unresponsive giantism—making the U.S. the most stagnant of all the industrial economies in the last decade, with productivity rates a third of those of Japan and Denmark, a quarter those of Finland.

▪ The rise of unemployment, unabating no matter who is in office or what laws are passed, amounting to a $3.5 trillion loss to the economy since 1953, according to Leon Keyserling, and producing an inevitable loss of revenue for public treasuries, the decline in social and municipal services, an increase in the levels of poverty, and mounting civil unrest.

▪ The increase in corporate criminality and scandal—estimated by even the U.S. Chamber of Commerce to amount to $40 billion annually—and apparently increasing year by year in response to the impinging pressures on every industry in the land, from milk producers to moviemakers.

In short, the capitalist "system," if that is not too grand a name for it, is simply, on its own terms, not working. Hardly surprising, then, that such a sober, if perhaps melancholic, observer as Robert Heilbroner, an economist at the New School for Social Research, has been led to declare: "I still believe that the civilization of business—the civilization to which we give the name capitalism—is slated to disappear, probably not within our lifetime but in all likelihood within that of our grandchildren and great grandchildren."

It is worth observing that the central reason for this immediate period of crisis—roughly since World War II—is the operation of the Law of Government Size.

As the American government in particular—though the same is true enough of other industrialized governments as well—has expanded its role, its influence, its economic power, and its personnel, it has inevitably produced economic and social misery in ever-increasing degrees, manifested most clearly in the high rates of inflation that serve to sap buying power from the general run of citizens and in the high rates of unemployment that leave more than a quarter of the productive citizens idle and impoverished. The way it works is this:

As the government takes unto itself more and more of the tasks of serving the citizens in their every waking moment, and not a few of their sleeping ones as well, it runs up an inevitable debt. Some of the value produced by all these government services is returned to the government through taxes, and ever-increasing taxes at that, but a great deal is siphoned off by private businesses in the form of their own profits—as, for example, when Washington builds a superhighway system and the trucking industry reaps the profits from it, or when the Atomic Energy Commission develops and promotes nuclear power for electricity at the cost of many billions and private utilities then use it for corporate gain. Some of these profits can be recovered by the government through corporate taxation, but proportionately very little, particularly since the tax system is deliberately designed to help corporations prosper, since both legal and illegal means exist by which corporations can hold onto their gains, and since out-and-out tax breaks of some $136 billion a year are handed out to businesses in a general policy of what's-good-for-General-Motorsism. Thus, merely to keep itself going year to year—servicing its debts, meeting the costs of inflation, and all the while expanding its operations—the government has to increase its expenditures and hence its debt. And hence the rate of inflation and the economic hardship of the citizens.

At the same time, government expenditures foster corporate bigness and encourage monopolies in a wide range of industries, creating an economic segment that is in many ways immune to normal market forces. Large corporations are in the happy position of being able to expand profits by using government services without having to spend very much on enlarging their plants or increasing their wages, and passing on to the consumers such increases as inflation may create. Nor do large corporations do much to lessen unemployment, since they are the ones that can afford automation, that can dictate labor conditions, that can generally do without unskilled workers, and that can accomplish growth when necessary by merger rather than expansion. The eventual upshot is an increase in overall social misery: urban blight, joblessness, pollution, poverty, sickness, crime, and so on. Where this is abated at all, it is through increased government services, and there we are again with more inflation. In short, another double bind.

BUT DIRE AS the current economic crisis is, it represents only one side of the malady from which American capitalism is now suffering. The other is worse, because it is inherent and fundamental.

It is in the *nature* of capitalism, of course, to be unstable—boom-or-bust it's called, and Keynesian fiddling doesn't seem to change it a lot—and that is bad enough, particularly when both boom and bust happen at the same time, as over this past decade when the rich have gotten the boom and the poor the bust. But, one is forced to say, it is also in the *nature* of capitalism to produce certain other ills, the accumulation of which in recent years has become serious enough to put the advanced industrial societies into the realm of what can only be called peril. Chief among them:

Exploitation of resources. There are no special penalties attached in capitalism to the profligate use of natural resources, and indeed the principle is that if you can do that cheaply enough you can even increase your profits: the more exploitation, the scarcer the resources, the greater the demand, and the higher the price. However, in the last decade we have come to realize not only that these resources are finite and irreplaceable but that our rapacious use of them—the U.S. uses up about 43,000 pounds of raw materials per person per year—has serious economic and social effects. Petroleum is the familiar example—it used to be you could stick a pipe into any part of Texas to get oil, but now we have to spend $10 billion to get it from the Alaskan North Slope—and though there is a lot of blather about just how much of the stuff we really have, the authoritative *Science* magazine has asserted, "There is no longer much argument with the conclusion that U.S. resources of conventional oil will be seriously depleted by the year 2000." Iron ore, too, has been rapidly exhausted in this country, our annual output down 20 million large tons since 1950, the number of producing mines reduced from 321 to 50 in the same period, and the high-grade Masabi range in Minnesota virtually depleted. Water is even more serious: North America now uses up *twice as much* water as is replenished by rainfall, and water tables all across the nation are dropping, particularly in the Southwestern states, where it is calculated that some aquifers may not fill up again for a thousand years and some may never. Forests have been consistently depleted, especially in the West, where the cutting of old trees well surpasses the planting of new, and in the South, where softwoods have replaced hard. All in all, the National Commission on Material Policy concluded as recently as 1973, the U.S. is in the throes of "serious" shortages now and by 1985 will face even more acute shortages in *two thirds* of the basic raw materials upon which the American industrial machine depends.

Waste. Similarly, capitalism encourages waste, there being no special benefits attached to creating durable or recyclable goods and considerable advantage to getting people to throw away their old products and buy your new ones. (Indeed, there are positive values in making waste, inasmuch as it costs more than $6 billion a year in the U.S. to get rid of it, and that money has to go to *someone.)* Large contemporary industries like the plastic and petro-chemical industry complex have been particularly culpable here, because their plants demand so much capital to build and so much energy to operate that they are forced to turn out their products in the largest possible numbers to make a profit; thus many plastic goods now appear everywhere in all kinds of guises not because there was any particular demand for them or because society made any rational decision that it wanted them but simply because the out-of-scale realities of the industry made them "necessary."

Ecocide. Not only is ecological damage an inevitable consequence of exploitation and waste, it is also built into the capitalist imperative of maximum profit: in economic terms it *makes sense* to pollute the atmosphere if it reduces production costs, to pour toxins in the rivers if safe disposal costs too much, to use fluorocarbons that threaten the ozone layer if there is a market that seems to want them. Probably most businesspeople are aware, as individuals, that the environment is being stretched to its regenerative limits by the industrial assaults upon it and will visit the dangerous consequences upon their grandchildren if not upon themselves, but as corporate officers they continue to put up the most toe-grabbing resistance to even the kinds of mild controls the government has proposed. So powerful is this capitalist ethic that, for example, immediately after the automobile and oil industries pressured the Environmental Protection Agency into reducing the already lax smog standards for American cities in January 1979, the American Petroleum Institute, representing the big oil companies, announced that it would file suit in Federal District Court to force an even lower standard.

Social burden. Say what you will about capitalism and its achievements, it has always exacted a certain social price, and that has become particularly high in recent years. Just the normal quotidian processes of

American business create serious distending pressures for every American family bound up in it; they force couples to move a continent away and children to relocate school districts at a whim; they insinuate anxiety into every employee, from the marginal laborers right up to the scrambling executives; they tend to make all but the very topmost feel powerless and sometimes worthless, reducing their human needs simply to a weekly paycheck. If the environmental pollution of the industrial economy is severe, the behavioral pollution may be even worse, and in ways that we have yet to discover. Anthropologist Lionel Tiger suggests: "The industrial system we take so much for granted is only several hundred years old. We are beginning, I think, to realize that it is an extremely demanding economic system under which to live." Perhaps the citizens are willing to pay such a price for such trappings of modernity as we may boast, discounting the disintegrative effects; perhaps, however, we may be having second thoughts about the trade-offs as the possibilities of impending social chaos become ever clearer.

Social irresponsibility. The corporation in a capitalist system existing essentially for profit rather than product, there is no particular incentive toward well-made, durable, safe, or aesthetic goods; the corporation being responsible to its managers and stockholders rather than its public, there is no especial inclination toward healthy, safe, clean, or balanced environments. In this context, inferior products may generate more profits because they must be continually replaced, unsafe products because they are cheaper to produce. It is *rational* to put poisons into food if they increase its shelf-life, to design cars with gas tanks that explode if they can be made cheaper, to offer drugs that maim or stupefy if they can be convincingly peddled. The corporation, with only tremulous limitations provided by law, acts pretty much as it wants to, regardless of broader social considerations: it may produce what it wants rather than what the society may need (another "air freshener" or $3 billion worth of pet food while rats bite infants and families starve); it may invest as it wants rather than as society might prefer ($70 million to develop and market potato chips in tennis-ball cans or $12 billion a year to build overseas factories for overseas employment while our hospitals deteriorate and unemployment festers); it may locate and relocate as it wants rather than as society might find rational (setting up a factory in an already congested area or abandoning a town after exhausting its resources). The only question is how long any society can, or should, suffer the consequences.

Overgrowth. Growth is the very wellspring of capitalism, the force that makes it succeed where it does, but the inevitable result of growth, as economists have long since discovered, will at some point be increasing concentration and eventual monopoly. In the U.S. this has now reached the point, as we know, where no more than 500 corporations control about 40 percent of all private production and, in most major industries, two thirds of the market is in the hands of four companies or fewer. The first effect of this is to develop increasingly overextended and overloaded systems both within and among corporations, so that, as economist Hazel Henderson points out, "when complexity and interdependence have reached such unmanageable proportions . . . the system generates transaction costs faster than it does production"—it costs more just to keep going than we can ever make from it—and the society goes broke. The second effect is to put these large corporations beyond the influence of *both* the traditional supply-and-demand market *and* the regulatory government: they can manipulate the market by virtue of their advertising, distribution, borrowing, and marketing power (in Galbraith's blunt words, "the big corporation eliminates or subdues market forces"); and they can circumvent government control by lobbying, tax breaks, bureaucratic interlocks, overseas plants, simple noncompliance, and the threat of loss of jobs. The society must therefore pay for the burdens of corporate complexity as well as the travails of corporate whim.

Instability. Inherent in the capitalist system anyway, this is exacerbated by the size and intricacy of the corporations and their national and transnational systems. In a small world, instability may have made little difference, but in a connected world it begins to affect everything. We find that a *managed* worldwide economy is simply impossible, given the logistical and political complexities; we also find that an *unmanaged* worldwide economy is simply chaotic, unable to make rational decisions about resources or markets or employment or anything else. It seems an inevitable working out of the Beanstalk Principle that the world economy in recent years has experienced, among other disasters, runaway inflation, unabating population growth, severe starvation in Africa and Asia, overconsumption of such scarcities as petroleum and copper, and the effective collapse of the international monetary system.

All of that—exploitation, waste, ecological and social disarray, instability—does not by any means exhaust the characteristics of capitalism that, even if benign enough in the past, are now of such proportions that they pose a serious threat to a sustainable future; but it does suggest the most pertinent ones. I wish to stress that I do not mean to seem to be talking about them in moral terms, as being good or evil—I am not describing what *ought* to be but rather what *is*. It is important to appreciate that capitalism behaves as it does not because it is evil, and the corporations as they do not because they are immoral—that is the *nature* of the system, that is what, by the rules, is *supposed* to happen. At the same time it is well to recognize that there may be limits to where such a system can lead us, limits to which in recent times it seems we may have come.

SO THE CAPITALIST CRISIS, then, is really made up of two facts: first, that the industrial system is not working, even on its own terms; second, that it is.

Of course none of this is really news. Many have realized the quite catastrophic nature of what we have come to, not only as it impinges on the present but, even more, as it threatens the future. It was this that led some rather more far-seeing capitalists to establish the famous Club of Rome, whose reports continue to make clear the serious nature of the long-term crisis, presumably to allow rational businesspeople and politicians to devise some remedies before it is too late. It is this that has led the United Nations at various times over the last decade to sponsor a number of sober and comprehensive conferences on such matters as pollution, climate control, law of the seas, income disparity, food, and population.

Some who have seen the crisis have argued that everything will eventually shake itself out and there's no real need to make substantive changes. This is the attitude of the standard American business sector, of which Herman Kahn of the Hudson Institute is a capable spokesman. He argues, in effect, that ordinary supply-and-demand pressures in the market will provide most of our answers and that high-technology projects run by the government will provide all the rest: "Our capacities for and commitment to economic development and control over our external and internal environment and concomitant systematic, technological innovation, application, and diffusion, of these capacities are increasing, seemingly without foreseeable limit." His optimism is touching, and persuasive to many businesspeople, but for all his charts and diagrams it rests on nothing more substantial than bare belief—that the technological solutions will somehow come before the catastrophes do, that the market will operate in a rational way when it is pressed to do so, that a government capable of controlling vast economic projects will be otherwise benign. And he is quite dense to the appreciation of most of those qualities of capitalism that I have mentioned above—which one is perfectly free to do when one is a confirmed believer in the magic of the technofix but which will hardly prepare one realistically for the future. (It is perhaps characteristic of his understanding of the future that in 1967 Kahn was maintaining that "the U.S. is practically independent of foreign oil supplies" and never considered the possibility of an oil shortage just six years later.)

Others who also take a somber but less optimistic view of the crisis have argued that it does indeed require thoroughgoing changes in economic and political systems, particularly in the direction of planning, especially governmental and intergovernmental planning. This may be said to be the standard socialist response, to be found in such thinkers as John Kenneth Galbraith and Michael Harrington and typified by Robert Heilbroner, who has warned of "the need for an unprecedented degree of monitoring, control, supervision, and precaution with regard to the economic process." And he adds: "This may require allocations of materials, prohibitions against certain kinds of investment or consumption activities, international arrangements, and a general sticking of the public nose into private life wherever that life, left to itself, threatens the very survival of the system." (This is very much like the response, too, of a certain segment of the pro-government Establishment of the Northeast, represented by one-time Treasury Secretary Michael Blumenthal on the national level and World Bank President Robert McNamara in the international arena.)

Aside from the fact that such a cure sounds as if it would be every bit as bad as the disease itself, several other objections come naturally to mind. In the first place it is hardly encouraging to imagine all this wonderful planning being done by a government that, for example, planned Vietnam, Medicare, urban renewal, and many other of our familiar disasters. Then, too, it would probably depend upon a government of such size, complexity, and bureaucratic latticework that could not possibly accomplish what it set out to do and might very likely create fresh catastrophes of its own. Finally, there would not seem to be any particular grounds for optimism in the experience of the nations that have attempted centralized planning: the Soviet Union, for example, which has tried it for longer than any other government in the world, exhibits almost all the maladies of our system (cyclical boom-and-bust, shortages, inequities, poverty, inflation) plus a few of its own (rationing, maldistribution, regimentation, low living standards, and tyranny); even the more benign Sweden, with the most extensive planning of any nation in the West, has not, despite its successes, been free of recession, inflation, foreign debt, devaluation, declining growth rates, declining productivity, lagging investment, fuel shortages, severe losses by state-owned businesses, and staggering taxation rates.

There is, however, a third response possible, and it is one to which a number of people, bidden and unbidden, knowingly and not, have been turning in the past decade or so. In place of optimistic drift and pessimistic planning, this response points rather to the reduction of all economic systems and organizations to the point where they can be controlled by those who are immediately affected by them. It emphasizes, for the individual, self-definition of the job, self-scheduling of time, work in small groups based on consensus and cooperation. It encourages the development of family-farm agriculture, the decentralization of industry, worker ownership and self-management of firms and factories, and alternative technologies and all that they imply. It seeks self-sufficiency and self-reliance for neighborhoods and communities, community control of industries, and regional cooperation where necessary. And overall, it favors an ecologically minded, people-oriented, small-scale, steady-state society.

This is, of course, the human-scale economy. Obviously it has no very extensive models to offer, there being no nation at this time completely devoted to this path. But there are, as we shall see, innumerable examples within many nations, present as well as past, of one or another of these attributes, and there is, as we shall also see, right within our own nation a very large and significant portion of the economy that represents this human-scale trend.

British writer Gordon Rattray Taylor stands out among those who have studied and advocated this third response. He says simply:

> The present system is biased in favour of goods as against other desirables; it neglects bads; it provides motives for anti-social and inhumane behaviour because it is mechanical and inhumane itself. The way to correct it is not to substitute vast monolithic publicly owned boards and corporations, which are equally open to inhumanity and distorted valuations. It is to substitute decisions made by that marvellous instrument the human brain, which alone can weigh all the factors, ponderable and imponderable, in a situation. And a 'personalized' economics can only be built from small-scale, face-to-face contacts: a 'polylithic' instead of a monolithic system.

2

The Myths of Bigness

In 1979, the Blitz Weinhard Brewing Company of Portland, Oregon, which had served the area for 123 years, was sold to the Pabst Brewing Company of Milwaukee, Wisconsin, for an unspecified sum of several billion dollars. William Wessinger, board chairman of Blitz Weinhard and the grandson of the man who had founded the brewery in 1856, said: "I just hate it. I hate the whole trend of consolidation that the country is in." But, he argued:

> The big guys have more and more leverage with their advertising dollars and big-order techniques. Pabst wanted some of our brands, like Olde English Malt, which is big in New York and Georgia, and there was just one item after another that we couldn't match them on—as when they are able to buy beer cans at 26 cents a case under what we pay.

At one time during the nineteenth century there were 121 breweries in New York City. As late as 1915 there were 70 of them, and even after Prohibition had killed off others there were still 23 before World War II. In 1976 there were none. A city that had been making beer since the Red Lion Brewery was established in Nieuw Amsterdam in 1660, and that had known Schaefer, Rheingold, Knickerbocker, Congress, Excelsior, Metropolitan, Eichler, Huppel, Von Hink, Diogenes, Trommers, and those hundred others, now gets its brew from New Jersey and Pennsylvania and Wisconsin, from Holland and Canada and Japan.

In 1935 there were 750 breweries, all regional, in the United States. By 1946 there were only 471, by 1960 there were 200, and by 1980 less than 40, a thirtyfold decrease in just forty-five years. An industry that had always been decentralized because of the limits imposed by batch production and high costs of distribution has succumbed to the processes of monopolization: just five nationwide companies now account for nearly 80 percent of all beer sales. It is assumed that before long there will be no more regional beermakers left in the U.S., the market having been taken over by fewer than a dozen national breweries, most in turn controlled by conglomerates with only a marginal interest in beer-making itself.

There is no pretense by anyone, inside the industry or out, that this development has been beneficial to the consumer or the public at large:

▪ The quality of the product is not improved when local breweries are taken over by nationals, and indeed the acknowledged idea is to market a beer that is insipid enough to appeal to the lowest common denominator, with blander taste, fewer raw ingredients, and somewhat less alcohol. Most national firms also have switched to "heavy brewing," a process in which a beer concentrate is mixed with water only just before it goes into the bottle, an obvious saving of brewery capacity but accomplished at the expense of texture and taste. Orion P. Burkhardt, an executive of Anheuser-Busch, the leading national brewery, suggested something of the beermaker's concern for quality when he told the *New York Times* that his firm's only worry is how much beer drinkers "are willing to give up in a taste sense" before they are likely to protest.

▪ Nor is the product any cheaper when sold by the national firms, in spite of the fact that they are supposed to be able to buy supplies in bulk, use modern technologies, and employ economies of scale in manufacturing and storage. Largely because they have immense advertising and promotion expenses and complex marketing and sales bureaucracies, national breweries actually have to charge more for their beer than local ones do (except when they first come into a region to undersell the local brands); and because customers' loyalties are based actually upon brand image rather than taste and quality, the brewers have to use more and more advertising and more and more gimmicks ("light" beer, eight-ounce bottles, flip-tops) to keep their share of the market. Indeed, because they are mostly beyond the reach of competition, they are actually free to keep raising their prices without fear of being undercut by smaller firms or having bars and supermarkets drop their product.

▪ To top it all off, the national companies are the ones that created and maintain the system of non-returnable plastic bottles, at an enormous cost to the national ecology, both from the use of petrochemicals when the bottles are made and from the non-biodegradable assault upon the countryside when they are thrown away. Naturally they vigorously oppose returnable bottles—despite the fact that this would save raw materials, improve the environment, lower beer prices, and increase employment—because it is far harder for them to recycle their bottles across the nation than it is for a regional brewer who operates over a much smaller territory.

And the social consequence of this process of consolidation? Well, the people of Portland are in no doubt. The Blitz Weinhard people were always important in the civic affairs of the area. Their main concern was always in making profits, to be sure, but they were rooted enough in their locality to have at least a residual sense of civic responsibility and community concern. Nowadays, it seems, the Pabst people somehow don't share that same attitude: the people brought in from Milwaukee know that there is some value in kicking in to the Community Chest and the like, but they don't know very much about the particular town they're operating in, nor is it very important in the corporate scheme that they should care. Oliver Larson, executive vice president of the Portland Chamber of Commerce, not normally the kind of officer given to criticism of big business, has observed:

> The biggest difference between now and the old days is in the response to community need. Then, some war-horse would call a meeting of the people who decided things and it would get done. Now, they all have to contact corporate headquarters, at somewhere like Gravelswitch, Kansas, and it takes forever to get an answer.

Which is usually unresponsive when it comes.

THE PROCESS OF consolidation is not confined to the beer industry, of course: for many other products—particularly automobiles, soft drinks, and most recently coffee—the story has been similar, and many other industries—supermarkets, motels, restaurants, even bookstores—have been transformed by chain operations. In fact, as we have seen, the condition of bigness has stamped itself on the entire American economy, with fewer and fewer firms controlling more and more power with each passing year. It goes by the appropriately ugly name of oligopoly.

Now the usual stance of economists and businesspeople is that all of this bigness is either inevitable or positively beneficial: small firms are often uneconomical, they can't keep up with the new technologies, and eventually they have to lose out to the larger firms, which are more efficient, modern, productive, innovative, and profitable. Such a development, which has been going on for at least a century and seems inherent in modern capitalism, may not have solved all the problems of living, but it is responsible for the cornucopia of labor-saving technologies and the extraordinary high standard of living we now enjoy in the industrialized world. And all of those who have recently jumped on the Schumacher bandwagon with talk of the virtues of small enterprises are most likely leading us back to the Middle Ages and a world of inefficient, wearisome workplaces, and condemning the poor to unabating poverty and everyone else to a life of drudgery and bare subsistence.

Perhaps. But that perspective may be as automatic and as errant as the one that first regards the forest floor as empty and uninteresting—and discovers, only after a few minutes of concentration, that with reoriented vision the dirt is actually teeming with life, ants and earwigs and mites and untold miniature creatures busily going about their lives.

A reoriented vision here might perceive that these homilies are, in fact, based on myths. Let us examine in some detail the five most pervasive of them.

Economies of scale. It is an article of faith in the industrial world that bigger plants are more successful than smaller ones because the cost per unit goes down as the total number of units produced goes up. It's called economy of scale, and it is based on the assumption that Giant Widget Corporation, which buys its raw materials at a lower price (because it buys in bulk, a less complicated process for the supplier) and processes them more cheaply (because it has larger machines and plant-space and more specialized workers), will be more successful than Mini-Widget Company because it can either sell its widgets cheaper or make a larger profit at the same price. We are all familiar with industries—book publishing, for example, where price goes down as printing orders increase, and clothes manufacturing, where ready-to-wear is always cheaper than tailor-made—that operate on this principle.

Obviously there is some truth to the idea, and some sizes *are* more efficient than others. But it is *not* true, as all businesspeople and most economists believe, that bigger is continually or progressively better, or that Worldwide Widget Conglomerate will be the most successful of all. For in fact if you plot out costs and quantities you find that there is a clear "U curve": costs will go down for a short time, but quite soon they bottom out as you increase production and then they start to go up again. The reason for this is those "diseconomies" that always creep in after a certain very minimal size: the added costs of supervision and control, of bureaucracy and communication, of transportation and distribution, of warehousing and inventory, of energy use and maintenance, of labor costs for specialized workers, and in the largest and most dehumanizing plants, of absenteeism, strikes, grievances, and alienation. Thus, as economist Barry Stein has convincingly shown, economies of scale are "generally achieved in individual plants of modest size."

That is why, in every industry in every country, as we have observed before, there are always hundreds of small firms that manage to compete even in areas where the giant firms predominate—so in the U.S., 98.5 percent of all companies (1967 figures) have fewer than 100 employees. That is why every large corporation in no matter what field is broken down into relatively small plants—in the U.S. the number of production workers per unit was only 49.5 in 1947 and declined steadily to 44.9 in 1967 and 43.8 in 1972. And that is why even the largest companies depend upon a whole series of smaller independent firms instead of trying to make a complete product themselves—as auto manufacturers rely on many smaller outside businesses to supply them with glass, tires, brake drums, water pumps, springs, fuses, dynamos, carburetors, ball bearings.

Moreover, there seem to be no clear connections between the economies of scale in a particular plant and the *success* of that company. Even where optimum scales exist, it turns out, there is no way to predict the profitability, durability, efficiency, or overall success of the firm. Barry Stein again has shown that many companies—in fact, in most fields, the majority of companies—operate well *under* the theoretical optimum with no apparent economic penalty. Stein has concluded that "technical economies of scale are not the primary determinant of either competitive ability or true efficiency" and indeed that "there is no strong case to be made for significant economies of firm . . . size."

Finally, economies of scale, even when they are reached, turn out to have relatively little to do with ultimate consumer costs, at least in the complex modern marketplace. Such economies can, at best, affect only the *production* process, and these days that makes up only a rather small part—at the very most a third—of the total cost of a product. Thus a small firm that isn't able to meet absolute economies of scale in its plant can nonetheless offer products more cheaply because it can economize elsewhere along the rest of the pipeline, taking advantage of its size.

Efficiency. Naturally American business has fostered the myth of efficiency with particular fervor, since if bigness can't deliver *this* then most of its other justifications won't mean a whole lot. In typical form it sounds something like the words of engineer-turned-author Samuel Florman who tried to lead an advance against alternative technology in *Harper's* a few years ago by proclaiming that, "like it or not, large organizations with apparently superfluous administrative layers seem to work better than small ones." However you want to measure "work better," that just isn't true.

Just as big corporations have inefficiencies of scale, so they also have other deficiencies of size. They are less efficient in energy use because they have greater areas to heat and cool, typically more energy-intensive machines to operate, larger transportation requirements (particularly in distribution), around-the-clock shifts requiring full-time heating and lighting, and lighting banks geared to whole floors rather than individuals. They are less efficient in the use of raw materials, because there is less control over how the materials are used and hence more wastage per employee, because large production processes can't make corrective adjustments that smaller ones can, and because any error will be on a larger scale and will deplete more before being corrected. They are also less efficient in using the most precious resource of all, the human being, simply because, as countless studies have shown, size and complexity tend to create feelings of anxiety and tension, lack of individuality and self-worth, and other psychological problems that are expressed daily in faulty work, inattention, laziness, delays, and even sabotage.

But the central problem of large corporations seems to be, as we should now expect, the larger bureaucracies they must carry. Coordination is more difficult with large organizations, planning is more intricate, scheduling more rigid, information-gathering more complex, decision-making more attenuated, and assessing consumer reactions more complicated. (Just getting a part from the storeroom can be a chore: I have an IBM "Operating Procedures Guide" for one modest IBM plant in upstate New York, covering 340 pages of tiny six-point type, and Procedure 32, a request for materials from another section, requires no less than forty-one separate steps to be taken among nine separate plant divisions and coordinated with an untold number of forms, slips, files, requests, punchcards, and printouts, all to be done in triplicate.) Economist Frederick Scherer, after his study of eighty-six companies, said flatly: "The unit cost of management, including the hidden losses due to delayed or faulty decisions and weakened or distorted incentives . . . do tend to rise with organizational size."

It may be even worse than that. As economist Kenneth Boulding has noted: "There is a great deal of evidence that almost all organizational structures tend to produce false images in the decision-maker, and that the larger and more authoritarian the organization, the better the chance that its top decision-makers will be operating in purely imaginary worlds." And if that may be dismissed as an outsider's view, here's one from an insider: not long ago a vice president of Union Carbide, one of America's giant firms, confessed that neither he nor his colleagues "had any idea how to manage a large corporation" and couldn't know enough about all the intricate corporate workings even to solve a clear problem when one was presented to them. Baseline evidence for such statements may be found in the fact that the much-vaunted conglomerate, combining many already large businesses into a single corporation, only very rarely shows higher profitability after its acquisitions than before—and such profitability is usually because it proceeds to *sell off* the acquired firm and thus become smaller. As the House Antitrust Subcommittee once reported, of the twenty-eight conglomerates it studied, only seven showed a profit after acquisitions, three remained unchanged, and "eighteen companies had ratios lower in the years after acquisition [and] it would be reasonable to conclude that these ratios reflect ineffective management."

For those who still may doubt, it may be pertinent to point to the corporate experience of the last decade, hardly one to inspire much optimism about bigness and efficiency. American business in general, dominated as it was by giant companies, went through one of its worst periods, with diminishing productivity rates, recurrent liquidity crises, steady drains on the dollar, and ultimately its own version of the "British Malady"—obsolescent plants, declining markets, uncontrollable wage increases, and unabating price increases. At the same time a remarkable number of the very biggest firms gave a demonstration of their efficiency: a half-dozen railroads, led by the mammoth Penn Central, went bankrupt, as did W. T. Grant, Food Fair, Tishman Realty, and, for 1975, a record-breaking 254,000 others; a number of the largest banks, including Franklin National in New York and U.S. National in San Diego, went under; Lockheed Aircraft was saved from collapse only by a massive government loan, and innumerable other defense companies were propped up with artificial orders and refinancing; real-estate trusts, including the REIT owned by giant Rockefeller-controlled Chase Manhattan Bank, defaulted on millions of dollars; and even on Wall Street more than a third of the investment houses went under, including two of the very biggest, Goodbody and Francis I. DuPont.

Finally, the last word to the U.S. Treasury Department. Its statistics show, unequivocally, that "when efficiency is measured by return on assets, smaller businesses are more efficient than larger businesses in every industry." In fact, the smaller ones are able to increase their output with, on average, one-third less financial backing than big ones—or, in other words, they are three times more efficient.

Innovation. The idea that the inventions of the contemporary world are the result of the innovative genius of big business organizations has always beguiled economists; even the Scotch-cold Galbraith was once inspired to rhapsodize: "A benign Providence . . . has made the modern industry of a few large firms an almost perfect instrument for inducing technical change . . . [providing] strong incentives for undertaking development and for putting it into use."

Were this true, one might be hard-pressed to explain why the U.S., with its many large firms, has declined sharply in innovation over the last decade—the number of patents issued to Americans for inventions between 1971 and 1979 fell by more than 25 percent—and why headlines like "The Crisis in Innovation" have become a staple in business papers and academic seminars. But it is not, not by a long shot. Big organizations are far more resistant to change than small ones, and big corporations have their own particular reasons for resistance: they already have a huge investment in the equipment as it is; they have such highly specialized operations that any little change would mean considerable expense and disruption; they are guided by bureaucracies whose inherent bias is always against taking risks; they operate in general by successful systems of control, and control is inimical to innovation; and they cannot encourage much change within any particular division without risking the cohesion of it all and ending with each component going off its own way. Jane Jacobs gives the telling example of the Minnesota Mining and Manufacturing Company. When it was a small stone-mining company it could easily branch off into making sandpaper, and when it was making sandpaper it could naturally develop certain kinds of adhesives, ultimately creating what the world knows as Scotch tape. But after it became a huge tape company and grew into the giant 3M Corporation, it became increasingly locked into a set production process and fixed market commitment that simply did not encourage change and invention and in fact could not rationally allow very much.

Galbraith notwithstanding, the evidence on this point is considerable. Most businessmen admit it: "Smaller companies are the best innovators," says William Norris, chairman of Control Data Corporation, once small. For Dr. Robert Noyce, chairman of Intel Corporation, creators of the semiconductor revolution, the current "crisis in innovation" is caused by too much "group-think" and work-groups grown too large: "The spirit of the small group is better and the work is much harder." Particularly telling is the testimony of T. K. Quinn, a former board chairman of the General Electric Finance Company:

> Not a single distinctively new electric home appliance has ever been created by one of the giant concerns—not the first washing machine, electric range, dryer, iron or ironer, electric lamp, refrigerator, radio, toaster, fan, heating pad, razor, lawn mower, freezer, air conditioner, vacuum cleaner, dishwasher, or grill. The record of the giants is one of moving in, buying out, and absorbing after the fact.

Several academic studies underscore the point. Economist Jacob Schmookler concludes that "beyond a certain not very large size, the bigger the firm, the less efficient its knowledge-producing activities are likely to be," with the concomitant "decrease per dollar of R & D in (a) the number of patented inventions, (b) the percentage of patented inventions used commercially, and (c) the number of significant inventions." Arnold Cooper in the *Harvard Business Review* has shown that "the average capabilities of technical people are higher in small firms than in large ones" and hence "R & D is more efficient in small companies." And a report done for the U.S. Department of Commerce in 1967 concluded that from one-half to two-thirds of all important inventions in all fields were the product of individuals, public institutions, and small firms rather than big corporations, specifically advocating the encouragement of "independent inventors, inventor-entrepreneurs, and small technically-based businesses."

Most interesting of all is the comprehensive study done by economist John Jewkes and his colleagues a few years ago, in which they drew up a list of the seventy-one major inventions of the twentieth century and then sought to determine where each came from. Twenty-two inventions, they found—including fluorescent light, television, transistors, and Scotch tape—were the product of corporate research laboratories (though some of them, like 3M, may have been relatively small at the time of the invention); ten were of mixed origin or impossible to categorize; and thirty-eight, or 54 percent—including air-conditioning, the ball-point pen, cellophane, cyclotrons, helicopters, insulin, jets, penicillin, Polaroid, safety razors, Xerox, and the zipper—came from individual inventors working entirely on their own or in institutions where their work was autonomous. The Jewkes team concluded:

> 1. The large research organizations of industrial corporations have not been responsible in the past fifty years for the greater part of the significant inventions.
> 2. These organizations continue to rely heavily upon other sources of original thinking.
> 3. These organizations may themselves be centres of resistance to change.

Cheaper prices. Ralph Borsodi, the genius of American homesteading and one of the rare economic minds of our century, fled New York City in 1920 to establish a small farm in Rockland County, upstate. One day, being shown by his wife a jar of tomatoes she had just finished putting up for the winter, he could not refrain from asking the incurable economist's question: "Does it pay?"

> Mrs. Borsodi had rather unusual equipment for doing the work efficiently. She cooked on an electric range; she used a steam-pressure cooker; she had most of the latest gadgets for reducing the labor to a minimum. I looked around the kitchen, and then at the table covered with shining glass jars filled with tomatoes and tomato juice.
>
> "It's great," I said, "but does it really pay?"
>
> "Of course it does," was her reply.

Borsodi was skeptical: how could she compete with a firm like Campbell's, which had a skilled management, labor-saving machines, quantity buying, efficient machines, mass-production economies? He sat down to figure it out. He calculated how much the tomatoes had cost to grow, how much time he and Mrs. Borsodi spent in the garden, how much electricity was used in the kitchen, how much the canning jars had cost, and all the other expenses; and then he compared that with what it cost to buy canned tomatoes in the supermarkets near his home. The results astounded him: *"The cost of the home-made product was between 20 percent and 30 percent lower than the price of the factory-made merchandise."*

It was, he felt, as if "the economic activities of mankind for nearly two hundred years had been based upon a theory as false as its maritime activities prior to the discovery of the fact that the world was round."

Eventually, of course. Borsodi figured it out, and it proved to be a devastating indictment of the consequences of bigness. Then, as now, the complications rose not in the plants but far downstream from the lathes and belts and assembly lines. First, distribution. The more goods that are produced, the wider the market area must be, hence the more expensive the costs of distribution (including warehousing and transportation) throughout that area; it is now an accepted standard in the U.S. that, particularly for consumer goods, the unit costs of distribution will be higher than those of production, and they will increase as the price of gasoline goes up. Second, advertising. Mass production naturally necessitates sufficient advertising to create a mass market, and the more extensive it is the more expensive—which is why name-brand items are always more expensive than generics. (The high cost of advertising also tends to keep smaller and cheaper firms out of a market—creating an "entry barrier," in economic terms—thus reducing the competition that might lead to lower consumer prices.) Finally, promotion and packaging. In markets that are saturated, and where Brand A is not especially different from Brand B, it is necessary to find gimmicks that make a product stand out—bigger boxes, added partitions, coupons, toys, contests—and lead to added costs.

There are a few additional wrinkles, though, that Ralph Borsodi did not consider in his time but that also help to explain why big corporations do not provide lower costs. For one thing, the major American firms have discovered that price competition is just plain *uneconomic*—too unstable, too difficult to plan for, too risky. After all, if you've spent tens of millions of dollars and half a dozen years creating a new product—a methane car or yellow toothpaste—you want to be pretty sure of what you'll be able to sell it for, and that takes certain long-range planning and certain boundaries of agreement with your competitors about prices. It is not out-and-out price fixing in most cases (though the cardboard box people were found guilty of that as recently as 1979), but in every industry there is constant communication among the largest firms, all of them realizing that it is in their own interests not to fiddle around too much by lowering prices.

In addition, Borsodi could never have foreseen the immense role of the Federal government in the market economy. In some cases the government actually insists on price-fixing and the elimination of competition—as in trucking, where all firms are required to charge prefixed prices over pre-determined routes; and in some cases it attempts to limit the number of firms in a particular field or encourage mergers so as to reduce "wasteful" competition (as a Federal Anti-Trust Task Force put it in 1968, "In the regulated sector of the economy, the bias and its enforcement is overwhelmingly against competition"). Moreover, the never-ending billions of dollars from Washington are, in most cases, passed out to the biggest firms, usually without competitive bidding, in "cost-plus" contracts that encourage high prices, with an inevitable effect on consumer prices in all sectors of the market.

It is no accident that during this recent period of the greatest corporate concentration in history—the rise of the multinationals, the wave of mergers and acquisitions, the accelerated growth of the largest firms—the U.S. has undergone its most extreme and prolonged period of inflation—that is, of rising prices. If there were really some connection between big business and low prices it is logical to think that we might have seen some signs of it by now.

Profitability. Finally, what other justification is more important than sheer profitability—as they say, the bottom line? For if a big corporation is not profitable, then what other excuse could there be for its size?

It should be no surprise that, for all of the reasons above—from diseconomies of bureaucracy to higher advertising costs—large firms tend to be simply less profitable than small ones. Nothing new about it—it is a fact that has been documented at least since the Twentieth Century Fund study of corporate profitability in 1919 ("the larger corporations earned less than the average of all corporations . . . and the earnings declined almost uninterruptedly, with increasing size"); again by the Temporary National Economic Committee in 1940 ("Those with an investment under $500,000 enjoyed a higher return than those with more than $5,000,000 and twice as high a return as those with more than $50,000,000"); again by California economist H. O. Stekler, examining profits of manufacturing firms in 1949 and 1955–57 ("For the profitable firm, there is a declining relationship between profitability and size"); and yet again by the U.S. Senate Antitrust Subcommittee in 1965 (in twenty-three out of thirty basic industries, profit rates either decreased consistently as firm size increased or else had no connection with increased size). The most graphic demonstration of small firm profitability is probably this table compiled by Stekler:

SIZE AND PROFITABILITY

Assets	Rate of Return
$0–50,000	137
50–100,000	130
100–250,000	120
250–500,000	118
500–1,000,000	119
1–5,000,000	113
5–10,000,000	108
10–50,000,000	105
50–100,000,000	107
100,000,000 +	100

As Barry Stein concludes from these figures: "It is clear that, per asset dollar . . . smaller firms are more efficient users of capital." Something to bear in mind when we next hear from Mobil and General Motors.

MYTHS DIE HARD, of course, particularly when they are fostered by very powerful institutions whose interests they serve. After all, it would be unseemly for this nation to admit the fact that the Boston Tea Party was an act born not of patriotism but of John Hancock's predicament in having a warehouse full of tea that was about to be undersold by the British shipment just then in the harbor—against which he naturally protected himself by hiring a bunch of dockside roustabouts to readjust the competition.

Still, it is always wise to recognize myths for what they are. Repetition may increase their persuasiveness, but it cannot alter the facts. And the facts behind the myths of corporate size are incontrovertible.

3

Standards of Living

THAT'S ALL WELL AND GOOD, but you can't just do away with the large corporations that are the basic sinews of our modern economy without totally destroying the American economic way of life and the prosperous society that stems from it. Even if the corporate world is actually inefficient, unproductive, stodgy, and all the rest of it, it has at least produced a standard of living that has placed the U.S. far above most of the other nations of the world.

Or so the argument goes.

It may be the most fallacious myth of all.

LET US TAKE a look at what this standard of living really is.

First, it seems logical that any true measure of that concept would have to include not only the *material* standards of a broad spectrum of the population but the less measurable *social* standards as well. Clearly one could take the Gross National Product and argue that as it rises so does our standard of living—and in fact this is the familiar tactic of most of our generals and captains of industry. But as we have now come to learn, the Gross National Product measures none of the real things that enter our daily lives—our relationships, our feelings of security, our friendships, our sense of community, our freedom from anxiety, our happiness. If I have a terrible accident driving home from work, and my car is demolished, and a dozen police cars and ambulances and towtrucks are called to the scene, and I'm taken off to the hospital to be attended to by a battery of surgeons for several weeks, and if my wife is forced to buy a new car, and several crews are called in to restore the dividers and lamp stanchions I have knocked over—all of this is just *wonderful* as measured by the GNP, since it gives added employment to all kinds of people in all kinds of trades. Lying there in traction, however, on the brink of death, I may be forgiven if I don't share in the general economic euphoria.

Similarly, so many personal and social "bads" are, in this system, economic "goods." Pollution, congestion, crime, waste, alcoholism, corruption, accidents, disasters—all of them end up in the GNP as plusses: they're *good* for the economy. Hence though we may be told that our GNP is ten times greater than it was three generations ago, and the sheer number of products available twenty times greater, it may merely mean that our problems are ten or twenty times greater; we would certainly be hard-pressed to say that as a society we are ten or twenty times better off, or happier, than our grandparents were. Indeed, in not a few respects—the health of the food we eat, say, or the stability of our families and communities—we may realize ourselves rather less well-off. I do not believe that anyone measuring the particulates in the air or comparing the taste of the chickens or examining the curricula in the high schools or studying the rates of teenage suicides or observing the litter along the roadsides or comparing the civility of salespeople could seriously argue that our "standard of living," in any real way, had *improved* as a result of mid-century industrialism.

BUT PERHAPS THAT is too unscientific an approach to this question. What if we were to stick to hard economic data and see what that says about how much better people are living today?

Of course one might say it's easy to tell that we live "better," in the sense that we have television and airplanes and computers that were not available three generations ago. And yet the most sophisticated economic analysis yet devised suggests that in purely material terms we are living now only a small bit better than we were fifty years ago. Economists William Nordhaus and James Tobin of Yale University, after long and laborious calculations, contrived what is acknowledged by the trade to be the premier analysis of the U.S. economic living standard, something they call the MEW—the Measure of Economic Welfare. It is based on a complex matrix, but in essence it is designed to take into account not just the GNP but such things as the value of leisure activities, non-market transactions, social and urban disamenities, government operations, and similar "non-economic" calculations. By the Nordhaus-Tobin measurements, correcting for inflation, the per capita MEW was $4,506 in 1929 and rose to $6,391 in 1965—in other words, the real purchasing power of the individual had been increased by only $1,885 in all those years; a pleasant enough increase, I suppose, even if it's only about one percent a year, but it doesn't seem to argue very much that when I got married I had $1,885 more to spend than my father did when he got married.

Moreover, we must remember that there is some difference between *general* economic welfare and real *individual* prosperity. True, the GNP has gone up a great deal since 1945, but the percentage of it that is devoted to consumer goods, the things that really best define our standard of living, has not changed much in all those years: 56 percent in 1945, 67 percent in 1950, 63 percent in 1970, 64 percent in 1975. And when we look at personal expenditures for *non*-durable goods—alcohol, tobacco, food, travel, recreation, domestic help, education, all the things that might be said to be evidence of the real luxuries of life—we find that (in 1972 dollars) per capita expenditures have gone only from $1,122 in 1955 to $1,537 in 1977, an increase of just 37 percent and a sum total of only $415 in all those twenty years, not much evidence of unbridled prosperity.

BUT MAYBE THE most accurate way to measure standard of living is not in simple monetary terms but by the kind of high technology the U.S. has developed and the labor-saving devices that surround us both at home and work.

There is a lot of it, no doubt. But of its worth? It is sobering to realize just how much of all that technology is not only dangerous and unhealthy, not only resource- and energy-intensive, not only costly and indulgent, but downright *unnecessary:* it exists, more often than not, to correct some other previous technological or social error. We have all these devices around us, in other words, mostly because we *created* the need for them.

We have the technical miracle of power steering in the modern automobile not because anyone thought it was necessary to turn a corner with the press of a forefinger but because one year the engineers at Chrysler had created a car so heavy and unwieldy that there was no way to maneuver the thing by conventional steering mechanisms and they had to devise something to make it sellable. We have an ingenious array of "hypo-allergenic" cosmetics and skin medicines because normal cosmetics do such damage to most people's faces (according to the Cosmetic, Toiletries, and Fragrance Association, a trade group, it is nuns who have the best skin because "they normally don't use cosmetics"). We have remarkable chemicals that can make gray meat look red and keep aging food from rotting—even though they make the food taste worse and poison our systems besides—solely because we developed vast supermarket chains with mass buying and mass marketing and mass delays that kept real foods in the pipeline for too long and caused much of it to spoil.

In case after case, we can trace the technological achievement not to success but to failure. Invention, it seems, has become the mother of necessity. As Fritz Schumacher used to say often, "Would the ancients be more amazed at the marvels of our dental technology or the rottenness of our teeth?"

And what is it that this labor-saving technology actually does for us, anyway?

▪ It does not seem to have created a quieter, more easygoing, more leisurely society, as one might have expected—anyone who has been to Antigua, say, or Botswana or Mexico or even Portugal or Greece will realize how much less hectic the pace, how much less intense the strain, in those low-technology places than in the U.S. or Germany or even Sweden.

▪ It does not seem to have made life better for the countless workers who have been displaced by it and usually end up as social burdens on the welfare rolls—as, to take just one minor example, the 27,000 workers in the tomato fields of California who were able to document that the introduction of mechanical tomato-pickers cost them their jobs, their social functions, and their standing in society. Nor, must it be added, has it worked all that well for the businessmen who have adopted it, only to discover that the initial costs are exorbitant, the replacement costs are punitive, and the energy costs in between are excoriating.

▪ It does not seem to have done much at all to increase the leisure time it is presumably there for. Although the official U.S. workweek has declined from sixty hours at the beginning of this century to thirty-five today, a major study by Sebastian de Grazia for the Twentieth Century Fund found that workers are not actually working any less. To make ends meet, or merely to satisfy the conditions of their employment, they are putting in overtime, moonlighting on another job, taking work home, working in their basement shops, or commuting to work, and then are forced to do the household chores and workaday tasks that fill up the rest of those twenty-five extra hours a week of apparent "leisure" time. "The time involved in activities off the plant premises but work-related nevertheless," de Grazia found, "this is not less than it was at the turn of the Twentieth Century. . . . The American is actually working as hard as ever."

▪ It does not seem to have provided any more free time than was enjoyed by that scorned symbol of primitiveness, the "cave man." A comprehensive examination of anthropological data by Marshall Sahlins in his fascinating book *Stone Age Economy* suggests that ancient Paleolithic societies and contemporary hunting-gathering cultures rarely put in more than four hours a day at "work," that is, hunting and preparing food—the !Kung, for example, typically spend less than three hours a day, Australian aborigines put in four or five hours on average, the Peruvian Machiguenga contribute a little over three hours a day, the Hazda tribe in Africa manage on less than two. The modern American worker, even if laboring for only thirty-five hours a week, might well envy this schedule, and the American farmer, known for sunup-to-sundown toil even with garage-fulls of machinery, ought to be positively green.

▪ And it does not even seem to have eased any real burdens in the place where it is most often thought of as being vital: the American home. Joann Vanek, a sociologist at Queens College in New York, examined in considerable detail the activities of American women in the 1920s and the 1960s and found, somewhat to her surprise, that despite all of the household gadgets that have become standard over those years, women who are primarily homemakers "in fact devote as much time to housework as their forebears did." In 1924 the average woman spent about fifty-two hours a week doing housework; in the 1960s the average woman not in the outside labor force spent fifty-five hours a week doing housework. (Women *in* the labor force in the 1960s spent only twenty-six hours on housework, apparently because, Vanek suggests, by bringing in a paycheck they "do not feel the same pressure" to justify themselves putting in long hours at home.) But, but . . . what about all those appliances—what about washing machines, for example? Well, in 1925 the housewife spent a little more than five hours a week doing laundry, in 1966 she was spending more than six—because, apparently, people have more clothes to wash and wash them more often. And blenders and processors and slicers and all the rest of the electric kitchen? The housewife of 1926 put in about twenty-three hours a week preparing food; the housewife of 1968 about nineteen, not appreciably different. "It appears," Vanek concludes, "that modern life has not shortened the woman's work day. . . . Indeed, for married women in full-time jobs the work day is probably longer than it was for their grandmothers."

INDEED, I THINK it is fairly easy to make the argument that our material standard of living might be *improved* if we were ever to free ourselves of the high-technology economy and move in the direction of a small-scale and self-sustaining world. I mean that in all seriousness, bizarre though it may sound.

Just think. Suppose we were to eliminate, at a stroke, advertising and promotion of every product sold in the United States—that would save the society approximately *$40 billion* a year. I do appreciate that it would deprive some people of their livelihood—around 100,000 people are employed in advertising these days—but there should be no great difficulty in finding jobs for that small a number of people, and even if we were to pay each one of them $30,000 just to sit around the house we'd still come out $37 billion ahead. I also realize that this would deprive certain media of a major source of income (while creating much relief for the rest of us), forcing certain readjustments upon them, but the effect is most likely to be the creation of more, smaller, locally attentive outlets, and any serious dislocations could be made up for with a tiny fraction of the money we now spend. Moreover, the elimination of advertising would mean that virtually every product would cost less, some considerably less, many unnecessary commodities would drop from sight, wasteful fashion and model changes would be almost eliminated, the print media would save paper by the forestful, and all in all a considerable savings of both money and resources—at a rough estimate, I'd say another $40 billion—would accrue to the society. Not to mention the psychic benefits of being free of jarring commercials, roadside ugliness, cluttered magazines, the greater portion of junk mail, the pressures of Jonesism, and the sexual and psychological manipulation that is a standard of the trade.

Or suppose that we could capture and rechannel the resources—both natural and human—that are now being wasted on the production of useless, shoddy, trivial, insubstantial, unnecessary, status-ridden, duplicative, meaningless, and undesirable products (one would not want to call them "goods"). Take cars. The process by which General Motors fractures itself into half-a-dozen divisions, each of which then produces a half-a-dozen separate but virtually identical models, and dozens of variations upon each model, and then puts out model changes every two or three years, is faintly comical and extremely expensive; one economic survey determined that the extra price of cars merely as a result of model changes alone cost consumers about $750 per car, or $5 *billion* per year. Add to this the whole familiar strategy of built-in obsolescence to assure that the cars won't last successfully beyond a couple of years—reinforced in GM's case by its refusal to manufacture replacement parts after a number of years—and the whole thing becomes so obviously undesirable by any true social or economic standard that it seems almost amazing that it is permitted to continue. True, this making of cars does generate a lot of jobs—perhaps a million, no more—and churn around $75 billion a year or so. But can that really be worth the incredible waste, year after year, of raw materials, of energy, of human labor, that might easily be channeled elsewhere? Could not a rational society—even one that wished to have some cars around—find more productive ends than this?

Or suppose that we decided to abandon the custom of stretching out dead bodies in elaborate boxes in a room full of flowers—that alone would save American families upwards of $4 billion a year. Or determined that the chewing-gum business was not an essential industry—that would provide the society with an extra $1 billion a year. Or concluded that moonshots and space shuttles had about petered out in usefulness to American psyches and sciences—that would provide us with an additional $6 billion a year. Or chose to abandon the elegant variation of packaging in all its infinitude—that would enrich the society by approximately $23 billion (1978 figures) a year.

It is almost impossible to get reliable money figures for the kinds of madness that we continue to perpetrate upon ourselves in the folly of what we think of as our "economy" (the word did once mean "thrifty") but various experts from time to time have tried, and a compilation of some of their estimates provides a rather revealing idea of how *substandard* our standard of living is, compared to what it could be were we not pouring money down these drains. For example, these are the approximate savings should a healthy society decide to eliminate or sharply reduce:

• built-in obsolescence, fashion decrees, model changes, and the like (using 1978 figures)	$64 billion
• intentionally shoddy work (as in car repairs, an $8–10 billion-a-year fraud, according to Senator Philip Hart), useless or redundant products (as $1 billion worth of drugs), and duplicative or unneeded services (as $2 billion worth of insurance premiums)	302 billion
• advertising, promotion, packaging, product differentiation	70 billion
• business cheating, corruption, white-collar crime, and consumer fraud (including what the *Wall Street Journal* figures to be $1.5–10 billion a year in shortweighting alone)	200 billion
• industries based on social unease and distress, such as gambling, alcohol, tobacco, drugs, pornography, prostitution, and spectator sports	82 billion
• crime prevention ($20 billion in 1976) and crime losses	42 billion
• accidents and illnesses due to pollution of air and water, inadequate safety measures on the job, and psychological dysfunctions brought on by rat-race stresses	12 billion
• government regulations attempting to restrain and supervise the multiple abuses and illegalities of private businesses (and consequent paperwork by those businesses)	103 billion
• government-sponsored and -created wastage through subsidies, cost-plus contracts, unbid contracts, boondogles, and so on	175 billion
• tax loopholes	77 billion
TOTAL	$1,127 billion

Of course these are only the crudest guesses, I don't wish any misunderstanding about that; and even using the *Wall Street Journal* and Senator Hart among my sources, I don't mean to suggest that anyone in this enterprise has anything but a general approximation of the ultimate truth. Still, such estimates can certainly indicate some idea of the magnitude of the money at stake here and the feasibility of the claim that, to some degree at least, our present economy actually throws away an immense amount of its resources, thereby impoverishing rather than enriching the social whole. An alternative economy that was able to make substantial savings here, therefore, would by any means be living at a lower standard—and probably would have little trouble reaching a higher.

And it is conceivable that such a society might think of contriving, with its new-found trillion dollars, some cathedrals, some palaces, some symphonies to enrich its age.

But wait a minute—how about all those people out of work? Won't a trillion-dollar dislocation be depriving countless people of jobs? In one sense, yes, if "deprivation" is used loosely. Clearly if there is no more advertising there is no more Madison Avenue, and if the defense industry is halved by doing without cost-plus then half its workers will be laid off. But we are talking about a rational society here, one in which the elimination of such jobs is a "deprivation" only in the crudest sense—it certainly does not deprive the worker of meaningful activity or the society of essential goods. This is "deprivation" that enriches. And would enrich even more were those people to be re-employed on projects that would be uplifting for them and valuable for the society—rebuilding the eroded railroad beds, for example, or teaching illiterate youths or writing intelligent television scripts or converting buildings to solar energy (that alone, according to the Congressional Joint Economic Subcommittee in 1979, could employ at least 3 million people). Salary? Well, we have this $1 trillion, see. . . .

Indeed, the simple fact is that most people in this nation are *not needed* in their work, are not performing any particularly useful social functions, and the sooner they found re-employment the better for them and all of us. Keynes long ago pointed out that hiring half the unemployed to dig holes and the other half to fill them up would solve the unemployment problem, and that was dismissed as a clever donnish remark, but he was righter than he knew. For that is really how much of our current economy operates right now: many millions of citizens are doing nothing of *value* for society, only performing the necessary function of getting a paycheck and redistributing it to the manufacturers, banks, and governments of their choice. The defense worker is almost certainly not necessary, considering that the U.S. has already stockpiled enough weaponry to blow the earth apart several hundred times over; the designers creating biannual fashion changes in clothes and the workers who make and model and sell them are clearly not necessary, their talents wasted in an enterprise remarkable for being without any socially redeeming features whatsoever, nor are advertising executives, casino dealers, pornographers, fast-food purveyors, people who make sleeping pills or Barbie dolls or annual reports or underarm deodorants or celebrity books or plastic flowers . . . but need I go on?

Figure it this way. Only 40 percent of the population is in the labor force. Of that, according to such calculators as Herman Kahn, only 2 percent is necessary for the "primary" industries that basically keep us going, agriculture and mining. Add to that the workers that would be necessary for the manufacture and production of essential goods—approximately 5 percent more of the labor force, according to the calculations of Scott Burns—and we'd arrive at approximately 7 percent of all American workers, or 3 percent of the total population, that is in any sense truly necessary to keep the nation going. I should think it would behoove the other 97 percent of us to find something else to do, and the sooner the better.

Freeing the society's resources from waste and stupidity, then, and restructuring employment toward the essentials of goods and services would seem likely to promise not a lowered standard of living but a higher one. The economist Leopold Kohr has explained in theoretical terms why this might be so. He has constructed a diagram that compares the growth of *consumption* with the growth of *society* as a whole, and shows how, beyond a certain point, total consumption will become geared not to provide more luxuries for the citizens but simply to add to the powers of the state and modify the very troubles that increasing size and density will bring:

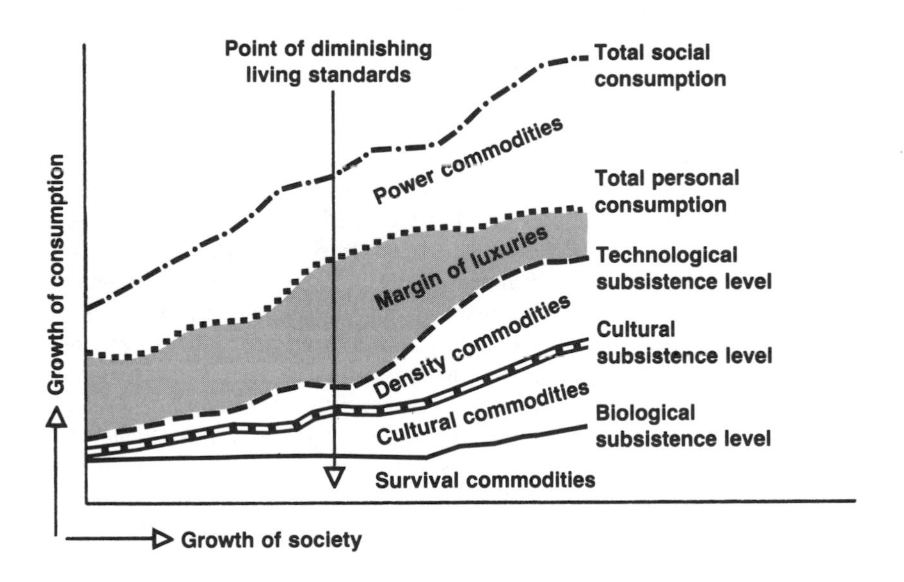

Living standards begin to diminish here at the point where the citizens must pay *less* for their own personal luxuries and *more* for the "power commodities" of the state—governmental expenditures at all levels, typically for defense, administration, and debts—and for the "density commodities" of the society—goods and services necessary simply to allow large numbers of people to function at all (commutation, insurance, all kinds of licenses, legal and medical fees, catalytic converters, burglar alarms, psychiatric consultation). Even with a rising GNP, in other words, a society that has gone beyond its optimum size will experience an increasing *decline* in its real standard of living—as measured by the "margin of luxuries"—the bigger it gets. And chasing after that elusive ever-growing economy, by technofix or otherwise, will only make it worse.

TOM BENDER IS an unusual man, one of the founders of *RAIN* magazine—the irreverent bible of the alternative-technology movement—and a person who combines a wealth of practical around-the-house mechanical ability with a wisdom and judgment I can only call philosophical. His notions of the possible economic future, though some may find them romantic, seem to me to be eminently reasonable:

> If the monetary savings of self-reliance turn out to be anywhere near what they look to be, a whole new world of options opens to us. Options to be more relaxed in our work and to restore the satisfactions of work well done and work done in more rewarding ways that we have given up in our forced pursuit of efficiency. Options to press less heavily on others and on our resources. Options to put our effort into creating a beautiful and satisfying world rather than a wealthy, efficient or powerful one. Options to create a new Golden Age with the freedom and energy to pursue the harmonious development of all our capabilities—that to love or to make music as well as that to make money.
>
> New dreams are possible. The Golden Age of almost every society in history has occurred *not* when all a culture's energies were focussed on increasing its wealth and power, but rather when the attainable limits of those dreams were reached and people realized that such goals had not left them with the quality, beauty or personal happiness they had envisioned. Freeing the vast resources of society that are now channeled into things that are no longer attainable makes new visions possible and attainable. Once we realize that greatness is not achievable through vast expenditures of resources but requires the development and refinement of our own personal abilities, we discover that our present wealth is more than adequate to achieve an equitable and golden age for the whole world.

That is surely the standard of living to which we strive.

4

Steady-State

IT WAS IN 1848, that notably fruitful year, that the idea of the steady-state economy was first put forth, by the political economist John Stuart Mill, in his path-breaking *Principles of Economics.*[1] Growth, as he saw, was inherent in capitalism, particularly the industrial capitalism then beginning to stamp itself upon the United Kingdom, and yet it was by definition a finite process: "the increase of wealth is not boundless." Capital will eventually cease to produce any sensible return for the capitalist, he posited, when the cost of extraction, manufacture, and disposal become great enough and when the redistributive tax systems of the state become extensive enough. At that point there will no longer be any point in being a capitalist: the breed will vanish. In place of the "progressive state," the "stationary state":

> I am inclined to believe that it would be, on the whole, a very considerable improvement on our present condition. I confess I am not charmed with the ideal of life held out by these who think that the normal state of human beings is that of struggling to get on; that the trampling, crushing, elbowing, and treading on each other's heels, which form the existing type of social life, are the most desirable lot of human kind, or anything but the disagreeable symptoms of one of the phases of industrial progress. . . .
>
> The best state for human nature is that in which, while no one is poor, no one desires to be richer, nor has any reason to fear being thrust back by the efforts of others to push themselves forward. . . .
>
> It is scarcely necessary to remark that a stationary condition of capital and population implies no stationary state of human improvement. There would be as much scope as ever for all kinds of mental, cultural, and moral and social progress; as much room for improving

1. Others before him had of course debated the idea of *growth,* especially economic growth, most notably the French physiocrats in the eighteenth century (they were mostly against it) and such Greeks as Aristotle (likewise). But Mill was the first to put the case of the "stationary" economy in positive, concrete terms.

the Art of Living, and much more likelihood of its being improved, when minds ceased to be engrossed by the art of getting on.

Like many ideas, the measure of its worth varied indirectly with the time it took to become recognized by others: for more than 150 years it was virtually ignored, and the few who noticed it—professionals, mostly—understood it little. With the revival of interest in ecology in the 1960s, however, and then with the energy crisis of the 1970s, a number of modern economists began to re-examine the notion of a stationary economy and reshape it for modern conditions.

Economist Kenneth Boulding in 1966 introduced the idea of "spaceship earth," by which he meant to suggest that we live within a basically closed system that must ultimately, like the spaceship, depend on its use of recycled materials and renewable energy. British economist Ezra Mishan in 1967 offered the idea of "growthmania" in his elegant and influential *Costs of Economic Growth* and suggested the rejection of "economic growth as a prior aim of policy in favor of a policy seeking to apply more selective criteria of welfare . . . to direct our national resources and our ingenuity to recreating an environment that will gratify and inspire men." In 1971 Rumanian émigré Nicholas Georgescu-Roegen, working out of the University of Tennessee, came forth with the principle of increasing material "entropy," an idea borrowed from the Second Law of Thermodynamics, to suggest that the excessive production and consumption of contemporary society was exhausting the finite material and energy resources of the earth to the point where soon the human species itself would be threatened. And the next year Jay Forrester, Donella and Dennis Meadows, and others at MIT hit the headlines with their pioneering computer models that became the basis of the Club of Rome's initial admonitory book, *The Limits to Growth.*

After that, a continual drumroll. Herman Daly's explorations, *Essays Toward a Steady-State Economy,* became available in 1972, Fritz Schumacher's path-breaking and deservedly popular *Small Is Beautiful* appeared early in 1973 as did Leopold Kohr's *Development Without Aid,* and the influential scholarly quarterly *Daedalus* came out with a full issue on "The No-Growth Society" in the fall of 1973. Other important writers joined in: Hazel Henderson, Rufus Miles, René Dubos, Barbara Ward, Robert Theobald, Howard Odum, William Ophuls. And by the end of the 1970s the notion of the no-growth, stationary, steady-state economy was sufficiently well established to be creating ripples throughout the circles of both academic and governmental economists.

Though the visions of the steady-state economy inevitably vary according to the particular proponent, they all agree on certain basic points. A growth economy uses up scarce resources, emphasizes consumption over conservation, creates pollution and waste in the process of production, and engenders inter- and intra-national competition for dwindling supplies. A steady-state economy minimizes resource use, sets production on small and self-controlled scales, emphasizes conservation and recycling, limits pollution and waste, and accepts the finite limits of a single world and of a single ultimate source of energy.

Herman Daly puts it this way: "If the world is a finite complex system that has evolved with reference to a fixed rate of flow of solar energy, then any economy that seeks indefinite expansion of its stocks and the associated material and energy-maintenance flows will sooner or later hit limits." Nothing can expand forever. By contrast, the steady-state economy is one "with constant stocks of people and artifacts, maintained at some desired, sufficient levels by low rates of maintenance 'throughput,' that is, by the lowest feasible flows of matter and energy." Or as Boulding says it more elegantly: "The essential measure of the success of the [steady-state] economy is not production and consumption at all, but the nature, extent, quality, and complexity of the total capital stock, including in this the state of human bodies and minds included in the system."

It is the art of living, not the art of getting on.

Perhaps the best way to envision the steady-state economy is in terms of ecological harmony. For in a real sense a steady-state economy takes the stable ecosystem as a *model,* since a properly balanced environment is in fact essentially stationary. Growth occurs in fluctuations from season to season and species to species, but overall any given system may be basically unchanged for eons; and however fierce may be the competition between some species, however preoccupied each may be with its own survival, none of the elements within it (except, in self-delusion, the human) has found any particular advantage in the growth of the system as a whole.

OBVIOUSLY THERE IS plenty of room for objection to the stationary economy, even after all the usual resistance of the myth-prisoners is disposed of.

The steady-state economy, say some, *is simply not necessary, it is a needless deprivation—for continued growth will provide the solutions to the admittedly severe but tractable problems we have now.* The idea that growth will provide technical breakthroughs is the old technofix response, and it amounts to recommending more arsenic for the patient dying of arsenic-poisoning. The variant that it will create money we can use for, in economist Henry Wallich's phrase, "The Great Cleanup," is something like the assurance of the miser who asks us to let him horde more money so that, someday, he may give even more of it to charity. And the Kahnian idea that it will force people to pay more for scarce goods and thus eventually limit plunder by price is akin to allowing the rich kid to buy up all the candy on the block in the name of conservation and letting him sell it back at any price he chooses. You will recognize the tone of the growth-at-all-costs argument, of course, for it is the standard American business response become familiar now through years of Mobil Oil ads. That source does not necessarily discredit it, perhaps, but it does not lend it much credence either—especially since it is precisely American business that has got us into the fix we're in. Fat men don't cure obesity by eating.

But the steady-state economy, say others, *would mean a permanent underclass of the poor because where there is no growth there is no way out of poverty.* Of course growth *hasn't* eliminated poverty in all the centuries it's been tried, not even come close, and there's not much reason to expect that it somehow miraculously will. The reason is simple: except for a few exceptional cases of up-from-poverty, the wealth of capitalism tends to go largely to the same old wealthy families and entrepreneurs who have always had it. The advantage of the stationary economy here is that—especially if confined to a particular community or region—it puts its primary value on the *limitation* of accumulation and consumption; it does not have to operate within the profit matrix, which always forces prices up, and therefore it has more wealth to distribute generally; and it explicitly values the human component in a balanced ecosystem, wherein the wastage of human potential would be considered as anti-social as the wastage of any other raw material. This is not to say that there might not be gradations of wealth within a steady-state economy—depending, of course, on the community's particular political and social ideals—but that these are not inherent, as they are in a growth system, nor are they necessary or desirable, nor do the mechanics of the economy work to support them.

But such an economy, say still others, *would lead to greater unemployment and labor unrest, maybe even social conflict, because without growth there are insufficient jobs.* Aside from the fact that the growth economy, even in the U.S., does not seem to have eliminated unemployment and labor unrest, it also has one insuperable built-in deficiency here. Because it depends on continual technical "progress," it always displaces the less educated and less skilled parts of the labor force with automated machinery and the skilled workers who understand it, leaving the former forever unemployed; and then it tends to restrict the chances for the necessary education and skills to the upper levels that already have them, the colleges remaining essentially the province of the upper-middle echelons. A stationary system, by contrast, with no careening interest in "progress," would welcome only those technical changes that remained consistent with its ecological values; it would also be labor-intensive rather than fuel-intensive, just out of simple conservation, and this would emphasize the employment of all able-bodied citizens; and especially if it were enmeshed in the small community, it would continually be generating countless occupations, from baby-sitter to health counsellor, for the less able-bodied, the aged, and the young.

Even so, a final set of critics says, *the steady-state economy, supposing it were achieved and running smoothly, would soon lead to exploitation and chaos, on the one hand, or excessive government control, on the other.* For if you deny people the opportunity to exploit nature the way the Western world has been doing these past five centuries, they are likely to seek their advantage by exploiting their fellow humans; yet any system that was strong enough to prevent that and at the same time to assure that minerals were not exhausted or waterways polluted would have to be excessively autocratic. Several ready answers to this come to mind: this posits an unacceptably crude and greedy view of human nature to begin with; it does not recognize that exploitation may not be a necessity in economic systems other than capitalism; and it does not, alas, describe a world very much different from the one we have right now, which has plenty of both chaos and autocracy. But there is a more substantial point here, and it has to do with the very nature of a steady-state economy. To arrive at, much less to sustain, such an economy, the citizens as a whole would have to be able to see that it is in their collective self-interest, and of course in the interests of their children, to live within limits. They would have to experience in their daily lives the benefits of limits—such as, say, the absence of advertising and rush hours. Such a perception can come about, it is fair to say, only within a fairly small geographic compass, where people still can see the effect of each of them on the environment and on their fellows, where there is not an adversarian but a communitarian perception of government and society. But where it does take root, neither exploitation nor autocracy would be much of a possibility, for that would be at such odds with the general perception of the common good that it would seem as foreign as an endotoxin to a healthy cell, and be as quickly discarded.

The spirit of getting on, in other words, would be replaced by the spirit of living within. The steady-state economy means stewardship instead of exploitation, enoughness instead of too-muchness, living instead of making a living.

SOME YEARS AGO Garrett Hardin, a professor of biology at the University of California, shocked the academic world with a parable he called "The Tragedy of the Commons." As he posited it, contemporary society is very much like the ancient commons ground upon which the village shepherds would graze their animals. At a certain point, inevitably, that commons will be unable to support any more sheep—there is just enough grass for each animal, and the addition of a single new animal will mean the overgrazing of the land and, eventually, as more sheep are added, the loss of all its ground cover, the erosion of its soils, and ultimately its effective destruction. And yet it is in the self-interest of each shepherd to add one more sheep to his flock because the added gain to him and his family, in terms of milk produced and meat to be sold, is considerably greater than any immediate loss from the eventual destruction of the commons. Thus each shepherd grazing sheep there, operating out of real and apparent self-interest, will continue to add more animals to his flock until the ultimate tragedy happens and there is no more commons for anyone.

This, Hardin suggests, is our present condition. The oceans continue to be overfished, particularly by Russian and Japanese fleets, because it is in the immediate self-interest of each nation to take as much as it can from the sea—particularly if there is a threat of eventual depletion—even though every nation knows that at some near time there will be no more fish at all for any to catch. American steel manufacturers continue to resist measures to limit the poisonous emissions they pour into the air each day, because it is in the interest of each company to avoid the expense of pollution controls and redesigned furnaces for as long as possible, although all steel company executives know that they and their families and workers and townsfolk have to breathe the resultant polluted air and that this is very likely to cause severe illness and early death. The American West is becoming desertified, losing perhaps 10 million acres of grassland a year, for the simple reason that the ranchers of the area, acting out Hardin's scenario, continue to overgraze their herds in this sparse territory despite a decade-long policy of the Bureau of Land Management to curtail them; as reporter Molly Ivins of the *New York Times* has noted, these are lands where, "the early explorers wrote, the grasses brushed the bellies of their horses. Nowadays, the sorry weeds would be doing well to brush a trundling armadillo." The examples, alas, may be multiplied.

No doubt this parable does suggest an important, and unquestionably tragic, truth. But it contains something else as well: an indictment of the getting-on economy and an appreciation of the steady-state one. True, those who are operating out of *self-interest,* who do not see and cannot feel a *communal-interest,* will almost inevitably proceed to their self-aggrandizement and the destruction of their surroundings, moderated only by the feeble constraints of church or government. But if there were a stationary economy it would have to be grounded precisely in the perception of communal-interest, quite unable to exist without a realization of the unity of the ecosystem and the simple centrality of ecological balance. The shepherd would know the limits of the commons and its importance to his family, to his children and their children, to his neighbors, to the past and future of his community, and that overriding communal-interest would easily outweigh the possible personal gain of putting another animal out to feed.

This, in turn, is more or less a function of size, both populational and geographic. There can be no "communal-interest" among 200 million people, or 20 million, or even 2 million, because there is no way for the human heart with all its limitations to perceive the interconnectedness of all those lives and their relevance to its single life; we cheat on our income tax and drive at 65 m.p.h. and ignore beggars on the street because we perceive no community at the scale at which we live. Nor can there be communal-interest over distances of 3,000 square miles, or 300 square miles, or even 30 square miles, because there is no way for the human mind in all its frailty to conceive the complexity of an ecosystem so large and its single place within it; we use plastic bags in the supermarket and wash clothes with phosphate detergents and drive untuned cars because we cannot understand ecology at the scale at which we function. Only when the shepherd knows his world and the people in it and feels their importance to his own well-being, only when he realizes that his self-interest *is* indeed the communal-interest, will he voluntarily limit his flock.

Only then will the looming tragedy of the commons be avoided.

IN THE MIDST of his arguments for a steady-state economy, Herman Daly tells the story of the village idiot who would be stopped every day by the townspeople and asked to pick between a nickel and a dime. The idiot always chose the nickel and the residents always went away saying, "There, you see what an idiot he is." Except that the idiot in later life explained: "After all, if I kept picking the dime they would have stopped offering it to me pretty quickly. This way I kept getting nickels every day."

And just so with the stationary economy. Its riches might not be so great at first, and we might seem to be idiotic for settling for them; but over time they would be sure to accumulate. Eventually they would far surpass those we know now.

5

Steady-State in Hiding

THE TRUE STEADY-STATE economy does not yet exist, of course, anywhere in the world today: the rich nations plunder and produce, the poor nations supply and buy, and within each of them the getters-on scramble over the getters-by.

But there is a most extraordinary . . . trend? stirring? perhaps "movement" is not too strong a word . . . in that direction within most of the nations of the industrial world, most pronounced in fact in the most industrialized: a loose, in some aspects subliminal, usually inchoate development that seems almost to be the acting-out of some deep and hidden longing on the part of a great number of people for something other than the conventional capitalist economy. The evidence for the trend is extremely diverse and some of it only fragmentary; some of it, as would be natural in a period of transition, is even contradictory. But taken together, and understood as manifesting something flexible and transitional, it gives unmistakable proof of what might easily be the groundwork, perhaps the substructure, of a steady-state economy in hiding. It is all the more important for being practically unknown.

The underground economy. Defined by *Newsweek* as "the vast and murky nether world of jobs, services and business transactions that never show up in official records," the underground—or, variously, "invisible," "subterranean," "black," or "hidden"—economy is such a chimerical and amorphous thing that it cannot be defined with exactitude. It includes baby-sitting and drug-running, bartering and tag sales, tool-sharing and street-vending, some practices ages old and purely capitalistic, others much newer and clearly expressive of some communal or ecological concern, and virtually all of it on a small scale. Its business is transacted without any official agency keeping tabs, its usual medium of exchange is barter or cold cash, it is not reported to or taxed by the IRS, and it is most often done among friends and neighbors.

Altogether the underground economy in the U.S. is thought to involve more than 20 million people and amount to about 10 percent of the official GNP, or an estimated $200 billion in 1978, according to Peter M. Gutmann, chairman of the economics department at Baruch College of the City University of New York. (A second economy exists also throughout Europe—as, for example, *travail au noir* and *schwarzarbeit* —and probably is even more pervasive; the chief tax collector of the United Kingdom has estimated an off-the-books turnover of up to $13 billion, and one Italian estimate suggests $27 billion.) That could be on the high side—the IRS officially estimates $100 billion—but whatever the figure, the phenomenon it points to is clearly widespread at all levels of the economy and in all walks of life.

Some of it operates for essentially capitalist motives, underground because it is illegal—as with the farms in California and Hawaii devoted to raising extensive marijuana crops (an estimated $300 million a year in one four-county section of northern California); this part reflects steady-state ideals only in the sense that it is small-scale, local, limited in ecological impact, and private rather than corporate.

Some of it represents an unquenchable individualism, reinforced these days by tight budgets—the retiree who stretches income as a tutor, the police officer who moonlights as a cabinetmaker, the high-school dropout who becomes a small-appliance repairman; this partakes more of the steady-state sensibility in that these are most often people outside the economic mainstream, who operate as individuals, who place only the most limited strain on resource-supply or waste-removal, and who find a particular social value in being their own bosses, setting their own hours, and using their own real skills.

And some of it, involving probably the greatest number of people, reflects a traditional spirit of neighborliness and a concern for simple living—as with the grandmother who teaches soapmaking and quilt-stitching to a young neighbor in return for rides to the supermarket twice a week, or the engineer who fixes his friend's car in return for a little help writing resumés, and tag and garage sales, volunteer hospital work, sharing tools, handing down clothes, informal apprenticeships in the local garage or carpentry shop; this is rooted in communitarian and often truly ecological attitudes, deliberately cooperative and non-exploitative, inherently tied to recycling and conservation, and generally opposed to keeping-up-with-the-Joneses growthism.

The household economy. Even larger than the underground economy, because it is entirely legal, is the "household economy"—that is, the goods and services produced in ordinary households that normally do not have a price put on them but that nevertheless have an obvious and real measurable economic value. This includes all the things that, if you weren't doing them yourself and had to pay someone for, would be immediately recognized as part of the "real" economy: child care, meal preparation, home decoration, furniture repairs, backyard improvement, gardening, canning, sewing, car transportation, laundry. And it includes all the value of the "plant and equipment" that, if translated into corporate terms, would easily rival the investment of all the rest of the economy: homes, property, lots, furniture, storm windows, cars, appliances, and the lot. The American family may think of itself as economically battered and worthless; actually, on average, it is an economic machine—counting the material assets, outside salaries earned, and unpaid work—of something like $59,000 (1978).

Though completely removed from the usual *market* economy, this *non-market* sector is enormous by any measurement. Yale economists Nordhaus and Tobin have calculated the labor-and-services part of it to be about a quarter of their "Measure of Economic Welfare"—amounting to as much as $295 *billion* in 1965 (the latest year of their calculations), a figure equivalent to more than 40 percent of the nation's GNP. But in addition, the plant-and-equipment segment may amount to total assets of perhaps $1 *trillion,* according to the most authoritative—and easily the most entertaining—scholar in this area, Scott Burns, in his seminal work, *The Household Economy*. Add it all up and it is an extraordinary sum for this part of the economy, all the more extraordinary in that it is essentially hidden, hardly ever factored into the considerations of either economists or politicians.

Scott Burns's figures, and those of other economists who have tumbled to this area in recent years, leave no doubt as to the importance of the household world. About 40 percent of the adult population is *not* involved in the standard labor force, most of it off doing the everyday things that never get noticed but are still economic. If all their labor was paid for, by going wage rates on the "outside," it would amount to an increase in the average family's income (according to one 1966 study) by 43 percent, and it would even be bigger than the total amount of all the wages paid in manufacturing ($212 billion as compared to $146 billion). And if all that income was actually toted up and accounted for, it would amount to about 48 percent (1964 figures) of the total disposable income in the nation. In sum, the household economy, "nonmarket" though it may be, represents *roughly half* of the real, year-in-year-out, economy of the land.

One particularly noteworthy part of the household economy that has developed with great speed over the last decade is what has come to be called "briarpatch businesses," working enterprises of essentially cottage industries operating out of individual homes and apartments or small storefront offices. The woman who takes in washing and the husband who becomes a full-time furnituremaker in the basement are familiar examples of this, but the increase in recent years has come about mostly from young people out of the 1960s who have chosen a consciously Spartan lifestyle with a deliberately small-scale business, as lawyers and carpenters, psychologists and veterinarians, auto mechanics and quilt-makers, restaurateurs and graphic artists. Any college town in America, almost every city of any size, has seen dozens of such briarpatch operators crop up over the last decade. It's hard to say how many, exactly, but clearly some significant part of the 6 million self-employed people in the U.S. (1977 estimates) would be counted among them, and some major part of the $43 billion worth of non-farm corporate assets (1977) would be in their hands. So numerous and successful are they in some cities—San Francisco, most notably, but also Boston, New York, Milwaukee, Chicago, Austin, and Denver—that they have established communitywide support organizations modeled on the Bay Area's pioneering Briarpatch Network, begun in 1974, and have published annual "People's Yellow Pages" to announce their services.[1]

Though the household economy is clearly consistent with such elemental tenets of the capitalist system as personal consumption and even Jonesism, in its essentials it is pointing to the nascent steady-state economy. It is all individual- and family-based, small-scale, and non-corporate; it operates mostly on non-monetary principles, with its labor outside the cash nexus; it stems from certain age-old homesteading and individualistic principles antithetical to the market system, especially to the rush-right-out-and-consume system; and it inevitably values, for sheer economic reasons, such qualities as durability, simplicity, safety, and efficiency in its goods. Scott Burns is quite explicit in how he sees the household economy as having the "necessary compatability" with a steady-state economy:

> The important difference between the market and household economies is that the former is committed to the idea of compounding, to the perpetual doubling and redoubling of capital, and the latter is concerned with the creation and use of capital in the present. The return on market capital is magical, Faustian cash, a substance that is infinitely exchangeable, compoundable, and otherwise flexible to meet the most extreme of human wishes. Capital in the household economy provides returns in services. These services are non-transferable and cannot be accumulated. I may own the goods and equipment that will provide me with a roof over my head, a dry shirt, a warm meal, or a visit to a friend. It remains that the return is a non-transferable service, something consumed in the here and now or not at all. A moment's thought will reveal the enormity of this difference.

1. "People's Yellow Pages" exist in Boston, San Francisco, Vancouver, Ann Arbor, Kansas City, Los Angeles, Madison, Montreal, Morgantown, Durham-Chapel Hill, Santa Cruz, Ogden, Seattle, Syracuse, Tucson, Tulsa, Washington, Minneapolis, Buffalo, Boulder, St. Louis, Berkeley, Cleveland, Cincinnati, New Haven, Honolulu, New York, Atlanta, Bristol, Ithaca, Tempe, Providence, Portland (Oregon), and Portland (Maine).

Moreover:

> The market economy is not a creature suited to an environment where materials and energy are increasingly scarce. The market is geared to "more and more," not to more from less. Yet the reality of our limited natural resources requires precisely that: more from less. The household, not the market, is the institution for such a condition.

Ultimately, he argues, *"the growth of the household economy and the arrival of the stationary state may be all that separates us from a social hell."*

The interstitial economy. In the nooks and crannies of the American economy today there is abundant evidence of a new mood that seems to have come over the average citizen in the last decade or so, betokening a profound dissatisfaction with the pattern of the growth economy and a shapeless but quite real impetus toward something more stable and more satisfying.

Its most obvious manifestation has been the widely acknowledged "loss of the work ethic," much bemoaned by corporate types in recent years, as with this financial consultant on the Op-Ed page of the *New York Times* in 1978:

> Any personnel manager will tell you that yesterday's generation of non-student is today's generation of non-worker. . . . These non-workers come from all the races and socio-economic groups. Many of them grew up as middle- and upper-middle-class children, but they all seem to be affected by the "welfare psychoses" of always getting and never giving. . . . Even more maddening is the absence of the work ethic. Work is not a means to an end; it is not a life's role; there is no fulfillment in work. Worse still, there seems to be a complete absence of ambition. It is as if Horatio Alger were some kind of subversive.

Absenteeism, wildcat strikes, tardiness, job refusal, on-the-job theft and sabotage, slowdowns, high turnover rates, low productivity—all these have been increasing over the last decade, in bad times as well as good, and all of them reflect a new and apparently pervasive unhappiness with the traditional American patterns of work. As the government's own *Work in America* report put it in 1972, a combination of the "blue-collar blues," the "white-collar woes," and "managerial discontent" has made all workers "perilously estranged" from the economic system; in fact only 43 percent of the office workers and 24 percent of the industrial workers could be found to acknowledge that their work brings them enough satisfaction that they would be in the same job if they could start all over again.

It has been a partial reflection of this that so many millions of Americans have opted for what the 1975 Stanford Research Institute survey called "voluntary simplicity"—a life of frugal consumption, ecological awareness, individual reliance, and limited income. The survey came up with an estimate of 4–5 million living "fully and wholeheartedly" in this way—which jibes well with the official Labor Department figure of 5.5 million between 16 and 64 who are not in the civilian labor force and are not housewives, students, or disabled—and another 8–10 million partially involved. By 1987, it predicted, there would be 25 million Americans living in voluntary simplicity, plus another 35 million "partial adherents."

And there are even more in this country who claim to hold to steady-state—or what is often called "post-industrial"—values, as revealed by a number of Harris polls over the last half-dozen years:

▪ 79 percent declared a desire for more emphasis on "basic essentials" of life rather than "reaching higher standards of living";

▪ 78 percent said they had or would change their lifestyles toward simplicity (by eating plainer meals, rejecting fashion and model changes, consuming less);

▪ 77 percent said "spending more time getting to know one another better as human beings on a person-to-person basis" was more important than "improving and speeding up our ability to communicate with each other through better technology";

▪ 76 percent wanted the nation to learn to rely on "non-material experiences" rather than "more goods and services";

▪ 74 percent opposed America's excessive use of the world's resources;

▪ 66 percent urged "breaking up big things and getting back to more humanized living" instead of "developing bigger and more efficient ways of doing things"; and

▪ 64 percent asserted that "finding more inner and personal rewards from the work people do" was more important than increased productivity.

It may be that such overwhelming responses are skewed—people having the tendency to be always on the side of the angels when responding to pollsters even if they're living with the devil—but they indicate at the very least that in theory a sizeable segment of the American public rejects the growth economy. And indeed, according to Ronald Inglehart's *The Silent Revolution,* a 1977 survey of attitudes and preferences in the West, a wholesale shift toward an emphasis on "quality of life" rather than "getting ahead" is going on in every industrial nation in the world.

There are numerous other signs, too, all pointing in the same direction: the increase in the number of cooperatives in the past five years (a trend that has been accelerated by the Federal Co-op Bank established in 1979); the vigorous development of the environmental (and particularly the anti-nuclear) movement; the rise in the number of major companies both owned and managed by the people who work in them; the rising popularity of books, magazines, catalogs, and encyclopedias devoted to homesteading, self-sufficiency, home gardening, personal development and self-help projects; the spread of community and public land trusts for conservation and stewardship of both urban and wilderness property; and the growth of corporate programs emphasizing "work enrichment," team production, "flextime," workplace democracy, worker participation, collegial decision-making, and the like.

And remember, these are all blades of grass that are forced to push their way up in between the cracks of a mighty concrete economy that is designed in every way to discourage them. That they exist at all, given the pervasive nature of the corporate economy, is surprising enough; that they have such power and popularity suggests that the development of the steady-state consciousness and a movement toward the steady-state society is already proceeding apace. "Getting on" is losing hold.

THOUGH HE WAS no doubt passed over by the Harris polls, Jonathan Swift once wrote:

> In all well-instituted commonwealths, care has been taken to limit man's possessions; which is done for many reasons, and, among the rest, for one which, perhaps, is not often considered: that when bonds are set to men's desires, after they have acquired as much as the laws will permit them, their private interest is at an end, and they have nothing to do but take care of the public.

I do not know of such commonwealths in the past, but that is clearly the spirit of the steady-state future.

6

The Logic of Size

THE SIZE OF hunting bands in early Paleolithic societies, as we have seen, normally seems to have been no more than five, that number apparently being found over the eons to be most harmonious for a human group around the campfire and on the trails and still efficient enough to bring down even the largest prey. From surviving Stone Age cultures, particularly in Africa and Australia, it appears that most of the early languages did not even have a word for any number over five.

It was part of the customs of the early Greek maritime cities that ship crews could have no more than five individuals, any number above that being considered sufficient to engage in piracy.

In 1305 Philip the Fair of France, engaged in a protracted struggle with the papacy, forbade unauthorized groups of more than five people to congregate anywhere within his realm, on the assumption that a gathering greater than that was probably a Romanist clique bent on conspiracy against the crown.

After he came to the English throne following the Cromwellian Protectorate, Charles II re-established the episcopacy, and his Parliament, in the English Conventional Act, decreed that all home assemblies for religious reasons of more than five people would be considered suspicious and punishable by law.

Twentieth-century researchers, particularly the psychologists who have carried out extensive experiments with small groups, have determined, according to the English management expert Charles Handy, "for best participation, for highest all-round involvement, a size between five and seven seems to be optimum."

C. Northcote Parkinson believes that most committees start with five members—before that there is not enough sense of cohesion for the group to consider itself a committee at all—and after a time, when they grow and become unwieldy with new members, they tend to revert to the original size by establishing an "executive committee" or "board of directors," or somesuch, of the five most powerful members.

One contribution to the basic *Handbook of Small Group Research* suggests that the optimal committee size is five, since below that number people tend to feel that getting together is "not worth it" and "won't prove anything," whereas much above that people start to participate less and eventually stop paying attention. This number, too, is good for decision-making, because of course it is odd and because it is large enough so that no one feels isolated when taking a stand (divisions are most often three–two) and small enough so that participants can shift roles and points of view without onus.

A. Paul Hare, editor of the classic *Small Groups,* introduces a variety of laboratory studies showing that groups between four and seven are most successful at solving problems, and that as groups increase in size they tend to take more time to come up with solutions, make less accurate judgments, produce fewer ideas, achieve less communication, and stand less chance of reaching an agreement.

One leading sociologist, John James, determined several years ago that "action-taking groups" average out to 6.5 people and groups that get to be as big as ten and fifteen people are "non-action-taking." He also took the practical route of surveying the subcommittees of various legislatures—the places where the real work of any body gets done—and found an average of 5.4 people on U.S. Senate subcommittees, 7.8 in the House, 4.7 in the Oregon state legislature, and 5.3 in the Eugene, Oregon, city government.

And Frederick Thayer, whose exclamatory *An End to Hierarchy! An End to Competition!* is a careful look at organizational forms, sums up: "While common sense would seem to dictate that there can be no 'magic' number, five appears so often in so many environmental situations as to carry persuasion with it."

Now I am, of course, fond of "magic" numbers, and though I'm not sure this is another one, it does seem obvious that through the ages people have decided that groups of about this size have an unusual power and competence about them. This is of some importance in considering a steady-state economy, because in such a system, where efficiency is inevitably of greater importance than it is in our profligate capitalist arrangements and where happiness and worker morale are more vital than in the getting-on world of making a profit, it would be important to determine the optimum size for effective and congenial work units, workplaces, work organizations.

Luckily there are innumerable studies here that can serve as useful guides.

As to the basic work unit, the small group, hovering somewhere right around the magic number of five, has proven time and again to be most effective in the workplace, easily outperforming both the isolated individual and the assembly line. There are literally hundreds of studies confirming that (particularly in small settings, but also in the larger factories) small groups tend to have higher morale, lower absence and turnover rates, fewer accidents and labor disputes, and—bottom line—greater productivity. Part of the reason for this seems to be that working without hierarchy, without managers and bosses wielding power, a work group sets its own tasks and pace most efficiently. Part has to do with the fact, borne out in innumerable cases, that the more you let workers run their own affairs, the better they will perform. And part is due simply to a group loyalty that tends in time to give individuals a cause, a sense of identity, a meaning—however limited—that inspires greater enthusiasm and productivity; research into wartime behavior similarly shows that it is the loyalty to the group—not to country or motherhood or democracy or the general—that produces the most successful soldiers and accounts for the exceptional feats of bravery and courage.

The classic case in point here is the "Relay Assembly" study made at the Hawthorne, Illinois, Western Electric plant in the late 1920s. Researchers from the Harvard Business School took a group of six women workers out of a department of a hundred and put them into a special test room, then variously altered their conditions of work—lighting, breaks, hours, lunchtimes, piece rates, and so on—to see how it affected productivity. Quite to the researchers' bewilderment, the women increased their output at *each* stage of the experiment, no matter which benefit was added or subtracted, ultimately with an 80 percent decrease in absenteeism, a 40 percent increase in hourly output, a clearly observable increase in morale on the job, and a new friendliness and cohesiveness after work. But the most extraordinary result was that the women's productivity increased even when, after many months, they were returned to the original conditions of work without any benefits at all. Consternation. The researchers were in a turmoil: what could explain it? Eventually, with academic ingenuity, the Harvardians determined it was something they called the "Hawthorne effect"—that is, simply taking people out of their ordinary routine and making them feel special will increase their productivity—and this became the standard business school line over the succeeding fifty years.

Clever, but it does seem something of a reach, doesn't it? What the researchers were observing, though they couldn't realize it, was simply and obviously the effect of two essential factors: the women were isolated in a small and coherent group with shared tasks and rewards, *and* within this group they were allowed an unusual degree of independence in setting the conditions of their work: they *saw* themselves as a small group, they *worked* as a small group, they felt themselves to *be* a small group. And small groups, particularly autonomous small groups, feel better and work better.

One might wonder why, after so many years of proof, corporate executives throughout America have not reorganized offices and plants into small work groups. In Europe this has been done in many industries in many countries—Saab and Volvo in Sweden, Daimler-Benz and Bosch in Germany, Fiat and Olivetti in Italy—with the predictable beneficial results. The Singer sewing-machine plant in Karlsruhe, Germany, switched its 1,200 workers in most operations from assembly line to small-group operation in 1977 because, according to general manager Hans W. Gilbert, the line moved only at the pace of the slowest worker and if people made mistakes they "blamed the next man down." Since then "workers get greater job satisfaction, while their output is improved and increased," and the savings to the company were put at some $370,000 in the first year of the change. But American firms, though testing small-group productivity in some instances, have tended to resist any wholesale shift, and most places that have tried it chose to abandon it within a year or so. Apparently managers are fearful of the kind of solidarity that develops within these groups, they worry about their jobs if it proves that supervision isn't necessary, and they basically distrust change of any kind. As between improved productivity and flux, and muddling-through and stasis, the corporate executives of America—particularly the middle management—will choose the latter.

SIMILAR DATA ABOUT small offices and plants show that those places with a limited number of people, where workers and managers know each other personally, where bonds of conviviality and cooperation are forged, are much more satisfying and productive than larger, more impersonal workplaces.

Take any measure you like. Studies of worker morale show uniformly that unhappiness increases as plants get larger. One researcher who studied sixty-six manufacturing and twenty-seven non-manufacturing businesses, with anywhere from 10 to 1,800 employees, found that "the overall level of satisfaction [with both the job and co-workers] decreased with size"; a study by the Acton Society in England found that as organizations become bigger the workers show less interest in the welfare of the business, know less about the administrators, are more inclined to accept rumors about the plant, have less devotion to "the work ethic," and display less cooperation with both managers and fellow workers. Absenteeism has also been found to be more prevalent in larger businesses; studies of airlines, metal manufacturers, engineering firms, coal mines, chemical plants, and package-delivery businesses all indicate that workers skip more days, come in more erratically, and take more sick days in the larger organizations. Labor disputes are less common in smaller firms, the rate of job turnover is lower, and the number of complaints about defective goods and inadequate service smaller. Accident rates are much smaller in smaller firms: a government survey in 1976 showed 3.9 injuries per 100 workers in places with fewer than 20 employees, 8.7 in places with 20–49 employees, 11.2 in places with 50–99 employees, and a high of 12.6 in places with 100–250 employees, before dropping again somewhat to 8.0 in giant firms of 1,000 or more employees.

And it is not only working stiffs. Even the managers, it turns out, are happier in small businesses. A five-year study of Stanford Business School graduates, who may be presumed to have settled into fairly cushy jobs and pretty much where they wanted to be, indicates that the "men in smaller firms were more satisfied with their work . . . had higher compensation and were able to participate more broadly in management"—better jobs and, surprisingly, more money. Two other surveys, one of 1,900 administrators and the second of 3,000, found that in both authoritarian and collegial managements executives in smaller organizations were significantly more satisfied both with their lives and with their jobs—though, as might be expected, those in the least authoritarian organizations were the most pleased.

In short, as sociologist Emile Durkheim discovered many years ago, "Small-scale industry . . . displays a relative harmony between worker and employer. It is only in large-scale industry that these relations are in a sickly state."

Fittingly, the logic of a steady-state society points exactly toward the "small-scale industry" and its harmony. For it would seem that any true steady-state enterprise would inevitably approach optimum smallness because such principles as minimal ecological intrusion, control over resource use, community involvement, and individual participation *necessarily* put limits on the extent to which any enterprise can grow. For example, a steady-state factory would not need a large sales and promotion staff, say, or an advertising department; it would not need to manufacture so many products, since it needn't create model changes or arbitrary differentiations, and would build for permanence rather than obsolescence; it would not want or need to compete against another firm simply to try to dominate a given market; it would have to keep production within the limits of finite resources and modest energy use; it would confine its output according to the declared needs of the community rather than trying to manipulate those demands; it would depend on alternative technologies that would not demand ever-larger machines; and, particularly where there was maximum participation by the workers in the firm's affairs, it would experience no need to expand operations beyond what was comfortable for the workers. The steady-state enterprise, in short, would have no *reason* to be big, every incentive to stay small.

Is IT JUST a haphazard happenstance that the stationary system leads to inherent limitations on sizes, having to do with that table about the number of contacts a person can make within different-sized committees? Or might there not be an underlying logic in all human behavior, allowed to play freely in such a stable economy as it is not now permitted in our exploitative one, that inevitably provides that small economic groups function better than large ones?

Mancur Olson, Jr., thinks there is. He is a professor of economics at the University of Maryland—a quite conventional, even conservative, professor of economics, I should add—and his book exploring this precise question, *The Logic of Collective Action,* has become a classic in the field because it demonstrates this reason so convincingly. Olson challenges the standard notion that a normal group in an economy will just naturally act in its own self-interest—that works for small groups, he shows, but larger groups cannot function with that same self-interest and are thereby less efficient and effective. Olson bluntly summarizes it thusly:

> Unless the number of individuals in a group is quite small, or unless there is coercion or some other special device to make individuals act in their common interest, rational, self-interested individuals will not act to achieve their common or group interests. . . . If the members of a large group rationally seek to maximize their personal welfare, they will *not* act to advance their common or group objectives unless there is coercion to force them to do so.

This seems a bit mysterious at first, since we all know of large organizations apparently operating in their own interests—unions, say, and professional societies. But Olson illustrates both theoretically and with real-world examples why *any* large organization cannot function optimally:

1. In the small organization, any individual's share of the common good that the group can obtain is larger, since there are fewer people and only with a small group can the rewards exceed the effort or money put into it. Conversely, *"the larger the number in the group, other things equal, the smaller the* [gains for the individual] *will be, the more individuals in the group, the more serious the suboptimality will be. Clearly then groups with larger numbers of members will generally perform less efficiently than groups with smaller numbers of members."*

2. In a small organization, because the rewards are clearer, there is a greater likelihood of one person, or a small subset, absorbing extra costs or making some sort of a sacrifice so that the organization will continue. Conversely, "the rational individual in the large group in a socio-political context will not be willing to make any sacrifices to achieve the objectives he shares with others." Thus a Typical American Family faced with inflation will not cut spending by itself, even though it realizes that if every family did, inflation would be halted, because it knows its *own* spending won't have an appreciable effect on the nation as

a whole and because if there is any ultimate lessening of inflation the family will automatically get the benefits without any sacrifice.

3. Similarly, in a small organization each individual realizes that if any one member fails to contribute, "the costs will rise noticeably for each of the others in the group" who may then "refuse to continue making their contributions" and threaten the continuation of the organization, thus eliminating all of the common benefits. "By contrast, in a large group in which no single individual's contribution makes a perceptible difference to the group as a whole, . . . it is certain that a collective good will *not* be provided unless there is coercion or some outside inducements that will lead the members of the large group to act in their common interest."

4. In a small group, the costs both of reaching an agreement and of carrying it out—communicating, bargaining, meeting, staffing, monitoring—are much smaller, sometimes quite negligible. But "the larger the number of members in a group the greater the organization costs, and thus the higher the hurdle that must be jumped before any of the collective good at all can be obtained."

5. In small organizations, social and peer pressures can operate to encourage participation, and face-to-face association assures that individuals are known and snubbed for misbehavior. "In any large group everyone cannot possibly know everyone else, and the group will ipso facto not be a friendship group; so a person will ordinarily not be affected socially if he fails to make sacrifices on behalf of his group's goals."

There. Simple, really—and inexorably logical.

The point I am sure need not be labored unduly, but Olson offers a number of interesting illustrations:

A *partnership,* for example. This works when small and the rewards are clear, but "when a partnership has many members, the individual partner observes that his own effort or contribution will not greatly affect the performance of the enterprise, and expects that he will get his prearranged share of the earnings whether or not he contributes as much as he could have done."

Or a publicly held *corporation.* Stockholders as a rule almost never act to get rid of a particular management, even when it is performing disastrously, because "any effort the typical stockholder makes to oust the management will probably be unsuccessful; and even if the stockholder should be successful, most of the returns in the form of higher dividends and stock prices will go to the rest of the stockholders," particularly those who hold greater amounts. (Occasionally, of course, there are serious proxy fights, but this happens when one or two stockholders with large blocks of shares have an immediate self-interest in replacing the present management with one of their own.)

Or a *union.* In most large labor unions—as opposed to small, single shops—the value of the large organization is either so negligible that the individual worker has no incentive to join or so pervasive that the worker gets the benefit regardless of joining. Thus it has been necessary ever since the creation of large labor organizations in the nineteenth century for the union to operate by outright coercion—compulsory membership, a "union shop" or a closed shop, wage checkoffs, job control—and even then it gets the month-to-month participation of only a tiny fraction of its members.

Or a *professional association*. These now exist in all sorts of occupations, but the most potent ones are surely the state bar associations, which by law have been given absolute power over who can practice law and who cannot, and on exactly what terms. It is extremely doubtful that lawyers would choose voluntarily to be a member of a state bar association, since again the rule applies: if there is any common good created by these organizations, all lawyers will benefit, whether they contribute or not, so there is no incentive for anyone to join voluntarily; that is why membership is obligatory.

In short, Olson demonstrates conclusively that the difference between the small group and the large one is not simply one of numbers. There is a *qualitative* as well as a *quantitative* distinction, and the benefits come out so clearly on the side of the smaller organization that it appears larger ones *would not even exist* in a country with true economic freedom and would not be able to survive without some continuing form of compulsion.

That is an extraordinary thing to say about a country that prides itself on being a "democracy." And it opens up what may be the most important element of all in a human-scale economy: democracy—real, felt, active democracy—in the workplace.

EVERY STUDENT OF economics knows the Law of Declining Productivity. Its unrelenting message: the increase in any economic variable (land, labor, capital, etc.) while others remain fixed will reach a limit beyond which further increase diminishes productivity.

It is simple. If two people can turn out one table a day, it may be possible that four can turn out three, and five can turn out five, allowing for greater efficiency with shared tasks and specialization of added workers. If the quality of work remains fixed, however, and the space and the time, the addition of other workers is likely to make only a marginal difference, with seven people able to put out only six tables a day and a complement of nine so complex that it can manage only five. At some early point, because the workers begin to crowd into each other and the communication becomes more difficult and no one is quite sure what anyone else is doing and planning and coordination break down, productivity—and productivity per worker—inevitably declines.

The law has been around, on at least the textbooks, for a long time. That it has been broken often does not change its eternal validity. It embodies the logic of size.

7

Workplace Democracy: Ownership

OVER THE ENTRANCE to virtually every workplace in America, it has been said, is an invisible sign saying, "ABANDON FREEDOM ALL WHO ENTER HERE."

It seems strange that a nation conceived in liberty, inspired by independence, and devoted to democracy should condemn the majority of its working citizens, for the greater part of the day, to conditions of unrivaled autocracy far more stringent than anything practiced by George III. Yet in fact, during the hours that most of us are employed, we forgo most of our basic democratic rights:

▪ We do not vote on nor have any say whatsoever in what our company will do or make, how it will raise money or invest its profits, or for the most part how it will operate the offices and plants in which we work.

▪ We do not have the rights of free speech, free assembly, or free press; we have no protection against personal or property searches or invasion of privacy; and we do not have the right to trial and due process within the workplace.

▪ We normally cannot modify or challenge the decisions that are made for us by a handful of distant people whom we do not elect to, and cannot remove from, office.

▪ We have no representation in the financial processes (excepting only as a union may intervene) and normally must accept the salaries, cuts, and raises offered by those over whom we have no control.

▪ And if we should displease this hierarchy for any reason we may be dismissed or disciplined at the whim of any superior (again except only as union power may intervene) and be out on the street without job or reference or—what's worse—recourse.

It seems somewhat stark when put like that, but no one really disputes that this condition—what Justice Louis Brandeis seventy years ago criticized as "industrial absolutism"—is the norm in American life. As one who should know, General Robert E. Wood, for many years the chairman of the board of Sears, Roebuck, has said: "We complain about government and business, and we stress the advantages of the free enterprise system, we complain about the totalitarian state, but in our individual organizations . . . we have created more or less of a totalitarian system in industry, particularly in larger industry." And T. K. Quinn, the former General Electric executive, has described American business thusly: "The directors were in every case elected by the officers. We had then, in effect, a huge economic state governed by non-elected, self-perpetuating officers and directors—the direct opposite of the democratic method."

If a human-scale economy means anything, it means the elimination of such conditions of totalitarianism and the creation of workplaces where individual humans can have a full and constant voice in the matters that affect all aspects of their working life. It means, in short, democracy in the workplace.

For all of its strangeness to American ears, the idea of workplace democracy in its various guises has been alive in much of the rest of the world for more than a decade now, and in the last few years it has begun to be taken up here in the U.S. The International Labor Organization in Geneva has established a comprehensive library on worker ownership, self-management, production cooperatives, workers' councils, and the like. Two international organizations have been established, the International Cooperative Alliance, in London, representing some 42,000 non-agricultural producer co-ops over the world with a total of 5.4 million members, and its allied International Committee of Workers Productive and Artisanal Societies, which sponsored the first world conference on industrial cooperatives, in Rome in 1978. Worker control and ownership is practiced in various degrees in every European nation: Italy is said to have 3,000 industrial cooperatives, Poland 2,700, Czechoslovakia 470, and France 522, including one firm of 4,000 that makes a third of all the telephone equipment in the country; Sweden has had a law since 1977 guaranteeing workers places on all boards of directors and consultation by management on "any important change" in work conditions; West Germany in 1976 passed a law providing "co-determination" (equal power for both workers and management on boards of directors) for all large firms; the United Kingdom has had a "common ownership" movement since 1958 (with thirteen member firms by 1975) and an Industrial Common Ownership Act promoting worker-owned and cooperative businesses since 1976.[1] The Third World has been particularly hospitable to worker-run enterprises: agricultural producers' cooperatives have been established in almost every country of Africa and Latin America; India alone is said to have 60,000 non-agricultural producer coops containing 4 million members; industrial collectives and self-managed firms are well established in Tunisia, Algeria, Sri Lanka, Argentina, Peru, and Venezuela.

Even in this country, the idea has borne fruit. An Association for Self-Management was begun in 1973 and now has a headquarters in Washington, a body of several dozen academic papers gleaned from four international conferences, and a membership of some five hundred scholars and practitioners. Self-management has become a legitimate academic specialty among economists and political scientists, and even among anthropologists and psychiatrists; Cornell University is perhaps the leading institution in the field, with both a Program on Participation and Labor-Managed Systems, run by the leading theoretician, Jaroslav Vanek, and a New Systems of Work and Participation Program under the direction of sociologist William F. Whyte. Among the think tanks and instigators and coordinators at work across the country there are such groups as the Center for Community Economic Development and the Industrial Cooperative Association in Cambridge, Project Work in New York, the National Center for New Alternatives and Strongforce in Washington, D.C., the Community Ownership Organizing Project in Oakland, and the Center for Economic Studies in Palo Alto. And the extensive body of literature now available includes several dozen books, thousands of papers, hundreds of pamphlets, and a steady spate of articles in such publications as *Working Papers, Co-op, Self-Management, The Journal of Economic Issues,* and *In These Times.*

WORKPLACE DEMOCRACY CAN BE—given all the books and pamphlets, *has been*—defined in countless different ways, but all working definitions embody in some degree three essential characteristics: *ownership* of the enterprise by the workers involved in it, preferably operating cooper-

---

1. I have a most charming letter from Ernest Bader, the founder of the Scott Bader Commonwealth, Inc., the first of Britain's common-ownership firms, who says of the 1976 Act, "We can say, without fear of contradiction, that this Industrial Emancipation Act would not have been possible without our successful example." He deserves every credit for that achievement, though it is fair to note that neither the act nor his company constitution actually provide for full worker *control* over the commonly owned firm; workers elect only two of the ten directors of the Bader Commonwealth, for example, and the directors operate through a conventional management responsible to them. When I asked Bader about this he wrote back a somewhat less charming letter with a quotation from the Bible: "As for the cowardly, the faithless, the vile, the murderers, fornicators, sorcerers, idolators, and liars of every kind, their lot will be the death in the lake that burns with sulphurous flames."

atively; *control* of the enterprise by the workers involved, through participatory machinery; and responsibility of the enterprise to the *community* in which it is located and in which its workers live.

There can be other ingredients and various mixtures, but these are the three indispensable characteristics, and the ones we will examine in these next chapters.

NOTHING DIFFICULT ABOUT about the idea of worker ownership: it simply means that the shares of an enterprise are bought and owned by the people who work in it, and when people cease working there they cease to be owners. Nobody on the outside, not brokerage firms or capitalists or little old ladies on pensions, can buy into it. In effect, this is like having labor "hire" capital, instead of the other way around as we normally do it, and it effectively reduces capital to a secondary element in the operation, behind the work itself. It also means that the workers have an equal share in any profits—and, similarly, losses—made by the company, and a "stake" in the firm's success proportional to the number of shares they own.

Thinking of janitors and lathe operators and secretaries as *owners* of the places they work in is startling, even frightening, for traditional types; as one consultant to a group of workers put it, "When we went to the big lenders, their first reaction was, 'Jesus Christ, the monkeys are going to run the zoo?'" But the fact is, according to the Whyte team at Cornell, there have been "over 500 cases" of worker- or worker-community-owned enterprises since the founding of this country and today there are at least thirty-five of them going strong—and that's not counting the small producers' co-ops, the health-food restaurants and the auto-repair shops and the women's bookstores found in most college towns and many large cities, about which Whyte has written: "There must be thousands of firms of this type around the country." Worker-owned businesses include a $60-million, 500-person insurance company in Washington, D.C.; two farmworkers' ranches in California; a furniture factory in Herkimer, New York; a printing press in Clinton, Massachusetts; the Chicago and Northwestern Railroad in Illinois; a daily paper in Madison, Wisconsin; a frozen-food operation in Texas and Georgia; a poultry-processing plant in Connecticut; and sixteen plywood corporations in Washington and Oregon—in fact, the whole range of American enterprise. And according to the surveys, there would be more such enterprises if legal and economic conditions were opened up: a 1975 Hart poll, for example, showed that 66 percent of the American public favored "worker-owned" companies, 20 percent "investor-owned."

Not only do they exist—they work, and work well. A 1976 study by Washington's Federal Economic Development Administration indicated that employee-owned companies have above-average profitability compared to regular firms operating in the same market. The University of Michigan's Institute for Social Research, in a study of thirty worker-owned firms, found that all "show a higher level of profitability than do similar conventional firms in their industry" and the "single most important correlate of profitability" is the percentage of the company's equity owned by the workers, especially the rank and file. Other research shows the same—perfectly obvious—thing: where workers feel that they have a real stake in what they are doing, they work with infinitely more care, concern, and commitment. Product and profit figures show the results.

The conventionalists can't believe that the people who actually do the jobs are more capable at running firms than the distant conglomeraters, but in any number of instances it's been shown that worker-owners actually do know what they're up to.

Take, for example, the Saratoga Knitting Mill in Saratoga Springs, New York. Owned by the Van Raalte Corporation until 1968, it was taken over by the Cluett-Peabody conglomerate at a time when it was doing some $72 million worth of business a year. But Cluett-Peabody, with the infinite wisdom of the conglomerate, decided to eliminate the factory's sales force and take care of selling with its own existing staff—they knew the market better—and then found that its own people knew nothing about the Saratoga product line and couldn't get it into the stores; by 1974 it managed to reduce the annual sales to $20 million, on which it lost $11 million. Naturally its instinctive reaction was to get rid of a loser, shut down the plant, and protect the bottom line, and the 140 workers would take the hindmost. By 1976, though, the workers had managed to raise some capital from a few local banks and a number of their own mattresses, and they offered to buy Cluett-Peabody out. They did, with dispatch, immediately restored the sales force, improved production methods throughout the plant, and by the very next year the mill had not only pulled itself out of the red but showed a profit of $300,000.

Or take the Byers Transport company in western Canada. A successful trucking firm operating throughout the American Northwest, it was taken over by a hungry conglomerate, Pacific Western Airlines, in 1973. It immediately started to lose money. After four years and a loss of $600,000, PWA wanted out and offered it to the employees, who finally managed to get the capital to buy it out. In the very next year it made a profit of $200,000.

Or take the Vermont Asbestos Group. When the GAF conglomerate chose to get rid of this abestos mine in northern Vermont in 1974, it was such a losing proposition that it didn't even seem worth putting in the new anti-pollution equipment the government was insisting on. But the workers wanted to protect their lives and livelihood, and with community support they managed to put together $2 million, take over the plant, put in pollution controls, and keep production going without a hitch. Within the first year they had paid off their major loan, paid out a 100 percent dividend, and still managed to increase their wages by nearly 20 percent across the board. Of course the fact that asbestos prices suddenly went soaring up by 65 percent didn't hurt, but the dramatic turnaround was ascribed by the plant manager simply to "a better attitude" and increased productivity among the workers themselves.

The monkeys, it turns out, actually know a lot about running the zoo.

ONE INCREASINGLY PREVALENT form of worker ownership is the producers' cooperative, a tradition in the U.S. at least as old as the 1790s, when the first shop—a smithy near Philadelphia—was bought out and run by its workers, a tradition still thriving as late as the 1880s when at least 200 co-ops were operating in trades as diverse as cigar-making and home construction.[2] In a cooperative arrangement the workers not only own the shares of the company they work in but have, theoretically, an equal voice in its operation, can select the board of directors on a one-person-one-vote basis, and take an equal share of the profits. Not nearly as common as *consumers'* cooperatives, such as the innumerable food co-ops, or the *marketing* cooperatives, such as most of the agricultural co-ops, *producers'* cooperatives are nonetheless becoming increasingly common, particularly among the young workers who formed their politics in the 1960s and have tried to make their working lives conform to their philosophical beliefs. No one keeps close track of such organizations, not even the North American Student Cooperative Organization, the leading American watchdog of the co-op movement, but NASCO ventures the estimate that there may be as many as a thousand, mostly small.

The most successful—perhaps overly successful, as we shall see—of the producer co-ops are the plywood companies of the Pacific Northwest, which have gained a certain fame among researchers of the Left in recent years as living, home-grown examples of the kind of labor power found otherwise only in places like Europe and China. There is nothing particularly ideological about these cooperatives, though—like many worker-owned firms—they were formed in response to the threat of plant shutdowns and unemployment. The first plywood co-op was

2. Derek Jones, a researcher working with the Cornell Vanek group, estimated after considerable research that a total of 583 cooperatives had been established in the U.S. since 1790, with a high point of 215 in the 1880s, 56 still operating in the 1930s, and 30 existing in the 1970s.

formed by 125 workers in Olympia, Washington, in 1921 and over the years as many as thirty co-ops have been started, though by 1978 only sixteen were still worker-owned. A measure of their importance, and profitability, is that at one point in the 1950s the co-ops' output accounted for a quarter of the total capacity of the plywood industry, and even today it is estimated to be around 10–12 percent.

The co-ops have differed in details, naturally, but most operated with the same kind of system: each worker owns a single share, each has a single vote, the workers as a whole elect the board of directors and the corporate officers (usually fellow-workers), and those officers hire an outside (non-owning but high-paid) general manager who is guided by general employee meetings held twice a year. All the workers get the same salary regardless of job, and in fact there is a good deal of job rotation and job sharing to assure that no one gets the cushy—and no one the nasty—jobs exclusively; similarly, they share equally in the profits. All this has gone to make the cooperative plywood firms far more productive than their conventional counterparts—*from 25 to 60 percent* more productive, as the IRS itself determined some years ago—and to enable them to give hourly rates and year-end profit-sharing well above the industry norms. Daniel Zwerdling, a young writer who has done more and better reporting on worker-run establishments than anyone in the country, quotes one thirty-year veteran of the 270-worker Puget Sound Plywood Cooperative as saying:

> The more effort you put into it, the more you'll get out of it. I've worked in seven plywood companies since I was 15 years old. At the other companies you're only interested in getting in your eight hours and taking home the check. But here it's altogether different. It's *our* money that's involved. There's always the possibility that we could work harder and make more money. The harder we work, the more money we make. And that's the beauty of a co-op.

Unfortunately, such success has wrought its own price at some of the firms. Because the cooperatives do so well, the market value of the worker's shares increases, and some shares bought for $1,000 thirty years ago may be worth $25,000 or $30,000 today. That means a young worker will find it almost impossible to buy in—who starts out with $25,000?—and the older member hangs on until he can find a large corporation to come along and offer him and his fellows that kind of price. Such scavenger firms exist: between 1954 and 1976 five of the oldest and most successful cooperative mills were sold to large conventional corporations, each one after about thirty years of operation.

This "single-generation malady," however, is not inevitable. Many newer co-ops elsewhere have taken steps to guard against it, as for example with a "passive trust" or a "business trust" that has non-marketable shares that remain in the hands of the active workers. David

Ellerman, an economist with the Industrial Cooperative Association in Cambridge, has come up with one of the most ingenious plans. In his scheme for the Colonial Cooperative Press, which ICA helped to set up in late 1978, the workers do not buy ownership shares but "membership rights" that are not for sale and are bought back by the cooperative at the original price whenever a member leaves. Thus when Mary Mechanic joins the Colonial cooperative she has to put up the original membership fee of, say, $1,000; thereafter, she has a capital reserve "account" consisting of the percentage of the firm's annual profits distributed among the workers each year, which may be used by the firm until her departure; and on leaving, she takes into her retirement the original membership fee plus whatever she has accumulated in her account. By this method the share doesn't have any negotiable value and cannot be sold to any outside investor: it is simply Mary's stake in her enterprise while she is there. Ellerman argues that the idea of worker-*ownership* is in fact a misnomer when it comes to describing true cooperatives, since members should no more be able to "own" a company than union members "own" their union or the citizens of Albuquerque "own" Albuquerque; the more appropriate concept is worker *membership*.

But even though the plywood cooperatives face a somewhat difficult future if they cleave to their current practice, they have already demonstrated beyond a doubt or denial certain crucial truths: that worker-owned firms in America can be successful, and over a long period of time; that people are able to work cooperatively, even to the point of absolutely equal pay, in a competitive business world; that forms of democracy and self-government are not antithetical to good industrial operation; and that worker ownership creates a sense of pride and grit that cannot be found in any other system.

ONE OTHER FORM of employee "ownership" that has become popular in recent years might best be labeled "ESOP's fables." Since 1974 the Federal government has worked to create Employee Stock Ownership Plans, a complicated scheme by which corporations are encouraged to give shares of stock to their workers in return for certain tax breaks and investment privileges. So far at least 3,000 corporations have tried it out.

As normally operated, an ESOP begins with a corporation setting up a trust for its employees, say with 10,000 shares of stock. The trust then borrows money from the bank against those shares, but it gets special beneficial rates and write-offs that the parent corporation would not be able to get. The trust then gives this money to the corporation, as payment for the stocks it has been given, and it pays off its low-interest loan to the bank with some part of the corporate profits that it is given every year—and that the corporation writes off as a tax deduction. And the employees get the stock or may cash it in when they retire. It sounds all very democratic, indeed downright subversive, but the fact that it is the pet project of Senator Russell Long, the millionaire who runs the Senate Finance Committee, suggests that perhaps it does not betoken a true workers' revolution.

For one thing, "ownership" does not mean any effective voice in the operations of the company. In most cases the shares issued the workers are of non-voting stock or stock that will become voting stock only after a decade or two or stock that is controlled by a management-appointed trustee; according to a survey by Michigan's Institute for Social Research in 1978, 73 percent of the ESOP companies do not let the employees actually *use* the stock they "own." Then, too, the number of shares turned over to the employees is typically a minority of the outstanding shares, so even if the workers did someday demand voting rights they would not have much say in what goes on. Finally, since the employee stock is distributed according to salary, the upper management always has the preponderant share and can usually be counted upon to resist any real challenge from below; according to the Michigan Institute's survey, even when all the ESOP voting rights now held in abeyance eventually accrue to the employees, production workers will have a majority of the stock in only 7 percent of the ESOP firms.

The experience of the South Bend Lathe company in Indiana shows the limitations of the ESOP approach. In 1974 the Chicago-based Amsted conglomerate announced it was getting rid of South Bend Lathe—500 skilled operators, $20 million annual production, and all—for $10 million. There were no corporate takers, though, since the company had been run into the ground by the unheeding parent corporation. So the city of South Bend, the employees, and the U.S. Economic Development Administration came up with a complicated but rather slick little arrangement whereby an ESOP employees trust would own the 10,000 shares of the company and borrow the necessary $10 million from the government and local banks, with the expected future profits going to pay off the loans.

The deal went through in the summer of 1975 and everything went well enough at first—Amsted was happy, the city was happy, morale at the plant soared, and right away the company started operating in the black for the first time in six years; local papers were full of stories with headlines like WHEN THE BOSS AND WORKER ARE ONE and WORKER CAPITALISM: A BOON FOR CITIES. But all the talk about worker-ownership soon turned to ashes. Nothing changed: the old managers were still there and still running the show. The employees did not have actual control over the stock and would not get control over all of it, they came to realize, until 1985; and even then, because of the salary differential, the managers would still control at least a third of the shares, a decisive amount. Morale began to fall and some of the once-enthusiastic workers began to grumble. "A lot of people felt that now that it's theirs, they'd be in the driver's seat, they'd be the boss," a local union official explains. Instead, "when you get down to the real meat of it, there really isn't much difference."

The potential is there, to be sure, at South Bend and at a few other firms in which the voting stock is scheduled to end up in the workers' hands. And it is possible that eventually the voting power will someday be translated into effective day-by-day decision-making power in the hands of—and *used by*—the shop-floor workers side by side with the front-office staffers. But until then—and so far no ESOP company has ever even approached that—it is probably wisest to treat this approach to worker ownership with considerable skepticism.

As one would the fable about the tortoise and the hare.

WITHOUT DOUBT the most unlikely country in all the world is the setting for the most extraordinarily successful model of worker ownership in all the world. The country is Spain, a nation that during most of the past forty years has been about as hospitable to workers' democracy as to democracy of any kind or shape; the model is the complex of more than seventy interlocking producer cooperatives based in the small Basque city of Mondragon (population 30,000), which altogether have a workforce of more than 14,000 people and annual sales of close to $600 million.

That this complex grew up over the last twenty-five years in the shadow of the Franco dictatorship is probably due to the strong mutualist tradition of the Basque region and the leadership of one José Maria Arismendi, a Basque priest who happened to be assigned to Mondragon by the Catholic Church in 1941. That it has been so unusually successful, though, is due not to fortuitousness but to design—careful, slow, labored design.[3]

The Mondragon idea began in a community-funded technical school that Arismendi helped to launch in 1943 to teach local youths both useful technical skills and the principles of worker participation. A number of graduates of the school, eager to put both into practice, decided to establish their own cooperative factory and by 1956 had managed to raise enough money from friends and families in the town to begin a plant, with twenty-three workers, producing stoves and cooking equipment. That cooperative, ULGOR, was a resounding success almost from the start, and upon its prosperity the rest of the complex was built, one firm at a time, two or three a year, first industrial, then agricultural, then

3. It is also true that the Mondragon participants generally downplayed Basque nationalism during the Franco years—to the consternation of some Basque separatists—and generally worked within Franco's Falangist institutions.

financial, then educational, every one operating on the same principles of cooperative worker ownership.

Today ULGOR is the leading manufacturer of refrigerators and stoves in all of Spain, and the Mondragon complex has more than sixty other industrial plants, six cooperative schools (with 3,500 students), five agricultural cooperatives, a consumer co-op, a research-and-development center, a welfare agency, and knitting them altogether, a community bank—also a co-op—that develops, finances, advises, and regularly monitors each enterprise. They must be doing something right. Net profits nowadays average between 6 and 10 percent of annual sales and 18 percent as a return on capital, which is even above the average for manufacturing firms in the U.S. Only one cooperative has had to be disbanded in these twenty-five years—and that failed largely because of ineptness by the Spanish government—and not one has ever defaulted on a financial obligation. In all that time there has been no unemployment and no one let go for lack of work—even if there should be no job immediately available, a Mondragon worker is guaranteed 80 percent of regular pay at all times.

Hardly a wonder that two researchers from Cornell's Work and Participation Program were moved to conclude, "Nothing remotely like the Mondragon system has ever appeared before anywhere in the world."

The system of worker ownership is quite simple. Each worker puts up some membership capital—it has varied over the years but now ranges anywhere from $2,000 to $6,000 depending on the enterprise—which can be paid in a lump sum or taken out of wages over the first two years; a quarter of that is automatically transferred to the collective for general operating funds, the rest remains as the worker's "capital account," on which annual interest is paid and to which the worker's share of the firm's annual profits is added. This account cannot be sold or transferred (and so in practice it usually remains available for the firm to use for reinvestment, thus creating new jobs), and it has to be withdrawn when the worker leaves, the full amount on retirement, around 80 percent if before retirement; in recent years, for workers who have been with Mondragon for twenty years or more, this account on retirement has come to as much as $20,000 and $30,000—in addition to a retirement pension equal to 100 percent of the final salary. The individual's stake in the company is therefore obviously quite considerable, and all observers agree that this identity of interests is what creates the particular sense of motivation among the Mondragon workers.

When a new enterprise is begun, approximately 20 percent of its working capital is raised in this way from the people who will be working in it; at least 60 percent more is put up by the co-op bank, which has both the corporate accounts of the member firms and the private savings of most of the workers (and non-member citizens of the community) to draw on; and if needed, another 20 percent is available in long-term loans from the Spanish government under a program promoting cooperatives. The bank takes special care with its investments: it makes sure that any proposed project is studied for a full two years, that there are managers available who know how to run the plant, and that there is an ample market for the products when they come on line; it also keeps a ninety-person "management service" ("*empresarial*") department to keep an eye on managers and plant finances and offer advice when needed or asked.

Wages are established in each firm at a base level equal to the prevailing wage elsewhere in the province, and a differential of 3:1—here a remarkable feature of the Mondragon experiment—is fixed between the lowest salary and the highest. (You can get an idea of the sense of equality and solidarity this provides by comparing it to a large American firm where the differential between new employee and chief executive is more like 50:1 or 70:1.) Profits of course are redistributed to the worker-owners into their capital accounts at the end of the year, with the only stipulation being that at least 20 percent must be held back in the firm's collective reserve fund and at least 10 percent used for social services and projects in the community. (Losses similarly have to be borne by the workers, except in the first two years of a new enterprise, and are deducted from their capital accounts, up to 70 percent.)

Altogether it is a system of remarkable financial self-sufficiency, since almost all the capital needed to establish, operate, and expand the firms is available right there within the community, either from the workers' capital accounts or from the savings bank; no more than 20 percent ever comes in from the outside, and that infrequently.

Control over plant operations is nominally in the hands of the workers, as well, with the employees electing from their membership a board of directors (the *control board)* that meets once a month and sets broad policy for the firm, including the hiring and direction of top managers. In practice, control boards have tended to let their hired managers have a considerable degree of independence, so that frequently the enterprises operate just as if they were ordinary firms on the outside, with aspects of hierarchy, autocracy, routinization, and pay disputes. Still, managers are always aware that their jobs depend on keeping the allegiance of the very people they are overseeing, and they know that their decisions may be overridden by the control board or general membership at any time. In one instance not long ago, the managers turned down the request of the evening-shift workers for a 5 percent pay differential over the daytime workers, arguing that it would cost too much money; those workers called for a general assembly of the entire plant, put their case, and won over a majority; management had no recourse but to agree to the raise. In addition, most Mondragon firms have a "social council" made up of elected representatives from each shop to hash over grievances about working conditions and salaries and to present complaints to the management or the control board.

Mondragon is by no means a paradise; it has its problems, some general (too many routinized assembly-line jobs, a lack of democracy and solidarity in the larger firms), some particular to its own methods (workers voting against shutting down uneconomic branches, domination of the control boards by an active few). But it has shown beyond question that an economic system of worker ownership with a cooperative base is not only possible but profitable, that it can thrive in large and complex industrial plants as well as in small workshops and offices, and that it can survive and prosper for many years, even beyond the departure of the first generation. Beyond that it shows that economic enterprises that are immediately rooted in a community covering a small area, that explicitly have the community's interests at heart (as with the profit-sharing scheme), can rely on both the financial support and the popular enthusiasm of the citizens to a degree unparalleled elsewhere.

Could it happen elsewhere? Is it duplicable? The officers of the Mondragon bank certainly think so; they say that starting producer cooperatives is actually pretty simple, once you draw up the plans carefully and find (or train) the managers who can carry them out. There's nothing about Basque blood or the Spanish sun that makes such places unique to Mondragon, the bank president says, and adds that the only necessary conditions to imitate are an industrial tradition and sophisticated workers, both of which are found throughout the West. A team of British observers (a businessman, a banker, and two financial journalists) observed Mondragon close up in 1976 and concluded:

> We can see no insuperable obstacles to the establishment of a cooperative sector run on Mondragon lines in Britain. . . . Indeed, we came away feeling that a network of regional cooperative groups, on the Mondragon model, each serviced by a local bank, would be socially and economically very advantageous to this country.

The Cornell researchers similarly: "It may indeed take a great man or woman to create and further develop a social invention of the magnitude of Mondragon, yet lesser human beings can examine that invention and seek to apply what they can learn from the structure and social process of Mondragon to organizations and/or systems they themselves seek to build within their own cultures."

8

Workplace Democracy: Control

THE TASK OF saving the Saratoga Knitting Mill was formidable. Workers borrowed from friends and relatives, dug into savings meant to send their kids to college, got second mortgages on their homes, borrowed on their life insurance policies. One woman, a loom operator, held a garage sale every week all summer long until she had the necessary cash; a front-office worker chose to sell his vacation home, over the sizeable objections of his wife and children. But somehow between the spring of 1975, when the 140 employees were told the plant was closing, and the summer of 1976, when the workers of the mill took title to their own operation, they managed to raise $7 million: 60 percent from their own pockets, 25 percent from banks and businessmen in the local area, 15 percent from outside investors. It was a considerable, and considerably daring, move on the part of people who had never even heard of worker ownership before, but there was a good feeling about it, too: "Sure, I'm saving my job," said one worker, "but I'm also becoming my own boss. Don't forget that."

Not quite. The managers that were there when the plant was run by Cluett-Peabody are for the most part still there, and most of them took the worker-owner opportunity to become the dominant shareholders in the new company, the plant manager himself becoming one of the largest, with the maximum permissible 100 shares. The work routine is still onerous, despite the introduction of more modern knitting equipment, and there is no more influence of the shop-floor workers on the production process now than there ever was. And ownership of the company hasn't changed the decision-making one whit: as plant president Donald Cox says bluntly:

> We have been able to make this employee-owner conversion without making any changes in the management group. My company is not a self-managed enterprise. I make all the decisions.

Hugh Hibbert, a knitting foreman at Saratoga, told writer Danicl Zwerdling the somewhat disillusioned reaction of the floor workers three years after their "takeover": "People expected a little more in the decision-making process than there's been. They were of the idea they'd have a little more to say at the managerial level." Today, outside observers report, cynicism is high, the plant morale as low as ever.

IN COUNTLESS CASES, from small, cooperative health-food restaurants to large manufacturing plants, from the Briarpatch Auto Works in Palo Alto to the asbestos mine in Vermont, from the Scott Bader Commonwealth in England to the metalworking factories of Stockholm, workers have found time and again that mere formal ownership doesn't necessarily have much to do with day-to-day authority. As the leading Swedish expert on workplace democracy has put it:

> With a few exceptions most experiments have very little to do with industrial democracy in a sense which includes forms for worker influence on larger economic and technological decisions. By and large these cases—important as they may be—represent only new ways to increased productivity through increased motivation to work. This means that this approach is not answering all the relevant problems involved—such as worker claims for a broader base in decision-making—and it also means that its potential is not fully realized.

It takes more than just worker ownership for the democratization of work—there's got to be some mechanism for worker *control* as well.

Control can come in a multitude of ways, and the range of experiments in employee democracy has included everything from meetings to decide how long the coffee break should be (at the McCaysville textile plant in northern Georgia) to weekly assemblies of entire factories to determine the following week's production quota (in a few self-managed plants in Yugoslavia). But certain features have proven to be basic:

▪ the regular and open *assembly* of all employees, empowered where necessary to elect (or, better, choose by consensus) whatever directors and officers seem necessary for the firm, with the right of immediate recall of any one of them;

▪ regular *decision-making,* on all matters deemed to be important to the workers, by either the full assembly of the workers or some representative body;

▪ the easy acquisition and sharing of *information* about everything that goes on in the workplace, including who earns what and how much the raw materials cost and what the plans for long-range development are; and

▪ the regular and complete *rotation* of jobs, allowing every employee to sit anywhere from the metal punch to the treasurer's desk, with special encouragement for those who are reluctant.

Strongforce, a private group in Washington devoted to worker self-management, has proffered a checklist of the kinds of things over which any workgroup should be able to have a direct, informed, decisive, and regular say:

1. Raising capital
2. Allocating profits
3. Determining investments in new plant and machinery
4. Arranging job assignments, rotations, responsibilities
5. Choosing the type of products or services, the markets, and the prices
6. Organizing research and development
7. Setting salaries, wages, and fringe benefits
8. Hiring and training new workers
9. Creating job security and layoff standards
10. Setting work standards and work rates
11. Establishing safety rules and practices
12. Overseeing physical working conditions

In sum, the worker should control the workplace, not the other way around. Anything less and what you have is an operation devoted not to increased democracy but merely to increased productivity.

ULTIMATELY, IT'S CLEAR, what all this is leading to is the abolition of the stratum in most workplaces known as management, and at first blush that seems a mad and foolish thing. After all, even the plywood firms hire outside managers, even the Mondragon cooperatives retain hierarchy and management. True enough—but might it not be said that to that extent they are not truly democratic, that in fact they have not altered the basic conditions of old-style labor by all that much? Would not the removal of such managerial aristocracy—outsiders, the highest paid, separate in habits of thought and living—be the necessary and desirable consequence of worker control?

Well, yes—but is it possible?

Indeed it is, and not so rare as all that, either. In some collective workplaces the manager's job is rotated among the workers, effectively demystifying the role while giving everyone an understanding of what it entails. In other worker-run shops managerial decisions are reached collectively, by either vote or consensus, with each employee expected to have the requisite information and expertise. And in a number of places a managerial committee is elected by the workers at large to take on basic administrative tasks for a year, subject to intervention by the workers at any time.

There are problems in such a system, no one denies. Most people in our industrial culture are not used to doing without leaders and taking responsibility into their own hands. There is a sort of dependency mentality that in subtle ways gets ingrained in most of us, creating an inevitable resistance to the thought of taking charge: I don't feel like getting involved; it's too much of a hassle; I don't have the time. Most people, particularly women and often the undereducated, have little experience with leadership, are usually afraid to assert themselves, and tend to clam up when made to sit on a serious committee. And yet experience has shown that in practically every instance where it's given a real try, enough time and enough practice with self-management will overcome these difficulties.

Examples of non-managerial successes are not all that hard to find. In the classic study here, Seymour Melman, professor of industrial engineering at Columbia University, laboriously compared twelve industrial enterprises in Israel, half of them run with traditional managements and half run cooperatively, without bosses. By every measure—productivity of labor, productivity of capital, efficiency of management, cost of administration—the cooperative plants performed at least as well as, and in most respects better than, the regular ones: cooperative sales per production worker were 26 percent higher, sales/asset ratios were 33 percent greater, profit/investment ratios were 67 percent greater, and profits per production worker were 115 percent higher. Melman concluded, in his careful academese, that "industrial enterprises of a modern technical sort" can indeed operate without "management decision making" and that "cooperative decision making is a workable method of production decision making in the operation of industrial enterprise." As to the reason, Melman suggests that "this capability is linked to the pervasive motivational and operational effects of cooperation in decision making and in production, pressing toward stability in operations, and thereby toward optimal use of industrial facilities." In other words, people work better when they care. Now it is only fair to point out that the collective enterprises Melman studied were on Israeli kibbutz settlements and therefore the workers were presumably conscious of the entire community that stood beyond, and depended upon, their work, a strong motivating condition that presumably did not apply to the standard enterprises. Nonetheless, the *fact* of their superiority is unquestioned.[1]

1. Cooperation, according to two psychologists in the basic *Handbook of Social Psychology* (Volume II, 1954), normally leads to "strong motivation to complete the common task and to the development of considerable friendliness among the members." In numerous clinical studies, "with respect to group productivity the cooperative groups were clearly superior."

An even more unusual example of dispensing with traditional management comes from Norway and involves ten ocean-going commercial ships—places where authority and rigidity have always been thought to be indispensable. David Moberg, writing in *In These Times* in 1977, described how it worked:

> On the good ship Balao, for example, the traditional hierarchy was replaced with work planning groups. The newly integrated work crews required increased education and mechanical training for many of the lower level sailors. They learned how to navigate as well as to repair the ship, so they could work in nearly all areas. The old bosun, or supervisor, was eliminated in favor of the new job category of ship mechanic.
>
> Safety hazards were continually discussed in work planning groups, with the ship's crew completely in charge of designing all operations to maintain safety. Everyone was put on a fixed annual wage and a flexible work schedule for staffing the ship was established.
>
> The key to the whole experiment's success was "for those who do the things to have the initiative and control" [according to Einar Thorsrud, the man who began the whole project back in the 1960s], "to put back into the job coordination and control." Now the captain on the ship is frequently overruled. There is no uniform relationship of superiors to subordinates, but an adjustment of rules according to each task. The four separate eating areas have been combined into one, and no one even sits regularly in the same chair. This uprooting of bureaucracy has increased the satisfaction and power of the crew. The ship owners now find it easier to recruit and hold workers.

And if it can be done on a *ship,* could it not be done anywhere?

A final example, closer to home, is the experiment run at the Rushton coal mine in eastern Pennsylvania from 1973 to 1975, in which one section of the mine, with three shifts of nine workers each, established an autonomous work group, working without managerial supervision. Every worker in the section was paid top union salary, each one was trained in all aspects of the job, and when they met at the beginning of each shift they would decide among them what jobs they felt like doing that day, their only condition being that they had to keep production up and not violate any mine safety rules. The foreman, who in other mines was typically a mini-dictator telling workers where to cut, where to lay power cables, when to eat lunch, was not allowed to interfere with the daily work process in any way, could not give orders to the group, and had to make any suggestions to the entire shift, which would meet and decide whether to honor them or not.

The results proved conclusively that self-management was successful, both at maintaining productivity and, far more important, at increasing worker satisfaction. The actual tonnage mined during this period was not increased—it stayed roughly the same—but, according to Gerald Susman of the College of Business Administration at Penn State, during the first year the time lost to accidents declined 500 percent, safety violations were reduced from eighteen to six, and equipment breakage was down over previous years. ("It used to be when a machine got busted," one miner confessed, "we'd just sit around, happy-like, until the foreman spotted it and called in a mechanic. But when a machine breaks down nowadays, whenever we can . . . we just fix it ourselves.") Cost figures of the autonomous experiment are somewhat confused, but one measure of the apparent economic gain is that after a year the management of the mine announced that they wanted to open up a whole new section under the same autonomous rules. But what was most noticeable about the project was that the workers liked it. Reporters from a dozen major papers and the business press who visited the mine all heard the same kinds of remarks: "Now it's like being in business for myself"; "It's like you feel you're somebody, like you feel you're a professional, like you got a profession you're proud of"; "I'm not as tired when I go home any more, and my wife . . . told me just the other day that I was a lot easier to get along with." The United Mine Workers reported that "over and over again, we heard miners say that they would continue to work autonomously even if the pay differential were taken away, and that they needed no incentive outside the additional satisfactions they received to continue working autonomously."

It is not really anticlimactic to add that, in August 1975, after twenty months of the project, the miners themselves voted, 79 to 75, to discontinue it. The reasons? The UMW itself was no longer much interested in the experiment, having concluded that its ultimate effect on mine safety was negligible, and more important, the workers in the *other* shifts, the non-autonomous shifts, were plainly angered both with the high salaries and with the independence their fellow miners enjoyed, and their votes were sufficient to end it.

The literature on worker participation is copious, and throughout it runs a single and almost unequivocal conclusion: when workers are given an opportunity, they work just as well if not better without bosses. There are studies in garment factories, insurance companies, medical laboratories, shoe factories, newspaper offices, and auto-accessory plants, among men and women and mixed groups, with production-line workers and salespeople and office workers and scientists, and they all indicate that increased participation, increased autonomy, increased control, invariably means increased satisfaction and more often than not increased productivity and efficiency. In the words of Paul Blumberg, professor of sociology at Queens College in New York and the author of the pioneering work, *Industrial Democracy:*

> *There is hardly a study in the entire literature which fails to demonstrate that satisfaction in work is enhanced or that other generally acknowledged beneficial consequences accrue from a genuine increase in workers' decision-making power. Such consistency of findings, I submit, is rare in social research.*

NOW I AM not suggesting—nor do the studies conclude—that there is no need for *structure,* for organization, planning, guidelines, quotas, goals, and systems of responsibility. The absence of managers does not mean, as some (mostly managers) contend, the absence of order. There has been enough experience in the last decade or so with groups trying to work with complete spontaneity and randomness ("without all that capitalist structure bullshit") to show that, at least when anything more complicated than crash-padding is involved, such disorder is not particularly useful or, in the end, very satisfying. But hierarchy is not the only form of order, any more than autocracy is the only form of decision-making. Cooperative structures, cooperative responsibilities, can easily be established and have been in countless enterprises.

Nor am I saying that *decisions* do not have to be made, on occasion in a hurry and sometimes by a single individual. Obviously certain minor problems should not require a full-scale general assembly, like the magazine collective that debated for a half an hour if the shades should be drawn and if so how far—that decision could probably have been taken care of by the one in whose eyes the sun was shining. And obviously certain major problems should have an immediate yes-or-no answer that one person may have to give, though with the understanding that the decision will be reviewed sometime hence by the rest of the group and either confirmed or modified with new guidelines for future occasions; just such a procedure was followed during the extraordinary journey of the Quaker ship *Phoenix* to Vietnam in 1967, when despite harassment by South Vietnamese gunboats and American planes the ship's leader did not once use his authority to make unilateral decisions but instead hammered out every procedure with other project members, *by consensus,* before acting. It is surprising how many matters of substance and duration can be decided, easily and rationally, by a coherent group properly informed.

Finally, I am not saying that there is never any need for *experts* of one kind or another in a complex business—it well may be that trained people are necessary as accountants, engineers, sales managers, and the like. Two points, however, must follow.

First, experts do not need to be *bosses*. The people with the expertise have no special reason to have the monopoly of power; they are workers just like the rest and should have the same say in the products and procedures of the workplace. They may have some special knowledge that the other people do not, but it is precisely in sharing that, and explaining decisions based on it, that others come to learn new roles and the group as a whole comes to make rational choices.[2] Just as an architect does not normally dictate the way a house is built but works with and adjusts to clients, contractors, suppliers, and inspectors, so the expert in a worker-controlled operation would be expected to consult and discuss with all the others in the group.

Second, experience shows that with almost all jobs, and most particularly those of the kind that upper- and middle-level management occupy in American businesses, on-the-job training and general corporate experience is many times more important than prior academic preparation, a fact that virtually every manager will acknowledge. That means that the ordinarily intelligent worker is probably capable, in time, of learning the necessary skills of any managerial position—and can be presumed in addition to bring to it a certain valuable up-from-under perspective that no one else could have. Experiments in full job rotation have been fairly limited, confined usually to the communes and collectives of the past decade that have been explicitly committed to destroying the usual patterns of expertise and specialization. But they suggest that generally, though both the expert and the neophyte usually begin with great trepidation, it does not take long for even the most arcane skills to be shared. In some cases, according to a networking group in Cambridge called Vocations for Social Change, which has had experience with a great many alternative organizations,

> when the task involves a difficult skill, there is often a dual rotation system. Two people do it at a time, but one person rotates off and is replaced by a novice half-way through the time period. Thus everyone is in training half the time and is the "expert" for the other half.

Sometimes, too, a committee of trained people is established to take the place of a single expert and meets regularly both for basic policy and day-to-day decisions.

In general, the tenor of our economy seems to be moving away from hierarchical control toward more collegial decision-making, even in the stuffiest, most conventional firms, and the benefits are now routinely

---

2. They also may *not* in fact have the special knowledge they pretend to. The supervisor, foreman, and engineer at one toy company reported by sociologist George Strauss confidently predicted disaster if they gave into the wishes of the assembly-line women and allowed them to adjust the belt speeds as they wished; they *knew* the women would shirk, the speeds would be too slow, and productivity would fall. The experiment was tried, though, and in fact the women ended up working at a faster (though more variable) rate, increasing productivity by 30 to 50 percent, and so thoroughly calling management's expertise into question that the supervisor unilaterally ended the whole thing and made the women go back to the old system. Such stories are by no means unique.

touted by management consultants and workplace psychologists. Scott Burns, the Boston economist, argues:

> What is lacking in public consciousness—or even in the literature of management—is recognition that the decline of authority is a consequence of far larger events than faulty child rearing, permissive schools, the power of the local union, and the other popular hobgoblins of disorder. The larger fact is that the economic drive which justified (or at least sustained) the hierarchic structure of industrial society has matured. As a result, all that would support the continued existence of powerful hierarchies, from the organizations within the market economy itself, is disintegrating. The day of large organizations and small elites is at an end.

That may be somewhat optimistic, but there is no gainsaying the trend to which he points. As the Vocations for Social Change people put it in the title of their book: *No Bosses Here.*

Now there are—it will come as no surprise—size limitations on worker control, as anyone who has worked in a plant or office will realize. The mechanics of economic democracy—general meetings, group decisions, information sharing, job rotation—demand a certain cohesion, a certain constriction. Just as it is not possible to know a million people or hold a million dollar bills (they would make a stack 364 feet high), it is not possible to have an effective system of democracy in a workplace with more than a limited number of people.

Stephen Sachs, an official of the Association for Self-Management and one of the leading American researchers in this field, has surveyed the literature and determined that self-management is "more extensive and more successful in relatively small, face-to-face organizations than in larger organizations, and in smaller units of large organizations than in the organization as a whole." His work in Yugoslavia, a nation that has done more than any other to foster worker control throughout the economy, has confirmed this, for there after a decade of experimentation the Yugoslav government concluded that self-management "had generally been found to work better in smaller firms than in large ones" and officially moved to reduce the size of factories and offices and to break up large companies into semi-autonomous work units operating by direct democracy. In general the Yugoslavs have concluded that when a workplace gets much beyond a hundred people its self-governing difficulties start: there is a significant drop in participation by the workers and a tendency to let supervisors and senior managers make all the important decisions, with ultimately a decline in morale and productivity.

The Israeli kibbutzim offer another good real-world example. There most of the service groups—the basic "branches" that do the agricultural work, run the kitchens, operate the laundry, and so on—are seldom more than eight or ten people, even in the larger settlements of five hundred and more. The industrial units tend to be larger, but half of them have fewer than twenty-five people, less than a third have more than fifty, and only a handful have more than a hundred (and then usually broken down into two or more autonomous shifts). Arnold Tannenbaum and a team of academic researchers who examined industrial organizations in five Western nations found that the small size of the Israeli factory was the crucial element in its success; large organizations, they determined, are always more hierarchic and authoritarian, no matter what ideologies they fly under, and are simply not conducive to workers' self-management.

Finally, there is the Mondragon example. "The success of producer co-operatives at Mondragon," the British team who went there in 1976 reported, "has been limited to small and medium-sized enterprises": more than half of them have fewer than a hundred workers, and these are the ones agreed to have the most harmonious conditions. The only factory over 1,000 is the huge ULGOR refrigerator plant with 3,500 workers, and this is the one that has regularly been plagued by disputes and dissension, the only one where there has ever been a strike (unsuccessful; the workers fired the ringleaders).

In sum, it seems safe to conclude from these and other experiences that a small workplace will have no particular difficulty in achieving worker control and that units of even forty or fifty can find ways to maintain the necessary communications and intimacy, though at that point the strains begin to show; up to a hundred is still apparently manageable, and over that the troubles mount and true control becomes more and more distant.

For real worker control to exist, you would think, a number of other conditions than size would have to be fairly special: the people would have to be relatively well-educated, for example, and used to an industrial culture, with some experience of democracy in their daily lives, and some training in the processes of group decision-making. So it would seem. The fact is, though, that probably the most striking example of worker control and democratic process in all of North America is to be found at Cooperativa Central, a ranch in California run by seventy-five Mexican-Americans, most of them once dirt-poor and very few of them educated beyond basic literacy.

Cooperativa Central was begun in 1973 by a community-action agency with $1 million from the Federal Office of Economic Opportunity and some farmland it bought in the Salinas Valley, hardly a prepossessing way of implanting democracy. Farmworkers from the area were invited to join the co-op, and though most hadn't any idea what was being thrown at them, some seventy-nine workers signed up, took charge of a couple of acres of land each, and began harvesting the fields of, as it turned out, strawberries. For the first two years you wouldn't have known that there was much of a cooperative at all: the old ranch manager and the former salesman were kept on, the same companies came in to buy the berries at the same old prices, and the same old patterns of sharecropping remained.

But gradually some time along in the second year the workers began to realize that this was in fact their own farm and that they had the machinery to run it as they wished. In 1975 they fired both the manager and the salesman, ousted most of the old board of directors, and began to run the place themselves with their own ideas of democracy, and that is the way it is operated today. All of the basic decisions, from what kind of machinery to buy to what kind of housing to build, is made by a nine-member board of directors elected annually *and* by general meetings of the entire membership meeting monthly. The manager is chosen from among the ranks and is replaced at least once a year and often more frequently than that, as the board decides—a revolving-door policy that the farmers particularly welcome; managers are expected to be facilitators, not bosses, and they check everything with the board. One production manager interviewed by Daniel Zwerdling described the process:

> Even though I can spend $500 on my own, I ask the Board everything. If I have to take a truck to the garage to fix the brakes I ask permission from the Board. I had to buy a new water pump for $250, so I decided I had better ask, to play it safe. This morning we needed to put some fertilizer over there, and we needed to hire four workers to put it on. I had to keep going back to the board members in the field, again and again, to ask each one who he wanted me to hire. It's sometimes hard to bring the board members together to make those decisions because they don't want to stop working in the fields.

The board, in turn, checks everything with the members, even questions that according to the by-laws don't even require a vote. Farmworker Javier Ruiz told Zwerdling:

> We are completely democratic. Everything is done by majority vote. The workers meet at least once a month in the conference room we fixed up in the barn—I'd say at least 50 or 55 of the 75 members usually attend—and they hire and fire, they decide how much acreage we should plant, who to sell our berries to, and they make a lot of financial decisions.
>
> For example, members vote on buying new equipment, like a tractor or expensive office equipment. Members voted on which freezer

> company to sell to, and which one of the trademarks we wanted to use. Recently the Board voted to accept a new member into the co-op, but the members overturned the decision at their monthly meeting because they felt he wasn't qualified. We had some old houses on our property which we were just using as storage sheds, and some members thought we should rent them out to families. The Board voted not to lease them, because the houses are not up to code, but the members overruled the Board because some of the members need housing very badly. We're going to try to bring the houses up to code soon.

Can it work? Can unschooled Mexican immigrants with a system of haphazard management rotation and niggling-decision democracy possibly run a real live ranch in the modern world? It looks that way. Cooperativa Central is an extremely successful farm, with annual sales of $2 million and a productivity rate greater than the acreage had as a private farm; in 1976, a good but not extraordinary year, the co-op members earned an average of $25,000, including profit-sharing and deferred payments, in an area where some Mexican workers are lucky to make a fifth of that; membership fees for newcomers now come to $14,000–$18,000 (though most of that need not be paid up front). The ranch has expanded out of strawberries with an additional 700 acres of land purchased in 1977 on which it is raising vegetables and cattle.

One has to believe that if worker control can work *here*—and so far it has, with a vengeance—it can work anywhere. Skeptics? "If they have any doubts," says Javier Ruiz, "send them over here and we'll show them."

9

Workplace Democracy: Community

ROBERT DAHL, the political scientist, offers a proposition in his analysis of democracy, *After the Revolution?*, which he calls the Principle of Affected Interests. It holds: "Everyone who is affected by the decisions of a government should have the right to participate in that government." It is a principle that would seem to be absolutely basic to any democratic system, and one that is very much in the American grain, something on the order of, say, "no taxation without representation." It would seem to be, moreover, a precept central to any democratic economy, speaking to the interests of the general public beyond the workplace, "everyone who is affected by the decisions" of the workplace: not just the workers but the people next door and the taxpayers who provide the roads and the couple who runs the luncheonette down the street and the accounting firm that does the books and the people downwind who breathe the air and every client and customer at its door. In short, what we may call—using the term in its very broadest sense—the community.

Just as workplace democracy cannot exist without worker ownership and worker control, so it cannot exist without some form of worker community. It is a triad: ownership represents its economic side, control its political side, and connections and communication with the community represent its social side.

FOR THE USUAL private business in the capitalist system, the community is essentially non-existent. No matter how public a firm's activities may be—and who could deny that the activities of General Motors, for example, have enormous public consequence?—it is protected by the fiction of being a private organization in the hands of private citizens. Yes, it might give to the Community Chest or outfit a local softball team or sponsor a public-service program; it might create a "public relations" division to mix into civic affairs while promoting company affairs; it might even, as a few "enlightened" businesses have done, put a woman or a black or a student or some other "community representative" in a token spot on the board of directors. It also presumably has to abide by certain regulations designed to insure public safety and pay occasional income and property taxes to offset the public services it uses. But for the most part the private firm is regarded as private, left to consider outside citizens only as they make themselves known through the market. The executives, who normally don't even live in the same town their offices are in, have no obligation to the community that sustains them, and a great many move from firm to firm with such rapidity that there wouldn't be much way for them to know the needs of their fellow citizens if they had. And the multinational firms are the most distant of all, because of course they have *no* community, not even a single nation, to which they have any loyalty or duty, nor any way to know about or respond to the interests of such a vast and worldwide public.

Above all, the private corporation is supposed to be concerned not with its society but with its profits, and that's quite a different and in most cases antithetical thing. Donald Conover, a director of "corporate planning" at Western Electric, has told what happened when he asked a group of managers to comment about a television film he had made on the problems of the city. One of them turned to him and said, "How do you want us to answer, as a manager or as a human being?" Extraordinary: in our system they *are* two different roles.

Things are no better, I hasten to add, under the "socialist" systems. There the idea of community is substituted for by the role of the state, which makes the laws and assigns the quotas and controls the markets, all presumably so as to evidence the public will. No matter how foolish or corrupt or anti-social, the state-owned firm is protected by the fiction that whatever it does will automatically be in the community interest—by definition. No doubt to some extent the socialist firm *does* respond in some way to community interest, in the sense at least that it cannot (as its capitalist counterpart can) flout government directives or discontinue a decreed product line or lay off its workforce or set up a runaway shop in the south. But it cannot represent the true wishes of the larger public because no one has any good way of knowing what those are, being determined in practice by diktat from the commissars in the capital—and if the commissars choose to divert a staggering amount of the national budget to armaments instead of automobiles (as in the Soviet Union), well, that is the public will no matter what the public wills. Nor does anyone regard it as important to serve the needs and appetites of the local communities where the plants and offices are located, for the system is relentlessly centralized and individual managers must listen not to what their neighbors but to what their commissars say. In those conditions even outfitting a local softball team might be hard to do.

It is a sorry comment that neither of these systems, as we know from long experience, really works to represent the community interest, and both are a pitifully long way from allowing "everyone who is affected" by economic decisions to participate in making them. But a worker-owned, worker-controlled system—is there any reason to think that it could achieve the Principle of Affected Interests and give the community a real voice in economic affairs?

ON A THEORETICAL LEVEL there certainly is. One would expect that workers, living closer to their workplaces and representing a broader spectrum of social strata than today's executives, would inevitably bring a far greater diversity and intensity of community viewpoints into a firm's decision process. Presumably also they would take their family's and their neighbors' concerns into the job with them, and since there are no distant shareholders or absentee directors to worry about, their processes and products would reflect those concerns. And it's reasonable to assume that the whole experience of self-management would spill over into the workers' private lives and inject itself into the homes and neighborhood organizations, developing democratic and participatory habits throughout the community. Finally, it would seem that a firm truly attuned to what the community desired would be in a very advantageous position to make rational and finely tuned decisions about what kinds of services to offer and how much of what kinds of products to manufacture.

But fortunately we needn't remain on a simply theoretical plane here. There is enough evidence from the real world to show how community-workplace ties are forged and strengthened when almost any kind of worker control is practiced.

On one end of the scale, take the limited experiment with an "open system" at the Proctor & Gamble plant in Lima, Ohio, begun in the late 1960s. Workers were given considerable power in setting their conditions of work, with control over hiring and firing, establishing pay rates, working out job rotation, and keeping the books. Predictably, productivity increased, quality improved and equipment costs decreased, and both salaries and profits showed a steady rise; but what was most surprising was the finding of Neil McWhinney, a psychologist from the University of California who helped to plan the Lima experiment:

> One of the striking features in our "pure" open systems plant is that workers take on more activities outside the workplace. The most visible involvements had to do with community racial troubles. Following major disturbances in the small city where they lived, a number of workers organized the black community to deal directly with the leaders of the city and of industry. . . . Blue collar workers won elections to the

> school board majority office and other local positions. Nearly ten percent of the work force of our plant holds elective offices currently. . . . We have noted that open systems workers join more social clubs and political organizations.

At the other end of the scale, take the extraordinary system of self-management in Yugoslavia, where the government has attempted to establish an entire worker-run economy, with every single workplace—industrial, agricultural, service—organized on self-management principles. Imperfections there are in the Yugoslav arrangements—particularly in the heavy roles played by the state apparatus and the Communist Party—but after thirty years of practical day-in-day-out experience, the success of it is beyond question: in fact, during the 1960s the Yugoslav economy grew at a rate second only to Japan's and in the 1970s, despite some slackening, remained among the world's highest. And during this time there has been a real and measurable increase in the involvement of workers individually, and their firms as businesses, with the wider community.

Stephen Sachs of Indiana University, who has done research in Yugoslavia on exactly this point, reported that he found a "strong indication that the participation by workers in management and organizational income does in fact increase the concern of the enterprise for community interests and problems"—and to a level that "seemed to exceed significantly that of business and business executives in the United States." He found in the region he studied that the workers councils regularly voted to give money to special village projects beyond their regular village tax—in one case for a new drinking-water system, in another for new road pavement—and normally sponsored concerts, soccer teams, chess clubs, dance companies, and the like. Facilities and services that a business had set up for its own employees—a bus service, for example, or a meeting hall—were automatically free and available to anyone else in the community, and even on-the-premises space and equipment—in one case an entire workshop—were open to students at nearby schools and technical colleges. And permeating most operations was the fixed idea that the purpose of the business was to benefit not simply the workers but also their families, their friends, their neighbors, their towns. It isn't all perfect harmony, to be sure—in all too many firms the workers have given over real decision-making power to the upper management, often less attuned to community needs, and in others the workers need to be prodded by "regulation from truly representative public institutions" to keep community interests uppermost. Still, in general, Sachs found, "the institution of workplace democracy significantly increases the community spiritness of an enterprise," and "maximum social responsibility of enterprises can probably only be achieved in a society in which firms are fully self-managed."

NOR ARE THE BENEFITS all one way, either: the involvement of a community in the affairs of a local business can help significantly to make that firm more harmonious and productive.

There are a multitude of problems with worker-controlled enterprises, of course, particularly among those who have little training for such a responsibility. But when the community makes itself felt, and autonomous workers realize that there is a constituency beyond the office walls, a body of people out there dependent in some measure on their efforts, that generally proves to be a real and potent solution to most of the difficulties. Among countless cooperative and worker-run firms in this country and in case after case in similar enterprises elsewhere, where community ties are strongest the businesses tend to perform better, stay in business longer, hold their turnover rates lower, and work with fewer internal conflicts. This is particularly true of retail businesses, of course, and people in food and bookstore co-ops and record-shop collectives are the first to say that community feedback is essential for success and the first to establish the means—regular open meetings, suggestion blackboards, Sunday-afternoon forums, local representatives in their councils—to insure it.

The most telling example of this kind of influence is the Israeli kibbutz, for there a sense of shared commitment to the kibbutz ideal is added to the powerful day-to-day input of a close-knit community. We have already seen that the businesses on the kibbutz settlements are more productive and efficient than other Israeli forms, and in explaining that Seymour Melman says revealingly:

> The people working in the kibbutz enterprise are motivated to feel needed and wanted within the context of the total community. Such feelings, among people who share common tasks, are powerful motivating forces for individuals to give their best in the performance of shared responsibility.

Since there are no wages for work, and since everyone gets an equal real income, it is status and admiration that normally act as the incentives for the kibbutz worker—and both of those are by definition *social,* dependent upon the wider community. Another researcher, Haim Barkai, has noted:

> Respect and esteem for a good day's work and for the success in managerial and entrepreneurial functions is undoubtedly an important factor in the attitude of individuals towards work and responsibility in the kibbutz environment. And per contra, the disfavor with which shirking is viewed is, in the closely knit communities, which even the largest kibbutzim still are, a powerful sanction.[1]

1. Obviously this doesn't apply to the growing number of outside "hired workers" who make up some 8 to 10 percent of the kibbutz labor force, one reason the hired worker has become a serious issue on many kibbutzim.

Moreover, the kibbutz enterprise, whether agricultural or industrial, has to depend in very real and practical ways on the community, for its resources are limited—particularly labor, which has always been in short supply, and land, which has to be wrested from the desert—and the settlement as a whole has to determine how they should best be allocated. Hence it is the general assembly of the whole kibbutz that actually decides what kinds of crops and products are to be produced, how big the workteams should be, how the annual productivity targets are to be set, and what the development and reinvestment plans for the future are to be. And though there are obviously wide differences among the 230 or so kibbutzim, the fact that they have all been economically successful—per capita income growth rates of an average 5 percent in the 1950s and 1960s, productivity rates higher than that of the rest of the country, a generally high living standard—suggest that this communal role has been unusually advantageous.

BUT I THINK it would be a mistake to assume that worker control and community self-interest will always coincide, as if by some communal magic. Enough evidence has come from the Yugoslav experience, as well as from any number of well-intentioned collectives and Community Development Corporations in this country, to show that there will be times when workers are going to put their own (or their firm's) concerns ahead of the specific considerations of the outside populace: they want a raise, the community wants a playground; they want to take off an extra week of vacation, the customers want the services continued uninterrupted; they want to make cars, the people would rather have buses. For such times, it pays to have systematic arrangements that guarantee some sort of mutuality.

The spectrum of possibilities is broad, but in general it has this four-part configuration.

Community representation through contractual obligation. Under this sort of plan, about the simplest, every self-managed firm by contract agrees to give back to the community in one way or another some of the rewards it gains from doing business there: by turning over a certain percentage of the profits beyond local taxes, or agreeing to the annual upkeep of one or another neighborhood institution, or maintaining a public facility on its premises. Most often such an obligation is voluntary, produced by groups that explicitly wish to demonstrate communal values—the collective bookstore that sets up a library available to any community person free of charge, or the Community Development Corporation that writes into its charter the percentage of profits to be used for day-care centers; but on occasion it has been written in as a proviso of loans from a local bank, trade union, progressive foundation, or cooperative credit union.

An extensive variant of such a plan is found in the Mondragon system, where the central savings bank, the institution that funds and monitors all projects, stipulates in advance that each venture must pay back at least 10 percent of its annual profits to the community, with a sliding scale by which the more money a firm makes the more it has to turn over. This is reinforced, too, by the Mondragon cooperative school system, which acts as both a programmer and a guide—and ultimately as a kind of conscience—to insure that each worker separately and the multitude of firms collectively do not shirk their social obligations. Yet finally the real guarantee of community interaction is the simple fact that Mondragon is a small city of 30,000, with a strong tradition of mutual cooperation and social cohesion, where people have seen in their everyday lives the practical benefits of shared resources. In the words again of the British economic team:

> It means on the one hand, that [the cooperatives] receive the support and backing (notably savings) of the local residents; on the other, it makes possible the integrated planning of industrial initiatives, housing, education, the training of skilled personnel, and community services (medication, and social services). In other words, the whole operation is run by the community in the interests of the community.

Community representation on workplace boards. Somewhat more formal and complex, though still quite accessible, this has been tried out by a number of the smaller cooperatives in the last decade and by many of the CDCs (though the latter are not necessarily worker-controlled) explicitly devoted to local, often inner-city, progress. Normally several of the established neighborhood civic and religious organizations are asked to name a representative or two to the firm's board or send delegates to the firm's general meetings, but in some cases mass meetings or even special elections are used to select representatives at large. The People's Development Commission in the South Bronx, like some other inner-city organizations, considers everyone who lives in its particular eight-block area as a full voting member of the group, though its legal form as a corporation requires it to have a board of directors as the nominal locus of power.

One form of this representative arrangement was concocted by the Washington-based National Center for Economic Alternatives in preparing its proposal for community ownership of the Youngstown, Ohio, steel mill that the Lykes Corporation decided to shut down in 1977. Under that plan, one third of the new plant's board of directors would be elected by a non-profit "community corporation" composed of "recognized community leaders"; one third by the employees themselves, through an ESOP scheme that would give every worker full voting rights and a slice of the profits; and one third by individual shareholders, both workers and residents in the Youngstown area. (Interestingly, when this proposal was actually put to the Youngstown workers they asked for two significant changes to enhance the community role: the community corporation should be elected from a broad range of political and civic organizations, not just "leaders," and it should have only a quarter of the voting stock, with a full half to go to individual shareholders.) With the Youngstown populace unable to raise the purchase money, however, this scheme has never been put into practice.

A more complex and far-reaching version of this idea was recommended by E. F. Schumacher in *Small Is Beautiful,* where he suggested that half the shares of any enterprise be allocated to a public board (a "Social Council") chosen by public democratic vote and half of the profits allocated to it (in lieu of income tax) at the end of the year. This Social Council would have no power to vote the shares or to intervene in basic company matters—with twenty years as economic advisor to the British National Coal Board, Schumacher was not full of optimism on the ability of the public hand to guide private enterprise—but it would act as a regular watchdog with a place on the enterprise's board of directors and could appeal to a special court to intervene in cases where it felt the public interest was not being properly safeguarded.

Community ownership, worker usufruct. The old Latin concept of usufruct—use and enjoyment, rather than ownership, of property and goods—can be used to modify the straight-out idea of worker ownership, as another method of guaranteeing community interests. Jaroslav Vanek, the Cornell theoretician, has suggested a self-managed system in which the community has ownership of all plant and equipment and workers have usufruct rights to it—thus workers have "the exclusive right to control and manage the activities of the firm" but not "the full ownership—in the traditional sense of the word 'ownership'—of the capital assets"; they "can enjoy the fruits of production" but "must pay for this a contractual fee—or rental, or interest"; but they cannot arbitrarily "destroy the real assets or sell them and distribute the proceeds," and the community thereby retains control over the basic productive instruments of its economy. Presumably in such a system the ownership rights would not mean much day-to-day interference but would give the community the right, when it chose, to criticize and instruct and recommend, and in certain extreme cases to withdraw the whole operation from those who were misusing it.

Not surprisingly—Vanek is a Yugoslav by birth and had a hand in the initial Yugoslav self-management system—this model bears a strong resemblance to the practice in Yugoslavia, except that there the practice has tempered the theory somewhat over the years. Officially all businesses in Yugoslavia are public, all assets of any firm are society's—not ownership by the state, as in the Soviet Union, but ownership by a rather abstract legal idea called the society. As it works out, the society is sometimes represented by the federal government, sometimes by the Communist Party, sometimes by the trade unions, and sometimes by the local government unit, similar to our county units, called a "commune." All of these impinge to one degree or another on the worker-controlled enterprises, on behalf of the public interest, with the exact mix and the total influence varying considerably from commune to commune and region to region. The two most potent public safeguards are the national government's "social accounting service," which keeps track of all resources and material goods in the country and can inspect the books of any particular firm to insure that the resources are being used properly, with any conflicts to be settled by a separate "economic court"; and the local commune government, which has the legal right to intervene in certain public aspects of any enterprise and to make recommendations in certain intramural aspects, with recourse to the courts in case of disputes. In practice there is a good deal of harmony between the enterprises and the communes, particularly since the same people tend to be prominent in both, but conflicts have arisen over the annual distribution of the firms' proceeds, which are supposed to be divided 70 percent to the workers and 30 percent to the commune and federal governments but which are not always so readily or accurately accounted for.

The Yugoslav system is similar in some respects to the sort of usufruct proposal that was put forward in this country by Henry George, one of our most extraordinary (and today most neglected) economists, as long ago as the 1880s. George argued then that people should not have ownership rights to a piece of land or extract rent from it if they didn't themselves *use* the property; rather the land would be owned by the society at large and each person, family, or firm could make use of it as they saw fit. The society's obligation would be to insure that the fruits of everyone's labors would be secure, and in return it would receive an appropriate rent on its property—in effect, a tax on the value of the land, "the taking by the community, for the use of the community, of the value that is the creation of the community." George wrote:

> It will be obvious to whoever will look around him that what is required for the improvement of land is not absolute ownership of the land, but security for the improvements.
>
> Nothing is more common than for land to be improved by those who do not own it. The greater part of the land of Great Britain is cultivated by tenants, the greater part of the buildings of London are

> built upon leased ground, and even in the United States the same system prevails everywhere to a greater or less extent. Thus it is a common matter for use to be separated from ownership. . . .
>
> It is not necessary to say to a man, "this land is yours," in order to induce him to cultivate or improve it. It is only necessary to say to him, "whatever your labour or capital produces on this land shall be yours." . . . It is for the sake of the reaping that men sow; it is for the sake of possessing houses that men build. The ownership of land has nothing to do with it.

The Georgist principles provide a way for a community to secure its financial interests in a rational economy of usufruct, but to guarantee its social and environmental interests something more may be necessary. One method that has been developed and put into practice in recent years is the *community land trust,* an arrangement by which a group of people can form a corporation to buy and hold land in perpetuity and then use and develop that land as they choose, guided only by their original contracts and whatever trustees they may elect—and by their consciences. Perhaps thirty such land trusts exist now in the U.S., the oldest one since 1968, and most have remained both small (with ten to fifteen families) and agricultural. But the principle seems to work, and there is no reason a full-fledged town with industries, farms, shops, and services could not operate in the same way: individual interests are secured by people acting to improve their own lives in their own settings, individual rights through contractual leases guaranteeing people the use of a certain amount of property; community interests are secured by contractual agreements and the election of trustees to oversee the well-being of the community, community rights by careful land-use planning and environmental allocation of land and by adherence to these plans and contracts.

Community ownership and direction. At the extreme, a community might simply take control of the entire economy of its area, allocating resources and assigning jobs, creating and guiding all enterprises and services, setting workplace conditions and quotas, and even controlling the allocation and distribution of goods. A truly self-sufficient community could decide whether or not it wanted to stick to standard market arrangements—you make, I buy, I service, you pay—for the distribution of goods or, in a more cooperative way, distribute according to work, or need, or even absolute equality—a system feasible, however, only on relatively small scales. It could decide as well whether to do away with currency (except for outside transactions) and replace it with other systems available to smaller units, such as scrip or (as in various nineteenth-century communes) "labor notes" or family-allowance credits, or even with simple barter; capital, after all, is an invention to assure the mobility of a product, but if mobility were unnecessary and products could be easily exchanged in a small area then capital would be essentially unnecessary.

Various more-or-less utopian experiments along these lines have been tried over the last century and a half, some with extraordinarily long-lasting success: the Warrenite communes of Utopia and Modern Times both lasted for nearly two decades with systems modeled on Josiah Warren's ideas of labor-for-labor exchange, and the Amana communities survived for well over sixty years on the basis of simple family allowances, until modern enticements and currencies intruded during World War I. The Shakers, who lasted nearly two centuries in small villages throughout the Northeast, operated essentially without currency (except for outside dealings) and for the greater part of their existence were able to distribute more or less along lines of from-each-according-to-his-abilities, to-each-according-to-his-needs. More prosaically, a number of small cities in Germany and Austria during the Great Depression abandoned the worthless national currencies in favor of their own local currencies, which immediately became accepted legal tender and were used for salaries, purchases, and the like within their conscribed areas. The town of Wörgl, Austria, a community of about 6,000 people, moved from widespread unemployment to nearly full employment within three months by the use of its own currency, since local people could understand that it had the real value of a day's work behind it and were willing to accept it as a means of exchange; naturally when the practice began to spread and threaten central bank holdings the government moved against it, and ultimately the Austrian Supreme Court declared it illegal.

Community control of community economies is by no means a farfetched idea: that is exactly how the Israeli kibbutzim have operated for seventy years. They make no pretense to being cut off from the wider Israeli (and European) economy, of course, and their exchange system and currency are meshed with the rest of the nation. But they are islands of collectivism in a capitalist sea. In the kibbutz system, all property is owned by the community, all productive assets are the property of the community, the land is distributed and the labor allotted by the community, what is to be produced, with what, by whom, for when, what is to be paid for raw materials, how much is to be expended in production, what is to be done with the proceeds—all that is determined by the community. And not just some economic czar, either, or a bureaucracy—the essential decisions that guide the economy of each kibbutz are made by the general assembly open to all the settlers meeting weekly. Of course the assembly is guided by the individual firm and the branch experts and the secretariat charged with coordinating the economy, but the fundamental responsibility and decision-making is its, and in most kibbutzim this is taken as a heavy and a serious burden, upon which the fate of the settlement may rest. The proof that it can work, that it can even flourish, in more than 200 quite differently populated and differently endowed communities—despite all the problems the kibbutzim have had and all the turmoil that Israel has gone through—is in the settlements' extraordinary economic record.

THESE FORMS OF community economic involvement, merely the representative wavelengths along a very broad spectrum, are all quite different but they all seem to work in varying ways in varying communities. The precise form is obviously a matter for the people in any given locality to determine, providing only that there are some regular democratic channels for that determination to be made and, if necessary, revised. Ultimately the exact form will not matter quite as much as the social climate that those forms have to operate in, the political goals and philosophical styles and psychological aspirations of the inhabitants—for if that climate is not right it is hard to see how any economic contrivances will work.

And from both theoretical and practical evidence, it seems that one crucial determinant of that climate is size.

The very concept of "community" suggests limits—it is not "nation" or "region" or "city," all words with other connotations—and whether or not we pick for it that "magic" range somewhere between 5,000 and 10,000 people, there is no doubt that it demands some sort of circumscription. It is simply not possible to forge an economic identity with 7 million other people so that plans of allocation or production or distribution can ever be determined, much less harmonized, by even the most sophisticated methods; nor, really, with 500,000 others, or even 100,000. Nor is it possible to achieve a coherent economy adjusted harmoniously to the ecosystem and the health of the people who live in it without some clear and perceived geographical limits.

The face-to-face unit, in which everyone is known in some degree to all the others, and in which the quality of other people's work can be readily judged, seems to provide at least one size range for effective community-worker cooperation. We have already noted the success of the kibbutzim, and not a single observer, academic or other, has failed to comment on the importance of their small scale. From the beginning of the movement there has been an explicit policy of *fission,* splitting and setting up new units when one kibbutz feels itself becoming too big for effective communal control, and in practice that has been somewhere around 400 people, the average size of the current settlements, though a few are as large as 600 and 700 and three are over 1,000. Other examples are to be found in the more than seventy "communities of work" that flourished in southeastern France in the decades just after World War II; Boimondau, the largest and most successful and the model for the movement, had 150 households, or roughly 600 people. This, then, points to an effective economic community roughly within our previous "magic" range of 500–1,000 that defined a neighborhood.

There is also evidence, though, that workplace democracy can operate in units of even larger sizes, corresponding to what we earlier typified as a community. The most complete evidence here is from the early years of the Spanish Civil War, when all effective national government had collapsed and hundreds of small towns throughout Catalonia, Aragon, Levant, Andalusia, and Castile organized themselves into independent communities, operating both agricultural and small-scale industrial enterprises as worker collectives under community direction and control. Most of these towns seem to have numbered between 2,500 and 6,000 people, none was much over 10,000, and I would calculate the average to be somewhere around 4,000—too large, generally, to have any but infrequent general assemblies of all the townsfolk but small enough to be easily coordinated by a collective committee elected by, responsible to, and often recalled by the assembly. (There were apparently some abuses by collective committees exchanging aristocratic hauteur with syndicalist disdain and running as roughshod over the peasantry as those they had replaced—but for the most part a rough democratic process seems to have remained intact.) In most day-to-day matters the individual enterprises were largely autonomous—the collective farms decided when and where to irrigate, the local plant set its own work conditions—but in all matters spilling over into the town—determining how agricultural machinery was to be shared or how the crops were to be marketed, and in many places how food and goods would be distributed—it was the community, through the collective committee, that made the decisions.

All of this seems to have happened more or less spontaneously in widely separated sections of Spain, mostly places that had had some dim mutualist tradition but well beyond those that had been consciously organized by the anarcho-communalists of the Spanish CNT and often among people who were illiterate, apolitical, and certainly untutored in self-management. And all of it seems to have enjoyed surprising success: for nearly two years, until overrun by Franco's troops or invaded by hostile Communist brigades, these collective villages ran the economy of a significant part of rural Spain, and by most accounts increased production, introduced new machinery, diversified local industries, redistributed the land, established new schools, created new welfare systems, and in general brought a prosperity to the regions, even in a time of war, beyond anything known previously. It seems to me telling that in one Castilian town the aristocrat who came to reclaim his expropriated and collectivized lands a few days after the Civil War was so astounded by the improvements made by the "ignorant" peasants—new irrigation systems, new fields under cultivation, a mill and a school and dining halls and houses—that he immediately turned over some of this property to the villagers and arranged for the release from jail of the man who had helped them draw up the plans.

The contemporary journals of Gaston Laval—an unabashed CNT partisan, but a careful and straightforward reporter—tell the story of some of these remarkable experiments in workplace democracy. Take Graus, a town of small industries and a population of 3,000 in the northern part of Huesca province, on the French border:

> There was no forced collectivization. Membership in the collectives was entirely voluntary, and groups could secede from the collective if they so desired. But even if isolation were possible, the obvious benefits of the collective were so great that the right to secede was seldom, if ever, invoked.
>
> Ninety percent of all production, including exchange and distribution, was collectively owned. (The remaining 10% was produced by petty peasant landholders.) . . . Each factory and workshop selected a delegate who maintained permanent relations with the Labor secretariat [of the collective] reporting back to and acting on the instructions of his constituents.
>
> Accounts and statistics for each trade and enterprise were compiled by the statistical and general accounting department, thus giving an accurate picture of the operations of each organization and the operations of the economy as a whole. The list that I saw included: drinking water, bottle making, carpentry, mattress making, wheelwrights, photography, silk mills, candy, pork butchershops, distilleries, electricity, oil, bakeries, hairdressers and beauty parlors, soap makers, house painting, tinware, sewing machines, shops and repairs, printing, building supplies, hardware, tile shops, dairies, bicycle repairs, etc.
>
> Everything was coordinated both in production and in distribution. . . .

Or take Binéfar, a Huesca town of 5,000 or so, where the townspeople first harvested the fields of the departed big landowners and then sat down to work out their economy:

> After the harvest, industry and eventually commerce were socialized. The following are the rules that the popular assembly of all the inhabitants approved:
>
> 1. Work shall be carried on in groups of ten. Each group shall elect its own delegate. . . . The delegates shall plan the work, preserve harmony among the producers, and if necessary apply the sanctions voted by the popular assembly.
>
> 2. The delegates shall furnish the Agricultural Commission a daily report of the work done.
>
> 3. A central committee, consisting of one delegate from each

branch of production, shall be named by the general assembly of the Community. . . .

5. Directors of labor for the collective shall be elected by the general assembly of all the collectivists.

6. Each member shall be given a receipt for the goods he brings to the Collective.

7. Each member shall have the same rights and duties. . . . All that is required is that members accept the decisions of the Collective.

8. The capital of the Collective belongs to the Collective and cannot be divided up. Food shall be rationed, part of it to be stored away against a bad year. . . .

11. The general assembly shall determine the organization of the Collective, and arrange periodic elections of the administrative commission.

It was a most extraordinary, and largely ignored, period of modern economic history, of which these accounts can give us only the barest bones. But they do suggest, even attenuated, the kinds of local triumphs economic democracy is capable of—at least in its collectivized version—and they do allow us to see some real-world evidence of the success of community-level populations in making it happen.

And if that is not the Principle of Affected Interests in action, I don't know what is.

10

Self-Sufficiency

ONCE UPON A TIME the greater part of the world's population lived in conditions that, as we view them from our contemporary perspective, could only be considered opulent. They spent their time—all of them, regardless of birth or beauty—in the closest thing to indolence, working only a few hours a day, sitting around and sleeping and making love for hours at a time, literally living off the fat of the land and gorging themselves when food came along with little thought of saving for the morrow. They surrounded themselves with beautiful objects, participated in elaborate useless rituals, devoted resources to nothing more substantial than jewelry and wall paintings. They ate well, with balanced diets, got plenty of regular exercise, were spared most serious diseases, and lived to relatively ripe old ages. They were for the most part free of poverty, privation, pollution, crime, and war.

They were, as you have no doubt guessed, the Paleolithic hunter-gatherers of prehistory—the "cavemen."

It is difficult to generalize about a period that lasted maybe 40,000 years and was at such a remove from ours, but the diligent researches and diggings of the last hundred years, reinforced by studies of contemporary hunter-gatherer societies in Africa and the Pacific, have given us a pretty careful picture of what the Paleolithic societies must have been like. They knew a mastery over their technology that produced an intricate array of tools—more than sixty types of knives, awls, burins, axes, scrapers, spearpoints, cleavers—including the lovely "laurel-leaf" stone blades of the Solutrean period, some of them as thin as four-tenths of an inch, a feat said to be beyond even modern machines. They were accomplished in the arts—as the Lascaux caves alone may attest—and in the basic sciences. They were socially advanced to the level of community solidarity, altruism, communal industry, dance and food rituals, and the elaborate burial remains found in the Shanidar Cave in Iraq—bodies that had rested on beds of soft boughs and elegant flowers—testify to their elegant ceremonialization of death.

However rough and simple may have been the people produced by these early societies, they were by no means savages, they did not live in hand-to-mouth indigence, they were not ignorant of the workings of the world around them, and their lives were in no sense solitary, poor, nasty, brutish, or particularly short. (Their life spans have been calculated to have been about 32.5 years, exactly equal to that of the U.S. average in 1900.) Especially not when compared to the peoples of modern Calcutta, say, or the Sahel or Manila or Tijuana, or even the South Bronx. In fact anthropologist Marshall Sahlins, whose *Stone Age Economics* does much to correct our Alley-Oop stereotypes, calls the Paleolithic "the original affluent society": "one in which all the people's material wants are easily satisfied."

True, those wants were not very extensive—they did not go so far as brocaded cloth and brigantines and Burgundy and electric blankets—but they were real, and felt, and they did not go unfulfilled. In our affluent society, by contrast, where everyone's wants are great, or made to be great, they are almost never fulfilled, even by the wealthiest among us, and they are hopelessly out of reach for the great majority. Yes, the Paleolithic peoples had what economists today might want to call a low standard of living, measured as the accumulation of material objects. But *they* didn't know that: they no doubt thought of themselves as "unencumbered."

Given their objective, given what they sought out of life, these early people were more often than not able to satisfy it and many times able to exceed it, and with a minimum of hardship and labor. Sahlins quotes Lorna Marshall, an expert on the Basarwa:

> They all had what they needed or could make what they needed. . . . They lived in a kind of material plenty because they adapted the tools of their living to materials which lay in abundance around them and which were free for anyone to take. . . . They borrow what they do not own. With this ease, they have not hoarded, and the accumulation of objects has not become associated with status.

These Basarwa, and their identical remote ancestors, led a life of *perceived satisfaction:* want not, lack not. Who would not call it affluence?

OF COURSE IT IS not the Paleolithic state of affluence that I suggest we all return to, but there is something in the Paleolithic understanding of the limits of material amassment that does seem pertinent. For what made them affluent was, in truth, their self-sufficiency, their ability to satisfy all of their needs within their own means. (And if that meant they had to regard those things beyond their means—mammoth-roasts every Sunday—as being unneeded, and hence unwanted, that was not a limit on their self-sufficiency but rather an improvement in their happiness.)

And that kind of affluence, I propose, is available to us today, modified by our own vastly greater knowledge and our own vastly improved technologies, would we but direct ourselves to some small part of that Paleolithic comprehension.

Affluence, you see, is always relative. It is not only that of the car and the house and washing machine and vacation; it has something too of security and harmony and quiet and friendship and freedom from pollution and from powerlessness. Compare the inhabitants of ancient Media, learned in astronomy and science and skillful in the arts of administration but knowing little in the way of riches, with the Iranian followers of Cyrus the Great, mighty warriors fat with the plunder of nomadic bands but totally untrained in either artistic or scientific accomplishments: one man's Mede, no doubt, is another man's Persian, but who would want to say that all the affluence rested with the latter? Is affluence the Manhattanite condition of million-dollar Fifth Avenue penthouse triplexes and heatless, rat-infested, paper-walled SROs on East 139th Street, or is it the Vermont village, the Iowa town, the Oregon community, where no great extremes ever seem to exist but everyone has employment, where industry and agriculture co-exist, and where the air, the water, the streets, the police blotters, and the movies are all reasonably clean? Might the trade-off of a small-scale economy of sufficiency-enjoyed-by-all against the intricate multinational economy of wealth-for-a-few-and-poverty-for-many be, in fact, no trade-off at all?

SELF-SUFFICIENCY MAY PERHAPS seem a foolish anachronism in a world of interconnectedness, an economic cul-de-sac in an era of mutual dependence. And indeed it may not really be obtainable in its most absolute form, since almost any society, no matter how cut off, will normally have some sort of contact and trade with the outside world and may be enriched thereby. But as a goal and an ideal, and as something to be achieved in even imperfect conditions, self-sufficiency has inestimable virtues.

It is a way for a community naturally to achieve a stationary society, in absolute harmony with the environment, assuring to itself a rational control over resources and productivity. It allows the free establishment of workplace democracy in all its manifestations, unimpeded by economic tugs and pulls from other unsympathetic—or hostile—forces. It makes a place expand instead of contract, create instead of borrow, use instead of discard: just as a man left on his own, thrust on his own devices, develops strengths and uncovers inner resources and becomes the fuller for it, so too a community. Above all, it establishes independence: a self-sufficient town cannot be the victim of corporate-directed plant closings or a truckers' strike or an Arab oil boycott or California droughts; it does not have to maintain lengthy and tenuous supply lines of any kind, nor pay the shippers and the jobbers and the middlemen who are clustered along them; it does not have to be the accidental victim of toxic fumes or industrial poisons or nuclear wastes produced by or passing through the town; it does not have to bow to (always rising) prices set by distant A & Ps and GMs and GTEs in disregard of what the local farmer is in fact growing and the local shop producing; and ultimately it does not have to sway in the winds of the hurricanes of boom and bust as regularly generated, as it were offshore, by distant and uncontrollable economic forces.

It is on breaking the terrible dependence upon imports and exports, and the economic vassalage that results, that self-sufficiency must depend. A community grows and becomes more textured, as Jane Jacobs has noted, when it "replaces" imports—that is, when it manages to do on its own what before it had to pay others to do: she offers the example of Tokyo, where innumerable little bike shops grew up to repair imported bicycles, then started manufacturing spare parts, and eventually turned out the whole product, doing away with imported bikes entirely.[1] Any nation that comes to depend on imports will find itself perpetually vulnerable, scrambling always to create the exports that will be sent off in return or else living in ever-deepening enthrallment. The U.S., as an importer of vital resources—more than 50 percent of its petroleum, 75 percent of its tin, 90 percent of its chromium, 98 percent of its cobalt and manganese, 100 percent of its strontium and sheet mica—is finding out just how painful and dislocating such vulnerability can be; other nations in economic trouble—Britain, Italy, Greece, Spain, Brazil, the Philippines—have found out before. The other side of the coin is no better, either: an overdependence on exports is usually a sign of nothing more nor less than colonialism. Most of the Third World today—Liberia exporting rubber grown on land that could be raising food, Ghana depending on cocoa exports and importing chocolate bars, Cuba devoting half its agriculture to the sugar export crop—lives in such a condition.

This is not to say that a healthy economy might not *include* imports and exports—that would seem only natural for any but the most isolated community—but only that it *depends* on such trade to its peril and trusts in it to its detriment. Obviously the percentages will shift from place to

1. Jacobs does assume that Tokyo then has to go on importing some *other* goods, and creating more businesses to make exports so as to pay for them—and that is a trap that she, Tokyo, and the capitalist world seem unable to escape, the trap of growthmania: cities and societies always must grow and develop and extend and consume. It certainly does not *need* to follow, as one of her own examples makes clear: Manchester, which grew to enormous size and prosperity by total dependence on imports of cotton and exports of cloth, was a one-industry town that collapsed in the twentieth century when other nations began to spin their own cotton; Birmingham, which had a multitude of little independent workshops, none with more than a dozen workers, produced for itself the bulk of the products it needed and remained a thriving city right into the twentieth century.

place, but when any city or community starts to use more than a small proportion of its jobs and its resources to create things that will never be used by its citizens it begins to get into trouble, as the history of Manchester or Mohenjo-Daro or Milwaukee suggests.

Apart from its economic virtues, self-sufficiency conveys social benefits as well. It is a way for a community to survive, and thrive, as a closely knit group, to create stability and balance and predictability, to learn its own reserves and become developed in the fullest sense. In a town dependent on its own resources, people necessarily come to know each other, to appreciate each other's strengths (and weaknesses), in a way they never get to do in an atomized city, or even in many atomized small towns. Children are given the opportunity to learn the complexity of reality—a multiplicity of small shops and offices and plants and farms provides an unexcelled laboratory of the real world—so that they do not grow up believing that tomatoes grow in supermarkets and electricity comes from a switch on the wall. And when grown, those children are more likely to stay in a place with its own developed businesses and opportunities, there being then no particular reason to go off to seek their fortunes elsewhere. Other citizens, too, should find inevitable opportunities and if they are forced to be plumbers as well as librarians, poets as well as farmers, this can only be enriching for the person, as for the community. The feeling of competence, of pride, of selfhood, of independence, that might attend the citizen of a self-reliant town could be duplicated nowhere else.

Fraser Darling, the sociologist, studied some remote Scottish villages a few years ago that were essentially untouched by the outside world. He found that far from being stagnant and lifeless backwaters they were in fact teeming and surprisingly active eddies, far more diverse in both social and economic activities and generally more "alive" than company or single-industry towns or places like bedroom suburbs tied inextricably to distant cities. Their necessary self-sufficiency actually operated as a positive force: people coped better with the vicissitudes of life, they were more neighborly, and they had a greater diversity of jobs and responsibilities. And they had a better understanding of the town's natural environment and its importance in their lives, and were far less likely to despoil it—or let some intrusive entrepreneur despoil it—for some short-term gain.

Finally, one would have to believe that a self-reliant community would inevitably be less violent, both within its borders and without. Schumacher put it this way:

> As physical resources are everywhere limited, people satisfying their needs by means of a modest use of resources are obviously less likely to be at each other's throats than people depending upon a high rate of use. Equally, people who live in highly self-sufficient local communities

are less likely to get involved in large-scale violence than people whose existence depends on world-wide systems of trade.

How many wars in history have been the result of one people's threatening to deprive another of some theoretically valuable resource, of one nation seeking to protect or expand its far-flung export-import markets; how many waged by peoples entire to themselves, self-developed and self-content, how many by the self-sufficient cities of the Middle Ages, the self-reliant Amish and the Mennonites, by Switzerland?

THAT A HUMAN-SCALE economy would work toward self-sufficiency seems only natural, since it is only on the smaller scale that the human can understand and work some control over the surroundings, can perceive and regulate the variants in an economy, can determine the artifacts and services useful for or detrimental to the community. The human brain, however elaborate an instrument the human may consider it to be, is limited, and even extended as it is by certain modern technologies it cannot truly comprehend the labyrinthine involutions of any very vast scale—as the bewildered machinations of the official government economists over just the last five years show all too pointedly. On the other hand, an economy on too limited a scale will not be able to provide the complexity and diversity that go to create a full material life, and trading an economy of chaotic complexity for one of hairshirt simplicity would hardly seem to be much of a bargain.

The balance that is to be struck of course depends on what it is that a community needs, or thinks it needs (though the Paleolithic precaution is that probably many of our needs are not in fact needed). A village of, say, 500 people could probably grow its own food, operate its own energy systems, create its own handicrafts, perhaps carry on some manufactures, much as the Israeli kibbutzim do; but it would be hard-pressed to go in for much in the way of extensive manufacture or construction, would not likely have much variety in its wares, would have to keep its services quite simple, and would have to accept fairly limited opportunities of conviviality and culture. Even figuring a labor force of 250 in such a settlement, somewhat high by current American practice, there would probably be no more than 100 people or so available for manufacturing and recycling, the rest employed in agriculture, energy and transportation, services, and handicrafts. That would certainly be sufficient for a dozen small manufacturing plants, since we know from current American manufacturing statistics that 65 percent of all the plants in this country operate with fewer than twenty people (in fact only 11 percent have over 100), and in those *the average number of employees is only 5.5;* and it would no doubt cover such basics as lumber, paper, and textile mills, a carpentry and brick works, and a few small factories (bikes, maybe, and hardwares). But that would plainly be insufficient to create a full range of metal products, electrical equipment, medical instruments, books, rubber products, soaps, and paints, to pick only the basic categories of contemporary manufacturing. A self-sufficient village at this level is certainly possible, but it would be needlessly constricted.

If several such neighborhood-size populations join together, however, the possibilities become far richer. A community of 5,000 or 10,000 takes on the stature necessary for real economic independence—as indeed, if we needed reassurance, the greater part of human history has demonstrated. At that size, as we have seen, agricultural self-sufficiency and community energy systems are most economical and efficient, and at that level the labor force available for the rest of the economy (if it approximated current American percentages) would amount to between 2,000 and 4,000, divided about evenly between manufacturing and services. Now if we take the figures for current American manufacturing, we can see how many people it might take (both front office and production) to operate a plant in the sort of basic industries an independent community might require:

INDUSTRY	WORKERS PER PLANT
Textiles	132
Apparel	56
Lumber and wood products	29
Furniture and fixtures	50
Paper and allied products	104
Soap, cleansers, toilet goods	43
Stone, clay, and glass products	39
Primary metal industries	163
Fabricated metal products	50
Machinery (except electrical)	45
Electrical and electronic equipment	135
Motorcycles, bicycles, parts	81
Instruments and related products	75
Total Manufacturing Employment	1002

In other words, using current standards, a thousand people could operate one plant in each of the thirteen basic manufacturing categories—and as we know, those current standards are far bigger than the optimum for either efficiency or humanity and they include some truly behemothian places. In a rational economy it would no doubt be possible to reduce those sizes by half, but even if it were by no more than a quarter, that would still mean a community of 10,000, with 2,000 factory workers, would be able to staff *three plants* in each of these basic categories—enough to supply a small population with practically all of its manufacturing needs and allow it considerable diversity as well.

As populations increase over this level, self-sufficiency in one sense becomes easier, since there are more workers and so more kinds of products and services can be created. As we saw in examining city sizes, a small city of 50,000 typically has just about all the service and production enterprises of any size city, and these are places that do not even consciously seek self-sufficiency. But problems begin to accumulate at this level as well, along the lines of what I am tempted to call the Rule of Eternal Dependency: over a certain minimum size, the greater the population, the more complex the institutions, the more elaborate and advanced the technical needs, the more difficult it will be to satisfy them locally or in any single place. For example: as transportation and distribution become more diffuse and widespread, you need to start thinking of trains and rails or cars and roads and traffic lights, of bridges and tunnels, of planners and coordinators; as agricultural areas expand to support such a population, new kinds of field and irrigation equipment may be necessary, storage problems will require warehousing and probably refrigeration, harvesting and distribution will take additional labor and coordination; as waste disposal creates a burden the land can no longer easily absorb, new collection and processing machinery becomes necessary, sewage lines and treatment plants might be required, composting and recycling will demand additional complex machinery. On top of this, the whole process of establishing the city's needs, deciding which have priority, coordinating production and services to meet them, and making sure that they are met and at the right time in the right way becomes significantly more tangled and demands human and technical resources of its own.

It is not that self-sufficiency is not possible at this level—I am quite convinced that it is, given the additional people there to cope with it, because experience indicates that this has been a size at which cities have been able to survive, healthily, for many decades, and even in interdependent economies operate most efficiently. But it is clear that the city would quite consciously have to make trade-offs that the smaller community would not have to face.

SELF-SUFFICIENCY HAS NEVER occupied much of a place in modern economic thought—the closest that traditional economists come is in their "economic base" and "central place" theories—so there are very few theoretical models, and as a consequence even fewer real-life experiments, that would indicate how and at what sizes it might work.

We do know that during most of humankind's settled existence—say from 6000 B.C. right through the nineteenth century—most people in the world lived in small towns seldom holding more than a couple of thousand people that were self-sufficient because there was no other way to survive. The Greek villages of the seventh to fourth centuries B.C., as Mumford notes, "were both small and self-contained, largely dependent upon the local countryside for food and building materials." Monasteries from the sixth century A.D. on were noted as enclaves where both physical and spiritual needs—the former probably more limited than the latter—were met entirely by the collective gathering of brothers, usually numbering less than a hundred souls. In medieval Europe before 1500, as Ferdinand Braudel tells us, "90% to 95% of the towns known in the West had fewer than 2000 inhabitants" yet normally enjoyed a full range of craftsmen and artisans and merchants and farmers; throughout medieval times, as the *Encyclopedia of Social Sciences* notes, "provincial, even local, self-sufficiency was the order of the day." Even in the eighteenth and nineteenth centuries, and in some places well into the twentieth century, the basically self-sufficient village was the norm in Europe: a typical English town of several thousand as late as 1880 or 1900 would have blacksmiths and shoemakers, plumbers and carpenters, bakers and butchers, brewers and millers, saddlemakers and harness-makers, tailors and seamstresses, dentists and midwives, a pub and a church and a reading room, an inn and a market and a row of shops, and the food it ate was the food it grew.

All of these places, and others like them, made it clear that, at least at quite simple levels of living, self-sufficiency was certainly possible in units as small as several hundred and fairly easy to sustain in units as big as several thousand.

For more self-conscious models, however, we have to hunt farther afield. There are the Spanish rural collectives that we have already seen, though their brief span of only two years argues some caution in taking them as models; there are the kibbutzim, though their linkages with the outside economy mutes their exemplification. There are the Chinese communes, with populations of 10,000 to 30,000 and occasionally more, self-sufficient to the point where outsiders have witnessed four old women mending the filaments on broken light bulbs, though they exist primarily to provide exports of food for the cities and they live under a fairly rigid centralized thumb. There is Paolo Solari's "Arcosanti" taking shape in the Arizona desert near Scottsdale, a huge, sprawling twenty-five-story building that ultimately is supposed to contain all the life supports of a community of 5,000. There are the Bruderhof communities, villages of 200 or 300 people who hold all goods and property in common and, with the exception of a toymaking business (Community Playthings), tend to confine themselves to a self-sufficient, deeply religious, communal life. There is the grand social experiment of Cerro Gordo, a planned ecological community of 2,500 people being built about 25 miles south of Eugene, Oregon, hoping to be independent in agriculture, energy, transporation, education, health, and waste disposal, with an economy built on local self-reliance. There are the Mennonite settlements in Pennsylvania, villages and towns of only a few thousand, based on completely self-sufficient agriculture and essentially self-sufficient energy, services, and manufacturing—without outside electricity or internal-combustion engines, cars or tractors, radios or television.

Perhaps the most useful examples, though, because they have existed in the everyday world for some period of time, and because they seem to bracket the population ranges possible, are the American communes of the nineteenth century and the British New Towns of the twentieth.

The more than one hundred communes that existed in nineteenth-century America, generally founded with the idea of independence and often ostracized by surrounding populations, established a self-reliance of considerable fortitude and yet, on average, with no more than 300 to 500 people. They were not all strictly self-sufficient, since all of them had some sort of outside trade, but almost all lived with their own subsistence agriculture, including vegetables, fruits, and grazing herds; most had craft centers and workshops for everyday manufacture (clothes, furniture, tools); communal labor took care of most construction, road-building, and harvesting. Such outside commerce as did exist was usually confined to one or two commodities that could be sold in order to bring in a little cash, often to pay off mortgages on the land, and it is one measure of the health of their self-sufficiency that such small trading produced such large reserves: Bethel, in Missouri, began in 1844 with $30,000 worth of assets and dissolved thirty-six years later with $3 million; Hopedale, in Massachusetts, began with less than $5,000 in 1841 and disbanded in 1856 with a quarter of a million dollars in assets; the Wisconsin Phalanx increased its assets from a bare $1,000 in 1844 to $33,000 when it broke up six years later; the Amana communities in Iowa when they dissolved in 1933 after ninety years of self-reliance had more than $2 million to share among them. Historian Ralph Albertson, in his survey of the Amana and other "mutualistic communities," has concluded:

> Few colonies, if any, failed because they could not make a living. . . . They failed to compete with growing industry and commerce in a new, unexploited country. But they did not fail to make an independent subsistence living—and pay off a lot of debts and help a lot of stranded people.

The Oneida colony in New York was probably as successful as any of the "utopian" experiments, and it never numbered more than about three hundred inhabitants. It was not precisely self-sufficient, because it eventually developed a successful business making and selling steel traps and, later, silverware (as a matter of fact, that part of the operation still survives), but it regarded itself as such an independent unit that it specially prepared all its members who were forced to go "outside" and ceremoniously cleansed them "from contamination by worldly influences" when they returned. Its agriculture was self-sufficient after the first few years—made easier by the fact that the Oneidans did not smoke tobacco or drink alcohol and ate little meat—and the colony had, in addition to trapmaking, factories that made silk and handbags and machine parts; a sawmill; a carpenter's shop; a furniture plant; a harness shop; a printing press; workplaces for tailors, shoemakers, tinsmiths, clocksmiths, jewelers; and a dentist's office. The settlement made its own wheels and spokes, chains, mop handles, plows, scuffle hoes, slippers, sleigh shoes, musical instruments, and kitchen utensils, and installed steam heating in most of its buildings, long before it was common elsewhere.

At the far remove from all of this are New Towns that have grown up in Britain over the last eighty years. They have been founded with the explicit idea of "self-containment" rather than self-sufficiency, but they have achieved rough degrees of independence with populations of 30,000 to 70,000. The inspiration for all of them was Ebenezer Howard's justly famous "Garden City," which he concocted in the late 1890s as a plan for the way that Britain could escape the ghastly consequences of industrialism then desecrating the nation from London outward. Garden City itself, though Howard assumed it would have regular trade with the outside world, was conceived of as largely self-reliant in food, services, and industry at a population of 32,000; Howard's map for his imaginary city shows a four-mile-long ring on the outer edge of the city containing such things as coal, lumber, and stone yards, furniture, clothing, and boot factories, a printing press, a "cycle works," an engineering center, and even a "jam factory." Frederick Osborn, the city planner who went on to put Howard's ideas into effect in the actual garden cities of Letchworth and Welwyn just after the turn of the century, concluded from his experience that Howard's population estimates had been about right: "A town which is designed for modern industry," he wrote in his classic *Green-Belt Cities,* "employing people living on the spot, ought I think to aim at a population of at least 30,000" but not "in excess of 50,000."

Displaying an enlightenment most unusual in statecraft, the British government adopted Howard's and Osborn's ideas just after World War II and so far, with fits and starts, has sponsored the creation of some thirty New Towns. In practice the idea of self-containment has been applied more to local government services—transportation, utilities, schools—than to employment and trade, and indeed for most such purposes these settlements have achieved considerable independence; economically, none has been designed to stand alone, although the later towns built farther from London have included a much higher percentage of local employment. Still, compared to unplanned cities, all of the New Towns show a much greater economic independence and imply that true self-sufficiency would not be difficult to achieve at populations of about 50,000 (the New Town average is 57,000); indeed, the towns built from 1947 to 1958 mostly in the hinterlands actually have proved to be quite self-contained with populations that average a little under 39,000.

The old communes and the New Towns—they do not give us very precise guidelines, I realize, but they may help at least to delimit the possibilities of self-sufficiency. They do suggest that it can be a reality, apparently, in places as small as 300 or 500 and, with somewhat less cohesion, in cities as large as 40,000 or 50,000. Taken together with the other examples through history, they suggest that a settlement of the size of our community—5,000 to 10,000—would find no difficulty with it, and something around that range suggests itself as the likely ideal.

SELF-SUFFICIENCY IS OFTEN scorned for being simplistic. Indeed, I find that I am often taken to task for wishing to oversimplify things: "Life is just not that simple."

But in fact I wish to *complexify,* not simplify. It is our modern economy that is simple: whole nations given over to a single crop, cities to a single industry, farms to single culture, factories to a single product, people to a single job, jobs to a single motion, motion to a single purpose. Diversity is the rule of human life, not simplicity: the human animal has succeeded precisely because it has been able to diversify, not specialize: to climb *and* swim, hunt *and* nurture, work alone *and* in packs. The same is true of human organizations: they are healthy and they survive when they are diverse and differentiated, capable of many responses; they become brittle and unadaptable and prey to any changing conditions when they are uniform and specialized. It is when an individual is able to take on many jobs, learn many skills, live many roles, that growth and fullness of character inhabit the soul: it is when a society complexifies and mixes, when it develops the multiplicity of ways of caring for itself, that it becomes textured and enriched.

Those, obviously, are the ultimate goals of self-sufficiency.

11

Lucca's Law

BUT WAIT JUST A BIT: can everything really be produced in a self-sufficient community, particularly one as small as 10,000? All very well to bring up the ancient Greeks and the Oneida colonists, but doesn't the modern world require something more, even for a tolerably smooth existence, not to mention the good life to which we have become accustomed or at least aspire? Can a small community support a steel mill, make refrigerators, provide electric lights, produce all of the things that, however crude, are the stuff and substance of our lives and that for the most part we do not want to be without?

Is a small community, at bottom, *rich* enough ever to be able to provide the conditions of a decent material life?

Fair questions, and they deserve careful answers.

Of course the community of 10,000 people cannot possibly supply all the gadgets and geegaws and gimcrackery that is to be found in our stores today. It would not be able to provide the elegant variation of contemporary supermarkets—where, according to the A. C. Nielsen Company, the market research firm, some 53,000 new brands, sizes, flavors, and varieties have been introduced *just since 1971* and the number of items on the shelves has increased by 400 percent since 1950. It would not be able to produce the 58 different brands and sizes of cigarettes, the 79 different versions of breakfast foods, the 300 models and styles of cars, the literally tens of thousands of varieties of women's dresses. Variations there could be, since with a small market a simple factory would be able to shift its styles from time to time, but basically there would have to be, one feels, a good deal of the Henry Ford principle, "Any color you want so long as it's black."

One man who spent the greater part of a very long lifetime considering such matters was Ralph Borsodi, the tomato economist and decentralist philosopher who championed self-reliant homesteading and proved by both theory and example that breaking free of marketplace dependence was a comparatively easy—not to mention cheap and healthy—thing to do. A few years ago, in an interview in *Mother Earth News* not long before his death at the age of 91, Borsodi said:

> The evidence is pretty clear that probably one half to two-thirds—and it is nearer two-thirds—of all the things we need for good living can be produced most economically on a small scale . . . either in your own home or in the community where you live. . . . You can't make electric wire or light bulbs, for example, very satisfactorily on a limited scale. Still, virtually two-thirds of all the things we consume are better off produced on a community basis.

Not simply "produced," notice, but produced "better" and "most economically."

And he's wrong about the wire and the bulbs. In America today the average electric lighting and wiring factory employs no more than 65 production workers, and that's in the present system where plants operate beyond their most efficient sizes, where product differentiation is almost as absurd as it is with dog food (I have just counted eighteen completely different kinds of light bulbs around my house), and where 75 million households are being serviced. Posit an efficient plant, limited differentiation, and a market of 3,500 households, and the factory could be many times smaller; then figure that the quality could be improved by careful work on a limited number of items (that's one reason the employee-owned Torch corporation in New Jersey is able to offer five-year guarantees on their light bulbs) so production quotas could be considerably reduced and the plant made even smaller; lastly, add in the effect of recycling on limiting annual production (maybe not with little old ladies repairing filaments, but reusing bases and remelting bulb glass), and the operation could be smaller still. I would calculate that the manufacture of light bulbs for a small community need not take more than a handful of workers, two dozen in the most ambitious circumstances, well within the resources of even a limited community.

Moreover, at this scale these two dozen electrical workers should be able to finish off bulb production in a matter of months and then turn to other electrical work, making that wiring for example, thus in effect becoming a succession of electrical factories. If we figure that a town of 10,000 people might use 30,000 bulbs a year (about the current U.S. rate of consumption) and that even a small plant can turn out 15,000 in a month (half of the U.S. average of 32,000 per month per plant), the plant would have to spend only a couple of months a year on bulb production (or, more efficiently, turn out a full five-year's supply in ten months). And all those figures would be factorially lowered if five-year bulbs and recycling were added in.

In fact, I think it could be shown that even Borsodi's "two-thirds" is an underestimate. I think a small town can without much difficulty provide for virtually *all* material needs on a household or community level—and make the goods more affordable, more durable, more aesthetic, more reparable, and more harmless, too.

Thusly.

1. By sharing. With a regular system of sharing goods and the creation of accessible neighborhood centers where they could be found, an enormous number of the artifacts, machines, and tools that we now produce could be eliminated. Is it necessary for every suburban garage to have a lawn mower, which is used at most once a week, and only in the summertime? For every third or fourth basement to have a complete set of power tools, some in disuse for months on end? For every breakfront to have the set of silver and the punchbowl that's used only once a year? (Designer Victor Papanek has cited a survey done in Canada in 1976 in which on a single street of twenty-five houses, twenty-eight vacuum cleaners were found.)

Various adaptations to communal living might be useful here, too. A neighborhood social center, in addition to its other inestimable virtues, might be able with a single instrument to replace a multiplicity of television sets and pianos and phonographs. Shared travel would mean fewer cars and trains, shared laundering fewer machines, shared newspapers and books fewer trees lost.

2. By recycling and repairing. The more you recycle, obviously the less you have to manufacture or import, so in addition to lessening waste and conserving energy you are reducing the manufacturing load and heightening self-sufficiency. Neighborhood-level recycling centers, which already exist in some 23,000 communities in the U.S., could expand their operations with the kind of additional public cooperation self-sufficiency might engender and probably be geared ultimately to a 90-percent-return level on all recyclable goods. Such centers could easily be coupled with repair workshops where volunteers (the high-school tinkerer, the retired watchmaker) could fix up the multitude of products needing only an expert's touch and a little bit of solder: how many appliances are sitting now in other closets, as in mine, no longer working right but still essentially functional except for a frammis or a widget, put away on a high shelf because they could not be repaired and it never seemed worth the trouble to mail them back to some far-off manufacturer with some long-lost proof of purchase? At only a slightly more elaborate level, it is possible to establish "remanufacturing" centers, where a variety of slightly larger items like diesel engines and refrigerators can be salvaged and reconditioned with a few new parts and a little rehabilitation of the old. This is the sort of process at an informal level that has been going on for years in the poorer nations not permitted the throwaway indulgence, where it was not possible to buy replacements and not feasible to wait for imported parts, and it is one well worth borrowing from them.

3. By depending upon handicrafts rather than manufactures. At the modest scale of the self-contained community, there are many products that would not need to be—or could not be economically—mass-produced, eliminating the need for many factories; but most of those could be supplied by individual artisans crafting everything from clothes and pottery to furniture and tools. The costs would presumably be higher, as they are reckoned in the short run, but the difference in quality and durability is likely to make up for it in the long, as our grandparents—who had opportunities for handcrafted goods far more than we—used to keep telling us. (In a contemporary handcrafted shirt, for example, which is always made out of natural fabrics, there are at least thirty stitches to the inch rather than the ten or twelve stitches found in factory-mades, and shell buttons instead of plastic.) And the difference in the way the products look and feel, in the attitudes of the people who style them on us, in the satisfaction we get from them is, though possibly intangible, no less real. (To take bespoke shirts again, the exactness of the fit is assured by at least eighteen different measurements, including wrists and shoulder slopes and neck length, whereas department store shirts now frequently run in such generalized sizes as "14–15-inch" necks and "32–34-inch" sleeves.)

4. By developing and using local products and raw materials instead of depending on imported ones. Glass is almost always a better material than plastic, easier to work, cheaper to make, safer for the environment, simpler to recycle, and it can be made to be even more resistant to shattering, heat, scratching, and aging; it is made from the one mineral, silica, which is available in abundance in each one of our fifty states. Wood and concrete are building materials as good as aluminum or steel, and they are ubiquitous. Earth may be the best material of all, and it can be used for construction employing only the simplest tools; it will keep interiors warm in winter, cool in summer, fireproof and soundproof; it is as durable as—well, the earth—and as cheap as—well, dirt—and it is available, without transportation charges, practically anywhere anyone might want to live.

And special regions should be able to develop special materials and resources. Even as unprepossessing a region as the American Southwest has two such: a giant desert shrub called the jojoba, largely ignored today, that produces an oil that can be used for cooking, machine lubrication, paint additives, and at least a dozen other known things, and that can be stored for years without turning rancid; and a plant called the guayule, similarly undeveloped, that produces a latex almost identical to the kind derived from rubber trees and that could supply all of the rubber needs for the whole region from Texas to Southern California. Both of these plants were in fact proved to be successful substitutes during America's most recent period of self-sufficiency—in World War

II, when the government consciously developed alternative resources for threatened supplies.

Local-generation, as we might call this, seems particularly applicable to food. Self-sufficiency would probably mean the loss of Iranian caviar and Polish hams and French *fraises-des-bois,* not to mention in most cases Idaho potatoes, and Wisconsin cheeses and Florida oranges, except insofar as any particular community would feel it necessary to establish trade relations for such things or itinerant traders would find a market for them. And yet there is no region of this nation that was not once self-sufficient in food, coping for decades without the necessity of Brazilian bananas and Ugandan coffee and Indian curry, no region that could not be again. And not to its detriment but to its enrichment: the bounty from its own acres is infinitely cheaper, incomparably fresher, far more nutritious, free of shelf-life additives, and plain better-tasting. Is not the basic "peasant" cooking of any country always its best, was not the French "nouvelle cuisine" based exclusively on ingredients Paul Bocuse could buy that morning from the local market? Does it make sense for me to have to depend upon the cows 5,000 miles away to supply me with delectable cheeses when I live in an area that once was, and could be again, one of the prime dairy regions of the nation? Must I import apples from Washington state, as the local supermarkets begin to do in November, when I live in the second most prolific apple-producing state in the country?

I am not suggesting that there need be any back-to-berries deprivation here—many tropical products that seemed particularly desirable could be grown in community greenhouses, not on a large scale but sufficient for any local market, and probably certain kinds of sharing-and-bartering systems would naturally arise among communities and cities within a given bioregion. And I can't help thinking about the meal that was made up solely of the most elemental mundane offerings of the year-end harvest of one of America's least bountiful regions—that is the bountiful feast that was the first Thanksgiving.

5. By local ingenuity. If necessity is the mother of invention, self-sufficiency is obviously the grandmother. Local ingenuity and back-against-the-wall creativity have always been nurtured in the American small town—it accounted for the extraordinary number of inventions from such intentionally isolated communities as the Shakers (brimstone matches, the washing machine, a pea sheller, clothespin, and a thresher) and the Oneidans (another washing machine, a rag-mop assembly, and a string-bean slicer)—and there's every reason to suppose such inventiveness would not only continue but flourish in an atmosphere that made its products welcome. It is said that the reason for the lack of American inventiveness these days has to do with the forbidding legalistic red tape and the slim chances of design ever seeing reality—both these barriers at least would be removed in a self-reliant community eager for new methods that would conserve its resources and replace its imports: the medieval monasteries were almost single-handedly responsible for the agricultural revolutions in productivity and technique from the tenth century on.

Hubert Ignatious Fernando, an unknown technician for the Department of Agriculture in Colombo, Sri Lanka, created a working model of a revolutionary new rice-milling machine in 1972. His model uses local machinery and resources instead of imported ones, it costs about one tenth as much to build, it uses almost no fuels beyond the waste-products from the very rice it processes, and it increases the amount of finished rice—rice-on-the-plate, as they say in Asia—by at least 15 percent. Fernando had the inventiveness and the Sri Lankan farmers' organization had the interest in getting the machine into production, but unfortunately Sri Lanka is not self-sufficient. The four nations that dominate such technology—the U.S., Japan, West Germany, and Formosa—have decided, for various reasons, not to grant the rice-miller a patent, thus effectively preventing it from ever going into production. The way things are, it may have to wait as long as Hero's steam engine.

6. By using general instead of specialized machines. For the mass-production thinker this often doesn't make economic sense, but for limited markets the installation of multipurpose tools and assemblies can be far more efficient and economical. Only a limited number of such devices are available at present, there being no particularly strong market in the present production system, but there are many prototypes and no one doubts that the technology is simplicity itself. A boring machine, for example, that can rout out a small wire or a copper tube can be used to form a drain pipe or a sewage line, if it is retooled with a simple increase in scale; a lathe that can form a nail or a screw can be used for bigger tools and machine parts. Murray Bookchin's analysis of the effect of just this simple application of technology concludes:

> A small or moderate-sized community using multipurpose machines could satisfy many of its limited industrial needs without being burdened with underused industrial facilities. . . . The community's economy would be more compact and versatile, more rounded and self-contained, than anything we find in the communities of industrially advanced countries.

7. By a similar use of multipurpose factories. Working within a small market, a plant can turn out several different kinds of allied products, as with our electrical factory, instead of methodically making the same thing day after day and year after year, as is common now. A furniture factory, with only minimal alterations, can produce first beds, then tables, then chairs, then bureaus, in easy succession; and a machine shop geared for electric vehicles could switch without much difficulty to making tractors or reapers or even bicycles. For certain high-volume items—nails, say, or pencils—it would be possible to get mass-production efficiencies by gearing a plant for a month or two (or longer) to a single product, which would then be stored and used as needed, and then retooling it to make some other allied product for a month or two. This would never be quite as cheap and efficient as straight-through mass production, of course, but it would enable the total community manufacturing capacity to double and triple and more, and thus to multiply the number of goods with the same limited number of manufacturing workers.

8. By adapting plants to the community level. Factories do not have to be as gargantuan as they are, and it turns out that it is possible to scale almost any kind of manufacturing down to community dimensions. It is being done increasingly, in fact, even within our current industrial plant, as a variety of manufacturers discover a variety of new efficiencies and cost-cutting processes in decentralized production.

No, a town of 10,000 could not possibly support a giant steel mill of the kind you can see disfiguring Gary, Indiana, say, or Pittsburgh—but it *could* quite easily provide the machinery and power, the workers, the raw material, and the market for what is called a "minimill," a small-scale, self-contained steel operation. As two economists reporting on the success of minimills put it in a 1976 issue of *Land Economics:*

> Minimills are "neighborhood"-oriented operations sustained by a locally based and often locally trained labor force. Their power and material requirements are generally provided by nearby suppliers. Historically, the mills represented investment by local producers interested in developing reliable, low-cost local steel sources.

Minimills have been operating in the U.S. for more than forty years now, and as of 1973 there were as many as forty-five of them functioning in twenty-six states in every kind of geological and economic region from coast to coast. Their output is obviously limited—some are profitable at only 45,000 tons a year, most average 145,000, compared to the average full-size furnace that produces 750,000 tons a year—but among them they account for a quarter of the reinforcing and hot-rolled bars produced in the U.S. The majority are located in towns with less than 30,000 people, and a number in quite small communities, and all depend on local, rather than regional or national, markets which they are able to service far more quickly and flexibly than big producers.

Minimills have significant advantages over conventional mills. They cost far less to build and run; they take a quarter of the lead- and construction-time; they are technically simple and can use mostly unskilled labor; they use electric furnaces, which are more efficient and far less energy-demanding; they could be converted to solar power with an array of parabolic concentrating collectors without much difficulty; and they depend upon scrap—which is to say, recycling—as their primary raw material, though they also harmonize well with small-scale iron ore pre-reduction operations. (As an added attraction, they happen also to be noted as a source of inventiveness, including development of the electric furnace and the first on-line continuous-casting process.) And they are profitable, so long as they retain their limitations: "Smallness," say the researchers, "is a vital asset."

9. By networking, where necessary, with other communities. With judicious linkages with nearby towns—a sharing on a municipal level like the sharing on a personal level—a community could enlarge its economic possibilities and diminish its manufacturing requirements without necessarily sacrificing its self-sufficiency. (And a city of 50,000 or 100,000 would automatically be such a network, one imagines, with political links to harmonize the economic ones.)

Perhaps all that would be wanted is a communications network, with a wire, phone, or computer link that could connect a wide range of people for daily communications or emergency consultations or regional conferences or just general access to information; an existing model for this is the Electronic Information and Exchange System based at the New Jersey Institute of Technology with more than six hundred members and forty-five groups on line from coast to coast. Perhaps several towns might share special personnel, one technician or expert traveling regularly among a half-dozen towns and making specialized services available as needed—as at least one man in California has been doing since 1978, working as a city planner for five small incorporated cities near Eureka that could not each afford such a technician but can do so collectively. Or several communities might get together for a major public-works project—a modest dam to control seasonal flooding (if, of course, more ecological solutions aren't available)—or a public health program or even a theater or orchestra. Or maybe nearby towns would want to establish some sort of marketing network, increasing the distribution range for one or two special products, or a slightly more complex system of resource networking, trading one community's coal for another's copper, say, or some extra aluminum scrap in one town for some leftover bushels of tomatoes in another.

The dangers of such networks are obvious enough, since dependence upon them will eventually cut into any community's ability to act for itself, but kept within the bounds of simple mutuality, such systems can usefully expand productivity and save labor time.

10. Finally, and simplest, by doing without what is not needed. Bombs. Cosmetics. Moon exploration. Greeting cards. Food additives.

Junk mail. Packaging. Cigarettes. Electric dishwashers. Plastic bottles. Costume jewelry. Microwave ovens. Artificial fertilizers. . . .

IN FACT, COMMUNITIES are always much richer than they think they are. It is not smallness that deprives them of economic power and the ability to sustain themselves self-sufficiently but, on the contrary, their ties to largeness.

We have already seen in the example of a McDonald's outlet how at least three quarters of the sales of a single chain-store franchise are exported out of the community; a similar study for a supermarket chain-store operation shows that a firm with $5.7 million in annual sales will export out of the local economy $332,160 more than a store owned locally. Multiply that for virtually every product and service, and you can see that the only difficulty with a community's wealth is that it doesn't stay there.

If a community of 10,000 in the current American economy were somehow to keep all of its wealth, it would be quite rich indeed. The amount paid out in all personal taxes is currently about $1,500 per person (1975 figures) or a total of $15 million that leaves the locality; the amount paid out for all personal expenditures is about $4,600 per person, or a total of $46 million dispersed for goods and services. Now how much of that quite considerable sum of $61 million actually benefits the community? Well, some is returned in the form of services from the local municipality—roads, schools, police—and a lesser amount from the state government with its services, such as they are. But since the real capital outlay of local and state expenditures each year amounts to only $208 per person, or $2 million for our community of 10,000, there's an effective net tax loss of $13 million. As to the personal expenditure, some of that stays in the community, going to local landlords and merchants and plumbers and doctors, where it has a multiplier effect as it changes hands around the town. But even more of it is going out to distant manufacturers and suppliers, to A & P's headquarters and Sears's and General Electric's, to corporations in New York and Chicago and Tokyo and Riyadh; it is figured that even in the best of circumstances no more than half a neighborhood's income ever stays within its immediate area, and in our present case that would be about $23 million. So in sum, a contemporary community that is rich enough to be able to generate $61 million is likely to get back only $25 million of that, the rest going off to distant bureaucracies and distant corporations.

A lot could be done with that $36 million should the citizens ever want to try to keep their money at home; even more could be done with the whole $61 million, should they ever want to try self-sufficiency.[1]

1. It should be emphasized that this condition is just as true for "poor" communities. One study in the Shaw-Cardoza area of Washington indicated the 80,000 people there paid out $45 million in district and Federal taxes and got back only $35 million worth of services

It is a phenomenon that I believe deserves formulation as a law; let's call it Lucca's Law:

Other things being equal, territories will be richer when small and independent than when large and dependent.

The little town of Lucca, a Tuscan community that lasted for at least eight centuries as an independent entity within the *tempesto* that was Italian society during those years, may serve as our model. Emerging as a free commune sometime in the late eleventh century, it became one of the fiercely independent republics in the thirteenth century—with a population then of perhaps 10,000 or 12,000—and for the next 400 years, surviving ups and downs and feasts and famines, it was one of the most prosperous places on the entire Italian peninsula, not to mention the entire European continent. It enjoyed a rich and self-sufficient agriculture, with fruits, grains, wine, vegetables, chestnuts, and grazing fields; it was a major banking center from the fourteenth century on (still with no more than 15,000 people), was famous for its velvet and other textile manufacturing, was the home of recognized artists, musicians, and writers; and it expressed its prosperity, century after century, in a magnificent array of churches (one, the celebrated cathedral of San Martino), *palazzi,* castles, fortifications, and townhouses that was extraordinary even for those extraordinary years. Curtailed by the dead Spanish hand that closed over most of northern Italy after the sixteenth century, Lucca nonetheless survived as an independent and thriving trading town for another 250 years, its glory diminished but by no means extinguished. Then came the Napoleonic armies under the man who thought of himself as the emperor and unifier of Europe, and Lucca became a French plaything for a dozen years. After the Restoration, in the sweep of "nationalism" that moved over Europe, it was forced into a merger with Parma and then became part of a Tuscan state; in 1860 it fell under the control of Sardinia, which eventually became the foundation of the Kingdom of Italy, and under that banner the peninsula was "united" into the modern nation we know today. Union with successively larger territories in successively grander economies served mostly to impoverish the once resplendent republic, and today it is a forgotten backwater of 45,000 people, with a shaky industrial base (confined largely to jute and tobacco) and an agriculture quite dependent on imports from the south, the inescapable recipient of all that happens to the Italian economy of compounded chaos and inflation.

---

and welfare; another survey of an "inner-city" area by the National Center for Urban Ethnic Affairs found that the 10,000 families there spent over $1 million in a single year just on household appliances, not counting the $200,000 in carrying charges on them. Maybe most telling is that one Community Development Corporation, the Zion Investment Associates in a poor black district of Philadelphia with only 6,000 people, managed to raise, from people chipping in $10 a month, the impressive total of $2 million in just eight years.

History is replete with examples of Lucca's Law. Scotland, Wales, and Ireland were all measurably more prosperous as independent entities than they were after uniting with England and finding their economies sucked into the London maw. Most of the separate principalities of Germany, as well as the other city-states of Italy like Lucca, were far better off as small and self-contained units then they were when gathered together in the nineteenth century and made to pool their resources in one apparently bottomless pot. The United States itself—and in similar fashion Austria, Iceland, Norway, Belgium, Denmark, Ireland, Czechoslovakia—demonstrates that countries that split off from larger nations, far from becoming economic flotsam, are forced to turn themselves into viable states all on their own, and usually with considerable prosperity; the process seems to be helped when, as with most of these nations, they choose at the start that deliberate form of international self-sufficiency called protectionism. As Schumacher has reminded us, if Belgium today were a small northern province of France, it would surely be thought far too small to be a viable independent nation, and the learned economists would be assuring us that its separation would only lead to isolation and stagnation; only it *has* been independent, for 150 years now, and in that time it has forged its own thriving economy, with a per capita GNP (1975) of $7,697 that outranks France's $6,661. So much for viability.

It makes one think of Appalachia. That is today a synonym for poverty—no, not just poverty but a kind of grinding, inhuman, intractable desperation—and yet it is, of course, one of the naturally richest areas of the world, certainly one of the most productive coal regions in history. That sad paradox is a result of the terrible powers of fusion and agglomeration. Tom Gish is an Appalachian, the editor of *The Mountain Eagle,* a little paper in Whitesburg, in eastern Kentucky, the heart of Appalachia. Not far from where he lives is a valley in the headwaters of the Kentucky River known as Rockhouse Creek, a 10-mile stretch of, as he describes it, abject poverty:

> I saw hillsides denuded of growth long ago by burning slag heaps from underground coal mines. I saw other hills where growth had been swept away under strip mine debris. I saw two- and three-room tarpaper shacks and second-hand trailers located on piles of coal waste, without running water, without toilets, but with outhouses sitting directly over the creek.

It is, Gish says, "one of the greatest contrasts between natural resources wealth and human poverty that exists anywhere":

> For within that little 10-mile stretch Bethlehem Steel Company has a huge coal mining operation, and there are large mines owned by South-East Coal Company and Marlow Coal Company, plus a dozen or so

> smaller "truck" mines and several strip mines. All in all, that 10 miles produces at least five thousand tons of coal per day and has been doing so every day for the past 30 years at least. At today's values, that is some $1 million a week in coal going out of this tiny valley. And I haven't even mentioned the few dozen oil wells feeding into the Ashland Oil Company system, or the numerous productive gas wells feeding into the Columbia Gas and other gas systems. I don't know the value of the oil and gas coming out of this small area, but I would be surprised if it didn't approach the value of the coal that is going out.
>
> It is certainly a conservative estimate to say that this small area has contributed over $1 billion worth of minerals to the United States economy, certainly more than $1 million worth for every man, woman, and child resident there. It is an equally obvious fact that little or none of that money has found its way into the pockets of local residents. . . .

In short, "Appalachian residents have been stripped of any control over their own destinies." Never having had a chance to claim its own independence, in thrall for all these years to the cruel master of a national economy, Rockhouse Creek is very real, if pathetic, testimony to the truth of Lucca's Law.

IN SWEEPING TERMS, with all the caution that they imply, it may be said that we are today at the threshold of the fourth great economic revolution of Western times. The first was the development of feudalism and the medieval arrangements that emerged after the fall of Rome and sustained the continent until about the fifteenth century. The second was the development of capitalism and the system of wage-based industry that took over in the Renaissance and held sway until the start of the Industrial Revolution in the eighteenth century. The third was the development of industrialism and the form of state capitalism that has increasingly marked the nations of the West in the nineteenth and well into our present century. And the last is the development of the decentralized "post-industrial" economy and its manifestations in the developing steady-state consciousness and the growing movement for workplace democracy that are beginning to assert themselves throughout the developed, as in most of the undeveloped, world. With it is coming the acute awareness of the sorry shortcomings of corporate giantism and capitalism's apparent failure to manage the resources or secure the stability of the world, and with it too is coming a new appreciation of the need for a truly ecological economy built upon the resources of self-sufficiency.

The fourth era, in short, may mark the beginning of the human-scale economy.

PART FIVE

POLITICS ON A HUMAN SCALE

The first test to be applied in judging an alleged democracy is the degree of self-governing attained by its local institutions. . . . Only local government can accustom men to responsibility and independence, and enable them to take part in the wider life of the state.

IGNAZIO SILONE
School for Dictators, 1963

I am calling also for an end to giantism, for a return to the human scale—the scale that human beings can understand and cope with; the scale of the local fraternal lodge, the church congregation, the block club, the farm bureau. . . . In government, the human scale is the town council, the board of selectmen, and the precinct captain. It is this activity on a small, human scale that creates the fabric of community, a framework for the creation of abundance and liberty.

RONALD REAGAN
speech to the Executive Club of Chicago, 1975

Foederis aequas Dicamus leges.

VIRGIL
Aeneid

1

"A System Incapable of Action"

ON THE NIGHT of July 15, 1979, exactly three years after he had accepted his party's nomination for the presidency, Jimmy Carter went on television to give the most important speech of his career. He had just come back from two agonizing weeks at Camp David, where, in response to a succession of mounting domestic problems including that summer's gasoline shortage, he had taken a most extraordinary reading of the American mood, inviting several dozen people of many different callings and stripes, not just the usual gaggle of politicians and lawyers but business and labor leaders as well, some religious and academic figures for leavening, and an odd lot of just plain private citizens.

Looking thin and pallid, his lips pursed and tense, and with less of the fake bonhomie than usual, Carter told his nationwide audience:

> I want to talk to you right now about a fundamental threat to American democracy. . . . It is a crisis of confidence. It is a crisis that strikes at the very heart and soul and spirit of our national will.
>
> We can see this crisis in the growing doubt about the meaning of our own lives and in the loss of a unity of purpose for our nation.
>
> The erosion of our confidence in the future is threatening to destroy the social and political fabric of America. . . .
>
> We've always believed in something called progress. We've always had a faith that the days of our children would be better than our own.
>
> Our people are losing that faith. Not only in Government itself, but in their ability as citizens to serve as the ultimate rulers and shapers of our democracy. . . .
>
> The symptoms of this crisis of the American spirit are all around us. For the first time in the history of our country a majority of our people believe that the next five years will be worse than the past five years. Two-thirds of our people do not even vote. The productivity of American workers is actually dropping and the willingness of Amer-

> icans to save for the future has fallen below that of all other people in the Western world.
>
> As you know there is a growing disrespect for Government and for churches and for schools, the news media and other institutions. This is not a message of happiness or reassurance but it is the truth. . . .
>
> Looking for a way out of this crisis, our people have turned to the Federal Government and found it isolated from the mainstream of our nation's life. Washington, D.C., has become an island. The gap between our citizens and our Government has never been so wide. The people are looking for honest answers, not easy answers, clear leadership not false claims and evasiveness and politics-as-usual. What you see too often in Washington and elsewhere around the country is a system of government that seems incapable of action.

Extraordinary words, these, but no one doubted that they were spoken out of conviction, not politicking, and very few there were in the land who disagreed with them. The next morning the pollsters went out to get the public reaction: an astonishing 77 percent in the *Times*-CBS poll and 79 percent in the AP-NBC poll agreed that, yes, "there is a moral and spiritual crisis, that is, a crisis of confidence, in the country today."

THAT CARTER WAS unable then, or any time later, to provide even a semblance of a cure for the malady does not lessen his quite accurate diagnosis of the disease: the American political system is not working.

It is not a new diagnosis, not even new to Carter himself, who was elected to the presidency originally on his perception of the American electorate's profound unease and bewilderment. For at least a generation, but demonstrably and palpably since November 22, 1963, the American public, first the fringes and then the majority, has become disillusioned with the national government and apathetic or cynical about its efficacy.

Voting, that most basic and simplest of civic tasks, has been ingrained into us since the first grade as the very essence of our system, and the regular recurrence of national political campaigns every two years is always accompanied by well-financed and well-publicized appeals for us to get out and do it, as if this one activity were more significant than any other possible public activity. And yet every year the percentage of voters gets smaller and smaller. In 1962, the turnout for congressional races was so slim—just 45 percent of the voting-age population—that official Washington became alarmed and Kennedy appointed a commission to see what could be done. Since then almost all restrictions against Afro-Americans have been removed and a Constitutional amendment has given the vote to 18-to-20-year-olds—and the percentage of voters has continued to fall, reaching a low of 34.5 percent in the 1978 congressional elections. In 1960, the turnout for the presidential election, after one of the most publicized campaigns in history, was only 62.8 percent, the poorest for any nation in the industrialized two-party world; since then residency requirements have been greatly relaxed and postcard registrations introduced—and the percentage of voters has continued to fall in every succeeding vote: in 1976 less than 29 percent of the eligible voters bothered with presidential primaries and only a little more than half came out for the general election, which meant that a mere 27 percent of the voting population, and 19 percent of the population as a whole, chose the man who would fill the most powerful office in the land.

Voting is as accurate a barometer as we have of the national political weather, and no one is under any illusions as to how to read it. Curtis Gans, director of the Committee for the Study of the American Electorate, declared that the 1978 results show "the decay of political and social institutions, most notably the political parties" and "the growing impotence felt by the citizen in the face of larger public and private institutions and increasingly complex problems." More succinctly Arthur Hadley, the author of a book called *The Empty Polling Booth,* says: "Voters voluntarily avoid the booth because they see no connection between politics and their lives." That is an absolutely staggering statement to be made about any reputedly civilized country.[1]

Moreover, I suspect that a lot of the people who *do* go into the booths have no great faith that what they are doing touches their lives in any important way, for as near as we can tell the infection of political cynicism is at least as great among voters as among stay-at-homes. I have already touched on the polls that have regularly shown a declining loss of faith in America's political institutions, and there is scarcely a professor or a pundit who hasn't found the same thing: "A decline in the credibility of legislative bodies," according to Wayne State's Louis L. Freidland; a "loss of confidence in the nation's leaders and institutions," according to *Newsweek*'s Meg Greenfield; "mutual distrust between government and citizens," according to Case Western Reserve's Alvin Schorr. Or as the reformist Common Cause lobby puts it: "The reason the United States Government cannot solve our pressing problems is because the United States Government is the *problem!*"

Indeed, it was the perception of just such a loss of faith that instigated the Carter speech originally. According to the *New York Times,* Carter was moved by the findings of presidential assistant Patrick Caddell and his pollsters that there was "a surprisingly sharp rise in the number of pessimists in America, even in elite groups" in 1978 and 1979,

---

1. In Louisiana in the fall of 1979, Luther Devine Knox, a candidate for governor, made the appropriate response; he petitioned the court to change his name to None of the Above.

a "growing cynicism and pessimism in the country," and "a new pessimism among Americans about the future—both about the future of the country and about their own personal lives." So moved that he actually became convinced that "not only his Presidency but the nation were in dire peril."

That diagnosis may be still premature—and yet it reflects the unquestionable malaise that has pervaded the land in the last decade or so. Does Barbara Tuchman, a wise and learned woman whom I had always thought an imperturbable type, put it too strongly?

> In the United States we have a society pervaded from top to bottom by contempt for the law. Government—including the agencies of law enforcement—business, labor, students, the military, the poor no less than the rich, outdo each other in breaking the rules and violating the ethics that society has established for its protection. The average citizen, trying to hold a footing in standards of morality and conduct he once believed in, is daily knocked over by incoming waves of venality, vulgarity, irresponsibility, ignorance, ugliness, and trash in all senses of the word. Our government collaborates abroad with the worst enemies of humanity and liberty. It wastes our substance on useless proliferation of military hardware that can never buy security no matter how high the pile. It learns no lessons, employs no wisdom and corrupts all who succumb to Potomac fever.

I NEED NOT elaborate the point. The plain fact is that government in this country, in all its levels and its manifestations, is simply not working.

But of course it isn't. It's too big.

I think I have sufficiently detailed in Part Two the enormity of the Federal government, its entrenched bureaucracy, and its bloated burdens, and its consequent and inevitable inability to either understand or respond to the needs of the ordinary citizens—of Rudd, Iowa, say. But lest there be some lingering doubts—some latter-day liberals who may think that my emphasis on size is misplaced—let me offer the example of the supposedly most democratic national institution we have, the United States Congress.

Let us first put aside the textbook pieties. It is plain impossible for a member of the House of Representatives to be *representative* of any constituency—in terms of the Burkean imperative "to prefer their interests to his own"—since each one is supposed to have a district of approximately 465,000 people and it would take about twenty years just to find out what these constituents wanted even if one worked ten hours a day every day of the year. (For Senators the time would be anywhere from twenty years to—in the case of California, allowing 10 minutes per constituent—192 years.) But it is also plainly impossible for a member of

Congress to live up to the other Burkean charge—"your representative owes you not his industry only, but his judgment"—because the magnitude of the task is beyond any human capacity. In the 1977–78 session, 22,313 separate bills and resolutions were presented to the two houses, and if they were all to be considered, none could get more than a single hour's contemplation even if the members were to meet twenty-four hours a day for the full two-year term. Perhaps some of those were unimportant, but 3,211 of them at least were considered of sufficient magnitude to be passed by one or both houses. So, assuming that Congress were in session six days a week for two years and spent every single working day devoted to nothing but those important bills, the maximum amount of time any one could get—including study, committee and subcommittee examination, hearings, committee amendments, floor amendments, parliamentary maneuvering, lobbying, voting, assignment to conference, debate, and final enactment—would be slightly under two-and-a-half hours. That is hardly enough time for proper consideration of whether to expand the Federal white-pine blister-rust-protection program, much less make any intelligent headway with the ordinary tax bill, 200 pages long and confusing enough to cross an owl's eyes.

But of course our representatives do not have a full working day six days a week to devote simply to the mass of legislation. They must also travel back to their home districts, read and answer at least some of the great volume of mail, greet constituents both big and small, attend meetings of regional, party, and special-interest caucuses, keep contacts with other legislators and Federal bureaucrats, oversee both personal and committee staffs (which averaged 35 people per Representative in 1979), run periodic election campaigns, and presumably gather information for and have a hand in drafting the new bills—an average of forty-two per member—that are introduced each session. Small wonder that one legislator has confessed:

> We are losing control of what we are doing here. There isn't enough time in a day to keep abreast of everything we should know to legislate responsibly, dealing with so many bills, having to attend so many committee and subcommittee meetings, listening to the lobbyists, having to worry about problems of constituents, and, of course, keeping a close eye on politics back home.

Smaller wonder still that one study of how Congress actually spends its time found that, on average, each member was scheduled to be in at least two places at once for a third of each day and had, at most, eleven minutes a day for *thinking*.

The result of all of this is twofold. On the one hand, a great number of bills are enacted—some with quite sweeping effect, but all, given the pervasiveness of the Federal presence, of some consequence—without any thought or consideration whatsoever beyond the impulses of the original drafter, and many times even that has been lost in the legislative maze. The end-of-term Capitol Hill sessions are famous for this, sometimes churning out three dozen or more bills, each with a Christmas-tree full of amendments, in the space of twenty-four hours. But ordinary week-in-week-out sessions are very much the same, if less hectic. Washington reporter Elizabeth Drew, asking reactions in the House one ordinary May Monday a few years ago, found several Congresspeople willing to say, "Of course I don't know what I'm voting on" and words to that effect. Ned Patterson, a New York Democrat, confessed, "This is a symptom of the incredible overload that we deal with every day. There's no way that a member of this body can honestly say that he fully understands what he's voting on." He described how he went in to a routine vote with the House's new card-operated computer system: "As I stuck my card in, I asked the guy in front of me how he had just voted. He said, 'Yes.' I asked him, 'What's it about?' He said, 'Damned if I know.'" And, speaking particularly about the House's special method of "suspension"—voting with limited debate and without amendments—veteran Representative Morris Udall told Drew:

> There's no logic to why these votes go the way they do. There's a momentum that starts. Some bills will go roaring through, and on other days if some start to get in trouble they'll go down. Members are always looking for an excuse to vote no, and so if they see that one of these bills is going down, everybody jumps to join the pack.

On the other hand, a great many bills are *not* enacted, and on matters that would seem to be of high seriousness for the republic. The sight of congressional dithering on issues the public declares itself as finding important—taxes, energy, defense, inflation, welfare reform, medical insurance, regulations—is so commonplace now that popular approval falls regularly with each annual Gallup poll, only 19 percent of the public in 1979 declaring any confidence in the U.S. Congress. And the sight of congressional ineptitude in the face of the major crises that have confronted our times—Vietnam, Watergate, nuclear power, CIA and FBI illegalities, presidential abuse of power, invasion of privacy—is so recurrent that it is not surprising to find a majority of the public agreeing with former Congressman Michael J. Harrington—whose resignation in 1978 typified the sorry mess—that "Congress is irrelevant." Take the energy issue. After the first "crisis" in 1973, Congress dithered and slithered, devising nothing at all, and when finally prodded by a Carter energy proposal in early 1977, it managed to spend the next two years fumbling and stumbling before coming up with a bill that changed almost nothing except a few restrictions the oil companies didn't like, ignored solar and alternative fuels, and addressed itself to *none* of the substantive matters—the OPEC hold, oil-company profits, decontrol, rationing, divestiture, nationalization—that might have had some effect on the nation's energy problem. Five months later the nation was in the grip of another gasoline shortage, worse than before.

BUT IN TRUTH the malaise goes deeper than this: the failure of Congress is not so much the cause as the symptom of the distress in the body politic.

The crucial fact—the never-spoken, ever-present truth—is that Americans have given up their citizenship.

Citizenship—the act, the right, of participating in public affairs, of making the decisions that affect one's life, of having a continual voice in civic matters, of exercising regular judgment on the daily business of the state—has been sacrificed, in return for the right to vote, the right to find someone *else* to participate, the right to reaffirm every other November the loss of participation in public life. In past times when it was the locality that controlled daily affairs, the American adult participated in politics, joined civic groups, stood for office, took battles to the city legislature and problems to the city hall, met and thrashed things out in town meetings and ward assemblies from coast to coast. But that has changed. Today the locality is merely an appendage of some larger government, most major matters are decided for us in Washington, and there is virtually no way that our voices can be heard at that level in any sustained and percussive sense, no matter how many special-interest groups we may join or how many Mailgrams we may send. (Indeed, as special-interest groups have begun to blossom in the last few decades they have been treated by our representatives as dangerous, and as more citizens have written to their Congress members—a 400 percent increase in the last decade—they have been treated as an unpleasant burden and responded to with form letters signed by machines.)

Citizenship has simply evaporated in American life, leaving a residue of felt powerlessness. A President can declare a war or violate the territorial rights of 111 maritime nations all by himself and take countless secret actions at home and abroad jeopardizing millions of lives without any reference either to us or to our representatives. A Congress can decide to increase the Social Security burden by 300 percent or give a tax break to the oil companies or establish a new agency or pork-barrel a new dam, and most of us will not even hear about it until after it is done, much less have a chance to be consulted and give our opinion on it beforehand. State legislators can increase taxes or censor textbooks or raise utility rates without any participation from the people who are to be affected other than their pulling a lever in a polling booth two years before. Robert Paul Wolff, a professor of philosophy at Columbia University, has analyzed it this way:

> Since World War II, governments have increasingly divorced themselves from anything which could be called the will of the people. The complexity of the issues, the necessity of technical knowledge, and most important, the secrecy of everything having to do with national security, have conspired to attenuate the representative function of elected officials until a point has been reached which might be called political stewardship, or, after Plato, "elective guardianship." . . .
>
> It suffices to note that the system of elective guardianship falls so far short of the ideal of autonomy and self-rule as not even to seem a distant deviation from it. Men cannot meaningfully be called free if their representatives vote independently of their wishes, or when laws are passed concerning issues which they are not able to understand. Nor can men be called free who are subject to secret decisions, based on secret data, having unannounced consequences for their well-being and their very lives.

The simple fact is that in a system as large as ours it is essential that the individual *not* have a regular voice in political affairs. To allow each of 220 million people, or even the 150 million over 18, to participate in politics in a serious way would simply be too unwieldy, too chaotic; not even the wildest of technofix schemes of telephone voting and computer tallying could solve the sheer logistical problems if every person were to behave as, for example the Greek citizen of Periclean Athens, demanding to know the issues of the day, judging them, debating them, determining which were capable of being effected and when and how and by whom.

But not being able to participate has its terrible price. No wonder we feel so apathetic about voting: we do not have authentic political *selves,* we do not act politically, we do not know what is happening in and cannot much change the affairs of the nation, so the meager act of voting hardly carries much weight. *We do not understand ourselves publicly,* as public beings, nor could we be permitted to; we do not have public duties and public rights and public responsibilities of any meaning; there is nothing in our extended system that binds us as individuals to the public weal as there is in truly democratic societies.

We have sacrificed our citizenship to bigness, slowly over the decades—more rapidly in the last half-century but still slowly enough so that we have hardly been aware that it is gone—so it is not surprising that we do not have the interests, the attitudes, of citizens. Thus we do not vote. We do not pay taxes voluntarily—at least 5 million adults are reckoned to be tax avoiders, and virtually everyone else trims and cheats. We do not support our government in time of war—in fact a majority declared themselves *against* the government during the most recent war. We do not obey its laws by habit but by force, and a great many of the most highly placed people both in government and business, including even our Presidents and our representatives and the executives of the largest firms, are regularly and increasingly seen to be disobeying these laws. We do not willingly serve its army—many millions have recently flouted its jurisdiction—or do so badly if forced. We do not honor, for the most part, its veterans or its fallen or its flag.

I do not think that any nation has long survived under such conditions, almost certainly none nurtured on the political traditions of ancient Israel, of democratic Greece, of republican Rome, of the egalitarian Enlightenment.

THE "CRISIS OF CONFIDENCE" that Carter detected, then, would seem to be a very real thing, and undoubtedly far more serious than he knows. A public having lost its citizenship is beginning to be aware that the instruments of the American Behemoth are beyond its control and unresponsive to its needs, that somewhere the vaunted legislature and revered executive became separated from the body politic, that the institutions of governance no longer *deserve* any confidence.

It is a mood that goes beyond the normal American attitude of kick-the-bastards-out that has always been the expression of anti-government suspicion in this land: this is something deeper, more profound, something that is seeing past and beyond the politicians and bureaucrats who inhabit the machinery of government right into the machinery itself. *Newsweek* quoted a White House "political operative" as saying after the 1978 elections, "The voters have a feeling that the bastards who want to kick the other bastards out aren't going to be any better bastards once they're in." It is not the people but the positions that are seen as failing, not the incumbents but the institutions. "People think it's the *system,*" reports California pollster Mervin Field. "They don't believe you can change it by voting for one candidate over another."

Many years ago—nearly a hundred, in fact—a publication called *The Rebel,* put out in Boston, carried this editorial from one Arthur Arnould:

> An individual eats some mushrooms and is poisoned by them. The doctor gives him an emetic and cures him. He goes to the cook and says to him:
>
> —"The mushrooms in white sauce made me ill yesterday! To-morrow you must prepare them with brown sauce."
>
> Our individual eats the mushrooms in brown sauce. Second poisoning, second visit of the doctor, and second cure by the emetic.
>
> —"By Jove!" says he, to the cook, "I want no more mushrooms with brown or white sauce, to-morrow you must fry them."
>
> Third poisoning, with accompaniment of doctor and emetic.

> —"This time," cries our friend, "they shall not catch me again! . . . to-morrow you must preserve them in sugar."
>
> The preserved mushrooms poison him again.
>
> —But that man is an imbecile! you say. Why does he not throw away his mushrooms and stop eating them.
>
> Be less severe, I beg you, because that imbecile is yourself, it is ourselves, it is all humanity.
>
> Here are four to five thousand years that you try the State—that is to say Power, Authority, Government—in all kinds of sauces, that you make, unmake, cut, and pare down constitutions of all patterns, and still the poisoning goes on. You have tried legitimate royalty, manufactured royalty, parliamentary royalty, republics unitary and centralized, and the only thing from which you suffer, the despotism, the dictature of the State, you have scrupulously respected and carefully preserved.

I do not claim that Americans in any great numbers are yet prepared to do away with the state that has made them so ill. But I do not believe I am wrong in detecting the beginnings, the stirrings, the growth, of a true anti-mushroom sentiment in the land: people who, perhaps only dimly, have come to realize—in a way that even Common Cause does not yet understand—that "the government cannot solve the problem because the government *is* the problem." Inchoate, to be sure, unexpressed except by a few, this sentiment is nonetheless found today in a thousand guises. It penetrates issues as diverse as laetrile and the metric system, nuclear power and Proposition 13. It lies behind the anti-tax revolt and the commune movement, the ecology and alternative-technology movements, the campaigns for neighborhood and small-town revitalization, the growing attachments to holistic health, natural foods, and organic gardening. It has been the bedrock source of the civil-rights and anti-war movements, it energizes every drive for local control and community development and decentralized government today, and when it collides head-on with all the old rubric of "government of the people," it produces the widespread cynicism, apathy, and bewilderment that Patrick Caddell found in the nation.

It is this that may ultimately point the way to the human-scale polity. I do not mean to suggest that it is around the corner or necessarily inevitable. But I do think it is possible to perceive everywhere the strength of the new localism at work—not just in this country, I should point out, but around the world—a new insistence on the values and potentials of one's own region, one's own ecosystem, one's own cultures and traditions. The national malaise, the "crisis of confidence," is not going to be corrected until this spirit of place is given some political expression, until there is a devolution of power from the state to the locality and a decentralization of institutions to the point where individuals may in fact control them. A return to the power and sovereignty of the community: this is politics on a human scale.

2

The Three Rs

THERE MAY HAVE been no time within the last century of our history when there has been such a fundamental distention of the bodies politic and social as in our present generation.

Other times have seen pressure groups, of course—but today we have special-interest groups on every conceivable matter, in every conceivable enterprise and industry, most of them well organized and well financed, many able to influence the course of legislation and elections. Other eras have had protest groups, for example on women's rights and liquor and vivisection, but today there is not one social aspect of our life—birth, sex, death, education, health, environment—and not one biological aspect of our being—age, gender, race, ethnicity, physical handicap—that does not have one group, and usually two or more and often competing, organized around it. Other ages have known minority pressures, but today there is not a single ethnic group, however marginal, that is not organized and vocal and as often as not directly opposed to the interests or policies of at least one or two of the others. Other decades have had regional hostilities, but laying aside the Civil War and its immediate aftermath, seldom has there been the depth of hostility and fragmentation that we see today, with sectionalism infecting practically every national issue, setting a number of regions against each other and all against the Federal government. Other times have experienced cultural segmentation, inevitable in a vast country before the development of mass communications, but at no time since the advent of the national media has there been such a pluralism of tastes and forms and accepted practices, such an acceptance of disparate styles according to age, race, region, or status.

Kevin Phillips, the conservative Republican strategist, has seen in all of this what he calls "the Balkanization of America":

> The United States has been divided and fragmented before, but—save for the Civil War—with the underlying trend pointing in the direction of unity, fraternity, and increasing federal authority. Now American society seems determined to pursue smaller loyalties—regional, eco-

> nomic, political, ethnic, and even sexual—rather than larger ones. Unless the trend reverses in the next few years, and no such prospect is apparent, it bespeaks a fundamental reversal in the American experience. The heterogeneity of America will become a burden, the constitutional separation of powers crippling, the economy threatened, the cohesion of society further diminished.

The liberal cultural critic Robert Brustein has discovered much the same thing:

> The separatist impulse now seems to dominate dozens of special interest groups, each trying to elbow the other aside for power and influence, and we have become a nation of political lobbies instead of a people, dissipating vital creative energies in the formulations of petitions and propaganda.
>
> At the same time our culture is in the process of becoming an extension of our politics. . . . Instead of a national community in whose embrace we are all enfolded, we have broken into a multi-faceted complex of isolated constituencies, each with its own advocates and publicists, each arguing for moral, social and esthetic supremacy. . . .
>
> Whereas American politics, in the past, might have featured a conflict between large forces like North and South, and American culture divided between high and popular art, today we have splintered into so many fragments that the shape of the country is now almost indeterminate.

These are not alarmists, nor radicals of any sort, just sober observers of a contemporary America in the throes of a very palpable crisis.

WHEREVER ONE MAY LOCATE its epicenter, whatever one may feel about its power and duration, however one may assess its ultimate consequences, about the fact of this disintegrative crisis there is no doubt. There are any number of ways to approach a description of it, but perhaps the most comprehensive is to measure first its depth, as captured in the phrase, the *rejection of authority;* then to survey its scope, as evidenced in the remarkable worldwide *rise of separatism;* and lastly to assess its ultimate political impact, as manifested in the *resurgence of localism.* And what we will see by the end is the very real pull, the dynamism, of the human-scale ideal on the political landscape.

THE REJECTION OF AUTHORITY

THE MOST TRENCHANT analysis of this comes from Robert Nisbet, professor in the humanities at Columbia University, a fellow of both the Academy of Arts and Sciences and the American Philosophical Society, and one of the few distinguished academic conservatives. His sober book, *The Twilight of Authority,* came out in 1975; to give its flavor let me select a few of its subchapter titles: THE WEAKENING OF POLITICAL PARTY, GOVERNMENT AS DECEPTION, THE NEW CORRUPTION, THE OBSOLESCENCE OF IDEOLOGY, THE SPECTER OF VIOLENCE, THE EROSION OF PATRIOTISM, THE NEMESIS OF POLITICS, ESCAPE FROM CULTURE, THE LOSS OF SOCIAL ROOTS, THE LOSS OF THE HERO. Somber enough? He goes on to talk about the "decline and erosion of institutions" in America, "a vacuum . . . in the moral order for large numbers of people," "the crumbling walls of politics," "the waning of the historic political community," and most particularly does he see "a gathering revolt . . . against the whole structure of wealth, privilege, and power that the contemporary democratic state has come to represent." And he sees with bitter clarity:

> What we are witnessing today, foremost perhaps in the United States but increasingly in other parts of the West, is the breakdown of that concept [of public order] and with it of confidence in political government and its ordinary police and judicial powers of protection. . . . What has happened is that political government, in taking on itself a great variety of social, cultural, and economic responsibilities, for many of which it is ill-fitted, has ended up derelict in performance of its timeworn function of preserving public order.

This is no wild radical, mind, but a wise and weathered Tory, and yet his case for the unraveling of the fabric of the state—for what the sociologists like to call its "delegitimization"—seems quite consistent with the most outspoken voices of the Left.

It is no secret, of course, that anti-authoritarianism characterizes much of the contemporary world, as much in work as in politics, in the schools as in the churches, in the arts and culture as in personal and social intercourse. We are already accustomed to it in the form of truancy and vandalism, rising rates of divorce and illegitimate births, the end of the draft and high AWOL rates, national and local marches and demonstrations, lower productivity and more strikes, increasing crime and decreasing security, even blaring radios and garish graffiti. Among the most interesting manifestations of it in the U.S.:

▪ The Daniel Yankelovitch polling firm in 1978 published the results of a survey of young parents under the title, "Raising Children in a Changing Society," in which it disclosed that fully 43 percent of these new parents represent a "New Breed," teaching their children a new set of values quite unusual for America, and 24 percent are raising their children to reject customary American ideas of hierarchy and competition. At least 65 percent no longer believe, and probably will not teach, that there's any validity in "my country right or wrong," and only 13 percent are raising their children to believe that "people in authority know best."

▪ The most important development in American religion in more than two decades has been the rise of the independent laity, a movement that *Newsweek* has called "the New Reformation" that threatens to "transform both the structure and the operations of most U.S. churches." These lay believers, including most of the much-publicized "charismatic" and "born-again" Christians, reject connections to any ordained clergy or church hierarchy and commonly establish their own prayer groups, "house churches," and religious communities. Anti-clericalism is as central to this movement as anti-paganism.

▪ Government recommendations on health and safety, even those with wide support from doctors, are increasingly rejected by the American public, suspicious and distrustful of any proclamation from Washington even when it affects their own lives. A two-to-one majority of people polled rejected the idea that saccharin was dangerous and opposed the government ban on it; government recommendations for massive swine-flu immunization were ignored almost totally by the public at large; laetrile enjoys widespread and increasing use in spite of an official ban on the substance by the Food and Drug Administration and a nationwide advertising campaign; more than two-thirds of pre-school children are not immunized by their parents against polio, measles, rubella, mumps, or tetanus, in defiance of years of government exhortations.

▪ Since 1970, citizens in every state have attempted to use recalls, referendums, and initiatives, where permitted, to challenge one or another governmental practice or official, with at least 126 initiatives filed between 1970 and 1978, more than three times as many as a decade previously. Proposition 13 in California was only one of twenty-two similar tax measures initiated by citizens against their governments, many of them successful, a fervor that reached such proportions that it has led to the first serious demand for a new Constitutional Convention in the nation's history. Other citizen-led campaigns have been directed against government policies on war and peace, nuclear power and waste disposal, homosexual rights, returnable bottles, legislative payrolls, city growth, electric rates, busing, and probably most everything else in between. "To a degree unparalleled in recent American politics," noted the *Reader's Digest* in 1979, "citizens . . . are demanding—and getting—a more direct voice in legislation and are deciding issues previously left to elected officials."

▪ The University of Michigan has measured public attitudes toward authority since 1958, at which time it found that 20 percent of the public expressed distrust of political authority. Every year since then the percentage has risen, and in the latest survey, in 1979, 55 percent, more than half the citizenry, pronounced itself against the institutions of authority in America. No amount of jingoism over Afghanistan seems likely to alter that, try as the politicians might.

There is no suggestion that all this is a passing phenomenon. Most commentators on the subject seem to agree with Norman Macrae, the deputy editor of the *Economist,* who concluded after a worldwide tour in 1968 that the "bandwagon against big government" would continue to roll at least through the end of this century. Some argue that there will be no end to the sweeping process of, as one writer has put it, "people hoarding power from the state."

THE RISE OF SEPARATISM

As A GLOBAL event, the current rise of separatism—or variously, regionalism, tribalism, parochialism, nationalism, localism, autonomy, or devolution—is quite without precedent in history. Eric Hobsbawm, the British Marxist and no great admirer of it all, has acknowledged that "the characteristic nationalist movement of our time is separatist, aiming at the break-up of existing states including—the fact is novel—the oldest-established 'nation-states,' such as Britain, France, Spain and even—the case of Jura separatism is significant—Switzerland." It is, he finds, "an unquestionably active, growing and powerful socio-political force." Michael Zwerin, from a quite different perspective, would agree. In his enchanting book called *A Case for the Balkanization of Practically Everyone*—which is in fact a serious collection of sketches of separatist movements throughout Europe—he puts it this way: "The pendulum which for centuries has been swinging towards larger political groupings is swinging back and just as exterior colonies broke away from empires to form the Third World, so internal colonies—the Fourth World—are now trying to break away from States." There is even, in Wales, a magazine called *Resurgence* whose subtitle is: "Journal of the Fourth World."

A survey, as brief as it is possible to make it:

In Europe, there is scarcely any nation without a separatist movement, and most often these days an organized, militant, and occasionally violent one. The dominant trend of the continent seems to be, as the *New York Times*'s C. L. Sulzberger reported in 1977, "visible fragmentation," for "a rash of separatist movements is in vogue"; even the supposedly unifying European parliament elected in 1979 was based for the most part on regional, not national, constituencies. France has separatists in Brittany (where they are particularly strong), Alsace, and Normandy, a Corsican independence movement that bursts into occasional warfare, and since 1970, a growing nationalist sentiment in Occitania (Languedoc). Spain's historic regional struggles have resulted in considerable victories in the post-Franco years, with Andalusia, Catalonia, and Basque all being given regional autonomy and separate provincial governments in 1978, though some Basque elements in particular have continued to press for total independence; a guerrilla independence movement has also emerged in the Spanish-controlled Canary Islands. In Britain, the strong nationalist movements in both Wales and Scotland suffered setbacks when voters rejected Parliament's limited offers of devolution in the 1979 elections, but both are too strong to be ignored in policies or polls and both are pressing for additional autonomy that is sure to come; the title of Tom Nairn's book *The Break-Up of Britain,* which came out in 1977, was as much predictive as descriptive. And the bloodiest campaign of all is that which has engulfed Northern Ireland in the last decade, since the separatist civil war began in 1969. Italy has its separatist movements in South Tyrol and Sicily; Yugoslavia has been rocked by the Croatian independence revolutionaries, whose violence has spilled over into bombings and airplane hijackings in this country; Belgium is still divided between the Walloons and the Flemish; Denmark has been forced to give Greenland home rule after an overwhelming vote for that island's autonomy in 1979. And Russia, which is nothing more than a country made up of separate nations kept together by whim and weaponry, has been forced to recognize the separatist movements in Georgia, Armenia, and Azerbaijan to the extent that Moscow in 1978 was forced to give each region new constitutions providing local language rights and increased regional autonomy.

In the Middle East, the central issue of our times is exactly what one separatist state-in-being, Israel, is to make of another wishing-to-be, Palestine. Surrounding that are the divisions that since the 1975 civil war have turned Lebanon into two nations, a Moslem one in the north and a Christian one (which actually did declare its independence in April 1979) in the south; the incipient revolts by Sunni Moslems in Turkey; the actual division of Cyprus into two hostile states; and the armed rebellions in recent years of the Kurds, the Turkomans, the Baluchis, the Pashtunis, and other ancient and distinct peoples against the various governments controlling them.

In Asia, practically none of the national boundaries corresponds to nationalist realities and that is inevitably the source of separatist movements. The fiction of a united Pakistan gave way to the creation of Bangladesh, though only after a senseless, bloody war, and both remaining parts are themselves rent with additional religious and caste differences. India is today, as it has always been, a pastiche of many dozens of separate peoples, divided by race, religion, sect, language, history, and custom, whose rivalries erupt into violence someplace in the country every year; in the east at least three of these peoples—the Mizos, the Nagas, and the Tripurese—have launched independence movements, the first even creating a Mizo National Front guerrilla force in 1979. Sri Lanka contains a Tamil minority that has carried on a sporadic movement for autonomy—sometimes violent, sometimes parliamentary—for almost two decades and still maintained a small guerrilla group as of 1979. Indonesia has succeeded as a national fiction by allowing a great deal of regional autonomy among its disparate peoples, but it has had an out-and-out rebellion in East Timor since 1976 and a simmering separatist movement in both Borneo and West Irian since the 1960s. Vietnam, Laos, Cambodia, and Thailand all have minority tribes and nationalities that have resisted their centralized governments in various ways during the recent tumultuous years of warfare, often with arms but to date with only minor success.

Africa is, even more than Asia, a colonial fiction: there is not one nation where the separatist pulls of tribes, sometimes a multitude of tribes, does not exist, does not in fact dominate the nation's politics, not one where the great majority of people do not find the tribal allegiances more compelling than the national. Coups, countercoups, rebellions, and depositions have been the regular manifestations of these divisions—particularly in Ghana, Nigeria, Uganda, Dahomey, and Upper Volta in recent years—and all-out warfare has attended the tribal independence movements of Katanga, Biafra, Eritrea, and Western Sahara, as well as the tribal civil wars in Rwanda-Burundi, Sudan, Chad, Zanzibar, Angola, and Zaire. The South African policy of apartheid, loathsome and inequitable in its practical aspects, is at least founded on the perception of these tribal realities, leading to the establishment of nominally independent governments in Venda, Transkei and Bophuthatswana so far, with eight other "homelands" eventually to come.

In Canada, the much-discussed separatism of Quebec is not the only wrench on the already distended confederation patched together by the British in 1867, though it is by far the liveliest one. Both the votes of 1976, which brought the Quebecois to power, and of 1979, which confirmed their isolation, have gone a long way to making the province a distinct entity in fact, and whether or not it ever achieves actual independence it has already taken unto itself considerable practical autonomy. That trend was confirmed by the victory of Joe Clark in the 1979 elections, a triumph for those who want only the most minimal federal presence (CANADIAN ELECTION: A TONIC FOR SEPARATISTS, as one headline put it); by the rise to power in Alberta of a new party that is a strong advocate of provincial autonomy; and by renewed actions for separatism by several diverse groups in British Columbia. The *New York Times* put it simply but not unfairly in its 1979 round-up of Canada: DIVERSITY TEARS AT SOCIAL FABRIC OF VAST LAND.

The world over, then—and I have really touched only on the high points—the impulse toward separatism, fragmentation, division, has shown remarkable resilience during this last generation, a global process that seems to have touched nations of every stripe and for the moment at any rate shows no sign of abating.

And *here*—does it exist here in the place, perhaps more in hope than in truth, we are still pleased to call the United States?

We can no doubt dismiss the actual votes for secession of such places as Martha's Vineyard and Nantucket from Massachusetts and Grande Isle from Vermont as the clever ploys of local citizens angered at being denied local representatives in the state legislature. We might also regard the organized movements for independence or separate statehood in such places as New York City, the Michigan Upper Peninsula, Baja California, the Alabama "black belt," and the "ecotopian" Northwest as being forms of some political gamesmanship. We could disregard, as merely the disgruntled citizens' way of blowing off steam, letters like this one Jerry Brown's office received in 1979:

> The District of Columbia is as much out of touch with California today as was England before that [American] Revolution. We don't have to take it, Jerry. We have enough of our own resources. Let the other 49 bargain with us. California is a great state—it would make a great country.

And we could think of the increasing number of lawsuits being brought by individual states against the Federal government—representing disputes over countless environmental, educational, occupational, and anti-discriminatory issues—as just the newest method of jockeying within our complex federal system rather than any real sign of separatism.

There is more than that, however, and it goes deeper. There is the strong and still-growing movement among American Indians for recovery and independence of their original tribal lands taken illegally from them by the Federal government many decades ago; so far the Mohawks in New York, the Narragansetts in Rhode Island, the Penobscots and Passamaquoddy in Maine, and the Sioux in South Dakota, among others, have taken their cases through the courts and some of them have backed up their demands with armed force. There is the drive for Puerto Rican independence that, despite considerable Establishmentarian and some popular resistance, keeps gaining strength with every year and seems to be fated to eventual success. There is the much less cohesive but quite apparent distancing of the races, black and brown and red from the white, and each from each other, exacerbated rather than eroded as minority races come to take political power into their own hands in a number of cities and regions, leading not to the integrative dream of the 1960s but the disintegrative reality of the 1980s.

Above all there is the marked increase in regional competition and hostility over the last decade or so, in defiance of all the superficial similarities across the land wrought by mass marketing and the mass media. As Colorado Governor Richard Lamm told *Newsweek* in 1979: "Regional politics are greater than at any time since the Civil War."

The primary cleavage, now as it has been for the last twenty years at least, is between the developing Sunbelt and the decaying Northeast. It is a reality, as I've tried to show before, that goes far beyond Texas bumper stickers (DRIVE FASTER: FREEZE A YANKEE) and Georgia editorials ("At last we have a President who doesn't have an accent"). It involves the rival regional development commissions and warring congressional caucuses that have been formed in the last few years—including the Southern Growth Policies Board, the Western Governors' Policy Office, the Coalition of Northeast Governors, the Northeast-Midwest Congressional Coalition—and that scramble with considerable intensity both for Federal dollars and for relocating corporations. At the congressional level, it divides and often paralyzes party and even national loyalties: "Regional differences," according to *New York Times* political specialist John Herbers, "along with the proliferation of special interests in the political processes, have been a factor in Congress's inability to agree on a strong national energy policy, a unified approach to urban affairs and other domestic matters." At the presidential level, it is what lies behind the White House inability to work with entrenched Eastern foreign-policy and civil-service interests, the defensive good-ol'-boys reaction in surrounding Carter with Georgians and Southerners, and the final emergence of Teddy Kennedy to champion the Northeast in the presidential lists.

Regional differences have also surfaced in a most notable way in what the press is calling the "Sagebrush Rebellion" of the Western states against the Federal government. Paranoid alienation has long been a familiar affliction of the South, often completely justified in view of the way the rest of the country short-shrifted it, and now the West is beginning to feel the same way. In its cover story—THE ANGRY WEST VS. THE REST—*Newsweek* found both the populace and the politicians in every state west of the Rockies in the grip of "the new strain of sectionalism" that has visited the nation, with surprisingly vitriolic feelings of isolation and political estrangement. The *Times* found similarly that "according to a range of evidence and persons interviewed across the country, the West has replaced the South as the region that feels itself most abused by the Federal Government and the least understood by the rest of the nation."

But regional hostilities don't end there. Particularly as the nation is buffeted by increasing inflation, increasing unemployment, economic stagnation, foreign competition, and the recurrent energy crises, competition among all regions has grown serious and intense. There is constant and bitter competition between regions for defense contracts, army bases, public-works projects, R & D money, tax subsidies, agricultural supports, energy boondoggles; both public and private forces are at work to lure businesses, venture capital, conventions, book collections, academic superstars, conductors, advertising agencies, ballplayers, and a wide variety of other perceived attractions from one section to another. New England competes with the South over runaway companies, Appalachia with the Rockies over coal development, the Southwest with California over water allocation, the Northwest with the Rocky Mountains over water power, Hawaii with California over immigration, Alaska with the West Coast over oil shipments, the Southeast with the Northeast over natural-gas regulation, the Grainbelt with the East over beef prices, the Gulf States with the mid-Atlantic over offshore drilling—and on it goes, a display of divisiveness without parallel in this century.

When James Schlesinger in 1978 warned the United States of the threat of "Balkanization" he was not wrong, but possibly a bit too late.

THE RESURGENCE OF LOCALISM

IF POWER IS a finite quantity, its diminution in one area should presently be followed by its expansion in another. And so it is that, as the large national governments of the world have shown themselves increasingly unable to provide for the needs of their constituents, small-scale self-help efforts and political organizations at a local level have begun to emerge to take the tasks into their own hands.

The *Christian Science Monitor* perceived this resurgence in the U.S. in 1977, sent two of their best reporters out across the country to document it, and in a lengthy series of articles demonstrated that there was a new "broad-based citizens' movement swelling in the land" from the inner city of Baltimore to the barrios of Los Angeles:

> The movement is non-ideological, racially integrated, fiercely proud, and rooted in the nation's neighborhoods. Its constituency is an unlikely coalition of "rednecks" and radicals, blacks and blue-collar ethnics, willing to use confrontation tactics to defend conservative traditions of "turf" and home ownership.
>
> Some call it the "new localism" and "backyard democracy." Others refer to it as "politics of the '70s." Whatever one calls it, it appears to be a reaction to the failures of big government to handle local problems; and it is demanding a transfer of political and economic power to the neighborhood level. . . .
>
> It is steelworkers and socialists, loggers and little old ladies in crisp cotton dresses. It is citizens in urban ethnic villages, tiny towns, and suburbs. It is hundreds of independent but cross-fertilizing community organizations. . . . It is a grass-roots upsurge of citizens who are fed up

with top-down government solutions and who call for decentralized decision-making.

This is by no means a movement without obstacles. Most Federal law and municipal regulation, as well as the encrusted political habits of generations, have been against it. Traditional politicians, while always quick with pieties about neighborhoods, have despaired of organizing it into the party of their choice and so have tried to dismiss it. The corporate world, bankers in particular, have been absolutely ingenious in erecting barriers wherever possible. And yet, the movement apparently will not be put down, and it continues to win its victories, some small and some rather larger, on its own blocks, in its villages, in the city halls, across whole regions, and even occasionally on the national scene.

At the local level, new grass-roots power has wrought significant changes in a number of cities, even some of the largest and most troubled. ACTIVIST NEIGHBORHOOD GROUPS ARE BECOMING A NEW POLITICAL FORCE is the way the *New York Times* put it in a front-page story in June 1979, pointing to the way these groups "have been winning legislative and policy battles" on a "broad range of issues" from education and health to police protection and nuclear-waste transportation. Atlanta, Pittsburgh, Birmingham, Fort Wayne, Houston, Providence, St. Louis, Boston, Chicago, Baltimore, Dayton, New York, Dallas—all have well-organized neighborhood groups and community coalitions that have come to be heard in the city halls, and in several instances (Atlanta, New York, and Providence among them) have succeeded in rewriting city charters to increase neighborhood power. In New York, where at least 10,000 of the city's 35,000 blocks have developed their own block associations, and where some sixty wider community-level organizations have been formed, much of what the city has actually enjoyed during its recurrent fiscal crises has been provided not by the municipal government but by the local groups: neighborhood associations have planted trees and flowers, started recycling centers, turned acres of abandoned lots into parks and gardens, and formed street patrols and crime watches; neighborhood organizations have taken over at least seventy abandoned buildings from the city, rehabilitated them, and turned them into serviceable rental or cooperative units; and at least fifty neighborhoods have created community-wide volunteer ambulance corps, complete with their own neighborhood-trained medical teams and mobile units. Formal recognition of this devolution was granted in the new city charter of 1975, which established fifty-nine "community boards" with considerable decentralized power over the allocation of city services and land-use and development policies; additional power will come in the 1980s under provisions that both municipal services and budget allocations have to be matched to, and kept under the supervision of, the jurisdiction of these boards.

At the regional level a wide variety of groups has emerged in the last ten years devoted to community organization and rural redevelopment, to citizen activism and consumer resistance, in statewide and multi-state contexts. Some are specifically regional—Save Our Cumberland Mountains, Kentucky Rivers Coalition, Carolina Action, Appalachian Peoples' Organization—and some are narrower—Massachusetts Fair Share, Illinois Public Action Council, California Citizens Action League—some are essentially rural—La Raza Unida in South Texas, ACORN (Association of Community Organizations for Reform Now) in the South and Mississippi River valley—and some more urban—the California Campaign for Economic Democracy, founded by Tom Hayden, which develops reform slates for city and community elections, and Community Jobs Clearinghouse, which is a kind of activists' employment center for the West Coast. In practically every sizeable state it is now possible to find statewide groups working for local control of health care, food-growing and marketing, housing, communications, alternative technology, energy development, education, and culture, an extraordinary and quite spontaneous development of localism that has taken politicians and pundits, not to mention the professors paid to predict such things, quite by surprise.

On the national level, the resurgence of localism has been responsible for a number of laws and regulations designed to promote neighborhood development and encourage local-level employment. Through such new nationwide coalitions as the National People's Action (based in Chicago and with chapters in more than 120 cities in almost every state) and the National Association of Neighborhoods (based in Washington and drawing together groups in sixty cities), the neighborhood movement has registered some notable victories: the passage of the Home Loan Mortgage Disclosure Act of 1975 and the Community Reinvestment Act of 1977 to curtail the practice of bank redlining, and the creation of a Federal Office of Neighborhoods and its program of "urban homesteading" grants that has so far begun housing rejuvenation in at least twenty major cities. The neighborhood movement was also largely responsible for the enterprising—if somewhat ineffectual—National Commission of Neighborhoods, which issued a major 1,000-page statement in 1979 detailing a variety of ways to "reorganize our society—away from the stranglehold of a relative few with a near monopoly of power and money in political, social and economic institutions to a new democratic system of grass-roots involvement that allows individuals to have control over their own lives"—and this from an official, national commission. So far the movement's most successful national measure is probably the creation of Neighborhood Housing Services, which promotes urban restoration and preservation by providing housing "facilitators," start-up grants, and mortgages to help neighborhoods organize locally managed housing corporations; from 1973 to 1979 NHS corporations were started in forty-six cities with loans of some $6 million, a model precisely because the power and accountability remain at a local level rather than in Washington.

Naturally any movement drawing its strengths from strictly local roots is not likely to achieve the kind of spectacular extravaganzas so dear to the media, nor can it point to sweeping nationwide exploits. Its accomplishments are deliberately small, perhaps the more durable for that, but they can be seen many places across the country at the community level, block by block, neighborhood by neighborhood; and wherever one finds evidence of local revitalization, it is likely to be as a result of the ordinary people in these neighborhood organizations, not of private developers or municipal politicians.

And there is no mistaking why these community groups have arisen or why they have achieved their success. In the words of *Times* political reporter John Herbers:

> By being small, unstructured and decentralized, they are in keeping with the times, which have seen a fragmentation of authority in institutions that held the country together for many decades. Every neighborhood is different, and the complexity and range of the organizations have permitted them to deal more specifically with needs than national or regional institutions could.

Exactly so: when the disease is giantism, health is "being small, unstructured, and decentralized."

AT ONE TIME our diets, our sex lives, our family relations, our education, and our jobs were under the serious and daily control of churches and kings. We are no longer under their especial thrall, prisoners as we once were to their every whim. Is it so fantastic to think that someday we may be as free of the regular interference of the government of the nation-state?

Certainly the idea of the all-powerful national government—the prevailing political notion of the past hundred years or so—is open to question and revision, if only because it has been amply demonstrated that such governments are no more capable of solving our abiding problems than were the churches and kings before them. As the world begins to experience the diminution of its resources in the decades to come, it may be that the nation-state will similarly be forced to diminish its role because it no longer has the same power at its command, forced to limit its intervention in local affairs because it no longer has the same riches—and the same economy based on those riches—to support it. It may be, as some economists are saying, that sheer necessity will force that state to reduce its activities and services and to thrust local communities upon their own capacities and resources. Over time, as the nation-state withdraws from this or that aspect of, say, health care or transportation or housing, localities will be forced to assume that burden, thus potentially enabling them to develop their full capacities and permitting the evolution of a stronger and more variegated society.

It has been said that the chief reason for the success of the mammalian class, when it emerged those many long eons ago in the world's evolution, was that, unlike the reptiles that until then had dominated the earth, it could regulate its own internal conditions. Mammals, being warm-blooded, were able to adapt to all kinds of diurnal and climatic conditions to which the dinosaur was not, and being self-regulating, though far smaller, were able to survive geologic conditions and, eventually, to dominate. It is not fanciful to think that we may be in just such an Eocene epoch now, again watching the emergence of new smaller forms to challenge the old, superior for exactly the same reasons of size and self-regulation.

3

The Decentralist Tradition

THE IMPULSE TO local governance, to separatism and independence, to regional autonomy, seems an eternal one and well-nigh ineradicable. The long experience of nation-states—in Europe going back several centuries at least, in parts of Asia somewhat longer—has not destroyed that impulse, not in those countries, such as Britain, say, or the United States, where the state has grown to be most powerful and ubiquitous, not those places, such as Iran, where it has been most overreaching and oppressive. Indeed what is remarkable during these long years is how this decentralist tradition remains so resilient—so resilient that every time the power of the nation-state is broken, as during wars or rebellions, immediately there spring up a variety of decentralized organizations—in the neighborhoods, in the factories and offices, in the barracks and universities—that reinstitute government in local, popular, and anti-authoritarian forms.

The historical evidence is unmistakable on this point. In Paris in 1871 the collapse of the empire gave birth to the Paris Commune and its popular assemblies, while every neighborhood began its own committees of governance and defense and most of the business of the capital went on as usual, only with the workers themselves in charge: in Hannah Arendt's words, "a swift disintegration of the old power, the sudden loss of control over the means of violence, and, at the same time, the amazing formation of a new power structure which owed its existence to nothing but the organizational impulses of the people themselves."[1]

In Russia in 1905, and then again more sweepingly in 1917, industrial workers organized themselves into committees that took over factories and shops in practically every industry after the owners and

1. Karl Marx himself saw in this neighborhood government system "the political form of even the smallest village," which looked as if it might be "the political form, finally discovered, for the economic liberation of labor." He was right, of course, but he forgot it.

bureaucrats had fled, and the real work of running most of the cities was done—until Bolshevik violence eventually put an end to them—by local *soviets,* popularly elected assemblies.

In Germany at the end of World War I, workers and soldiers in a number of cities organized themselves into local councils—*Räte*—in defiance of the Social Democratic regime in Berlin, demanding a new German constitution based on local autonomy through a nationwide *Rätesystem,* in Munich even establishing for a time a *Räterepublik* of Bavaria where, an *un*sympathetic observer noted, "every individual found his own sphere of action and could behold, as it were, with his own eyes his own contribution to the events of the day."

In Spain in 1936–37 the collapse of the national government was followed by the emergence of independent collective governments in hundreds of smaller towns, as we have seen, as well as in a number of the larger cities—Barcelona and Alcoy, particularly—where entire industries were run by self-management, municipal services such as the electric works and streetcar systems were operated by independent collectives, and political and economic affairs were in the hands of "technical-administrative" committees elected within each industry.

In Hungary in 1956, the uprising that toppled the Communist regime led immediately, even in a country little used to popular government, to the formation of an extraordinary array of local councils, in neighborhoods and coffee houses, offices and factories, among writers and soldiers, students and—*mirabile dictu*—civil servants, indeed everywhere in the society, gradually coalescing within days into a rough network capable at least for a time of running the entire country.

In Iran in 1979, after the fall of the totalitarian government of the Shah, local institutions long suppressed suddenly appeared overnight, independent *ayatollahs* took control of their own provincial towns, neighborhood *komitehs* (even their name reminiscent of the Paris Commune) emerged to control their own territories in the cities, and at least four of the ancient minorities—the Baluchis in the southwest, the Kurds and Azerbaijanis in the northeast, and the Arabs of the Persian Gulf—asserted their independence with armed rebellions and demonstrations in the streets.

The same sort of thing, too, can be found in America after 1776, in France in 1789, in several European capitals in 1848, in many parts of Italy during the 1850s, in parts of occupied France and Italy in 1918–19, in Ireland throughout the civil war of the 1920s, in China in 1949, in Cuba in 1959, in Czechoslovakia in 1968, in Chile in 1970, in Portugal in 1974–75, in Angola and Mozambique at that same time, in fact as near as I can tell wherever and whenever a central government loses its hold (and before some new centralizing force, as often as not proclaiming itself revolutionary, takes over). Hannah Arendt has studied this phenomenon, noting with some wonderment how local councils and societies "make their appearance in every genuine revolution throughout the nineteenth and twentieth centuries":

> Each time they appeared, they sprang up as the spontaneous organs of the people, not only outside of all revolutionary parties but entirely unexpected by them and their leaders. They were utterly neglected by statesmen, historians, political theorists, and, most importantly by the revolutionary tradition itself. Even those historians whose sympathies were clearly on the side of revolution and who could not help writing the emergence of popular councils into the record of their story regarded them as nothing more than essentially temporary organs in the revolutionary struggle for liberation; that is to say, they failed to understand to what extent the council system confronted them with an entirely new form of government, with a new public space for freedom which was constituted and organized during the course of the revolution itself.

Everywhere it seems to be the case that the absence of government does not lead to bewilderment and confusion and disorder, as might be imagined if all of government's claims for itself were true, but rather to a resurgence of locally based forms, most often democratically chosen and scrupulously responsive, that turn out to be quite capable of managing the complicated affairs of daily life for many months, occasionally years, until they are forcibly suppressed by some new centralist state less democratic and less responsive. They seem to be, as Arendt says, "spontaneous organs of the people," expressive of the natural human scale of politics and inheritors of the long tradition of decentralism.

IT IS STRIKING to re-read history with eyes opened to the persistence of this tradition, because at once you begin to see the existence of the anti-authoritarian, independent, self-regulating, local community is every bit as basic to the human record as the existence of the centralized, imperial, hierarchical state, and far more ancient, more durable, and more widespread.

Obviously for the 3 million years that humanoids were becoming human they lived in small clans and groups, as we have noted before, and for the next 10,000 years that they were becoming "civilized" they lived in small communities and towns, needing none but the most limited kinds of governmental structures. Throughout the era of oriental empires—Persian, Sumerian, Egyptian, Babylonian—the greater part of the world's people still lived in independent hamlets, ever resistant to the imposition of outside authority, and even within the empires themselves local self-governing communes always persisted. Later, the Essenes, the people of the Dead Sea Scrolls who established an egalitarian community in opposition to Romanic Jerusalem in the second century before Christ, were only one of myriad tribes and sects that lived deliberately outside the Roman imperial influence. And still later the Christians themselves often lived in democratic and independent communities, sometimes in secret and sometimes openly but always apart from and hostile to whatever state might claim sovereignty.

The settlements of Greece were typical of such resistant localism: for many centuries they clung to a fierce independence, city upon city, valley after valley, no matter what putative conquerors might intrude, in time achieving that Hellenic civilization that is still a marvel of the world. As historian Rudolf Rocker has written of them:

> Greece was politically the most dismembered country on earth. Every city took zealous care lest its political independence be assailed; for this the inhabitants of even the smallest of them were in no mind to surrender. Each of these little city-republics had its own constitution, its own social life with its own cultural peculiarities; and this it was that gave to Hellenic life as a whole its variegated wealth of genuine cultural values.
>
> It was this healthy decentralization, this internal separation of Greece into hundreds of little communities, tolerating no uniformity, which constantly aroused the mind to consideration of new matters. Every larger political structure leads inevitably to a certain rigidity of the cultural life and destroys that fruitful rivalry between separate communities which is so characteristic of the whole life of the Grecian cities.

Even to call it "Greece" is indeed to employ a modern fiction: the citizens of that ancient culture thought of themselves as Athenians and Spartans and Thebans, not Greeks, alike in language and civilization but not in political stamp or rule.

Traditional historians write of the European period from the fall of Rome to the Renaissance as if nothing much were going on outside of the consolidation of feudal families into the monarchies of the subsequent nation-states. But that is like talking of the night as the presence of stars or the ocean as if it were only waves. What was going on throughout the continent from the Atlantic to the Urals, what kept European civilization alive for better than ten centuries, was the maintenance and development of small, independent communities—here in the form of Teutonic and Russian and Saxon villages with their popular councils and judicial elders, there as the medieval city-states with their guilds and brotherhoods and folkmotes, and over there as the chartered towns spread by the hundreds over France and Belgium with their special instruments of sovereignty and self-jurisdiction. Characteristic of the look of the continent were the divided cantons of what became Switzerland, beginning with the first democratic commune in Uri in the 1230s, a form that spread through dozens of villages in the fourteenth and fifteenth centuries and lasted until dominance by Napoleon at the end of the eighteenth century; at its height a typical canton, the independent Swiss Republic of the Three Leagues, covering about the area of Dallas, consisted of three loosely federated leagues, twenty-six sub-jurisdictions, forty-nine jurisdictional communes, and 227 autonomous neighborhoods—and, as an eighteenth-century traveler put it, "each village . . . each parish and each neighbourhood already constituted a tiny republic."

That Europe did eventually evolve some families designating themselves royal, and that some of those conquered vast areas of land they liked to call nations, and that the whole became a system of border-drawn nation-states such as we know today, does not mean that this was the tide and trend of that long era. Indeed, as between the statist tradition and the decentralist, these thousand years were clearly the period of the latter, into the fifteenth century in western European territories, in some places into the nineteenth century in eastern. No one has understood this period better than the Russian scientist and anarchist Peter Kropotkin, whose careful researches into its long-neglected intricacies, built upon the absolute explosion of interest in village government by scholars everywhere in the nineteenth century, have given us a telling picture of those centuries:

> Self-jurisdiction was the essential point [of the commune] and self-jurisdiction meant self-administration. . . . It had the right of war and peace, of federation and alliance with its neighbors. It was sovereign in its own affairs, and mixed with no others.
>
> In all its affairs the village commune was sovereign. Local custom was the law, and the plenary assembly of all the heads of family, men and women [!], was the judge, the only judge, in civil and criminal matters. . . .
>
> [In medieval towns] each street or parish had its popular assembly, its forum, its popular tribunal, its priest, its militia, its banner, and often its seal, the symbol of its sovereignty. Though federated with other streets it nevertheless maintained its independence.

This was the rule, mind, not the exception; it was exactly this self-governing community, through pestilence and war, the vicissitudes of nature and of kings, that sustained the many tens of millions of people of Europe for a millennium and more.

Nor did the tradition end with the rise of the nation-state. In many places it persisted quite a time—France did not outlaw local folkmotes until 1789, and Russian communes continued to exist in countless places until finally gutted by Stalin as late as the 1930s—and even in the age of nationalism it is not difficult to find, just below the surface, the roughly independent peasant village, the headstrong town, the self-minding neighborhood, in almost any country of Europe.

AND IN AMERICA. The decentralist tradition, manifested in a persistent anti-authoritarianism and a quite exuberant localism, is basic to the American character. (I am thinking of the European element here, but of course before that was the culture of the original Americans, almost everywhere communal, non-hierarchic, anti-institutional, and carefully localized.)

The Plymouth settlers, after all, were a proud and independent people who made the journey precisely to escape the press of the authoritarian state, and their original village was egalitarian enough, at least in its first two years, to have communal farming, the equal distribution of clothes and food, and cooking and laundry done in common. And when that first village tried to assert its control over such free spirits as Anne Hutchinson and Roger Williams, they simply moved on and started their own independent colonies—the beginning of a long, regular, native pattern of settlement that marked this land for at least three hundred years, until the Pacific stopped the march.

Others, too, who came here were anti-authoritarian by temperament, or tempering—the Quakers and the Mennonites, escaping state persecution, the freed convicts and indentured workers, the entrepreneurs and political free-thinkers seeking fresh territory. Even such modest governments as the colonies represented seemed to chafe such people, and their resistance climaxed in the "insurrections" of 1675–90, in response to which William and Mary granted new and more lenient colonial charters. That did not halt the movement toward independence, however, and even the kinds of concessions later offered by George III and his ministers—and they were generous—did not succeed in abating the strong separatist spirit. Resistance to unwanted laws and the flouting of colonial authority were common well before the Revolution itself, and riots and rebellions—the Regulator movement against the governments of the Carolinas in the 1760s, for example, and the Green Mountain Boys against the government of New York in the 1770s—were recurrent. These fledgling Americans wanted to be left alone, to sink their roots how and where they pleased.

The Revolution was precisely in this tradition, and the document that began it is permeated with the principles of the sanctity of community borrowed from the philosophers of the Scottish Enlightenment, of the primacy of the people over the state plucked from Rousseau, and of the inviolability of local governance that was largely ingrained in the Americans themselves. No better confirmation of these principles was needed than the experience of the colonies in the early years after the Declaration, when most of the British institutions had collapsed and many of their leaders fled, and yet the citizens went right on administering their own affairs, and successfully too. Largely through town meetings, common from Massachusetts to Virginia and not alone in New England, the settlements of the new country raised money and volunteers for the new army (in which, incidentally, the soldiers elected their own officers), organized militias for self-defense, and took care of the plowing and planting, the road-repairing and bridge-building, the schooling and policing. As Tom Paine was later to write:

> For upwards of two years from the commencement of the American War, and for a longer period in several of the American states, there were no established forms of government. The old governments had been abolished and the country was too much occupied in defence to employ its attention in establishing new governments; yet during this interval order and harmony were preserved as inviolate as in any country of Europe.

The government that eventually did take shape over these lands, under the Articles of Confederation, was little more than an extension, a federation, of these existing forms. The Articles, much maligned by statists and regularly misconstrued by textbooks written from the viewpoint of a later age, were "weak" enough, as conventional opinion has it, if by "weak" is meant that the affairs of the country would continue to be the stuff of the daily chaffer-mugger of the village square and the town meeting and not matters exclusively for professionals in some inaccessible capital; "weak" if by weak is meant that, in the words of the Articles' first and most basal provision, that "each state retains sovereignty, freedom, and independence"; "weak" if popular government be weak, if local control be weak, if direct democracy be weak. Such matters will always be murky, but there is excellent evidence that the greatest part of the free population supported the Articles wholeheartedly and was little interested in the drive for a stronger government that such people as Hamilton and Madison began pushing after the war. And even when the centralists and commercial interests pushed it through, the Constitution was approved only after the state legislators, "in order to prevent misconstruction or abuse of its powers," demanded that a Bill of Rights be added to it: the danger of the central government it was that was uppermost even in the minds of those who were constructing it.

Certainly those citizens who quickly came to feel its pinch had reason enough to look with suspicion on the new government. We remember them now as authors of the series of revolts—such as Shays's Rebellion in Massachusetts and the Whisky Rebellion in Pennsylvania—by local communities, mostly rural, that had just finished fighting for the right to run their affairs without the taxations and surcharges of an unrepresentative government and would, by God, do it again. Common to these revolts was the publicly stated desire to escape the hand of usurious banks, to establish local control over currency and taxation, and to select local officials (in the words of Shays's followers, to allow "each town to elect its own justices of the peace, each county its judges, and each military company its officers")—all demands that seemed reasonable enough in the light of the principles of the Revolution and the experience of the Articles. But such resistance was inevitably met with force and put down with considerable ferocity by Federal troops and state militias, with jail terms for the leaders who survived.

NOT THAT THE youthful United States government was a particularly autocratic one, not that at all. In fact it was run by men who saw themselves as truly libertarian, in service to those principles of federalism and republicanism that did much to spread power out from the center to the state capitals and the counties and towns. But it did not take more than a few dozen years before the acute Thomas Jefferson, who had done so much to assure that the new government would be restrained, began to fear that even this much centralism was beginning to rot the republic and remove the essential affairs of state from those who should of right be tangling with them. Around 1816, after having served his stint in the presidency perhaps not wisely nor too well, he began to revive an idea that had long been part of his political creed: ward government. A system of small "elementary republics," he began to feel—units of perhaps a hundred men or two, populations of 500 to 1,000 in all—was essential to the salvation of the American state, and a better alternative than his earlier notion of recurring revolutions ("a little rebellion, now and then"). What he urged on all who would hear him was "small republics" by which "every man in the State" could become "an acting member of the Common government, transacting in person a great portion of its rights and duties, subordinate indeed, yet important, and entirely within his competence." "Divide [government] among the many," he declared, so that each citizen may feel "that he is a participator in the government of affairs, not merely at an election one day in the year, but every day; when there shall not be a man in the State who will not be a member of some one of its councils, great or small, he will let the heart be torn out of his body sooner than his power wrested from him by a Caesar or a Bonaparte." Thus a nation of face-to-face democracy, of town meetings, of neighborhood government, where "the voice of the whole people would be fairly, fully, and peaceably expressed, discussed, and decided by the common reason" of every citizen, every day, everywhere.

The Jeffersonian formula was never tried, not even seriously debated in the land—Jefferson himself had effectively retired from public life and chose not to enter new lists at this late date, his passions not quite up to his convictions. And even as it was being voiced, the large shadow of the Federal and state government was moving slowly out to dim and extinguish the small lights of self-government that existed: one after another the towns and cities of the mid-Atlantic region abandoned town meetings or made them ritualistic annual affairs; the corporate form of city government, with mayors and councils, was pushed by conservatives as a way to keep the peace and put decisions into "responsible" hands; and the township system was downgraded in favor of greater power to the state and, to a somewhat lesser degree, Federal governments. It was, as historian Merrill Jensen has put it in his authoritative *The New Nation,* "the counter-revolution."

But the Jeffersonian ideal remained, in one form or another, the philosophic pole at one end of American politics throughout at least the first half of the nineteenth century. It was the guide for Thoreau ("That government is best which governs not at all") and for Emerson ("The less government we have, the better—the fewer laws, and the less confided power"), for Calhoun and the Carolina Nullificationists, and for both white abolitionists and black insurrectionaries who repeatedly defied state and Federal laws. At the time of the Civil War there were many who would say with Thoreau:

> This government never of itself furthered any enterprise, but by the alacrity with which it got out of the way. *It* does not keep the country free. *It* does not settle the West. *It* does not educate. The character inherent in the American people has done all that has been accomplished; and it would have done somewhat more, if the government had not sometimes got in its way.

The Civil War and its centralizing aftermath—wars are always centralizing; that's why governments have them—brought a temporary halt to this tradition, but it broke out again with real ferocity once more in the latter part of the century. Against the increasing monolithicity of government, industry, and political party, there sprang up a variety of movements diverse in cause but similar in resistance to the centralists: the Greenbacks, the Grangers, Oklahoma Socialists, Knights of Labor, Georgists, feminists, anarchists, communists, utopians, and above all the groups that fused to become, around 1880, the Populists. The Populists seemed almost to be that party Emerson had dreamed of—"fanatics in freedom: they hated tolls, taxes, turnpikes, banks, hierarchies, governors, yea, almost laws"—except that they added to this anti-authoritarianism a profound regard for communalism, cooperation, federation, networking, and localism, and actually developed an extraordinary variety of ventures to foster those. From Texas to the Carolinas, the Populists represented a major part of American politics, winning over whole towns in the South and West, achieving electoral victories in several states—in fact, gaining control of the North Carolina legislature in 1890 and passing laws for local self-government through county autonomy. It may in the end have proved a failure, but Populism was American to the grain, built upon the small farmers and artisans of the still-frontier settlements, rooted in the values of the local community and those enterprises—grange, church, school, newspaper, local shops—that gave them expression, set against all those Eastern institutions—industrial trusts, railroads, big-city machines, national banks—that in fact in time were to suffocate those enterprises.

With the first two decades of the twentieth century the triumph of Federal power was made manifest. The central government was acknowledged as supreme, its authority over its population's pockets (the Income Tax Amendment of 1913) and habits (the Prohibition Amendment of 1919) and even lives (the Selective Service Act of 1917) fully established. Those who resisted, on whatever grounds, were given a show of raw Federal power: the Espionage Act of 1917, the Immigration Acts of 1917–18, the Sedition Act of 1918, the Red Raids of 1919, the Palmer Raids of 1920, and countless little sins of commission in between. What happened then in the 1930s and 1940s, with the familiar events of New Deal consolidation, seemed only a natural extension of the past.

Even then, the decentralist spirit did not disappear. It found expression in the Agrarian movement of the 1920s and 1930s against corporate giantism and the growing industrialization and urbanization of the South; in the cooperative movement, both agricultural and consumer, that set roots then that are still in place today; and in a variety of home-grown radicalisms—of Ralph Borsodi's homesteaders, Arthur Morgan's communitarians, the Catholic distributists, the Technocrats, the folk-schoolers, the Black Mountain anarchists, and so on. Many of these eventually joined to found the journal *Free America,* which for ten difficult years from 1937 to 1946 represented, as nothing in this century has, the voice of decentralism in America, summed up in its founding creed:

> *Free America* stands for individual independence and believes that freedom can exist only in societies in which the great majority are the effective owners of tangible and productive property and in which group action is democratic. In order to achieve such a society, ownership, production, population, and government must be decentralized. *Free America* is therefore opposed to finance-capitalism, fascism, and communism.

IN RECENT YEARS the decentralist tradition has shaken itself awake again after the deadening hand of the Cold War years, aroused first by the black and student insurrections of the 1960s, expressions of a generation that sought, as the *Port Huron Statement* of the Students for a Democratic Society put it, "a democracy of individual participation, governed by two central aims: that the individual share in those social decisions determining the quality and direction of his life; that society be organized to encourage independence in men and provide the media for their common participation." The convincing disillusionment with government power as symbolized by Vietnam and Watergate of course added to this spirit, and then evidence of FBI and CIA repression, congressional illegalities, bureaucratic failures, and, with Carter, government ineptitude, added further, producing what ex-Congressman Michael Harrington in his farewell statement to the House in 1978 called "a tide of revulsion and rejection" that was "unique to our time, at least in its intensity." And on the other—its more positive—side, it has, as we have seen in some detail, produced a quite extensive resurgence of localism and a lively renewal of the quest for the authentic community.

The decentralist tradition, no matter what, will not die, for it is as wide in the American soul as the country is wide, as deep in the American psyche as the riches are deep. One may well wonder, with historian C. Vann Woodward, why, given its steady opposition to centralization and authority, the American environment nonetheless "should have proved so hospitable to those same tendencies in government, military, and business. A huge federal bureaucracy," he taunts, "a great military establishment, and multinational corporations, not to mention big labor, seem to have successfully surmounted all the handicaps to centralization." But the answer is as easy as it is revealing. The centralizing tendency has always existed in this country alongside the decentralizing—for every Anne Hutchinson a Governor Winthrop, for every Jefferson a Hamilton, for every Calhoun a Webster, for every Thoreau a Longfellow, for every Debs a Wilson, for every Borsodi a Tugwell, for every Brandeis a Frankfurter, for every Mumford a Schlesinger, for every Schumacher a Galbraith. And obviously this century, not only in this country but around the world, has belonged to the centralists and all their totalitarian machinery. But the decentralist movement, if it has not triumphed of late, has not disappeared either, and it seems to have survived Woodward's decades of bureaucracies and multinationals quite intact. Indeed, one gets the sense that these next few decades may provide its chance again; and more: that these decades offer the opportunity for it to establish its patterns—of localism, self-sufficiency, ecological harmony, and participatory democracy—for a long time to come.

I HAVE JUST picked up a copy of the local newspaper in the small community where I am privileged to spend half the year. Across six columns of the *Putnam County News and Recorder* ("We Are One Hundred Twelve Years Old But New Every Wednesday"), in what may be the largest headline I have ever seen it use, are trumpeted these words over the story of a recent public meeting with state and Federal officials:

PHILIPSTOWN TO THE BUREAUCRATS: LET US DECIDE WHAT IS BEST FOR US!

Not many papers nowadays still use exclamation points in their headlines, but somehow here the old tradition (LINCOLN SLAIN!) seems altogether appropriate. The issues at the meeting were in their way trivial—whether New York State could take more lands off the tax rolls, whether the Federal government could run its Appalachian Trail wherever it wished—but to county residents they seemed consequential, for in them were not only matters of pocketbook but matters of local pride. The officials were full of reason-whys and it's-best-for-yous, quite misapprehending the audience—"Congress has mandated the relocation of the trail," "this study was done under paragraph 242 of the 1978 law providing for a Hudson River Level B study"—while the residents time and again rose to ask why they hadn't been consulted, what right the bureaucrats had, when would the state stop squeezing tax monies, who in the state offices ever paid attention to letters and petitions. The byplay was ultimately inconsequential, the sense being that the bureaucrats would try to go right on doing what they had been doing, only holding their cards a little more closely to their chests, the residents resolved to keep on the watch and from now on take a peek over as many shoulders at as many hands as possible.

This is the voice of the decentralist tradition resonating once again, in words as old and familiar as "Don't Tread On Me": *Let us decide what is best for us.*

4

Society Without the State

In a healthy society of non-human primates, there are no nations, no extended organizations, no supraterritorial forms that might resemble a state. Troops of primates do have a social cohesion about them, creating various temporary leaders or hierarchies according to the function at hand—one for fighting, say, another for vigilance, a third for feeding—and establishing customs and groundrules of permissible behavior. But nowhere do these basically autonomous troops combine into larger associations, nowhere do members of the same species, though clearly knowing themselves to be essentially related, attempt to create large-scale units of social control. "Superorganizations, alliances made up of two or more troops," as anthropology expert John Pfeiffer notes, "have never been observed among baboons or any other nonhuman primates."

In a healthy brain, though there are many major processes operating at once, there is none, either physical or psychological, that is dominant. In the words of neurologist Gary Walter:

> We find no boss in the brain, no oligarchic ganglion or glandular Big Brother. Within our heads our very lives depend on equality of opportunity, on specialisation with versatility, on free communication and just restraint, a freedom without interference. Here too, local minorities can and do control their own means of production and expression in free and equal intercourse with their neighbours.

Only in the diseased and malfunctioning brain does one process ever become dominant.

In a healthy ecosystem, the various sets of animals, whether themselves organized as individuals, families, bands, or communal hives, get along with each other without the need of any system of authority or dominance—indeed, without structure or organization of any kind soever. No one species rules, not one even makes an attempt, and the only assertion of power has to do with territoriality—the claim of one or another species to a particular area to be left alone in. Each community in the system has its own methods of organization, its own habits and styles and food supplies, and none attempts to impose these on any other or to set itself up as the central source of power and sovereignty. Predation there will be within such an ecosystem, and some basic wariness by the fly of the frog and the frog of the snake and the snake of the hawk, but there is no *inter-special* warfare, no pseudo-Darwinian war of all against all; on the contrary, there is balance and adjustment, the broad cooperation of nature's communities with each other and with their particular environment. Independence, complexity, variety, flexibility—these are the characteristics of the healthy ecosystem, and, among all creatures, only the human has ever tried to transgress that principle.

IN A WORLD in which the decentralist tradition was allowed to proceed to its logical endpoint, in which separatism and localism were permitted to run their courses, the importance of the centralized state would of course diminish rapidly. It is not even difficult to see, in time, the beginning of that process long held dear by philosophers both of the right and the left, the "withering away of the state." For as local units took unto themselves the regulation of internal harmony and democratic governance, the point and purpose of the traditional nation-state would rapidly decline, its functions absorbed by local populations with a more precise perception of the problems and a more realistic understanding of their solutions.

Examples of societies that have lived, and lived long and well, without the trappings of the state are surprisingly common, once one begins combing through the scientific literature. In fact they are so common, occurring right throughout the Indian societies of both North and South America, through much of North Africa and almost all of the great region from the Sudan to the Kalahari, and throughout the islands of the South Pacific from Sumatra all the way to Polynesia, occurring among patrilineal as well as matrilineal societies, settled and pastoral as well as hunting and nomadic, large and scattered as well as small and cohesive, isolated and ingrown as well as confederative and cooperative, occurring in such variety and profusion that it comes to seem from the anthropological evidence that this is indeed the basic natural organization of human societies. As British anthropologist Aidan Southall has said about the historical spectrum, "People with state organizations were exceptional."

This is the way most humans have organized their societies since the very earliest beginnings of anything that could be called human, so long ago that there is a sense in which its patterns may be encoded in our genes. This is the way even more developed societies must have formed themselves since the beginnings of settled bands and tribes some 15,000 years ago. This is the way the greater part of all humanity must have lived even after a few isolated peoples began forming fixed hierarchies and chieftaincies and states some 5,000 years ago.

Clearly any method of living that has been so widespread and so long-lasting must have something going for it, must be considered in some way successful, even if largely for "primitive" people in "prehistoric" times. Its success in almost every instance seems to be due to a very simple mechanism: the social control exerted by those who feel themselves part of a single, cohesive, multi-entangled social group such that the transgression of one is likely to threaten the well-being of all. It needs no parliament to decree that my act of theft or murder is going to be wrenchingly disruptive of a community, and that if I have even a minimal sense of self-preservation I had better not commit such acts. It needs no chief or ruler to tell me that I must at times tend my neighbor's cattle and help him in his harvest, for it is perfectly obvious that my own survival depends on his doing the same for me. It needs no policeman or soldier to prevent me, nor prison to scare me, from acts of social disruption, for the social well-being that I shatter, the social peace I destroy, is my own. There is nothing to keep me, strictly speaking, from mayhem; yet there is nothing to propel me either, and everything to restrain me.

As Southall puts it in his important entry in the *International Encyclopedia of Social Sciences:*

> Fundamental responsibility for the maintenance of society itself is much more widely dispersed throughout its varied institutions and its whole population in stateless societies. . . . In stateless societies every man grows up with a practical and intuitive sense of his responsibility to maintain constantly throughout his life that part of the fabric of society at which at any time he is involved.

Or, more bluntly, Peter Farb about the Basarwi:

> They have an intuitive fear of violence because they know the social disruption it can cause in a small group. And they know that because of the poisoned arrows always at hand, an argument can quickly turn into a homicide. This explains why the Bushmen attach no honor or glory to fighting and aggression. Their culture is without tales of bravery, praise of aggressive manhood, ordeals of strength, or competitive sports.

What's missing from such societies, and what makes them seem so strange to our eyes—as to the eyes of the first Europeans who encountered and usually misreported them—is the concept of *power,* and hence of hierarchy, control, obedience, all the elements that are necessary to contrive a state. No one here has power, no one wants it, no one even thinks about it—or has a way to think about it. Competence, yes, ability, skill, these are all desirable, even the talent to lead in battle or in dance or in harvesting or in magic. But never does it imply power. French anthropologist Pierre Clastres has described the pathetic turn of Geronimo's career, after that Apache warrior had been a successful leader of troops in battle and tried to make himself into a chief with political power, demanding that the Apaches join him in further wars against the Mexicans: "He attempted to turn the tribe into the instrument of his desire, whereas before, by virtue of his competence as a warrior, he was the tribe's instrument." Quite naturally, the Apaches would have nothing to do with him, and he spent the next twenty years in silly, futile battles with a handful of followers, becoming a chief, and heroic, only in the eyes of the white myth-makers who never understood. Clastres's conclusion for the Apache, as for the dozens of Central American cultures he has studied: "One is confronted, then, by a vast constellation of societies in which the holders of what elsewhere would be called power are actually without power; where the political is determined as a domain beyond coercion and violence, beyond hierarchical subordination; where, in a word, no relationship of command-obedience is in force." Even those tribes that create leaders of a kind, people we might call chiefs, do not invest them with power:

> The chief has no authority at his disposal, no power of coercion, no means of giving an order. The chief is not a commander; the people of the tribe are under no obligation to obey. *The space of the chieftainship is not the locus of power,* and the "profile" of the primitive chief in no way foreshadows that of a future despot. . . .
>
> Mainly responsible for resolving the conflicts that can surface between individuals, families, lineages, and so forth, the chief has to rely on nothing more than the prestige accorded him by the society to restore order and harmony. But prestige does not signify power, certainly, and the means the chief possesses for performing his task of peacemaker are limited to the use of speech . . . *the chief's word carries no force of law.*

Even the word "chief," it should be noted, is European. Such cultures as these have neither the word nor the concept.

HOW DOES SUCH a society work, how *can* it work? What is to prevent, as Hobbes envisioned life for such peoples, "continual fear and danger of violent death" and "a war as is of every man against every man"?

Let us take a look at one such society, the Dinka of the Southern Sudan. At the time of the most careful study of these people (by anthropologist Godfrey Lienhardt, between 1947 and 1951) they numbered about 900,000—proving stateless societies can encompass very sizeable populations—with a common language and common cultural traits but no sense of being in any way united, of being a "nation." They were divided into some twenty-five loose "tribal groups" averaging about 35,000 each (though ranging from 3,000 to 150,000), each with its own clear but fluctuating territory, but so far from there being any union among these groups, there simply wasn't any organized way for them to *relate* to each other, and if any two had a dispute they weren't even organized in such a way that they could *fight* each other. Within these tribal groups were identifiable tribes, with populations that averaged around 10,000 or so, and this was the basic unit both for defense—the limit of the people who would come to your aid if attacked from the outside—and for the mediation of disputes—the limit of those who could agree to a settlement by compromise and compensation.

But even these tribes had no political shape, no permanent bodies, no structured rules. It was in still smaller units, "subtribes" that would contain about 1,000 people living in the same village, swelling to 3,000 at the most, that anything representing a political office existed. Each subtribe would have a priest ("spearmaster"), who was the interpreter of local magic and tradition and the mediator in interfamilial disputes, and a warrior, who was expected to lead the village in the occasional fights against other subtribes (which were fought with clubs only, nothing lethal). But though these positions were hereditary, attached to a particular lineage that was thought to supply the best of each officer, the process was fluid, and if someone in the job turned out to be an inept priest or a timid warrior they were simply bypassed, with no ceremony about it, in favor of someone who could measure up.

Loose as it was, this system seems to have worked with persistent success over many generations and going back who-knows-how-many centuries. Allegiance to family, village, tribe, was strong enough to establish a sense of community such that the disruption was so anguishing it was felt by all and avoided as much as possible. If disputes did arise between families—perhaps over cattle, for the Dinka were herders—and could not be settled by the family heads, they could be taken to the spearmaster to hear his opinion; and though that would only be a recommendation and not a judgment, it would often be followed as a simple face-saving device for all. Disputes between villages, the subtribes, were likely to be less frequent but more heated when they came, occasionally escalating to retributive theft and even murder; such feuds, seriously disrupting the valued harmony of the region, would normally occasion the gathering of the tribal elders—not any fixed officers or legislators or judges, merely the oldest males around, meeting informally—who would hear both sides, offer opinions, try to set compensation (normally payments of cattle), and do what they could to get everyone back to the tenor of their ways. They would have no body of laws to refer to, of course, merely the accumulated wisdom and accepted tradition of the tribe, and they had no method of enforcement other than the abiding morality of the community. (Plus, to be sure, the abiding custom that the aggrieved villagers would go and grab the recommended number of compensatory cattle from the neighboring village's fields.)

Thus government for the Dinka was self-government in the very best sense. Common rules and practices were observed because they were seen to be, in normal daily life, the most harmonious for the individual, the family, the village, the tribe; they did not have to be legislated and codified and policed because they had everyday meaning. Transgression was viewed as disrupting the safety, security, and harmony of the unit, leading to a threat to the life not only of any one individual but potentially of the whole tribe. Such disputes as might arise could be handled through local machinery brought into temporary operation for that single occasion and then disbanded, and warfare was such a rarity—absolutely unheard-of between villages, for example—that to have kept a standing army would have been egregiously wasteful. Thus, in Dinka eyes, a state would be superfluous: with a system as neat as theirs, what earthly use could there be for lawmakers and kings and sheriffs and soldiers?

Obviously such a system must have a well-tuned sense of balance, of limits, in order to keep working, and that in turn was based on two other essential features of the stateless tribe: self-sufficiency and limited size.

In the Dinka economy there was no place for the individual accumulation of property or for the market system that extends from it. Aside from the most minimal personal possessions—a few clothes, some jewelry, maybe a piece of furniture—no Dinka ever had, or seemed to want, additional property: what, after all, could anyone *do* with such additional baggage, especially when moving around to find suitable watering holes in the dry season? Similarly, since there was no exchange of goods with any other society and each village wanted only enough cattle and food to guarantee security in lean years, why would anyone want to build up a surplus of goods, what could any village *do* with it? The economy was entirely local, entirely self-reliant, and thereby entirely free of the pressures that come into play with the purposeful and unending accumulation and exchange of properties. And more: a society (hardly what I would call "primitive") that does not amass personal property does not normally build up an economic hierarchy, one that does not accumulate surplus goods for trading and "becoming richer" does not so easily go to war, coveting other people's lands and goods and "riches." Self-sufficiency, as we noted before, inherently tends toward stability.

Similarly, in the Dinka system as in most stateless societies, there was the conscious and regular limitation of the size of the constituent units. Since order was to a large degree dependent on a sense of communality—one person being known to, and the concern of, all—

there was an obvious need to hold the basic population below the point where that sense became diffuse and fragile. Basic to the Dinka polity, therefore, was the easy fragmentation and fission of villages and tribes; as Lienhardt notes:

> It is part of Dinka political theory that when a subtribe for some reason prospers and grows large, it tends to draw apart politically from the tribe of which it was a part and behave like a distinct tribe. The sections of a larger subtribe similarly are thought to grow politically more distant from each other as they grow larger, so that a large and prosperous section of a subtribe may break away from the other sections, with its own master of the fishing spear, its own separate wet-season camp, and eventually its own age-sets. In the Dinka view, the tendency is always for their political segments . . . to grow apart from each other in the course of time and through the increase in population which they suppose time to bring.

Or as the Dinka say with sophisticated simplicity when asked to explain why a village divides: "It became too big, so it separated." And this system pertained, mind, among quite large populations—the Dinka numbered 900,000, the Tiv, in Nigeria, at least 800,000, the Lugbara of Uganda perhaps 240,000.

Absolutely fundamental to the stateless tribe, in other words, is, in both demographic and economic terms, the human scale.

NOR SHOULD ONE think that it is only among such archaic peoples that forms of statelessness exist. One can encounter much the same sort of thing, sometimes with a few more touches of formality, often enough in the histories of peoples of acknowledged sophistication.

The Greeks, for example. From the eighth century B.C. on to as late as the fourth, the great majority of Hellenic villages and cities operated without kings, without ruling priesthoods, without fixed aristocracies, organizing their daily affairs through assemblies of citizens and popularly elected leaders. The Greek *polis* may have created more civic officers than the Dinka did, and it may have relied to a greater extent on codified law, but it had the same avoidance of authoritarian forms, the same distrust of chiefly powers, the same dependence upon popular assemblies for the settlement of disputes. Even Athens, probably more encumbered than most Greek cities because of its size and prominence, studiously avoided the governmental trappings of the riparian empires that had preceded it. It was governed not by imperial wizards and pharaohs but by a public assembly, the *ecclesia,* a regular gathering of all (male) citizens who wished to participate, which made all important decisions and set all guiding policies and then elected a rotating body of fellow-citizens, the Council of 500, and an assortment of citizen committees to carry them out. So far from having a permanent monarch, Athens until Periclean times generally selected a new "president"—in effect the mayor—every single day by lot from among the Council of 500, and his duties were totally symbolic and ceremonial. And none doubts the sophistication of *that* culture.

Similarly, the medieval cities were without anything that could be thought of as a state, certainly nothing recognized beyond their narrow borders. They did have their forms of governance, to be sure, somewhat more complicated than the Dinka, with town officers and magistrates and written charters. But they knew no authority beyond their fortified walls, no lord or nation or king (which is why even today the British monarch theoretically has to ask permission to enter the City of London). They were entirely self-administering, self-governing, self-adjudicating (in most places, until the fifteenth century, local priests and often local feudal lords were subject to the decisions of the city's folkmote or juries). They recognized only the authority of their own citywide folkmotes, open to all citizens, which in turn generally selected the military *defensor* and the judicial *magisters* for the town (and it is precisely because they did not have such folkmotes that certain cities—Paris and Moscow for sure, and possibly London and Lisbon as well—were chosen by certain feudal lords and bishops to be the capitals of their growing dominions). And they regarded themselves as completely sovereign, even granting their own citizenship (as Braudel points out, each city "was the classic type of the closed town, a self-sufficient unit, an exclusive Lilliputian native land").

Even then, the average independent city hardly resembled what we would think of as a "state," so decentralized were the effective functions of municipal life. Especially in northern and western Europe, each neighborhood or parish normally had its own forms of governance for most day-to-day affairs, its own assembly meeting once a month or more, its own tribunal, priest, militia, and emblems. In addition each trade had its own guild to see to the welfare, protection, and health of its members, with its own assembly, its own courts and punishments, and its own military force armed with its own bows or guns. It was only disputes between the neighborhoods or disagreements among the guilds that went up to the consideration of city officers, and most of these were settled by schemes of local custom and adjudication as honored and respected as those among the Dinka.

There are many other examples of stateless societies to be found in more contemporary history. In sixteenth-century Ireland, we are told, there was "no legislature, no bailiffs, no police, no public enforcement of justice . . . no trace of State-administered justice." The Swiss Confederation in the eighteenth and nineteenth centuries was so formless that it had no central government at all but rather an obedient secretariat of two regular officials and a few aides that moved around the country from canton to canton (the entire apparatus of the "state" was once stuck in the snow in a single railway coach near Mellingen), and even though those cantons enjoyed almost total internal autonomy most matters of substance in them were in fact decided by "sovereign villages." Such religious communities as the Quakers, Dukhobors, Mennonites, and Hutterites have lasted for centuries without any trappings of the state—nor do most of them formally recognize the civil states around them—and with their own law and governance as established by religion, custom, community, and popular democratic assemblies.

But it is probably among the classic New England towns of the eighteenth century that the best modern example of the essentially stateless society is to be found.

ALTHOUGH IN STRICTLY legal terms the towns of the province of Massachusetts were part of a considerable state, consisting of a governor and legislature in Boston and a king and parliament in London, in any practical sense the provincial charter of 1691 established a territory without a state, without effective central authority of any kind. Such governance as there was—and it was not much intrusive into most citizens' lives—was grounded in the towns, the small, scattered settlements, very few with more than a thousand people in all, that stretched from the Atlantic to the Berkshires. As Michael Zuckerman describes it in his excellent *Peaceable Kingdoms* (an awkward misnomer that, for they had nothing to do with kings), these towns "were free enough of both Boston and London to enjoy an autonomy which, if never total, was always genuine," enabling them to establish "local custom and usage as guides for the management of most facets" of their politics; and they were even able to "set the law aside, without ever a word that went beyond the town boundaries, when the law proved inconvenient for local purposes," so that "effective law in the provincial community was ultimately only what each particular place would abide by."

> Their operative ideas of appropriate authority and meet relations among men were those they drew from daily conduct in the communities . . . The town became the only governmental agency with more than a sporadic impact on the lives of its residents, and as it came to provide most of them with the only essential experience of public authority they would ever know, it encroached ineluctably upon the traditional prerogatives of the central government. As the Revolution eventually revealed, the very maintenance of order in the province had come to depend, in the eighteenth century, upon the towns.

Talk about the decentralist tradition.

Not only did these towns manage their own affairs for upwards of a century, they did so with extraordinary peace and harmony. The local church, the local market, the town schools, the town meeting, all taught concord and control, all emphasized harmony and homogeneity. And with such success that Zuckerman figures on average only one town out of the 200 in any given year had a quarrel serious enough to take to the legislature in Boston to settle for it.

And of course there was no need for official agencies of public control and public order. There was, to be sure, a constable for each community, but he was a person elected annually, seldom in office more than a year at a time, almost always serving with great reluctance, and having no enforcement powers whatsoever beyond those sanctioned by the town as a whole: "Constables . . . could command compliance only when almost everyone was prepared to give it anyway." There were in certain centers courts of law, but they were seldom used (it is said that there were probably no more than a dozen lawyers in the whole province in 1750, and most of those in Boston), and they were without any real powers of enforcement; most parties in dispute preferred to establish their own arbitration committees, call together the church elders, or even ask in impartial men from neighboring communities. What kept the peace, what maintained order, were the communal bonds and religious ties, and behind it all, the town meeting.

The town meeting, despite its look of formality, was an instrument hardly more statist than the council of elders of the Dinka. It was nothing more—though nothing less—than the regular occasion for the expression of community solidarity through popular decision-making, about everything that affected the town, from the upkeep of the roads and the tax rates to the amount of firewood to be cut for the minister. All decisions were normally unanimous—what, after all, would be the point of a victory by a majority that left unhappy a minority of neighbors and friends?—and since "the recalcitrant could not be compelled to adhere to the common course of action," as Zuckerman notes, "the common course of action had to be shaped so as to leave none recalcitrant, and that was the vital function of the New England town meeting." Even more than in the Greek or medieval city, this was "government by consent of the governed"—always remembering that women were regrettably, except in their lobbying function, excluded—for this depended on the open consent, active, perceived consent, by all, not merely a majoritarian segment of, the governed. One can read over and over again in the records of town meetings throughout Massachusetts, and similarly throughout the rest of New England, recurrent variations on this theme: votes "by the free and united consent of the whole," "by all the voters present," or this especially felicitous one from Weston, "By a full and Unanimous Vote that they are Easie and Satisfied With What they have Done." It was this unit, this expression of the real general will, that was the essential political embodiment of the New England people for generations on end.

Here at the very beginnings of American society, here at the fount of the American soul, we find the most developed, the most settled, the most reasonable demonstration of the worth and happiness of life without the state.

WE DO HAVE to go back for some time in history—except for those few remaining examples of the archaic state in Africa and Australia—to find examples of societies without the state, that being the particular burden of the contemporary world. And yet when we do, through anthropologists and archaeologists and historians, we almost never fail to find an extraordinary record of stability and equilibrium that suggests, and goes a long way to proving, that the human animal, without the patterns of the state and the pillars of authority, tends to peace not war, to self-regulation not chaos, to cooperation not dissension, to harmony not violence, to order not disarray. Indeed, looking at the long human record, it is hard not to find an increase in all of the *latter* characteristics *with* the development of the state.

It took me many years to understand it, but I do now—the remark of Proudhon that liberty is not the daughter of order, but the mother.

5

The "Necessity" of the State

BUT "LET US put away our infantile fantasies, the yearning to return to an infancy of the species that never was, where mankind existed in small and totally autonomous units like tribes or villages and practiced primary democracy and knew peace and harmony. It may be that if most inhabitants and the technical know-how already existing on humanity's fragile and crowded spaceship are consumed in a thermonuclear Judgment Day, the survivors might be able to exist, for a time, in tiny, fully sovereign, widely separated units. Only a madman would think that a solution. No, because the species *is* interdependent, giant political systems are a necessity."

So scolds Robert Dahl, and he is such an eminent political scientist that it behooves us, I suppose, to listen. Of course, as we have just seen, it's not true that a small village of peace and harmony "never was," nor is there any reason to believe that the world's population would have to be reduced to a handful to accommodate separate sovereign units: remember, the entire population of the world could fit into the United States with a population density less than that of England (and onto our prime agricultural land with a density less than that of Malta). And it is not easy to see in what way this "interdependence" works—the Manchurian peasant and the Chilean barber? the Ghanaian professor and the Alaskan fishermen?—or why it should necessitate "giant systems" even if it did pertain.

Still, let us put these quibbles aside, for Dahl is a learned man and he may indeed be expressing those resistant feelings any American might experience on being told that it is possible to live without a state:

> As for making all large political systems vanish into thin air, when the silk scarf is pulled away there in full sight are matters that cannot be handled by completely autonomous communes, neighborhoods, or villages: matters of trade, tariffs, unemployment, health, pollution,

> nuclear energy, discrimination, civil liberties, freedom of movement, not to say the whole tragic range of historically given problems like the threat of war and aggression, the presence of great military establishments around the world, the danger of annihilation. These problems inexorably impel us to larger and more inclusive units of government, not to small and totally autonomous units.

Again, we must take the man seriously, though again I do sense some confusion. Aren't the problems of war and aggression and nuclear energy and tariff barriers *caused* rather than solved by big governments? (They certainly *haven't* solved them, in any case, have they?) Surely if there were no giant governments there would be no "great military establishments" around the world and a good deal less "danger of annihilation." And if we have in fact been impelled to "larger and more inclusive units of government" because in that way we would solve the problems of unemployment, pollution, and discrimination, it does not take any special genius to show that it seems to have been rather futile, or at the very best inadequate; if to provide us with high levels of health care and a profitable balance of trade, it does seem to have failed, or at the best proven highly erratic.

I would be readier to see the end of those "infantile fantasies," as Dahl calls them, if he did not put up against them so many adult ones of his own.

BUT WE SHOULD confront the philosophy that stands behind the Dahlian attack, because in truth the line of criticism is common enough and deserves examination. Historically there have been four general arguments for the necessity of the state (and, in modern terms, for the necessity of large governments):

1. To provide defense in case of attack and to guarantee peace and security.
2. To provide economic regulation and development.
3. To provide public services beyond the competence of the community and the individual.
4. To assure social justice and protect the rights of the individual.

These arguments have a persuasive feel to them—they should, they've been drummed into us for centuries—but it is fair to ask: do they stand up? are they sufficient justifications for the existence of the state? and do they refute the desirability of small-scale politics?

DEFENSE

TRADITIONALLY THE STATE'S initial claim to necessity was that it prevented warfare by erecting firm barriers of defense. As we have seen, this is tommyrot: the Law of Government Size shows that the duration and severity of war has always increased with an increase in state power. Larger states, far from providing peace, merely provide larger wars, having more human and material resources to pour into them. War is the health of the state, as Randolph Bourne has said, precisely because it provides the excuse for increased state power and the means by which to achieve it. (Victory, by the same token, is incidental: many losing nations emerge from wars with more powerful governments.) In fact what goes by the name of the defense of the citizens by the state may really be better thought of as the other way around.

Insofar as "defense" implies security, the state is the instrument least capable of providing this. Indeed, in the contemporary world of massed nuclear weapons, there simply *is* no defense of the citizens from a hostile aggressor—if Russia, for example, should unleash all of its missiles against this country, the only "defense" we would have would be to launch ours back again, guaranteeing the population nothing but nuclear holocaust. The very presence of the state and its huge pile of threatening weapons endangers rather than protects the populace, by inviting pre-emptive strikes; the fact that the U.S. is now sitting on something like 8,900 hydrogen bombs, the equivalent of 600,000 Hiroshimas, cannot be said to provide the ordinary person with the sense of security and safety implied by the word "defense." (Particularly since, according to a 1977 Brookings Institution report, between 1945 and 1976 Russia and America threatened military action not just once or twice, not just every other year or so, but 330 times, an average of nearly once a month.)

Moreover, in the course of attempting to provide its defense the state exercises its own forms of coercion and violence. It conscripts young people at will and takes to itself the right to jail or kill those who resist conscription or desert after induction; it forces the wider population to pay for warfare and its preparation through increased taxation, which has never in history been paid voluntarily; it amasses an army that denies individual rights and freedoms through accepted methods of authoritarian control. And such a state, preoccupied with defense, begins to justify all acts, however dangerous, in its name (as Presidents justify all crimes with "national security")—when all you own is a hammer, as Mark Twain once said, all problems begin to look like nails.

I am not positing a polyannalysis of human nature that argues all warfare will disappear and all defense become unnecessary in a world of small societies. I am only suggesting there are a number of reasons to believe that the occurrence and severity of warfare would be diminished.

To begin with, it is likely that a small society, particularly if concerned with a steady-state equilibrium, would not amass the kinds of glittering riches that would attract some predatory outsider; the riches of the stable community are likely to be in its air, its range of foods, its quiet streets, its heightened participation, not the sorts of things inspiring envy and conquest. By the same token, a small society already enjoying stability and some prosperity might well be reluctant to sacrifice that for a war whose economic benefits would be so unclear; as the University of Chicago's famed economist Henry Simons once put it, "No large group anywhere can possibly gain enough from redistributing wealth to compensate for its predictable income losses from the consequent disorganization of production." Nor would such a society particularly want to divert from its economy the kind of human and material resources that would be necessary for it to amass a strong fighting force capable of successful aggression.

Yet all that supposes rationality, and since the history of warfare is marked just as much by madness as by calculation, we would have to expect that somewhere war is bound to erupt even when it seems most illogical. Still, the very nature of that warfare depends on the size of the society waging it: with limited populations and limited resources for weaponry, wars cannot have the sweep and severity they do when waged by powerful nation-states. Nuclear warfare would certainly be almost impossible, not only because of the expense of such weaponry would be beyond any stable economy but, at bottom, because no small society could think of sacrificing millions of people and still surviving in the way current superpowers can.

Historical evidence is abundant to show that even the wars that do take place in small societies tend to be—well, if not pacific, at least constrained. A typical account in the anthropological literature is the one by Australian scientists C.W.M. Hart and A. R. Pilling, in *The Tiwi of North Australia,* describing how two bands declared war, amassed their armies, met around a clearing at dawn, and then began throwing spears at each other—but with the *oldest* men throwing the spears, none too accurately, the oldest women careening around trading insults, and as a result "not infrequently the person hit was some innocent noncombatant or one of the screaming old women who weaved through the fighting men, yelling obscenities at everybody, and whose reflexes for dodging spears were not as fast as those of the men." Stateless or not, such small bands and tribes rarely engaged in warfare for plunder or conquest, and territorial aggression and enslavement is practically unknown. Similarly, we have evidence from early Greece that at least some towns and cities worked out ways to make wars more like athletic events than battles-to-the-death, fighting to "maintain honor" or display manliness with games and contests; as early as the seventh century a "war" between Khalkhis and Eretria was fought out as a kind of wrestling contest, apparently quite vicious and serious but managed without spears, arrows, or slings and with the only stakes being honor, pride, and boasting rights.

And even the small regimes of the late medieval period, though undeniably and elaborately "stated," fought their wars at scales much smaller than our Hollywood conceptions would have led us to believe. Three or four knights and several dozen crossbowmen would make up a standing army, battles would commonly rage for no more than an hour or two before everyone was exhausted, and whole wars would be over in a month, the casualties rarely more than one or two percent of the fighting force. The Duke of Tyrol, it is said, declared war on the Margrave of Bavaria over some real or pretended slight, their war went on for two weeks, one man was killed, one village overrun, one tavern emptied of its wine, and in the end the Margrave paid a hundred thalers in reparations. And however many minor feuds and wars there may have been in a Middle Europe of duchies and principalities and republics and palatinates, that was as nothing compared to the wars wrought when they were all united into the *reichs* of Germany.

In a world of small societies, in fact, size itself would likely act as some kind of deterrent to aggression. Who could picture an American continent composed of even fifty different nations deciding to pool its resources to fight a little nation around the world in the jungles of Southeast Asia? Would Texas, say, or Ohio, be sufficiently rapacious—or, let's say, heroic—to want to mobilize its forces and send them across the Pacific to save Saigon from Communism? It is hard to see how the citizens of, for example, Bavaria, no matter how bellicose they might be, could have such a grievance against the citizens of Brittany that they would want to travel half a continent away to fight about it—though as we know when those citizens are combined into the larger states of Germany and France they have no trouble going at each other's throats with some regularity. There is something in the very size of the modern nation-state that leads it not into tranquillity but into temptation, and even when it builds up power that is supposed to be purely defensive it always seems to have a way of turning it offensive without much trouble. The Germans have a saying that fits: *Gelegenheit macht Diebe*—opportunity makes thieves.

Yet suppose that somehow a single large superpower should appear (or remain) in a world of smaller communities. What is to be done with the legitimate fear that, in Robert Dahl's words, "in a world of microstates, let there arise only one large and aggressive state and the microrepublics are doomed"?

Historically the response of small states to the threat of such large-scale aggression has been temporary confederation and mutual defense, and indeed the simple threat of such unity, in the form of defense treaties and leagues and alliances, has sometimes been a sufficient deterrent. The record of medieval Europe in fact, during a period of small territories confronted by rising states, is absolutely teeming with leagues and federations and compacts and associations among towns and cities that joined together to establish peace and to resist the martial incursions from the outside. The Lombardian and Tuscan Leagues in

Italy, the Westphalian, Rhenish, Swabian, and Hanseatic in Germany, the Lyonnais and the Marseillaise in France, the Flemish, the Swiss, the Dutch, everywhere small towns came together with independent cities, the cities united with each other throughout a region, and though experiences naturally differed, the rule was of long periods, several generations, of unperturbed calm that were broken seldom, if ever, by the federated small communities themselves.

Moreover, the difficulties for any large power trying to subdue a host of smaller societies are truly formidable and would be additionally so if those societies, in a human-scale world, were efficiently governed, harmonious and homogeneous, and concertedly self-protective. The problems that Nazi Germany had controlling a Europe of large nation-states were bad enough, but they would have been infinitely greater if each little community had been independent, without connections to centralized systems of administration and control, with effective traditions of local autonomy and defense. The material game of conquering—and controlling—a small society that offered a great deal in the way of resistance and very little in the way of exploitable riches would hardly seem worth the military candle.

As recent history, indeed, suggests. Despite the existence of a number of enormous states, the events of the contemporary world give little support to the idea that Gargantuas are out there dying to subdue all the Lilliputs. Lithuania, Tibet, Goa—yes, there are some such instances, though in each of those cases the usurper had some legitimate prior claims to the territory. But in general the pattern has been one of the division and independence of states, an increasing number of smaller states, not of conquest and subjugation by the superpowers. Scores of quite tiny states exist, most under the economic pull of one large state or another but politically autonomous and without much fear of being taken over. And some of the smallest states in the world—Andorra, San Marino, Liechtenstein, Switzerland—have existed for many centuries quite untouched and unthreatened.

In sum, the long human record suggests that the problem of defense and warfare is exacerbated, not solved, by the large state, and that smaller societies, especially those without governmental apparatus, tend to engage in fighting less and with less violent consequences. Indicating that a world of human-scale polities would not be a world without its conflicts and disputations, but would likely be a world of comparative stability.

ECONOMIC DEVELOPMENT

THE JUSTIFICATION FOR the state in classical economic theory is that it is supposed to stabilize economies over wide geographic areas, standardize currencies and measurements, establish protective tariff barriers and regulate international trade, prime the productive pumps for sustained growth and steady employment, and ultimately promote economic development and stability. To which we may add a contemporary responsibility, the regulation of businesses on behalf of the public so that it is not recklessly overcharged and endangered, fleeced and fummeled.

Of course the easy response is that if *this* is what the state is so vital for, it does not seem to have proved its worth. What has characterized all large national governments, particularly in the twentieth century, is their clear *inability* to provide economic stability, security, or employment, to secure people against the dangers of depression and inflation, in either capitalist or "socialist" worlds. A nation like the United States, assumed to be the most powerful of all, though it has a truly mighty GNP and has recently enjoyed a period of high imperial prosperity, has not been able to forestall regular depressions, prevent inflation, maintain the dollar, regulate tariffs to the benefit of national industries, provide either jobs or income security, create a stable retirement system, or manage any but the most meager protections against the corporate contaminations of food, air, water, waste, or soil.

Still, it is perfectly true that national governments do issue and protect a standard currency—the only trouble is that this is of itself one of the major causes of inflation, and always has been, and the advantages of it could be as equally well enjoyed by a congeries of cooperating communities accepting certain common tokens—much as the cities of Europe agreed to accept the coins of Florence (hence the florentine) in the thirteenth and fourteenth centuries. Nor do these governments ever manage to keep their currencies stable for very long, as theory would dictate—as the worth of the dollar would testify. Moreover, as long as there is mutuality it is possible to have a great variety of currencies in one very small area: nothing collapses in northern Europe in spite of the fact that there are *nine* major currencies (plus such international currencies as the dollar and drawing notes) operating within a 400-mile radius of Copenhagen, an area about the size of the state of Texas.

Well, then, what about tariffs—are not nations necessary to maintain them? Right enough—only the reason that tariffs exist in the first place is to protect the state, enable it to develop while new industries grow within its boundaries, for its own aggrandizement; the advantages to the consumer are negligible, if existent. Like armies, tariffs are there to safeguard the state, not vice versa.

International trade? Nothing that *couldn't* be done without the state, as is amply demonstrated by the history of Europe between 1000 and 1500, a period of the most extensive and elaborate sort of trade relations, fundamental to the development of capitalism, quite separate from any national machineries. In fact, *is* being done—for the operations of the current multinational corporations, as any number of studies have shown, are essentially beyond the control of any serious governmental regulation of any kind.

Pump-priming? Governments have become better at this since the advent of Keynes, but there is not one of them, including those that Keynes advised, that has succeeded in creating unabating growth without regular severe downturns, not one of them, indeed, that has succeeded in directing the flow of those pumps out of the front yards of the rich and toward the backyards of the poor. And might a process efficient at sustaining growth—even if the nation-state could achieve it—be outmoded in a world where it is not growth but stability we wish to sustain, where growth has in fact become a threat?

Shortages? The idea that big governments are good for monitoring and allocating precious resources—as with gasoline shortages, for example—so as to distribute them fairly to all has a nice socialist ring about it and is John Kenneth Galbraith's favorite idea for the 1980s ("more big government"). But those who remember the government's skill at monitoring and planning Vietnam and swine-flu shots and public housing, and—need I go on?—may not relish the prospect of its handling irreplaceable resources.

Regulation? Aside from added costs to consumer and taxpayer, the trouble with governmental regulation is that it is always a catch-up operation, fighting the problem at the wrong end *after* the damage has been done. If the only criteria in the economy are those that capitalism dictates, as they are, then of course there will be those problems that no amount of government law or supervision can correct; if the primary purpose of the government is to protect the smooth workings of the corporate system, as it is, then of course its duodenary attempt to restrain its excesses by patchwork edicts is always doomed to failure. Thus the famous actions against cartels and monopolies have proved insignificant, as the existence and growth of conglomerates and oligopolies makes clear; the regulation of public utilities has been almost useless, as the existence of telephone gougers and electricity polluters, the awful failure of nuclear plants, makes obvious. But wouldn't the consumer be in ever more trouble *without* government regulation, however expensive and inept? Perhaps so. But probably not if at the same time the myriad government supports favoring large (and therefore largely unresponsive) businesses were withdrawn and the small firms in the market had to offer—as they did in an earlier age—"good goods at a fair price" in order to compete successfully. (It is not competition but the *lack* of it that fosters abuses, not the neighborhood butchers but the large meatpackers that produced the fetid *Jungle.)* Certainly not if in each office and shop the full range of workers made self-managed policies, in each community the full range of citizens set democratic decisions, as to what products and services would be sold.

One needn't claim that economic relations among small and self-regarding communities are going to be untrammeled bliss to reject the notion that state supervision will solve more difficulties than it manufactures. Yes, the kinds of cooperation within a community, the kinds of federations among communities, would, in the absence of a state, take skill and savvy. Yes, careful balances would have to be wrought between upriver towns and downriver cities (the word "rivalry," after all, comes from the Latin *rivus,* for river); between the community that had surplus steel and the community with surplus grain; between the city with all the copper and the other with all the zinc. But it need not be all that complicated, nor need it entail some *deus ex civitas.* After all, the stamps of all 152 independent nations of the world are recognized by all the others and the mail goes around the world without any state to guarantee it; the railroad cars of thirty-one sovereign nations travel the same rails for 4 million square miles throughout Europe without any state, not even any central agency, to oversee them. International telephone and telegraph systems, worldwide maritime regulations, international airline agreements, global currency exchanges, and dozens of other forms of cooperative arrangements all prove to be successful without any universal state, or anything near it, to supervise them. Would the cooperation of a half-a-dozen small communities in a single limited bioregion be all that hard to effect?

PUBLIC SERVICES

As Robert Dahl puts it, the state is ultimately necessary on grounds of the "criterion of competence": "Only the nation-state has the capacity to respond fully to collective preferences," only the state can supply public goods in the common interest, which individuals and communities won't do by themselves. Who but the state can control population growth, say, or feed the starving, secure public health, build highway systems, provide welfare, offer disaster relief, distribute scarce supplies? Who but the state—to pick the most commonly cited example—can control pollution?

This is the way Dahl poses the problem, given a polluted Lake Michigan:

> Let us assume, without arguing the matter, that the capacity to reduce pollution in Lake Michigan is, up to some threshold, a function of size. A citizen who wants the lake cleaned up might well reason as follows: If the authority to deal with pollution is left to local governments, the job is unlikely to be accomplished even if a majority of people around the lake—not to say in the whole country—would like it cleaned up. For those favoring vigorous pollution control may be in a minority in some towns. Consequently that town will probably continue to pollute the lake. Thus only a regional or perhaps even a federal agency, responsible to regional or national opinion, can be effective. Of course I am bound to have only negligible influence on the decisions of such an agency,

whereas I might be highly influential in my own town. Nonetheless, on balance I prefer the agency to the present system of decentralized and uncoordinated units.

The hidden assumption here, of course, is that the Federal government is in fact a solution. In real life it doesn't quite work out that way. The first problem is to get the government to take any action at all, to rouse the politicians or the bureaucrats somehow, to force the courts to intervene, and that may be an expensive and time-consuming task for any individual, especially since the object of the appeal is probably 600 miles away in Washington, D.C. The next problem is to see to it that the action will be appropriate, that the decisions of all these people will be wise.[1] Immediately following comes the problem of making sure that the wise decision can actually be carried out, that the competence to do the job is in human, or at least bureaucratic, hands in the first place; the inability of any government to keep the various oil spills from polluting Texas beaches in 1979 suggests that there are some matters as yet beyond our skills to solve. And making sure that it *is* carried out, and carried out properly, by government officials whose interest in the task may be minimal and whose budgets may be small, without any unintended side effects. And the final problem is keeping the government always around to make sure that the elegant solutions stay in place and the errant polluters don't backslide when no one is looking—since the solutions have been imposed on them by a higher authority rather than drawn from them by cooperation and negotiation. Oh—and paying for it all.

Still, even if we concede that the Federal government *is* a solution, and a benevolent one, wouldn't it be more rational for Dahl's citizen to figure out why that pollution exists in the first place and trace the problem to its source? The cause may be an accident, or ignorance, in which case the polluter doesn't need the heavy hand of Federal regulation but a little help, or information, or some money, any of which can be supplied far more easily and cheaply by the concerned majority around the lake than by Federal intervention. If the pollution is found to be deliberate, it is most likely to come from some anti-social corporation—the battery company, the detergent manufacturer—passing its "external" costs on to the public, which it has been allowed to do by provisions of governmental law for a hundred years or more, and

1. The experience of San Francisco, during its 1976–78 water crisis—to pick just one example at the large city level from a bulging file—is not encouraging. To conserve water, the city government asked people to cut consumption by 25 percent, in return for which the public utilities commission granted the water department a 43 percent rate increase so its revenues wouldn't fall. But the residents actually went and cut consumption by 40 percent, confounding the experts as usual, so the department asked for an additional 22 percent rate hike—and then publicly suggested that people ought to find a way, in the middle of this drought, to use *more* water so the rates wouldn't have to go so high.

ignoring any but its own sheer profit considerations, which it is explicitly *protected* in doing by the same long line of laws. Or if the pollution is from a municipal agency, it is probably because of Federal provisions that defy ecological balance—for example, Washington's funding complex sewage systems instead of septic tanks and composting toilets, or giving tax credits to utilities for expansion but not for conservation, or spending billions to develop nuclear rather than solar power. In other words, Dahl's citizen would find that the state may be the underlying *cause* of the pollution Dahl would ask it to *cure.*

To find the government as the root cause of such problems, of course, should not surprise us by now: it is in the nature of the state, we have repeatedly seen, to create the problems that it then steps in to correct and uses to justify its existence. But there is a further point to that process that is pertinent here; in the words of British philosopher Michael Taylor:

> The state . . . in order to expand domestic markets, facilitate common defence, and so on, encourages the weakening of local communities in favour of the national community. In doing so, it relieves individuals of the necessity to cooperate voluntarily amongst themselves on a local basis, making them more dependent upon the state. The result is that altruism and cooperative behaviour gradually decay. The state is thereby strengthened and made more effective in its work of weakening the local community.

This is important: it is exactly this that accounts for the inability of the Lake Michigan communities to regulate their pollution problems in the first place. Communities that were in control of their own affairs, whose citizens had an effective voice in the matters that touched their lives, would almost certainly choose not to pollute their own waters or to permit local industries to do so, out of sheer self-interest if not out of good sense—particularly if they were small, ecology-minded, economically stable, and democratically governed. (And if by some chance a community or two did go on polluting, resistant to all appeals, their toxic effects would likely not overstrain the lake's ability to absorb them.) It is this process, moreover, that accounts for the failure of the concerned majority to have cleaned up the pollution once it existed. Individuals and communities conditioned to cooperative and federative behavior, particularly those whose interests are greatest (in this case fishing villages, towns with bathing beaches, beach clubs, marinas, lakefront hotels, boardwalk businesses), would almost certainly work out, and pay for, a way to restore the lake—especially if there were no Federal or state governments to siphon off the locally generated money through taxation.

As with pollution, so with the other public services of the state. There is not a one of them, not one, that has not in the past been the province of the community or some agency within the community (family, church, guild) and that has been taken on by the state only because it first destroyed that province. There is not a one of them that could not be re-absorbed by a community in control of its own destiny and able to see what its natural humanitarian obligations, its humanitarian *opportunities,* would be. Invariably when the state has taken over the job of supplying blood for hospitals, there is a shortage, even when it offers money—the U.S. now gets much of its blood from overseas; invariably when a community is asked to do it voluntarily, and when the community perceives that the blood is to be used for its own needs, there is a surplus. This is not magic altruism, the byproduct of utopia—this is perceived *self-interest,* community-interest, made possible (capable of being perceived by the individual) only at the human scale.

Indeed, there is not one public service, not one, that could not be *better* supplied at the local level, where the problem is understood best and quickest, the solutions are most accessible, the refinements and adjustments are easiest to make, the monitoring is most convenient. If it be said that there is not sufficient expertise in a small community to tackle some of the complicated problems that come along, the answer is surely not a standing pool of Federal talent but an appeal throughout neighboring communities and regions for a person or group who can come in to do the job. (This is in fact what the Federal government itself most often does today, hence the great reliance on contract firms and $200-a-day consultants.) If it be said that some problems are too big for a small community to handle alone (an epidemic, a forest fire, or some widespread disaster), the answer is clearly not the intervention of some outside force but the ready cooperation of the communities and regions involved, whose own self-interest, even survival, is after all at stake. And if it be said that there is not enough money in a small community to handle such problems—well, where do you suppose the government *got* its money in the first place, and how much more might there be in local pockets if $500 billion of it weren't spent by Washington, $200 billion by state capitals, every year?

I cannot imagine a world without problems and crises, without social and economic dislocations demanding some public response. I see no difficulty, however, in imagining a world where those are responded to at the immediate human level by those who perceive the immediate human effects and control their own immediate human destinies.

SOCIAL JUSTICE

THE FINAL ARGUMENT for the necessity of the state is that it alone can provide social justice for all its citizens and guarantee civil rights and liberties to every individual.

To be sure, it *doesn't,* not anywhere in the world. Even in this country, so concerned with these matters, the areas of social injustice and individual repression are wide, and certain people—American Indians, poor blacks, prisoners, chicanos, homosexuals among them—are particularly ill-served by the state. Indeed the case could easily be made that over the years as many inequities and injustices have been *caused* or fostered by the government with its left hand—let's mention only slavery, Indian genocide, union suppression, the Palmer Raids, Japanese prison camps, FBI and CIA illegalities, and Watergate—as are *prevented* by its right. Not that the U.S. should be singled out in this regard: over the world, states have been claiming to offer social justice and equality for five hundred years now without ever having achieved it, or an approximation of it, not even the most benevolent of them. And even the well-meaning attempts to do so, which have proved extremely costly and in many instances socially rebarbative, have very often been accompanied by iniquitous or punitive or violent ends, the state taking more power to itself in order to make its interference more successful—perhaps the long British campaign against slave-trading and the American Civil War and its aftermath may stand as sufficient examples of this.

But still, the state presumably *means* to protect its individuals, and presumably they would be worse off without state intervention. Didn't the American government abolish slavery, establish civil rights for blacks, outlaw segregated schools, and work to end racial discrimination in housing, hotelling, hiring? Hasn't it treated minorities—the Inuit in Alaska, the chicanos in Texas—better than they would have been treated in racist local communities?

Perhaps. The record is by no means perfect even here on its more positive (right-hand) side, and many would say that alacrity was not among its notable features—but yes, the American government has effected certain reforms and has maintained certain freedoms. But who exactly would want to say that this is *because* of the state, that the government accomplished these goals by sighting a problem on its own, tackling it boldly, wresting a solution? I cannot think of a single victory in this area that has not been extracted by force from the government, has not been achieved by the affected parties themselves, pushing, cajoling, petitioning, suing, marching, demonstrating, usually quite on their own; and as often as not—as with the abolitionists of the 1840s, the labor organizers of the 1890s, the strikers of the 1930s, the McCarthy victims of the 1950s, the initial civil-rights demonstrators of the 1960s—they did so *against* the agents and stated policies of the national government. When we look at the great accomplishments here, what we find is the state against the victims of injustice—against Zenger, Thoreau, John Brown, Nat Turner, Susan B. Anthony, Eugene Debs, Margaret Sanger, W.E.B. DuBois, Norman Thomas, Martin Luther King, the NAACP, the ACLU, the Communist Party, SDS, the *New York Times,* the *Progressive.*

What we generally have, in other words, is another example of the state, having taken power into its own hands, sitting on those hands until somebody shoves it off—at which point one of the hands goes to congratulate the shover and the other points grandly in a gesture of "Look-what-I've-done!" That minorities are protected as much as they are is due mostly to minorities; that individuals have the opportunities they have is due mostly to individuals; that the press has its freedoms is due mostly to the press. The Bill of Rights, we must not forget, was put there not as an instrument of the state for the citizens but as a means of protecting the citizens *from* the state.

It is also worth noting that even after all of these gains, it is fairly said that the freedom of the press is enjoyed only by the person who is rich enough to own one. Individuals still as a rule have very limited use of freedom of the press, very circumscribed access to free speech beyond their own shouting range, very restrained possibilities of free assembly that the police do not sanction; they still have very fragile real-world protection against searches, seizures, brutality, and cruel and unusual punishment by police agents of the state; and they still have only the most tenuous guarantees in practical terms of truly due processes of law, speedy trials, impartial (or any) juries, or freedom from cruel punishment in prison. And what is true for the ordinary, God-fearing, white, well-dressed, comfortable, and respectable individual is, sadly, even more true for the tinted, tattered, timid, tasteless, or tipsy one.

But is there an alternative: is it possible to think of a community achieving justice for individuals and minorities it did not find congenial?

Historically the community is noted for its failure in this regard: small towns, neighborhoods, even homogeneous cities, can be extraordinarily narrow-minded and cruel, inhospitable and downright vicious to people and opinions and customs they don't happen to like. But it is also true that they can be very receptive to minorities that they do not think of as threatening or that they regard as enriching, and they have a well-known capacity to tolerate a great variety of eccentric individuals—the town drunks, the Elwood P. Dowds, the Tom Edisons—who accept them in return. Should we posit communities in true harmony with their surroundings, with some democratic and cooperative operation of their economy, and some form of participatory governance, then it would be reasonable to suppose that much in the community would already operate to soften personal animosities, to accommodate differing minorities, to harmonize a wide range of opinions and beliefs. Without outside pressures, either legal or coercive, such a town could no doubt mute if not solve the essential problems of injustice and intolerance.

But even should this not be possible, there are the recourses known to centuries, indeed millennia, of human history: migration and division.

For the individual, the most successful relief from injustice, should the community be unresponsive, is not the appeal to the state, whose workings are always lengthy and uncertain, but migration and relocation. This is not easy, especially for people burdened with families and homes and investments, but it is obviously the most immediate and effective solution; and in a nation where one out of five families changes residences every year anyway, it cannot be all that difficult. If the individual cannot get along with the community, and the community cannot tolerate the individual, what real good will state intervention produce—wouldn't separation be, in any world, the rational, non-coercive, non-violent solution? Yes, it might be possible to contrive a state process that would force a Jewish Community to accept the Nazi Individual, or a White Community the despised Black, or a Fundamentalist Community the threatening Atheist. But it needs only for the principle of free travel to be observed—to the advantage of both the leavers and the stayers—and the Nazi, the Black, the Atheist can all find congenial communities of their own. The virtue of a multi-communitied world would be precisely that there would be within its multitude of varieties a home for everyone.

And for the aggrieved minority, the obvious way to win relief is not to use the state to demand rights (which only inspires hostility in the majority) or to interpose restrictions (which insures implacability in the majority), but to divide and resettle. As was true in the Dinka village and the New England town, and as we shall see in more detail in the next chapter, the eternal human solution to dissension has been division, the separation and sometimes relocation of the disgruntled minority. It is possible for a minority to use the machinery of the state over many years and, at the cost of considerable anguish and enmity, secure minority rights, but it is also possible for it to relocate with peace and good will and to create its own society on its own terms somewhere else. The commodious solution is not minority *rights,* but minority *settlements.*

Clearly I am not supposing a communitarian world without problems, even its instances of inadequacy and injustice. No one could expect that. I am only positing a world in which the community is empowered to seek the solutions right where the problems erupt, to deal within its bosom with the people it knows, and to seek migration and division when all else fails. Compared to the record of what the state in all its glory, all its centuries, has accomplished, this, for all its difficulties, seems simpler, safer, cheaper, surer.[2]

IN A HIERARCHY of necessities, the things provided by the family, the neighborhood, the community, the small city, would certainly come first:

2. It used to be argued that only national governments could achieve justice because they were free of the corruption of small towns—though anyone familiar with pre-Mao China or modern Africa or Asia might have doubted. Since Watergate, we hear that less often.

love, fraternity, security, cooperation, sex, comfort, order, esteem, above all *rootedness*. ("To be rooted," as Simone Weil has shrewdly noted, "is perhaps the most important and least recognized need of the human soul.") Those provided by the state—taxation, standing armies, police, regulations, bureaucracies, courts, politicians, nuclear power, corporate subsidies, moonshots—especially when taken in balance with its deficiencies, would no doubt come last. Looked at that way, they just may not even be necessities at all.

"Giant political systems are a necessity" in the mind of someone like Robert Dahl, and who is really to blame him? But in the light of what would seem to be the irresistible decentralist tenor of our times, particularly in the United States, of the documented and apparently unavoidable failures of the encrusted state, "a system incapable of action," it is not irrational to disagree. Even such an arch-conservative as Herman Kahn has been forced to acknowledge that the institution of the state may be—though of course he probably would never use the term—"withering away." The nation-state, in his view, historically had two essential reasons-for-being: to wage war and to foster economies; it will shortly find, particularly in the developed world, that in a nuclear age the dangers of doing the first are too great to risk, and in a multinational age the second is unnecessary, so its role and importance will decline in the decades to come.

That does not seem so fantastic. If humans lived for the first 3 million years without a state, and most of them for the next 8,000 years without one, and the experiments with the nation-state as we know it are only a few hundred years old, there is clearly nothing eternal about it. It may have been a serviceable device for one small period of human history, but as we move out of that period it may begin to lose its value and its meaning in the daily life of the planet, even eventually disintegrating from disuse, to be remembered as we now remember the wooden plow or the sundial—a quaint tool, useful no doubt for its time, but of course no longer important, no longer . . . necessary.

6

The Importance of Size: Harmony

WE SEE IT in the archaic villages, the Greek city-states, the medieval municipalities, the New England towns, the religious communities: the regulation of size. It is the instinctive, eternal application of the Beanstalk Principle to human affairs: for every harmonious, self-governing human unit there is a size beyond which it ought not to grow. And, one might add, a fairly conscribed size, too, as the long chronicle of human settlements, regardless of culture, complexity, or continent, amply shows.

In each of these societies, and in countless others, the regulation of size was most often achieved not through warfare or infanticide or starvation, though all were certainly known, but through *division,* whether the process be separation, segmentation, fission, resettlement, partition, or colonization.

In the Dinka system, villages regularly split when they became "too big." A man who for some reason was bypassed in the village systems of power (the younger brother of the man to whom the office of spearmaster descended, for example) would be permitted to round up his family and a few followers in other families and start a new village nearby with its own grazing areas and water holes; or families whose feuds with the rest of the village became so fierce that daily community life became tense would be encouraged to move elsewhere, perhaps with the parting gift of an extra cow or two.

In the eighteenth-century New England towns, the process of fission was so essential and regular that it made up the greatest part of the business that communities had to do with the colonial government in Boston, it being necessary to get state approval to establish a new town. Even separation into quite tiny settlements was preferred to continuing disputes or quarrels: "We acknowledge we are but small," ran a fairly typical petition from the town of Dunstable, "but we apprehend a small society well united may more easily go through such business of building a meeting house and settling a minister than a greater number when there is nothing but discord and disaffection."

In Switzerland, where regions have been dividing and subdividing for at least six centuries, the guiding principle has always been to provide minority territory rather than go through the endless struggles of minority rights. The canton of Appenzell, for example, while nations all around it were fighting bloody religious wars in the aftermath of the Reformation and would continue to persecute religious minorities for centuries more, peaceably agreed to divide itself: the new Lutheran minority cut itself off as the discrete community of Ausser Rhoden, the Catholics stayed in the new community of Inner Rhoden, and they lived in peace for the succeeding 400 years.

And so we find it elsewhere, too, practically everywhere that people are free to move and no coercive state insists on unity, the ancient and rational practice of *harmony through division.* The Beanstalk Principle at work.

AND HERE WE come to the last and perhaps the most fundamental reason proffered for the "necessity" of the state: that it provides social harmony and prevents criminal disruption. Ecological harmony and workplace democracy and all that are fine, so it is said, but given human nature we cannot assume that people will always be good or curb their impulses to destruction. It is therefore up to the state, and the state's apparatus of law and police and courts and jails, now as it has always been, to control criminal, deviant, and disruptive behavior.

The first answer, of course, is that, on the evidence, the state has *not* controlled or prevented crimes in any society where it has become powerful, particularly not in those modern nations where it has become most powerful of all. The United States, easily the mightiest in the Western world, has easily the highest crime rate and the greatest prison population. Despite the expenditure of approximately $25 billion in 1978 at all levels of government—that is truly an enormous sum, more than twice all the U.S. gold reserves—and the employment of more than *one million* police and agents, and the imprisonment of more than 450,000 people judged to be criminal, the crime rate has not decreased, the number of criminals has not diminished, the public streets are not safer, and the populace does not feel itself appreciably more secure. It might be regarded as perhaps too extreme to say that the state actually has less interest in *eliminating* crime than in justifying its existence and expanding its power by *allowing* crime to continue, but that at any rate is the effect of its performance, not only in the U.S. but around the world. The laws don't work, the police don't work, the courts don't work, the prisons don't work—this is not any contrivance of mine, this is the accepted judgment of criminologists of many stripes—and yet what we hear as the solutions, and what we are forced in fact to pay for, is more of the same. Can this be serving society's interests in any real way—or only the state's?

The second answer is that the state *cannot* control criminality—not at that size. For what we have to realize is that there is always and everywhere a connection between population size and social harmony, and the large state is beyond the point of effective stabilization.

LET'S LOOK AT it from the standpoint of "human nature." There are no certainties here, and the issue continues to generate arguments among all kinds of scientists—ethologists, biologists, psychologists, anthropologists—and no doubt will for some time to come. But whether one wants to side with the aggressivists—Lorenz, Dart, Ardrey, et al., who argue that humans are naturally warlike and disputatious, bloodthirsty and competitive—or the pacifists—Montagu, Leakey, Dubos, et al., who find that humans are essentially cooperative and neighborly, peaceful and altruistic—there seems to be this common ground: humans have the capacity for great evil and great good, great aggressiveness and great generosity, and a given set of circumstances tends to promote one and discourage the other. Thus a human male defending his family might, like the baboon male in similar straits, become a raging, killing maniac, shorn of mercy or reason; or a human male enjoying a feast might, like the chimpanzee in similar circumstances, share his food with any fellow creatures who happen along, the very soul of generosity and brotherhood. A person surrounded by subways and blaring radios, beset on every side with clamor, confusion, filth, boredom, and hopelessness, might well strike out in rage when provoked; a person surrounded by calm, decorousness, purposefulness, and comfort is less likely to. Behavior basically depends on the setting, the time and place and society in which it happens. Obviously the desired effort is to find those settings in which the benevolent side of our human nature is encouraged and the malevolent side dampened: that will not *ensure* constant and unvarying harmony, it seems safe to say, but it certainly will go a long way to permitting it and fostering it.

And without question such settings, as our examination of stateless societies has shown, are those of the small community. In small places it is difficult to commit malevolent acts without being seen and identified, and therefore known and punished, and this sense of being "in the public eye," whatever its unpleasant consequences, certainly discourages misbehavior. Moreover, if the result of being identified is the disapprobation of friends, or ostracism, or ultimately exile, this puts a heavy and very real social weight against disruption. Self-interest also enters in as a deterrent, since if your acts are likely to create serious turmoil in the community and prevent it from being able to protect itself or carry on its commerce or harvest its crops, that is going to adversely affect you, too.

In a more positive vein, living at a small scale generally creates a sense of friendship among neighbors, or at least of trust and honesty, and it becomes very difficult to violate that, even if individual self-interest didn't argue for its reciprocity. Likewise can a sense of cooperation and mutuality arise among those who continually rub shoulders—it is difficult to think of "cooperation" among millions, but very easy to conceive of it among hundreds—that works against aberrant behavior both practically and psychologically. Psychological health, indeed, is essential for social harmony, and as we saw in discussing city sizes the evidence is strong that the small, cohesive community is by far the most beneficial unit because it provides a spirit of belonging, of place—what the Spanish call *querencia,* suggesting the inner well-being that comes with knowing a particular spot as your home. It also works psychologically in providing predictability and order, and thus cushioning against the dislocations of "future shock" that make city dwellers anxious, and in supplying the systems of support and friendship that encourage a sense of individual security and self-worth. (Michel Foucault has even argued that the concept, if not the very fact, of madness was born in the sixteenth century, after the decline of the stable and supportive small medieval community.)

The increased participation in all aspects of life that is fostered by the small community is also beneficial. In social terms this allows an individual to ventilate grievances and to make changes, escaping the pressures that tend to build up in people who feel powerless or useless or ignored. And in political terms participation allows people to see that the process of decision-making, having included them, is fair, and that the decisions themselves, having been influenced by them, are fairly arrived-at—the very two reasons, not so incidentally, that historically people have obeyed laws and honored social norms.

What we know about human nature, in short, suggests that its better side is more likely to emerge in smaller settings than in larger ones. And if we then suppose those settings to incorporate some of the human-scale features we have already noted, the likelihood of social harmony is even further increased. To the degree that a community can incorporate into its society a regard for the human being's natural and interlocking place in the biosphere, it will encourage people not to do unnecessary violence to any part of that world, their fellow creatures presumably included. To the extent that it can provide organizations and systems at sizes where individuals may feel some sense of control, it diminishes the kinds of psychological and social dislocations that arise when people face large, depersonalized, and violent ones. And to the degree that it can develop a community-controlled, self-sufficient, and participatory economy, it is likely to remove many of the causes of crime, because that should provide everyone with not only the general economic satisfactions but the fullest and most rewarding kinds of employment.

THAT ALL may discourage malevolent behavior, right enough, but even in such a world, I think we should be prepared to say, there will be deviates and "criminals," people who for reasons sane or insane may cause private harm or shatter public harmony. What is to be done with them?

Obviously any community may make its own rules and regulations—will have to make them, unless it wants half the people stopping on green lights, the other half on red—and can set its own patterns as to how its transgressors are to be judged and treated. And I'm afraid the record here is not unblemished: we know that "witches" were burned in small towns, popular justice can sometimes get inflamed into lynch-mobbery, and Shirley Jackson's "The Lottery" is not wholly farfetched. But for the most part these were the aberrations, not the rule—as they would have to have been, if the town was not to lose half its population at the stakes and hanging trees, as we know from the accounts of myriad small societies around the world whose justice, however rough, did not often descend to this sort of irrationality. And for the contemporary self-governing community there are so many examples of other fair and reasonable means there is little reason to think that the bizarre ones would ever prevail.

The first thing the commodious community would do would be to put an end to executions: official violence always begets violence rather than deterring it, it creates severe psychological tensions in any close-knit population where it is sanctioned, and it ultimately puts a disastrous social weight at the wrong, the punitive, end of the process of justice. The practice of imprisonment would also seem to be counterproductive, since the whole bizarre notion of sending someone to jail to get rehabilitated is as sensible as sending someone into a rainstorm to cure a cold; and the financial costs of building and maintaining prisons—it now costs an unbelievable $20,000 a year to maintain a prisoner in New York State, money that would be better spent if it was just given to the inmate—clearly outweigh whatever they offer in temporary security. (Interestingly, the institution of the civil prison is unknown through most of history, becoming common only with the rise of the nation-state in the nineteenth century.) Even those individuals driven so mad that they need to be kept away from others lest they do more harm—and I must add that they are quite rare in the records of village life—do not have to be put into prisons to safeguard the populace—they can be treated therapeutically in hospitals, kept under the watch of family or therapists in controlled settings, or, *in extremis,* banished and sent into hermitage.

For the normal run of wrong-doing, a variety of non-violent options is open to a community conscious of the fact that it is dealing not with an unwanted monster but a friend and neighbor who is part of its daily life. Some sort of therapy, or "re-education," can be sanctioned in the worst instances, where there is some thought that misperceptions or ignorance are at work; or a system of reparation and restitution, requiring the transgressor to repay the injured party or the community at large through money, property, or labor; or community ostracism or humiliation, which can be particularly painful in small societies; or the withdrawal of mutual aid and of the normal forms of cooperation. Each of these has been used by self-regulating communities and all of them have been shown to be generally successful in deterring and in punishing, depending of course on the effective social cohesion and interdependence of the population. Compared to what these can achieve, the recourses of the state apparatus in being today seem crude indeed.

Essentially all that such a method of justice requires is a sense of *responsibility:* to the individual transgressor, to the customs and patterns of the community, to the collectivity as a whole. That is not a great deal to ask, nor in the self-governing community is it rare to find. It is only in the giant systems of the contemporary world that we see so little of it, for it is here, where the state takes over responsibility, that the individual is asked to shoulder none. As an individual living in New York City I have no responsibility to deter a crime or stop a robber running down the street or see that a suspect is properly processed or guarantee that justice will be meted out or assure that punishment will not be cruel: these are the things that the state is supposed to be responsible for and there is very little way I could (safely or effectively) have anything to do with them even if I wanted. Hardly surprising, in a system in which all power and responsibility lie with some remote government, that there are Kitty Genovese stories, that subway cars submit mutely to being mugged, that even the police and prosecutors and judges shut their eyes to abuses up or down the line and mutter, "I'm only doing my job."

I HAVE SO FAR begged the question of *what* size it is that we might expect to find most desirable for achieving social harmony and stability. But it will be apparent from my use of the words "neighbor" and "community" that I think it is to be found in populations in the range of several hundreds and several thousands.

Of the general connection between the size and social control there is no disagreement whatsoever: the larger the group, the more difficult it is to keep peace. Anthropologist William Rathje, who has studied a variety of ancient societies in the Americas, puts this as a general theory: as the size of a population *doubles,* its complexity—the amount of information exchanged and decisions required—*quadruples,* with a consequent increase in stress and dislocation and a consequent increase in the power and sweep of the mechanisms of social control. That seems perfectly plausible when we consider that in a group of 25 people there are theoretically 300 possible conflicts between two people at any given moment and in a group of 10,000 no fewer than 50 million possible conflicts.

Anthropology and history both suggest, as we have seen, that humans have been able to work out most of their differences at the population levels clustering around the "magic numbers" of 500–1,000 and 5,000–10,000.

For the first, John Pfeiffer notes that anthropological literature indicates that it is when a population reaches about 1,000 that "a village begins to need policing," and as we have seen, the Dinka villages, like villages in most stateless societies, hold about 500 people on average and almost never more than 1,000. (Rough figures for village sizes in some other stateless societies: 100–1,000 for the Mandavi, 50–400 for the Amba, 300–500 for the Lugbara, 200–300 for the Konkomba, 400–500 for the Tupi.) Evidently in these face-to-face societies, where every person is known to every other—and presumably every idiosyncracy, sore spot, boiling point, and final straw—it is comparatively easy to keep the peace and comparatively easy to restore it once broken. Confirmation comes from the New England towns, the great majority of which were under 1,000, where harmony was the regular rule and "concord and consensus" the norm; from the Chinese villages of all periods until the most recent, with rarely more than 500 people, where traditional law of many varying kinds operated independently of dynastic decrees; from Russia, where the traditional *mir,* with seldom more than 600 or 700 people, was the basic peace-keeping unit for more than a millennium, each with its own version of customary law and all without codification or judicial apparatus.

Even in the archaic societies, however, wider "tribal" units of several thousand people, though rarely more than 10,000, were common—this being apparently the level best suited for the settlement of those disputes that would break out among the villages, the level found effective for the combined defense of the villages against outside threats. Common customs would prevail among groupings of this size and, as with the Dinka, common agreements could exist as to how disputes could be settled peaceably; beyond this level, where no formal agreements existed nor even any machinery for accommodation, disputes when they arose tended to escalate to violence. The reason is conjectural but reasonable: at such levels, up to about 10,000, it is not possible to be on a first-name basis with everyone, but there would be enough interaction at ceremonial occasions, on market days, during seasonal migrations, and in occasional wartime camaraderie so that the elders of any one of the dozen or so villages would be known to all the others and some rough assessment of their capabilities and temperaments made by all. This is, interestingly, the maximum level at which the basic Greek criterion for successful harmony would seem to operate—"that the citizens should be known to each other," as Plato put it, that "they must know each other's characters," as Aristotle said—for much beyond 5,000 or 8,000 and, even in a densely knit society, it is humanly impossible to recognize the faces, much less the characters, of most citizens.[1]

There is also some evidence that tends to confirm the stability of small populations in many different settings. On the whole the more successful city-states of Greece, Italy, and Germany—peaceful, as a rule, in their internal affairs no matter how many defensive wars they were forced to fight—ranged from as small as 5,000 to nearly 20,000, though to be sure the smaller were the more peaceful. Some of the most pacific societies in the world—San Marino, the world's oldest republic, dating from the fourth century; Andorra, which has been a quiet and stable state since the thirteenth century; Liechtenstein, virtually without a police force for most of its 250 years and never once having an army—lived most of their lives with populations of 8,000–12,000, though today they are all closer to 25,000 (and beginning to display the resultant difficulties).

Or take the undeniably pacific nation of Switzerland. This land of small villages and small cantons has, by staying atomized, always offered stability; in the words of the sixteenth-century Venetian diplomat, Giovanni Battista Padavino: "Traveling in Switzerland is very secure; one can travel the roads day or night without any danger and can halt in woods or mountains, and every class and family enjoys its own profound peace and unbelievable security." Nothing much has changed to this day, and outside of the major cities, crime rates are among the lowest in Europe. All this with populations of only rarely more than 10,000; even as late as 1970 the average commune *(Einwohnergemeinde)* in a typical canton like Ticino had only about 5,000 people, though 18 percent had fewer than 100 and only three had more than 10,000.

Or take modern Greece. It has the lowest crime rate of any nation in Europe and one of the lowest in the world, and both peasant and pundit realize this is connected to its strong continuing tradition of small-scale settlements. Outside the big cities the average town has about 6,000 residents, and the social ties at that level remain effective bonds throughout the citizen's daily life. In the analysis of C. D. Spinellis, a lecturer in criminology at the University of Athens Law School:

---

1. The difference in society sizes between the Greeks and Dinkas has to do with the former regarding only free males as citizens, the latter in effect shaping their customs to include people of both sexes beyond puberty; thus with 5,000 "citizens" the Greek city could grow to be 20,000–25,000 people, the Dinka not much more than 10,000.

> Greece remains a traditional society, where family and community ties are still very strong. There is little anonymity here. When you know your neighbor you don't harm him because you need him. Studies in criminology show that informal controls, such as the family and the community, are the most effective controls.

Or let us take our own country. Crime rates are consistently and uniformly lower in cities of 2,500–10,000 than in any other city-size category. Taking 1977, a recent year as representative as any, the average community of that size had a total of 3,906 reported crimes *per 100,000 population* (which means in fact closer to 390 actual misdeeds) as against 6,006 for cities of 50,000–100,000 and 7,819 for cities over 250,000, and 10,742 for San Francisco, with the highest number; of those crimes only 230 were violent and by far the largest were thefts, the simplest category of crime and easiest of repair.

In sum, I do not think it is a reach to conclude once again that humankind seems to have worked out another problem—the one of keeping the peace—in fairly small units, roughly corresponding to the neighborhood and the community, and moves beyond those limits not to its swift and immediate, but to its eventual and ultimate, detriment. Just as certain circumscriptions must be placed on population levels so as to achieve ecological balance and economic self-sufficiency, so—and in direct and overlapping ways—must they be placed for harmony and stability.

GANDHI ONCE SAID it was foolish to dream of systems so perfect that people would no longer need to be good. That may be true, though I must say I despair at the thought of constructing systems where people *do* need to be good: that seems like a difficult goal, no matter what your view of human nature might be, and one that various saints have failed at time and again over many thousands of years. I would rather contemplate a system so simple that people would no longer need to be bad—that is to say, a system of support and sustenance, of rough equality and comfort, that would so guide and goad, chide and chivvy, prompt and protect, that the individuals in it would be inclined out of sheer self- and community-interest toward morality and harmony. The small community has provided such a system—not molded through any special design, nor guided by any millennial genius, nor organized by any party or sect, but simply by working out the rough, hard problems of existence as they have come along for many thousands of years. I see no reason to think that, left alone, it could not do so again. The mistake is probably to try to devise systems that revise human nature, reforming people and making them moral and upright; the better part of wisdom is to take people as they are and determine under what conditions they are by themselves more likely to perform the moral act than the immoral—to reform the conditions, not the people.

"You are a very bad man," Dorothy told the Wizard.

"No, my dear, I'm a very good man," the Wizard replied. "I'm just a very bad wizard."

As I see it, the idea of the human-scale community is that it can provide the opportunity for there to be very good men without the necessity of there being very bad wizards.

7

The Importance of Size: Democracy

IT IS A MATTER for some amusement in our time that people should wish to have any voice at all in the political affairs of their lives: a September 1979 issue of *The New Yorker* carried a cartoon of a man in the Washington office of his congressional representative saying to the receptionist: "Just tell him one of 'the governed' is here to see him!"

I suppose bitter laughter may be the appropriate response to the condition of the contemporary American government, for the fact of the matter is that the idea of "the consent of the governed" has become essentially ludicrous in a land where there are 220 million of them. We do not, as I have made clear, any longer have "representatives," people who *represent* our views or feelings or even interests or lifestyles; nor do we have the elemental qualities of citizenship that are essential in a democracy; nor do we even have that democracy itself. Whatever its patrons may find to say in favor of our present system of government, there is no pretense, except perhaps in some unrevised sixth-grade civics text, that it is designed to evince the popular will, or allow even a majority of the people to establish national policies, or let the public en masse behave as truly sovereign. It might be better to call our system—a system in which some of the people select between two candidates who are already beholden to other interests and are in no way bound to listen to these voters—an oligarchy of the elite, which we have had the good fortune to experience as essentially benign during most of its duration.

I do not especially wish to debate the merits of democracy, since that is a matter that has been amply proved by innumerable political philosophers over the last several thousand years and, one would think, should be axiomatic by now. Suffice it to say that, whatever else its problems, in its uncorrupted forms democracy provides more benevolence, stability, participation, responsibility, productivity, efficiency, diversity, justice, fairness, freedom, and happiness than any other known system of government.

Nor is there much point debating what democracy is. Democracy means the direct one-person-one-vote popular assembly of every citizen. It does not mean the bill-of-rights freedoms, it does not mean republican government, it does not mean federalism or pluralism. Above all, it does not mean *representation:* representative government may be a desirable expedient in a government of great size, but as we have clearly seen it has nothing to do with citizen participation, popular decision-making, or democracy. Rousseau may have been a great waffler on many questions, but about this he was Alpine clear:

> Sovereignty . . . consists essentially in the general will, and the will cannot be represented. Either it is itself or it is something else; there is no middle ground. The deputies of the people, therefore, are not nor can they be its representatives; they are merely its agents. They cannot conclude anything definitively. Any law that the people in person has not ratified is null; it is not a law. . . .
>
> The instant a people chooses representatives, it is no longer free; it no longer exists.

The only true democracy, therefore, is direct democracy.

MANY DISPARATE TYPES of theorists have analyzed the nature of democratic government, but virtually all are agreed on one point: a true democracy requires a small society. The human mind is limited, the human voice finite; the number of people who can be gathered together in one place is restricted, the time and attention they are capable of giving is bounded. From simply a human regard, there is a limit to the number of people who can be expected to know all of the civic issues, all of the contending opinions, all of the candidates for office.

The Greeks in general, whether partisans of democracy or not, agreed with Aristotle that the well-run *polis* had to be small: "If citizens of a state are to judge and to distribute offices according to merit, then they must know each other's characters; where they do not possess this knowledge, both the election to offices and the decisions of lawsuits will go wrong." European thinkers, likewise, though not all of them democrats, assumed with Rousseau and Montesquieu that populations and territories had to be kept circumscribed. "A fundamental rule for every well-constituted and legitimately governed society," as Rousseau said, "would be that all the members could be easily assembled every time this would be necessary," and therefore "it follows that the State ought to be limited to one town at the most"; and though he is never specific as to the size of its population—indeed, he argues sensibly that it depends on the geography and fertility of a region—he refers at one point to maximum freedom in a state of 10,000.

All subsequent democratic theory has proceeded from like assumptions. The triumph of the American and French revolutions recast much of this theory into national molds, and some there were who tried to argue that large-scale representative or republican systems retained "the essence" of democracy, but even a man like Madison acknowledged that a "pure democracy" was "a society consisting of a small number of citizens, who assemble and administer the government in person." Even John Stuart Mill, who was dealing with an England of millions, agreed that "the only government which can fully satisfy all the exigencies of the social state is one in which the whole people participate," and that, he said, cannot take place "in a community exceeding a single small town."

The twentieth century—and with undoubted good reason—has had occasion to reiterate that view in the face of mass parties, mass politics, and mass governments claiming to be democratic. John Dewey may have spoken for his generation—"Democracy must begin at home, and its home is the neighborly community"—as Lewis Mumford for his—"Democracy, in any active sense, begins and ends in communities small enough for their members to meet face to face." More recently, the eminent Robert Dahl: "Any argument that no political system is legitimate unless all the basic laws and decisions are made by the assembled people leads inexorably to the conclusion that the citizen body must be quite small in number." And Leopold Kohr puts it in this delightful perspective:

> A citizen of the Principality of Liechtenstein, whose population numbers less than fourteen thousand, desirous to see His Serene Highness the Prince and Sovereign, Bearer of many exalted orders and Defender of many exalted things, can do so by ringing the bell at his castle gate. However serene His Highness may be, he is never an inaccessible stranger. A citizen of the massive American republic, on the other hand, encounters untold obstacles in a similar enterprise. Trying to see his fellow citizen President, whose function is to be his servant, not his master, he may be sent to an insane asylum for observation or, if found sane, to a court on charges of disorderly conduct. Both happened in 1950. [And times subsequently.] . . . You will say that in a large power such as the United States informal relationships such as exist between government and citizen in small countries are technically unfeasible. This is quite true. But this is exactly it. Democracy in its full meaning is impossible in a large state which, as Aristotle already observed, is "almost incapable of constitutional government."

This *is* exactly it.

THE ACTUAL EXPERIENCE of direct democracy over the ages seems to have confirmed these theoretical insights—was no doubt the source of many of them—and suggests the possible population sizes at which it may operate. The results will not surprise you.

The cradle of direct democracy, of course, was Greece, from about the seventh to the fourth centuries B.C., and the hand that rocked the cradle was quite small indeed: the *Encyclopaedia Britannica* may even be a little generous in asserting that Hellenic democracy operated in areas that were "generally confined to a city and its rural surroundings, and seldom had more than 10,000 citizens." Athens itself may have outgrown those limits at several points in its career, and possibly a few other cities as well, but the Greek experience overall indicates that about 5,000 people would be the upper limit for regular and sustained participation in daily or weekly matters. The Athenian assembly at its best periods seems to have numbered around 5,000—one record suggests a quorum may have been 6,000, and Plato speaks of the ideal number of citizens as 5,040—and though that seems to us a large number for debate and decision, it seems to have worked. Obviously certain constraints have to apply at that number. Not everyone can speak on every issue, for example, because if they met for ten hours a day and each one talked for as little as ten minutes apiece it would take them eighty-four days to debate a single issue.[1] Not all issues can be brought to the group for discussion, because the maximum number of decisions that could be taken, even assuming a fairly rapid rate of one a day, would be no more than about three hundred a year. That in turn means that some degree of cohesion and agreement has to exist beforehand in the community at large; it needs the refinement of many issues to a limited number of viewpoints (this, of course, is what the *agora* and *gymnasia* were all about) and the acceptance of a limited number of spokespeople to put forth a particular cause; and it requires a willingness to let minor decisions be taken by functionaries (chosen by lot or election) operating outside the assembly. But given these restraints, and they seem to have come perfectly naturally to Greece, the Hellenic system was beyond question, despite its occasional flaws and lapses, one of the finest that humankind seems ever to have crafted.

Interestingly enough the other extraordinarily successful experience of direct democracy took place in another mountainous country, and the record there is even longer. The Swiss mountain cantons, whether as completely independent entities in the earlier centuries, or as parts of various loose alliances and federations later on, used a system of regular popular assemblies, referendums, and initiatives from the thirteenth well into the nineteenth century. Even now, according to Cambridge University historian Jonathan Steinberg, "underlying the provisions of a Swiss constitution is the assumption that ultimately the ideal state is the direct

1. This is what Bertrand de Jouvenal has called "the Chairman's Problem" *(American Political Science Review,* June 1961), and pertains in a group of any size.

democracy or the *Landsgemeinde,* the assembly of all free citizens in the historic ring," and he notes: "This, the pure form, not the clauses of a constitution or its preamble, is the truly venerable element in Swiss political life." The ancient "fundamental law" of Canton Schwyz gives some notion of what the *Landsgemeinde* must have meant:

> The May *Landsgemeinde* is the greatest power and prince [in the old sense of that word, meaning principal body] of the land and may without condition do and undo, and whoever denies this and asserts that the *Landsgemeinde* be not the greatest power nor the prince of the land and that it may not do and undo without condition is proscribed. Let a price of one hundred ducats be set on his head.

Even the punishment is typically Swiss.

Because each canton was a federation of districts, and each district was divided into communes, each commune was made up of sovereign villages, it is not possible to describe the "typical" democratic system in Switzerland. But in the main the *Landsgemeinde* covered a population averaging 2,000 to 3,000 people, of whom only the adult men were allowed to debate and vote (hence an assembly of around 500), and would meet roughly once a month or in some places once a year. The meetings were wide-open affairs, with plenty of horse-trading and even some vote-buying going on beforehand, and any number of factions would appear in the course of debate; but somehow after the decisions were taken, the divisions healed—they could hardly be allowed to fester in such small populations—and implementation was normally accepted and shared by all. In between meetings, for any particular matter of even the remotest seriousness, the *Landsgemeinde* officials would submit referendums to the citizens and accept as a matter of course the direction of the vote; at the same time the citizens whenever they wanted could force an initiative with a small number of signatures on a petition, and if the initiative passed it had the force of law. So well-entrenched are both the referendum and initiative that they are active parts of the politics of a number of modern-day Swiss cantons, and of the federal system as well.

Modern Switzerland has found, though, that increasing populations—and increasing pressures from the outside (especially corporate) world—have forced changes in their traditional democracies. There are still five cantons (out of twenty-five) that run their affairs through annual cantonal meetings of all the citizens, but the turnout tends to drop as the size increases, and the cantons average about 30,000 people now, ranging from 13,000 to 50,000. Town meetings show similar effects: the city of Grenchen, with 20,000 people and 12,000 voters, has found that only about 400 people show up for town meetings these days, whereas in nearby Wangen, with 4,300 people, 90 percent of the citizens may turn out for votes and assemblies. If there is a "tipping point" for Swiss democracy to work effectively, it would seem by my calculations to come at around 10,000, perhaps slightly earlier.

The last historical example of direct democracy in action is one we have already touched on: the New England town meeting. As a rule the towns in which they took place traditionally held no more than about 1,000 people, meaning that the assembly itself would attract upwards of about 200 people (only property-owning males were theoretically eligible), depending on the issues to be discussed or officials to be elected. At first, in the seventeenth century, the meetings would be monthly affairs, sometimes even weekly, but gradually by the eighteenth century the practice was to have them quarterly or annually and to let the elected officers of the town—typically there would be more than forty of them, from selectman to meeting-hall-sweeper—and the various designated committees—usually a dozen, from finance to roads—carry on the town business in the interim. These meetings, however infrequent, left little enough initiative to the town officers, though—they would declare on everything from whether the town should have a new bridge to which bushes marked the town boundaries, and the officials were entrusted merely with carrying out their wishes. And even when a town official made bold enough to propose a new course, he would not act on it until authorized by the town meeting, no matter how urgent, because he knew full well that it would never be carried out unless it had the meeting's sanction.

Town meetings still exist today in many parts of New England, though they are a dying institution as states intrude to take over more local functions and the Federal purse looms behind practically any project of size. But they still decide the laws that are to govern the town, the budget to be followed, the local taxes to be paid, the policies to guide the town officers, and who those officers are to be, much as they did 300 years ago. It is no longer local control in any real sense, what with the press of outside polities, but it is still direct democracy and, as political scientist Jane Mansbridge has determined, there are "citizens still directly controlling important decisions that affected their lives." Her reaction, in one town of 500:

> I left the town meeting grinning. . . . These people had debated energetically the practical and the ideological sides of issues vital to their town. They had taken responsibility for the decisions that they would have to live with. Votes had been close. Farmers and workers had spoken out often and strong. The town had no obvious "power elite."

Most such towns that still have annual meetings are small, averaging perhaps 4,000 or so, though they include many places of 500 and quite a number with 7,000 and 8,000. Joseph Zimmerman, a professor of political science at the New York State University at Albany and one of the leading academic experts on the town meeting, believes that there are definite limits on how well they can operate before things get so large that only special-interest groups bother to turn out:

> A lot of studies claim that the New England town meeting is undemocratic, because only special interest groups show up and only a small percentage of the voters come out, which is certainly true once a town gets up to 8,000 or 10,000 residents. But below that level it is a sort of informal representative government, where the people who don't go in effect elect the people who do go to act for them. If they don't like the results, they turn out in force the next time.

Again, the upper limits of the community.

THE MOST ENTERPRISING modern examination of the connection between population sizes and democratic government that I know of is a slim volume called *Size and Democracy,* by Robert Dahl (an absolutely indefatigable toiler in the theoretic vineyards, as we see) and Edward R. Tufte of Princeton University. As befits a scholarly study, it is barely able to offer itself of any firm conclusion, and it is evident that the authors are quite bewildered by their own evidence, which in the end they choose mostly to disregard. But the evidence is clear enough, and hardly surprising: in smaller units, people are more politically active, can understand the issues and personalities far more clearly, participate more in all aspects of government, and regard themselves as having some effective control over the decisions of their lives.

There have been any number of surveys of citizen behavior both in this country and around the world, and Dahl and Tufte have surveyed most of them. Their summary: "Citizens tend to believe that their local government is a more human-sized institution, that what it does is more understandable, that it handles questions they can more readily grasp, and thus is more rewarding, less costly, to deal with." As to power:

> Citizens saw local units as more accessible, more subject to their control, more manageable. In the United States, only one citizen in ten thought he would stand much chance of success in changing a proposed national law he considered unjust; but more than one out of four thought they could succeed in changing a proposed local regulation they considered unjust.

As to participation:

> In a number of countries, including the United States, levels of political participation other than voting are higher at local than at national levels. . . . Two to four times as many people said they had tried to

> influence their local government as said they had tried to influence their national government.[2]

As to equality:

> Only in smaller-scale politics can differences in power, knowledge, and directness of communication between citizens and top leaders be reduced to a minimum. . . . Larger-scale politics necessarily limits democracy in one respect: the larger the scale of politics, the less able is the average citizen to deal directly with his top political leaders.

In short:

> The relative immediacy, accessibility, and comprehensibility of local politics may provide many citizens with a greater sense of competence and effectiveness than they feel in the remoter reaches of national politics. What defenders of local government have contended throughout an epoch of growing centralization and nationalization of political life may prove to be more, not less, valid in the future: the virtues of democratic citizenship are, at least for the ordinary citizen, best cultivated in the smaller, more familiar habitat of local governments.

As the centerpiece of their findings, Dahl and Tufte offer evidence from what they call "the largest and most careful study bearing on the relation of size to democracy" ever undertaken, a $1-million survey by the Local Government Research Group of Sweden and the departments of five Swedish universities, from 1966 to 1970. In a close examination of the populations of thirty-six different-sized localities carefully selected to elicit the maximum information about citizen political participation and feelings of power, the group found that political awareness, political discussion, membership in political and voluntary organizations, and involvement in local government was far greater in units under 8,000 people than in any larger sizes. As Dahl and Tufte say, with some italicized astonishment: "The major finding of the study is that in Sweden the values of *participation and effectiveness are best achieved in densely populated communes with populations under 8,000.*"

EVEN A DEMOCRACY at its optimum size can have its problems, however, and it seems pertinent to confront the two most common: that

2. The reason that participation in local voting is usually lower than in national is that it is seen as a weak method of influence in the immediate setting where there are other more potent means of getting one's views across, but it is all there is when it comes to the remote world of national politics. The desire of a town for stability, the reluctance to create serious fissures, the built-up intimacy with an incumbent, and the lack of local media hype also tend to reinforce this.

it tends to operate with either-or voting systems that do not represent accurately the true popular will, and that it thus tends to promote factionalism, especially of the majority against the minority. The criticisms are valid enough, and experience even in small communities has shown them to be all too likely. The problem lies, however, not with democracy, but in the method of decision-making.

Majoritarian voting—win-lose, binary, or zero-sum, as the games theorists who have studied it like to say—is both imperfect and unrepresentative. The famous "voting paradox," first formulated by the Marquis de Condorcet in the eighteenth century, shows that majority-rule voting, despite our grade-school teaching, in fact has no necessary relation to the actual preferences of the majority of the voters. In one form we see it in the presidential pairwise elections, as for example a case in which (as we would know from polls) Ford could beat Reagan, Reagan could beat Carter, and Carter could beat Ford, so the true attitude of the electorate is unknowable. In a more sophisticated form, we might imagine that Reagan was actually favored by 44 percent of the voters, Ford by 30, and Carter by only 26. The Reagan supporters, however, are split 22–22 among Republicans and Democrats, while the Ford people are 24–6 Republicans, so in a primary runoff Ford would win 24–22. In the general election, however, Ford goes up against Carter, but having alienated the Reagan people he pulls none of those Democrats and only half of those Republicans, 11 percent, for a total of 41 percent with his own supporters counted in; Carter, with some of the alienated Reaganites, perhaps an 18 percent, and his own 26, gets a total of 44 percent. Thus the man actually preferred by a distinct minority of the public can win in a series of pairwise elections.

The same sort of thing is often seen, too, in legislation, where a bill will pass or fail depending solely on the order in which the amendments to it are considered; indeed, it is possible to arrive at any one of three different outcomes depending purely on what part is up for a vote at what time, which pretty much plays havoc with "will of the people." In schematic form, the paradox looks like this:

One-third of the legislators prefer	A to B to C
One-third	B to C to A
One-third	C to A to B

Therefore, if A vs. C, C wins, then C vs. B, B wins
if B vs. C, B wins, then B vs. A, A wins
if A vs. B, A wins, then A vs. C, C wins.

Therefore *any outcome* is possible, theoretically representing the "majority will," depending merely upon the order of the vote.

Majoritarian voting also generally leads to confrontation of one sort or another and hence to divisiveness. Life is very seldom either-or, but voting is, and that tends to cause a great many in-between possibilities to get lost and makes people cluster unnaturally around one or another pole. Particularly on a larger scale, but also in face-to-face democracies, such factionalism can get formalized into parties, which represent the rigidification, one might almost say the ossification, of politics. Too many sociological studies show how even fairly small units can break into bitterly opposed camps over electoral matters, creating divisions that can move into social affairs as well.

An alternative that normally avoids both these deficiencies of majoritarianism, and one that has been studied extensively in recent years, is the process of consensus. (Not to be spelled concensus and confused with poll-taking; it comes from the Latin *con,* together, and *sentire,* to feel.) Despite the contemporary misuse of the word, it means the achievement of an agreement with which all the people present can feel comfortable and none disagree strongly enough to blackball—not total agreement, merely agreement not to obstruct—and clearly where it can work it obviates the problems of both voting and factionalism. And it also solves the difficulties surrounding what are perceived to be "immoral" or "unjust" majoritarian votes—to go to war, to enslave a person, to deny homosexual rights, to permit leg-hold animal traps—since such actions cannot be taken where there is even a single person morally opposed and willing to speak out. To be sure, that will tilt consensual communities toward conservatism, since a lack of consensus will mean inaction on any given measure; but it will by the same token make them more stable, more predictable and more "comfortable," and less prone to ill-considered decisions.

The *process* of consensus is nearly as important as its result. The emphasis is on a search for agreement rather than, as in majoritarian assemblies, the clarity of divisions, on compromise and cooperation, that is, rather than maneuvering and competition. In order to try to get a whole meeting over to your position, it will not help to score debating points or ignore opponents' criticisms, and if you want to get out before dawn it is best to work out some arrangement with the other points of view rather than simply to solidify your own. As often as not the compromise, because it is a synthesis of a number of positions, turns out to be stronger and more durable than any original position, as iron gains strength when manganese and tungsten and carbon are added in the process of making steel. And because it has the assent of all the people who will be affected by it, it stands a better chance of commanding obedience and being implemented the day after—whereas in towns with majoritarian voting it is not uncommon to hear of decisions that remain null after a vote because it would cause too much friction to carry them out or there aren't enough volunteers to do the job. (In big cities, of course, they are carried out by duly appointed officers whether anyone likes them the next day or not.)

When it works, consensus has some remarkable effects. Hallock Hoffman is here trying to sell the system and may not be completely objective, but he is describing the process as he experienced it during many decades of Quaker meetings, all of which are governed by consensus:

> Although it may take longer for the meeting to act, when it acts it acts as a whole. No protection for minorities is necessary. There is no minority. Decisions need no refurbishing, as they often do under majority processes. Members develop the habit of questioning their own opinions—the individual conscience is uncoercively subjected to the examination and illumination of the group. In complement, the meeting questions itself and checks itself as a whole from suppressing the conscientious declaration of any member. Perhaps most important, when the meeting achieves its ideal, discussion is remarkably free and candid. The commitment is to truth, to light, and to divine inspiration. The love of all for each and for the meeting permits disagreement, inquiry into reasons for opinions, and mutual probing into matters elsewhere considered personal rather than public. In psychological terms, each member, secure in the affection of each other member, feels free from defensiveness and anxiety. The result is not victory for any part of the group. The only victory lies in the common achievement, without repression to anyone, of a decision satisfactory to all.

It is, self-evidently, a process that is not without problems. It relies upon a certain common understanding, a shared commitment. It works only when the group begins with some points of unity, in religion, geography, purpose, or philosophy, and best with all of these. It demands more time than up-and-down voting, more time in give-and-take, than most people are used to spending. It may not work in emergencies—though the story of the Quaker ship's journey to Vietnam tends to belie that—and it may mean delaying actions for weeks until differences can be reconciled. It depends upon some minimal participation from everyone, even the wallflowers not used to unburdening themselves in public gatherings. It requires a certain forebearance on the part of the meeting toward the individual dissenter who refuses to go along, just as it does a certain temerity on the part of the dissenter before opposing the meeting. And ultimately it means that the perpetual dissenters must recognize that it is in the interests of everyone for them to leave the community with which they are at such odds and go and find a more congenial one somewhere else.

Impossible, you say? Too demanding for real life? Evidently, not so.

Quaker meetings for at least the past 300 years have operated by consensus, not only in local meetings of from a dozen to several hundred, but in state and regional meetings where as many as a thousand might gather. Some of the Northeast American Indian tribes, certain Chinese and Japanese communities, and the villages of pre-colonial Java used to use consensus; even in modern Japan, it is said, "there is still a deep feeling in many quarters that it is immoral and 'undemocratic' for a majority to govern, for decisions to be reached without compromise with the minority." Some nineteenth-century intentional communities, usually those with religious underpinnings, had consensual governments for many years, though commonly in groups of only a few dozen and rarely of more than a hundred. Consensus was, and is, the goal of the villages in India that over the last generation have embraced Vinova Bhave's *sarvodaya* movement—the movement that helped to push its own J. P. Narayan to the prime ministership in the 1970s—and at least a few thousand settlements, ranging in size from a few hundred to several thousand, have found it possible to live by consensual government for decades. The settlements of eighteenth-century New England would operate by a kind of informal consensus, not dictated by parliamentary rules but simply called forth by the need for small-town harmony; and this is not so far from the practice of town meetings today that, as Mansbridge has found, do have votes and divisions but "still . . . prefer to make decisions unanimously rather than by majority vote," and "when deeply held agreement seems impossible, they usually strive at least for absence of conflict."

A great many contemporary "alternative" groups—communes, typically, intentional communities, co-ops, political organizations—have also attempted consensus government, with greater or lesser success depending upon their prior degree of agreement. Lew Bowers, a skilled and sensitive "facilitator" of many hundreds of consensual meetings, believes that the process can easily be assimilated into any group with shared interests that numbers as many as thirty or even fifty, but is skeptical about unanimity over that number. The Movement for a New Society, a federation of a dozen or so activist groups around the country, similarly recommends (and uses) consensus for small groups, although in at least one instance some of its members were successful on a larger scale: when 700 anti-nuclear protesters being held in an armory were asked to come up with a unified policy on bail-or-jail, MNS coordinators got the crowd to break into about sixty small groups, each of which thrashed out a consensus and then sent word to a central point, the center shuffled the various consensuses back and forth, and within two hours they had all agreed on a decision. This same strategy is also used by various anti-nuclear coalitions, including the New England Clamshell Alliance, which originated it in 1976 and has used it with fair success ever since. Clamshell is based on small "affinity groups" of ten to twenty people, rarely more (the unit is, incidentally, borrowed from the Spanish

Civil War), who meet with some regularity and make all decisions by consensus; these are then conveyed to a central committee by non-voting delegates—"spokes"—and this committee then coordinates the results into a fixed policy, sending word back and forth to various affinity groups until general agreement is reached. The Clamshell experience has been varied, and some groups have begun to feel that there are too many basic political differences to maintain a long-lasting coalition, but it has proven that federation of consensus-taking groups is certainly a workable organizational form.

It is a form, indeed, that might apply to a larger community of 8,000 to 10,000 people. Neighborhoods of, say, 500 or 1,000 people, asked to decide on some communitywide issue, could achieve consensus in an open popular assembly much as the New England towns once did, and then send a delegate—not a "representative" with an independent vote but merely a "spoke" to convey this consensus—to a community coordinating body; this body, with a workable size of a dozen or so (above the ideal five, it is true, but with never more than twenty in a community of 10,000 divided into neighborhoods of 500 each), could then harmonize the various neighborhood positions, easily referring back to neighborhood assemblies or committees where necessary, and then act as a kind of secretariat in carrying out the agreed-upon actions. This is a process of some complexity, to be sure, and each community would need to develop its own styles, but what we know of political affairs suggests that most aspects of governance are simpler at these levels—communications are easier, information is more readily and reliably gathered, feedback is faster and more reliable, participation is greater, personalities and competences are known, and agreements are more easily reached. Perhaps best of all, there are no fixed hierarchies here, no presidents or legislators or parliaments, no grand windy institutions beyond the citizens' control, no roles of such complexity and specialization that they cannot be filled by practically any citizen; which is not to say that there could not be "leaders," people whose intelligence or experience gives them special stature within their neighborhood—indeed, one would expect, and welcome, such people—but only that there is no statutory sanction, no official power, given to such figures, and therefore no abuse of it.

The process with a city of 50,000 is obviously somewhat more complex, and here one might imagine standing committees or full-time officials (their power limited by rotation and recall) necessary to coordinate and carry out consensual policies. But it could still work at that level, even though attenuated—either through a two-level system of neighborhood assemblies and a citywide delegate body of fifty or so (that would be large—Dahl and Tufte show that countries of 20,000–50,000 have parliaments of eighteen to twenty-five people—but not impossible) or a three-tiered system with delegates from the community level to a citywide body, which would then be around five or ten (if five communities of 10,000 or ten of 5,000). At this level one could not expect shared principles and ideologies to any great degree, but on the other hand since most real matters of the day would be taken care of at the neighborhood and community level anyway, the numbers and kinds of decisions that need to be citywide would be minimal. All that need be borne in mind—engraved on city monuments, flown from city flagpoles—is the maxim that decisions should be reached at the level where people, ordinary middling-competent people, can have control over them, and merely coordinated after that. *Vox populi, vox civitas.*

IT IS EASY ENOUGH to prove that small size is a *necessary* condition for the proper functioning of a direct democracy—even more for a consensus one—but could it also be a *sufficient* condition?

In truth, I do not think it would really be obligatory in a harmonious world that every community be a democracy, if only it remained of human-scale proportions. I could imagine each community going for its own singular form of governance—some might choose a republic, others a monarchy, some might want an oligarchy to rule, others an elected triumvirate, some may prefer a socialist dictatorship, others a cooperative federation, and only a few of the finest and most harmonious opt for a consensual democracy—and as long as none of them tried to impose upon the others, the conditions for a stable, ecological world would be met; and as long as the citizens of each had a free right in the choice of government, and the free right to leave the community if that government palled, then the conditions of justice and freedom would be met. The essential underpinning of a sound and stable society, I am convinced, is the community that is built to the human scale in all its proportions and cleaves to the human scale in all its institutions, not necessarily one that is democratic.

And yet, to my reading, history and logic both argue that a small community will tend toward the democratic, whether or not it expresses it formally, simply by virtue of the fact that individuals are known to each other, interaction is common and regular, opinions are freely exchanged, and every ruler is also a neighbor. In a small society even the prince will probably be accessible—as in Liechtenstein—and every parliament familiar; where the government is inherently limited in scope and accumulation, it is extremely difficult for any individual or set of individuals to dominate and overpower the populace at large, and extremely unlikely that the citizens will permit them. As Leopold Kohr has put it:

> In a small state democracy will, as a rule, assert itself irrespective of whether it is organized as a monarchy or republic. . . . Even where

> government rests in the hands of an absolute prince, the citizen will have no difficulty in asserting his will, if the state is small. The gap between him and government is so narrow, and the political forces are in so fluctuating and mobile a balance, that he is always able either to span the gap with a determined leap, or to move through the governmental orbit himself.

Moreover, any small society that sought stability and permanence, efficiency and rational governance, would most likely tend toward democracy almost automatically, as it were. Governing by diktat may look easy, but it does not permit reliable information coming in from below in conception, it does not allow diverging opinions to be heard in deliberation, and it does not encourage smooth and willing cooperation in execution. A community that wanted to be sure it knew what all its people were thinking, what the gripes and problems were, that wanted to hammer out the best solutions to the difficulties as they arose and wanted to be sure its suggestions were carried out and its regulations obeyed, would inevitably work toward some form of direct democracy. Likewise a community that wanted to create the maximum participation in the political process so as to give outlet for grumbling and dissension, that wanted to develop feelings of self-worth and effectiveness for the citizens' own psychic health, that wanted to insure loyalty and cooperation through common understanding of the political machinery rather than through coercion, would instinctively move toward some kind of participatory democracy. Healthy not only for the individuals in it but for the community itself, democracy would be likely to come to the fore in any rational community kept at a manageable size, no matter what its trappings may look like.

THE GREAT ENGLISH biologist J.B.S. Haldane, the man who gave us our earlier analysis of the beanstalk giant, was once asked, in a group of distinguished theologians, what he could conclude about the nature of the Supreme Being out of his immense store of knowledge of the nature of the universe. The old man thought for just a moment, bent forward and replied, "An inordinate fondness for beetles."

And indeed the scientist's perception was accurate: of the somewhat less than a million animal species that have been identified and named, almost 75 percent of them are insects, and of these insects about 60 percent are beetles.

Whether or not the secret to God's plan is in fact the beetle, as I must confess myself reluctant to believe, two indisputable truths seem to be revealed in the natural world. The first has to do with diversity, an incredible diversity, that generates so many hundreds of thousands of insects, and something like 400,000 kinds of beetles, more different kinds than of any other known animal species—spotted and striped, checkered and solid, green, yellow, purple, and rose, some living in sand and garbage, some in trees and roses, some quite minuscule and almost invisible to the unaided eye and some at least a foot long, some unisexed and some multi-sexed, some in the tropics, some in the Arctic, some indeed everywhere in the world. The second has to do with size, for the great preponderance of the many billions of insects are smaller than a human finger, and yet there are many times more species of small animals than of large ones, by a ratio of at least ten to one.

Does nature by any chance have a political message for us?

8

Deficiencies of Community

THE PARISH OF St. Denis, as described by a careful and not unsympathetic sociologist some years ago:

> The whole parish is always divided between the "blues," or Conservatives, and the "reds." Party affiliations follow family lines and family cliques and antagonisms. . . . Constituents of each party have a genuine dislike for those of the other. . . . Election time is one of great tension, of taunts and shouting. . . . Insults are common, and many speaking acquaintances are dropped . . . Campaigns reach their climax with the *assemblée contradictoire,* at which both candidates speak. Characteristically at these meetings there are organized strong-arm tactics, drinking, and attempts to make each candidate's speech inaudible. . . . The chicanery of politicians is a byword in the parish. Factional strife threatens the life of every organized association.

True, this small settlement (in Quebec) does not enjoy direct democracy and never thought of the idea of consensus; its economy is largely controlled by outsiders and has never known any such thing as workplace democracy; ideas of ecology, limits-to-growth, and self-sufficiency are completely foreign. Nonetheless, it is a small unit, roughly of the neighborhood size that I have been discussing—and it doesn't seem to have quite the characteristics of harmony and sagacity, fellowship and cooperation, that I have indicated the human-scale community ought to have.

The fact is, and we must face it here, that the history of the small community is by no means an unblemished one, and anyone who has lived in a small town for any length of time can probably find a variety of objections to life at such a level.

Small communities tend to be close-knit, homogeneous after a time if not at the start. They are suspicious of outsiders, actually xenophobic on occasion, and not quick to welcome outside ideas and customs, of whatever value. They are conservative, parochial, resistant to change, and accustomed to fixed ways and patterns that do not allow much room for the excitement of novelty or the unexpectedness of variation. They tend to make the individual second to the collectivity, breaking down walls of privacy, subjecting all to a kind of tyranny of public opinion. They generally demand conformity and punish intrusive eccentricity, can make scapegoats of those they find threatening, and may force what they regard as "deviants" into closets or sublimations. They do not necessarily prize culture at any but the simplest levels and seldom choose to divert time and effort to the creation of artistic institutions. They can put short-run economic interests over long-range environmental protection or bend to the power of a single company for the sake of jobs, and in many there is great and entrenched inequality. They do not always evince much egalitarianism and can sometimes fall prey to the authoritarianism of a single leader or tacitly accept a well-placed clique, avoiding the difficulties of participation and involvement, the arduousness of freedom.

This is by no means the pattern of all small communities, and for a town with any one of these deficiencies it is possible to find, in every time and culture, dozens of examples that would neatly show the opposite. Still, small settlements are undeniably prey to such deficiencies, and it is well to confront them.

Obviously the self-governing community is capable of doing practically anything it wants to do, including errant, xenophobic, irrational, and intolerant behavior. And given the variegated nature of the human animal, we could hardly expect to construct a world of such holiness that such deviance and malfunctioning would not exist. But the logic of the well-performing community is easy enough to see, the models are plentiful, and the virtues for the people in the aggregate as in the particular are clearly demonstrable, so it is only reasonable to suppose that the great preponderance of communities—given their free choice—would operate so as to maximize their own comfort and capacity, minimize their confusions and conflicts. Each of the elements of the human-scale community as we have seen it so far has been developed and practiced by people over the long years of human shoulder-rubbing precisely because it eliminated or eased the deficiencies of small-town living.

Thus the virtue of the community of *direct democracy* is that it does not easily succumb to governing elites and is able to offer systemic resistance to autocratic leaders. It provides the forum where new ideas may at least be considered and, through free debate, the means of making their virtues known. It produces in time an openness and tolerance in political matters and processes that can extend into the social sphere. It encourages participation, not simply because it is everyone's self-interest to show up at the meeting and keep from being elected dog-catcher but because the whole range of community problems tends to become as real as—in effect to *be*—personal problems. If it is in addition *consensual,* its workings will militate against the unjust treatment of any individual—individuals presumably having a say in their own fate—and thus against the "Lottery" and witch-hunt dangers. Consensus, too, if a regular process, works to round off the edges of minority opinions after a while, so that in time they fit in more smoothly with—or at least are accepted between—the opinions of the majority.

Similarly, the virtue of the community that *controls its own economy,* whether approximating to self-sufficiency or not, is that it accepts—indeed, values—the outside idea insofar as it makes immediate economic sense, and it searches for that kind of innovation that can comfortably fit in with the production and market arrangements it already has. It is likely not to want a newcomer who is an economic deadbeat or the individual whose skills are in pyramid-construction or SSTs—call it xenophobia if you like—but it will probably welcome and tolerate a good deal of bizarre behavior from the woman who shows skills at keeping the books of the town's enterprises or the man who supplies the much-needed expertise in windmills. It will probably be inclined to divert community resources away from monument-building toward culture, from aggrandizement toward pleasure, and to appreciate the sheer economic good sense of developing orchestras and theater companies of regional attraction, of contenting the populace with movies and museums, of keeping the young in town with concerts and entertainment. The community of *self-sufficiency,* moreover, will necessarily fall back on its own resources and develop its own latent cultural talents. And if in addition it enjoys *workplace democracy* it will have developed cooperation and involvement almost into habits, making great extremes of wealth in the town highly unlikely and making resistance to political rigidity highly likely—the autonomous economic individual is disinclined to succumb to being an automatous political one.

If, further, the community is guided by the tenets of *ecological harmony* and *steady-state* equilibria, it is hardly the type to despoil its environment or readily admit the toxic or polluting industry (which, being in control of its economy, it is free to reject). Conscious of the way it relates to the ecosystem, it would likely establish, and value, its connections to other communities within the bioregion—the city upriver, the town downwind—and keep in check its tendencies to isolation and insularity. Conservative it would certainly be, in the best sense of that word, for that is precisely what recycling and resource recovery, precisely what self-sufficiency, is all about; yet only the most shortsighted would overlook the new biogas invention that came along from elsewhere.

Likewise, the virtue of the community whose buildings and institutions are kept to the *human scale* is that this works against bewilderment and insecurity, toward self-competence and self-esteem, and thus toward tolerance and patience—deviance is far easier to accept and overlook when one's own stability and security are not in doubt. Small-scale institutions—schools, hospitals, offices, factories—where one's participation and influence can be felt allow for the kind of psychic health that can make of close-together life not an unnerving but an enriching experience. A community whose individuals tend to feel necessary and competent, not at the mercy but in control of the systems around them, will achieve some rough psychological balance, perhaps the capacity for most of the "Christian" virtues most of the time. It is even possible to imagine such a society being healthy enough to overlook the neighbors' deviations and respect their privacy—as well as too introspective, busy, tired, and preoccupied to care.

I am not describing an other-race here. All of the elements of the human-scale community above are actual and known precisely because they have been forced from us in our long experience in human groups—and the successful elements, as in all evolutions, survive. They have not often been found all together in one place in the past, I readily admit; but they are all known to human societies everywhere, and most have been known not merely for centuries but for millennia. What is required of us is that we find them, examine them, and craft them, each for ourselves and with our companions, into the kinds of places that we would wish to live in.

It is worth pursuing the question of the parochialism of the independent community, the dangers of its isolation, lest it be thought, despite all that I have said to the contrary, that I am positing a world of Neanderthal simplicity and backwardness.

Far from it. To be sure, I am certain that democratically run ecological communities in control of their own destinies will not choose to construct SSTs and DC-10s, MIRV missiles and nuclear bombs, fast-breeder reactors and cyclotrons, CAT scanners and Saturn rockets, Cadillac Sevilles and Astrodomes, TVAs and Alaska pipelines. (They could of course if they wanted to, within their own ecological restrictions, but there might not be much left over for their plows and pitchers.) But I am also certain that the concentrated use of alternative technology (including the cannibalization of elements of existing high technology), coupled with the accelerated energy of individuals in control of their own economies—and absent the absorptive debilitation of the state—will enable communities of quite modest size to achieve all that could be asked for in basic human comfort.

Not backward—but on the other hand, quite frankly, perhaps too small, too self-regarding, too precious of independence, to provide the large agglomerations we have grown to be accustomed to today. What is to be done, it is fair to ask, with the kinds of very large organizations we now prize—research libraries, for example, or multi-departmental universities—or the kinds of geographically dispersed systems we now expect—telephone networks, for example, or railroad systems?

The first question to be asked about such large agglomerations is whether they are really necessary. (Most of them, as we have seen time and again, are not, but I am speaking here of the benevolent ones.) To what extent is it more valuable to have the treasures of New York's Metropolitan Museum stored away in its catacombs, many of them molding and decaying away to be lost forever, with only a tenth of the artifacts ever on view at any one time—that is not art, as one curator says, that's *hoarding*—than to have them spread out to ten or twenty different museums, each capable of maintaining and displaying them properly, to the enjoyment and edification of additional tens of thousands of people? To what extent is it better to have a large collection of departments and colleges massed in a single mega-university, available only to those few students who can meet the price that a large institution demands and can travel to that single point, than to have them spread out across the land, each college in a different community where interaction of all kinds is really possible, accessible in both price and location to many thousands more people?

The next question is to what extent it is possible for some sizeable institutions, where they may be deemed necessary, to operate as self-contained communities rather than as constituent parts or larger, more cumbersome, and more fragile organizations. A college, for example, is a natural community all by itself, unnaturally joined in the American system to a variety of other dissimilar colleges mostly for the sake of "prestige," and might easily stand alone, the whole of it—student body, faculty, support systems—adding up to no more than a few thousand people, with the educational virtues of such a size that we have already enumerated, plus all the human and communal ones. One might also imagine a self-sufficient cultural center, devoted essentially to developing theatrical and operatic companies, orchestras and dance troupes, rock groups and movies, developing as a concentrated and isolated community, enriched by the diversity within its walls, and then sending individual units traveling throughout the nearby region.

The last question is to what extent a *network* might offer the same advantages in accessibility with none of the disadvantages of size and centralization. Many libraries already work this way, sharing and cooperating in the exchange not only of books but of manuscripts, theses, records, tapes, and other materials, proving that what is needed is not necessarily any one large institution but simply a large enough network of small ones. (My local library in upstate New York is quite tiny but it is connected through the mid-Hudson Valley system with a variety of municipal and college libraries from White Plains to Albany, and I have not yet found it to fail, in a day or two, to turn up any book I was looking for.) An electronic version of this, allowing the exchange of any kind of information or the holding of any kind of conference, is the computer network such as EIES (Electronic Information and Exchange System), which operates nationwide, and there are various citywide versions of a similar kind. Another sort of network, using only alternative technologies and existing telegraph systems, could be used to share medical information and permit a network of small hospitals to draw on more opinions and experts than could ever be gathered into one large medical center, or to keep any small community abreast of new developments elsewhere in any discipline or industry.

This last point inevitably leads us to consideration of widespread systems of any kind. Presumably many independent communities and cities, particularly those within a single bioregion, would want to maintain some kinds of links among themselves—roads, rails, telegraph lines, radio communication. Such linkages, kept to the human scale, would require no complex Amtraks or AT&Ts or RCAs (which would in any case be out of place in a small-scale environment)—they would of course be much reduced in size and wear because of the reduced amount of traffic that self-sufficient settlements would want or need, and reduced in complexity because they would need to be built and maintained in keeping with alternative technologies. But where wanted, these systems would not be all that difficult to forge, needing nothing more than the kind of minimal cooperation that existed once among the cities of the East Coast to maintain their trolley systems, as we will remember, or exists today in Europe to maintain its complex multinational rail network. Whether or not such links would be maintained from coast to coast would depend on the perceived self-interest of a sufficient number of communities—I would suspect that local and regional contacts would be the normal limit—but it is certainly possible, particularly with radio transmissions. It should also be possible for a number of communities to maintain telephone contact (and thereby computer and television links), should they wish to, with the same principle of self-interested cooperation—just as the 1,600 independent telephone companies in the U.S. do today among themselves and with the Bell system. Again the extent of such networks would depend on the communities and how much of their resources they were willing to divert to them, and almost any kind of interlock would be possible.

Self-sufficiency and self-regulation, in short, needn't mean insularity and backwardness. Each community could extend its horizons just as far as its citizens wanted—yet none need be drawn into elaborations it would rather avoid.

I HAVE, it will be noted, used the idea of "self-interest" repeatedly. That seems to be precisely the ingredient necessary—paradoxically perhaps—to overcome the deficiencies of a communitarian world. Lucky it is in such abundant supply.

I feel sure that in the long run, the very long run, the experience of an essentially communal society, whether in neighborhoods, communities, or small cities, will diminish the sort of selfishness and egoism familiar to us today. But I see no reason to think that it will happen anytime soon, and since I would rather consider realities than utopias, I assume we should take individual self-interest into account.

Very well. It is in the self-interest of all but a handful of us to stop pollution, overbreeding, overcrowding, crime, anomie, alienation, urban decay . . . and the rest of it. It is in the self-interest of all but a demented few of us to find social forms that will permit the greatest amount of personal participation and effective control over institutions, economic forms that allow the greatest amount of real accomplishment with the benefits of comfort and security, political forms that encourage the greatest amount of individual freedom and public happiness. And it is in the self-interest of all to do what is necessary to preserve those forms when once they are achieved or approximated, to maintain the limits at which they can be protected, despite the blandishments of such kings and colonizers and centralizers as may arise.

The difficulties of any human-scale society should not be minimized. But it needs only for each person to confront those difficulties in context—to measure the anguish of face-to-face political negotiation against the frustration of powerlessness, the deprivation of around-the-world vacations against the satisfaction of household artifacts handsome and durable, the nuisance of neighborliness against the cold loneliness of abstract cities—and then to make the rational and, precisely, the self-interested choice.

The truly self-interested individual, in caring about individual needs and satisfactions, would ultimately have to care about the people across the way, the neighborhood around, the community and the city at large, for in the larger contexts the particular needs are satisfied best—especially, as we have seen, those truest and most basic needs, of love and friendship and roots. Self-interest in the long run may be only communal interest, ecological interest, planetary interest. Theodore Roszak, at the first Schumacher Society Lectures in Wales in 1978, explored exactly what this does—what this *can*—mean, and his words stand as a fitting assertion of the resonance of a politics on the human scale:

> As the scale of industrial activity mounts, so also (at least along one important line of contemporary dissent in Western society) do our expectations of personal freedom and fulfillment. This, in turn, be-

comes an obstacle to the further expansion and integration of the system. So the system begins to *dis*-integrate, a fitful process that gets registered in the news of the day as truancy in the schools, the soaring divorce rate, declining morals and rising turnover in the workforce, the demise of military conscription, a growing reluctance to compete and conform, a general distrust of leaders, experts, official ideals, public institutions . . . in brief, the spreading ethos of cynicism and recalcitrance that social theorists refer to as "the twilight of authority," "the crisis of legitimation," etc.

But this disintegration is essentially creative, for, in our rising sense of personhood, we find a peculiarly postindustrial quality of life that is wholly incompatible with the mass processing of superscale systems. So we are moved instinctively to assert the human scale that will give us attention, respect, tender loving care. *In asserting the human scale, we subvert the regime of bigness. In subverting bigness, we save the planet.*

And, not in the least incidentally, ourselves.

PART SIX

CONCLUSION

In speaking about the definition of the ideal, we of course have in mind the definition of only four or five prominent features of this ideal. Everything else must inevitably be the realization of these fundamental theories in life.

PETER KROPOTKIN
"The Ideal of a Future System," 1873

Once utopia becomes a goal, the long and difficult undertaking of defining the content for a new society can begin.

RUDOLF MOOS and ROBERT BROWNSTEIN
Environment and Utopia, 1977

There *is* a means of realizing various microsituations through the voluntary actions of persons in a free society. Whether people *will* choose to perform those actions is another matter. Yet, in a free system any large, popular, revolutionary movement should be able to bring about its ends by such a voluntary process. As more and more people see how it works, more and more will wish to participate in or support it.

ROBERT NOZICK
Anarchy, State, and Utopia, 1974

Parthenogenesis

THE TENTH CHAPTER of Leopold Kohr's *The Breakdown of Nations* is an elegant summation of the arguments he has adduced in the previous 187 pages to show that, the large states of the present being unwieldy and disadvantageous, it would be possible to divide and reorder them, eliminating the great powers, liberating the small ones, matching future human politics exactly to those of small, democratic, humanitarian states. "The condition of a small-state world," he concludes, "could be established without force or violence," taking only "the abandonment of a few silly, though cherished, slogans," plus "a bit of diplomacy, and a bit of technique."

The eleventh chapter is the shortest in history. It is entitled: "BUT WILL IT BE DONE?" Its single word: "No!"

I HAVE NO easy answer to that hard question nor any sanguinity about the future. But obviously I would not have taken us all this far if I agreed with him. Not only do I believe, and I hope I have shown, that a human-scale world is *necessary* and *desirable,* I absolutely believe that it is *possible.* (And in moods when I am struck by its apparent improbability I recall what Polish philosopher Leszek Kolakowski has said: "It may well be that the impossible at a given moment can become possible only by being stated at a time when it is impossible.")

Not that it is in any sense inevitable; I have no illusions about that. We are, during these decades toward the end of the century, at a point where there will come a vast change, at least in the lives of the populations of the "developed" world—that seems obvious enough. But what form it will take, in whose hands and in what directions, I do not know. It may well be, as Karen Kodner has written in the *Progressive,* "that ever-larger structures will consolidate their power and, invoking the energy 'emergency' and the need for 'order,' restrict our freedom even further than they already have." Indeed, she says, "so long as our resources, our land, and our capital are controlled by huge corporations and self-perpetuating bureaucracies, inevitability lies with bigness, not smallness." Perhaps. But I think we have seen enough by now of the dangers inherent in that kind of world, and I for one don't believe it would be able to sustain itself—not politically, not economically, not socially—from the strains imposed by systems stretched too far, organizations grown too big, cities stuffed too full, environments picked too clean, technologies made too uncontrollable, populations become too ill-served and restive, economies grown too chaotic and poisonous.

The other possibility, the countervailing one of smallness, is less easy to imagine, of course, but it is not thereby less probable. Thomas Schelling has sagely warned against "a tendency in our planning to confuse the unfamiliar with the improbable," thusly: "the contingency we have not considered looks strange; what looks strange is thought improbable; what is improbable need not be considered seriously." It is exactly that syllogism used by those who find themselves in control of the dominant organs of society, so that they may ridicule what they do not understand, as surely as they must have once mocked Columbus, and Copernicus, and Darwin. They mock in order to dismiss, that they will not have to consider. And if forced to consider they often cannot see—as the ancient Greeks were essentially insensitive to color and did not distinguish between green and blue—and so misinterpret, that they will not have to change. They can, they need, to find ultimate refuge in that nice old English saying, "It won't work and we know because we haven't tried it."

I hope that what I have tried to set out at such considerable detail in the preceding pages offers a way to begin thinking more coherently, more rationally, about this alternative possibility, even for those who may have come to mock. Out of these details, these theories, these visions, it is possible to construct a coherent idea of what a human-scale world, for ourselves particularly but also for the rest of the planet, might look like, how it might work through redesigned settings and institutions to remold and refurbish our lives. No utopias, no panaceas: rather the nuts and bolts of what is needed, what has been done, what might be done, how best the human animal can contrive a world of harmony and plenty, of dignity and freedom, and the scale at which that might be done.

VERY WELL, THEN—that suggests where to go. How do we get there?

It is a question I am asked often, by friends and foes. It is one that many of my mentors in the past and many of my colleagues in the present have tried to respond to. But it is one that cannot—that *must not* —be answered.

I have no blueprint for how the human-scale world is to come about, and obviously if I did it would be ripped asunder and trampled on by free individuals and communities deciding for themselves what they wanted to build and how they wanted to make it. I can suggest a goal, I can certainly urge the necessity and dangle the desirability of that goal, but I could not suggest the way to get there—and if I did suggest one way, one answer, then immediately it would not be a human-scale solution, diverse for diverse people and local conditions, rooted in people's own conditions, wrought by people's own communities. (Anyone who can lead you into paradise, Joe Hill used to say, can lead you out.) If there is to be any realization of the goal, it will come not as someone dictates a path but as people work out for themselves a great variety of ways of taking control over their lives, varying as times and peoples and necessities and settings differ. I do not, as a matter of fact, imagine that it is all that difficult to accomplish, should the need be perceived and the will exist, and I have a sense that in any case the process itself is essential for every group and community to go through in its own particular way.

By saying that, I do not mean to abdicate an authorial responsibility. It is not equivalent to sitting back and saying, in that old barroom phrase, Let's you and him fight it out. It is just that I am as certain as tomorrow that there is no mass party, no first or second or third party, no vanguard or elite, no leader or guru, no treatise or formula, that is going to bring about the human-scale future or could possibly set for us the way it has to come about. That is an animal, as the Germans say, which there is not any of, and it would be irresponsible, however comforting, to suggest otherwise. There is nothing more here than the clean, hard task of showing what the needed and preferable future is and helping anyone who asks in the long, complicated, exciting process of reaching it.

Nothing more—and nothing less.

I OFTEN THINK, in this context, of the man who, when asked directions on how to get to the post office, thought a bit, wrinkled his forehead, and replied, "Well, I wouldn't start from here."

Indeed, I wouldn't. But I am also not as convinced as many that the *here* from which we start is as far away from the *there* we want to get to as many people think it is. Of the durability of the human-scale tradition, even through long periods of antagonism, I do not need to say more. Of the willingness (at least professed) of the bulk of the American people to accept change in this direction, as we have repeatedly seen, almost every poll will attest. Of the number and diversity and skill of people who are right now working in this direction, in practically every locale and every kind of enterprise, there is, as we have also seen, no doubt. The tradition is well-fixed, particularly in this country, the tendency is welcomed in many quarters, and the actual trend is unmistakable.

But there is more. I have seen people change, in ways I wouldn't have thought possible, within my own lifetime. All that we now mean by "The Sixties" changed not only a significant percentage of the young but at least some part of the old, and not merely in ways of style and dress, in ways of thought and perception as well. All kinds of people from all walks and stations have been drawn to the environmental movement, which was born out of nothing more than a vague discontent in the 1960s and is arising to a power that today affects legislation, corporations, individuals, and—environments. A consumers' movement has been born almost from nothing, built of disgruntled shoppers here and unhappy housewives there, a few young researchers and a few old lobbyists, and within the space of a decade or so has become strong enough to have at least some impact on corporate and political worlds. Those who came to see that the Vietnam War was connected to racism and sexism and capitalism in other parts of American life were changed in themselves, by the millions as near as we can tell, and some proportion went on to change others as well.

Perhaps most sweeping of all, the women's movement in all its many guises effected a change, still going on, of quite unprecedented proportions in the attitudes, styles, thoughts, feelings, lives, of people of all races, all ages, all sexes. One measure of the capacity for change of even the most apparently recalcitrant is the simple fact that, opened to this women's movement, all kinds of people in all sorts of circumstances—in the tiny village where I spend one half of my year as on the Greenwich Village block where I spend the other, among my friends the unregenerate Goldwater supporters as among my friends the sophisticated *New York Times* readers, and men nearly as much as women in these places and, as far as I know, around the country—have changed their behavior, their language, their understandings of the world. It suggests itself as being, within the brief breath of a decade or so, one of the most profound and extensive examples of people undergoing nonviolent alteration of beliefs that I can think of in contemporary history, certainly in this century.

And then there are those few nagging nuggets out of history. . . . The Roman Empire, which engaged in the wholesale and violent persecution of Christians in the third century of what we now call the Christian era, had embraced Christianity soon after the beginning of the fourth century and had established it as the official religion of the Roman world by the fifth. . . . Quaker slave-holders throughout the South and in the West Indies during the eighteenth century were persuaded by their brethren, merely on the rightness of the cause, to give up their slaves. . . . In Japan in the 1870s, after the penetration of capitalism, a number of the old feudal families, eventually fifty-five in all, inspired by Prince Hirobumi Ito, determined that the national welfare would be better served if they gave up some major part of their lands and wealth and power to the common cause of building up a nation, and they actually accomplished this divestiture voluntarily. . . . The American South, in the space of twenty years, has changed by its own efforts in many superficial and a number of penetrating ways, becoming a society of, let's call it accommodation if not affection, with the beginnings of a partial reordering of power, that would have been beyond the imagination of most people in the 1950s. . . .

It is purely fanciful, I know, but when I think of the possibilities of change in the states and institutions of the present, I think of Haydn's elegant *Abschiedssymphonie,* in which the musicians deliberately one by one pack up their instruments and leave the stage, until no one is left and the music comes to an end.

WE ARE AT—in the middle of, if you will—a turning point in American, and probably world, history. I know no better than you what is to come. But the choices are clear: drift, distention, or decentralism.

We can go on as we are, trying to muddle through (rather more muddle than through), patching up disintegrating and propping up decaying states, squabbling and warring incessantly over depleting resources and the last few tolerable environments, and coping and groping with increasingly anxious and uncertain lives. Or we can hope for rescue in ever-larger and ever-more-complex systems—757s and 797s after 747s, Models 1199 and 2199 after Model 499—and ever-stronger and more grandiose governments, giving up our liberty for an anticipated security, our initiative for an anticipated welfare system, and all the while moving closer to nuclear and environmental disaster. Or we can work to achieve systems and organizations of a size where we may regulate them, to reshape our landscapes to permit ecologically sound and locally rooted settlements, to create for ourselves a world in which our societies, our economies, our politics are in fact in the hands of those free individuals, those diverse communities and cities, that will be affected by them—a world, of course, at the human scale.

Pentagon . . . Pyramid . . . or Parthenon.

Acknowledgments

I wish to thank the following people for their aid and cooperation: Maurice Ackroyd, Gar Alperovitz, Ben Apfelbaum, P. Max Apfelbaum, Ernest Bader, Donald Barthelme, Philip Bereano, Peter Berg, Myrna Breitbart, Brian Carey, Don Carmahan, John Case, Robin Corey, Richard Cornuelle, Bob DaPrato, Rhoda Epstein, Jürg Federspiel, Annie Fitzpatrick, F.X. Flinn, Gil Friend, David G. Gil, Bertrand Goldberg, Phillip A. Greenberg, David Gurin, Chuck Hamilton, Cynthia Oudejans Harris, Neill Herring, Don Hollister, Lee Johnson, Marion Knox, Milton Kotler, Hal Lenke, Mildred Loomis, Dick and Pat Mackey, Jenny Mansbridge, Michael Marien, John Tepper Marlin, John McClaughry, Linda McDermott, Cynthia Merman, Griscom Morgan, Thomas Morgan, Lynn Nesbit, Robert Nichols, Constance Perrin, John Pfeiffer, John Ramsey, Ray Reece, Pam Roberts, Joyce Rothschild-Whitt, Norman Rush, Kalista Sale, Rebekah Sale, Michael Schaaf, Gertrude Schafer, Daniel Schneider, Ronald Schneider, Neil Seldman, Edward Sorel, Leo Srole, Barry Stein, Bruce Stokes, Bob Swann, Lee Swenson, William I. Thompson, Jaraslov Vanek, Charles Walters, Jr., Matthew Warwick, Bill Whitehead, Langdon Winner, Margaret Wolf, and the staffs of the libraries of Cold Spring, New York, New York City, and the U.S. Military Academy.

A special debt within words is owed to my editor, Joe Kanon.
A special debt beyond words is owed to my wife, Faith Sale.

NOTES

ABBREVIATIONS:

GPO = Government Printing Office; *NYT* = *New York Times;*
SA = *Statistical Abstract of the United States;* UP = University Press.

Part One: Toward the Human Scale

CHAPTER 1

Page 14
Parthenon destruction: *NYT,* 6/10/75, 6/23/78, p. A2; *Chemistry,* July 1975; *New Yorker,* 1/31/77; *Time,* 1/31/77; *Nation,* 6/9/77.

Page 16
Parthenon scale: William Bell Dinsmoor, *The Architecture of Ancient Greece* (Norton, 1975); Le Corbusier, *The Modulor* (MIT, 1954); Kent C. Bloomer and Charles W. Moore, *Body, Memory, and Architecture* (Yale, 1977); Roger Ling, *The Greek World* (Elsevier-Phaidon, 1976).

Page 18
Mumford, *The City in History* (Harcourt, Brace, 1961), p. 124 and Chs. 5 and 6 generally.

CHAPTER 2

Page 22
Suicide: *SA* 1979, No. 298; *Newsweek,* 8/28/78; *N. Y. News,* 11/15–16/77.

Page 23
Alcohol: *NYT,* 12/4/72, p. 28, 10/28/79; Ed and Jovita Addeo, *Why Our Children Drink* (Prentice-Hall, 1975); *Politicks,* 1/17/78.

Murders: *SA* 1979, Nos. 298, 300; *NYT,* 10/28/79, p. 1.

Mental health: President's Commission, 9/15/77 report; *NYT,* 9/16/77, p. 1, 4/28/78.

Poverty: *SA* 1979, Nos, 757–65; *NYT,* 7/17/77, p. 1.

Page 24
Polls: see *Assessment of Public Opinion,* Joint Economic Committee, U.S. Congress, 6/22/77; *Newsweek,* 4/12/76; *American Political Report,* 8/17/79; *NYT,* 11/12/79.

Page 25
Harris, testimony to Senate Government Operations Committee, 1976, and *New Times,* 10/1/76.

Hubbert, *NYT,* 12/3/76.

Page 26
Club of Rome, Donella H. Meadows et al., *The Limits to Growth* (Signet, 1972, 1975).

Page 27
Bateson, *Steps to an Ecology of Mind* (Ballantine, 1972), p. 241.

CHAPTER 3

Page 32
Gutkind, *Twilight of Cities* (Free Press, 1962).

Toynbee, Introduction to Walter Prescott Webb, *The Great Frontier* (Texas UP, 1951, 1952).

Page 33
Harrington, *The Twilight of Capitalism* (Simon and Schuster, 1976), p. 340.

Nisbet, *Twilight of Authority* (Oxford UP, 1975), p. 6 and Ch. 1.

Arendt, *New York Review of Books,* 6/26/75.

Club of Rome, Mihajlo Mesarovic and Eduard Pestel, *Mankind at the Turning Point* (Signet, 1974), p. 146.

Forbes, 11/15/76.

Page 34
Taylor, *Rethink* (Penguin, 1972), p. 21.

Barzun, *The Use and Abuse of Art* (Princeton UP, 1974).

Page 35
Seaborg, quoted in Peter Schrag, *The End of the American Future* (Simon and Schuster, 1973), p. 259.

Page 36
Kahn, *New York,* 8/9/76, and *The Next 200 Years* (Morrow, 1976).

Page 37
Beckerman, *Two Cheers for the Affluent Society* (St. Martins, 1975), quoted in *New York Review of Books,* 6/26/75.

Page 37–38
Human scale: see, generally, Le Corbusier, *Modulor* and *Toward a New Architecture* (Praeger, 1978); Constantine Doxiadis, *Ekistics* (Oxford UP, 1968); Paul D. Spreiregen, *Urban Design: The Architecture of Towns and Cities* (MIT, 1965); Edward T. Hall, *The Hidden Di-*

mension (Doubleday, 1969), esp. Ch. 13; John Mitchell, "A Defence of Sacred Measures," *CoEvolution Quarterly,* Spring 1978; and see Part Three, Ch. 3, below.

CHAPTER 4

Page 43
Lights and back-to-land movement: *NYT,* 6/9/75; author's interview, 12/19/79.

Page 45
Alternate religions: *NYT,* 11/18/76, 6/22/77.

Harris, June 1977.

Cadell, *NYT,* 4/19/77, p.1.

Back to nature: *SA* 1979, No. 398.

U.S. News, 10/22/79.

Page 46
Times, 1/3/77.

Cooperatives: *SA* 1979, No. 920; *Cooperatives at the Crossroads* (Exploratory Project for Economic Alternatives, 1977).

Gardening: Gallup poll for Gardens for All (VT), in *New Roots,* 1–2/80; *NYT,* 6/22/77.

SRI, Duane Elgin, Business Intelligence Program, Guidelines No. 100, 6/76, and *CoEvolution Quarterly,* Summer 1977.

Do-it-yourself: *Building Supply News,* 1978.

Page 48
Shannon, *NYT,* 9/18/76.

Page 49
Dubos, lecture to Human Ecology Conference, NYC, 4/24/77.

Sullivan, *Our Times* (Scribner's, 1935; reprint 1971).

Page 50
Contributors: Milton Kotler, *Nation,* 10/30/76.

Bell, *The Cultural Contradictions of Capitalism* (Basic, 1976).

Page 51
Needleman, *NYT,* 6/22/75.

Part Two: The Burden of Bigness

CHAPTER 1

Page 56–57
Haldane, *"On Being the Right Size," Possible Worlds* (Harper, 1928); and see D'Arcy Wentworth Thompson, *On Growth and Form* (Cambridge UP, 1942); J. M. Smith, *Mathematical Ideas in Biology* (Cambridge UP, 1968); Rousseau, *Social Contract,* Book II, Chs. IX, X; Galileo, *Dialogues* (Macmillan, 1914); St. Augustine, *City of God,* Book III.

Page 58
Higher Education, Algo and Jean Glidden Henderson (Jossey-Bass, 1974), Ch. 11.

Page 59
Studies, and Bernstein, in *Individualizing the System,* ed., Dyckman W. Vermilye (Jossey-Bass, 1976), pp. 18–28; and see Jonathan A. Gallant and John W. Prothero, *Science,* 1/28/72.

CHAPTER 2

Page 62
Radio City: quoted by Saul Maloff, *Commonweal,* 4/28/78.

Cars: *NYT,* 7/21/79.

Page 63
Bainbridge, *The Super-Americans* (Doubleday, 1961), p. 19.

Page 65
DeSales and German journalist, Bainbridge, *Super-Americans,* p. 18.

Page 66
Mishan, *The Costs of Economic Growth* (Praeger, 1967), p. 3.

Page 67
Adams, in Ralph Nader and Mark Green, *Corporate Power In America* (Grossman, 1973), p. 142.

Times (Arthur Burck), 3/4/77, Sec. 3.

Page 68
Van der Ryn, *Smithsonian,* 7/76, p. 47.

Adams, in Nader and Green, *Corporate Power,* pp. 138–39.

Page 69
Parkinson's Law, C. Northcote Parkinson (Ballantine, 1957).

Page 70
Stein, *Size, Efficiency, and Community Enterprise* (Cambridge: Center for Community Economic Development, 1974), pp. 20, 13–14.

Page 71
Mumford, *City,* p. 239.

CHAPTER 3

Page 73
Braudel, *Capitalism and Material Life* (Harper, 1973), p. xv.

Page 75
Goodwin, *New Yorker,* 1/28/74.

Page 76
Doxiadis, *Ekistics,* 8/75; Ritchie-Calder, *After the Seventh Day* (Simon and Schuster, 1961), p. 241; Mumford, *City,* pp. 107–8.

Page 77–78
City sizes: Lauro Martines, *Power and Imagination* (Knopf, 1979); J. C. Russell, *Transactions of the American Philosophical Society,* No. 48, 1958; William Petersen, *Population* (Macmillan, 1961).

Page 78
Russell, quoted in Leopold Kohr, *The Breakdown of Nations* (Routledge & Kegan Paul, 1957, American edition, Dutton, 1978), p. 127.

Page 79
SA 1979, Section 18.

Page 80
Corliss engine: *The Life and Work of George H. Corliss* (American Historical Society, 1930).

CHAPTER 4

Page 82
Kohr, *Breakdown,* pp. xviii–xix.
Page 83–87
Con Ed: *NYT,* 7/15–31/77, 8/1–4/77, 8/10/77, 8/25/77, 8/31/77, 9/1/77; *Newsweek,* 7/25/77, 8/1/77; *New Yorker,* 8/8/77; *Progressive,* 11/77.
Page 84
Inquiry: Con Ed report, *NYT,* 8/31/77.
Page 85
Systemantics (Quadrangle, 1977).

Hauspurg, *NYT,* 7/21/77.
Page 86
Newsweek, 5/7/79.
Page 87
Sanders, *NYT,* 9/9/76, p. 46.
Page 88
Marriott executive: *Wall Street Journal,* 12/2/77, p. 1.

ILSR study: *Self-Reliance,* 11/76.
Page 89
Car recalls: *SA* 1979, Nos. 1104, 1100; *NYT,* 4/29/79, p. 8, 11/12/79, p. A16.
Page 90
Durability of goods: Victor Papanek, *Design for the Real World* (Bantam, 1973), pp. 50–51.

New products: *Business Week,* 3/4/72; Alvin Toffler, *Future Shock* (Random House, 1970), p. 65; Papanek, *Design,* p. 123 and Chs. 5, 6.
Page 91
Diplomat: *NYT,* 6/3/77.

Peterson, *NYT,* 5/13/77, Op Ed.
Page 92
Mayhew and Levinger, *American Journal of Sociology,* 7/76.
Page 93–95
Books: *Authors Guild Bulletin,* 6–8/77, 4–6/78, 4–5/79; *Publishers Weekly,* 7/31/78; *ABA Bulletin,* 7/16/79; *NYT,* 10/23/77, Sec. 3, 8/19/79, Sec. 4; *Village Voice,* 1/24/77.
Page 96
NASA, Gall, *NYT Magazine,* 12/26/76.

Watters, *NYT,* 8/31/76, Op Ed.

CHAPTER 5

Page 97
Kohr, *Breakdown,* pp. 62–64.
Page 99
Government sector: Eli Ginzberg, *Scientific American,* 12/76.

Federal Advisory Commission, *Regional Decision-Making* (GPO, 1973).
Page 100
Federal regulations: *NYT,* 12/23/77, 1/22/78, Sec. 4; *US News,* 6/20/77.
Page 101
Goodwin, *New Yorker,* 1/28/74.

Rudd, *NYT,* 10/29/77.
Page 102
Times, 8/3/77, p. D15.; drought relief, *NYT,* 7/25/77, p. 10.
Page 104
McNamara, *NYT,* 11/28/76, Op Ed.

Michels, Weber, and Nisbet, in Nisbet, *Twilight,* pp. 52–60.

Kennedy aide, quoted by Reeves, *New York,* 12/6/76; Carter, *NYT,* 7/26/77, p. 11; Nisbet, *Twilight,* p. 55.
Page 105–06
Pentagon: *NYT,* 10/18/77, 3/8/79; *Newsweek,* 11/14/77.

CHAPTER 6

Page 107
For other governments, see James Cornford, ed., *The Failure of the State* (Totowa, N.J.: Rowman & Littlefield, 1975).
Page 108
Federal Commission on Paperwork, *Final Report,* (GPO, 10/4/77); *NYT,* 10/5/77.

Schultze, Godkin Lectures, 1976, quoted by Lewis, *NYT,* 12/2/76.
Page 110
Jubal Hale, *NYT,* 5/14/75, 5/28/75, 6/18/75.

Proxmire, *NYT,* 8/21/77.
Page 112
Congress study, Joint Economic Committee, *Economics of Federal Subsidy Programs,* 1/11/72; a Common Cause study, 5/78, found $136 billion.

Corruption: *NYT,* 7/6/77, 2/13/78 and 3/2/78 (Safire), 4/16–18/78.

CHAPTER 7

Page 116
Regulation costs: Murray Weidenbaum, *The Future of Business Regulation* (N.Y.: Amacom Press, 1979), and *The Costs of Government Regulation of Business,* Joint Economic Committee (GPO, 1978); *NYT,* 11/4/79, Sec. 4.
Page 117
Advisory Committee, *Forging America's Future* (GPO, 1976), pp. 8–9.
Page 118–22
South Bronx: Jack Newfield and Paul DuBrul, *The Abuse of Power* (Viking, 1977); *Fortune,* 11/75; Women's City Club of New York, "Report on the South Bronx," 1977; *NYT,* 10/6–14/77, 4/13/78, 10/10/78, 2/7/79, 2/11/79, Sec. 4, 11/30/79, 12/19/79.
Page 119
Federal government: see, e.g., Bernard J. Frieden and Marshall Kaplan, *The Politics of Neglect* (MIT, 1975).

Urban Institute, *Search,* Spring 1977.

Page 123
Mumford, *City,* p. 355; see also Peter Kropotkin, *Mutual Aid;* Henry Sumner Maine, *Popular Government;* Hilaire Belloc, *The Servile State;* Kohr, *Breakdown.*
Page 124–26
Mail service: John Haldi, *Postal Monopoly* (American Enterprise Institute, 1974); Bob Black, *United States vs. Alternate Systems, Inc.,* privately printed, Pittsburgh, Kans., 1976; M. J. Bowyer, *They Carried the Mail* (Washington, D.C.: Robert B. Luce, 1972); *NYT,* 5/30/77, 4/15/78, 6/15/78, 7/15/79, Sec. 3, 11/28/79.
Page 126
Times editorial, 1/5/78.
Page 126–27
Education: Lawrence A. Cremin, *Traditions of American Education* (Basic, 1977); *Sa* 1979, No. 215.
Page 127
Long Island board president, *NYT,* 7/18/77.
Page 127–28
Micronesia: Fox Butterfield, *NYT Magazine,* 11/27/77.

CHAPTER 8

Page 129
Warfare: Wright, *A Study of War* (Chicago UP, 1965), esp. Tables 22–54; and see Pitirim Sorokin, *Social and Cultural Dynamics,* Vol. II (American Book Company, 1937); Lewis F. Richardson, *Statistics of Deadly Quarrels* (Quadrangle, 1960).
Page 130
Toynbee, *A Study of History* (Oxford UP, 1947), pp. 244, 553.
Page 131
Mumford, *City,* pp. 42–45, 197, 214, 239, 242.
Page 133
Phelps Brown and Hopkins, *Forbes,* 3/1/75, 11/15/76.
Page 134
British historian, Thomas Frederick Tout, in Mumford, *City,* p. 353.
Page 135
Nash, *Perspectives on Administration* (California UP, 1969).
Page 136
French Convention, quoted by Kropotkin, "The State," e.g., in *Selected Writings,* ed., Martin A. Miller (MIT, 1970).
Page 137
Tocqueville, *The Old Regime and the French Revolution* (Anchor, 1955), p. 209.
Page 141
Tocqueville, *Democracy in America,* Vol. II (Knopf, 1951), p. 268.

Part Three: Society on a Human Scale

CHAPTER 1

Page 145
Malaria: *NYT,* 8/23/77, p. 20, 11/9/77, p. 2; *Newsweek,* 6/26/78.
Page 147
Estimates: Lee R. Dice, *Natural Communities* (Michigan UP, 1952), p. 7.
Page 150–51
MacLeish, "America Was Promises," 1939; Starr, speech to Electric Power Research Institute symposium, San Francisco, April 1979; Corps of Engineers, see Arthur Morgan, *Dams and Other Disasters* (Porter Sargent, 1971); Marx, *Acts of God, Acts of Man* (Coward, McCann & Geoghegan, 1977), p. 43; and see generally, William Leiss, *The Domination of Nature* (Braziller, 1972); Barry Weisberg, *Beyond Repair* (Beacon, 1971).
Page 151
McHarg, *Design with Nature* (Natural History Press, 1969).
Page 152
Will, *Newsweek,* 7/25/77.
Page 153
Federal board: *NYT,* 7/13/78.

Hutchinson, *Scientific American,* 9/70.
Page 154–155
Groundnut scheme: Ritchie-Calder, *Seventh Day,* pp. 302–8; Lord Hailey, *An African Survey* (Oxford UP, 1957), pp. 844–46, 906–7, 1296, 1559.

CHAPTER 2

Page 156
Ellul, in Robert Hunter, *The Enemies of Anarchy* (Viking, 1970), pp. 155–56.
Page 157
Dickson (Universe Books, 1975), p. 39.
Page 159
Smithsonian, James K. Page, 7/76.
Page 159–61
Alternative technology generally: Dickson, *The Politics of Alternative Technology,* op. cit.; Langdon Winner, *Autonomous Technology* (MIT, 1978); P. D. Dunn, *Appropriate Technology* (Schocken, 1979); Ivan Illich, *Tools for Conviviality* (Harper, 1973); Godfrey Boyle and Peter Harper, *Radical Technology* (Pantheon, 1976); *RAIN* editors, *RAINBOOK* (Schocken, 1977), and *Stepping Stones* (Schocken, 1978); N. Jequier, ed., *Appropriate Technology* (Paris: OECD, 1976); Lewis Mumford, *The Myth of the Machine* (Harcourt, 1970); Murray Bookchin, *Post-Scarcity Anarchism* (Ramparts, 1971); and of course E. F. Schumacher, *Small Is Beautiful* (Harper, 1973). Also: *Village Technology Handbook,* Volunteers in Technical Assistance

(Md.), 1970; *Appropriate Technology Sourcebook,* Volunteers in Asia (Calif.), 1976; *Appropriate Technology in the U.S., an Exploratory Study,* NSF (GPO, 1977); *Research in Philosophy and Technology,* Vol. I (Conn.: JAI Press, 1978); *Futurist,* 2/75; *Resurgence,* 1/78; *Science for the People,* 9/76; and *RAIN* and *CoEvolution Quarterly,* regularly.

Page 163
Banks, *CoEvolution Quarterly,* Spring 1975, and *Soft-Tech* (Penguin, 1978), p. 35.

CHAPTER 3

Page 165
MIT project: Robert Gutman, *People and Buildings* (Basic, 1972).

Pruitt-Igoe: William L. Yancy, in John Helmer and Neil A. Eddington, *Urbanman* (Free Press, 1973).

Page 166
Newman, *Defensible Space* (Macmillan, 1973).

Baker, in Nisbet, *Twilight,* p. 36.

Deasey (Schenkman-Wiley, 1974).

Page 167
Architecture and human scale: Heath Licklider, *Architectural Scale* (Braziller, 1965); Bruno Zevi, *Architecture as Space* (Horizon, 1957, 1974); Alexander Tzonis, *Towards a Non-Oppressive Environment* (Boston: i Press, 1972); Jonathan Lang, et al., eds., *Designing for Human Behavior* (Stroudsburg, Pa.: Dowden, Hutchinson, and Ross, 1974); Rudolf Moos, *The Human Context* (Wiley, 1976); Geoffrey Scott, *The Architecture of Humanism* (Gloucester, Mass.: Peter Smith, 1965); Georges Gromort, *Essai sur la théorie de l'architecture* (Paris: Vincent, Frere, 1946); Mumford, *City;* Spreiregen, *Urban Design,* op. cit.

Page 168
Zevi, *Architecture,* p. 76.

Page 169
Progressive Architecture, 6/65.

Kira, *The Bathroom* (Penguin, 1976).

Page 170
Diffrient, *Design Quarterly,* No. 96, 1975.

Dreyfuss, *Humanscale* (MIT, 1974), and *The Measure of Man* (Whitney Library of Design, 1966).

Page 172
Maertens, *Der Optische-Maassstab, oder die Theorie und Praxis des ästhetischen Sehens in den bildenden Künsten* (Berlin: Kgl. Baurath, Wasmuth, 1884); Hegemann and Peets (N.Y.: Wenzel and Krakow, 1922); Blumenfeld (MIT, 1967).

Page 175
Blumenfeld, *Metropolis,* Ch. 23.

Page 176
Bloomer and Moore, *Body, Memory, and Architecture* (Yale, 1977), pp.ix., 77.

Gibson (Houghton-Mifflin, 1950), and in Bloomer and Moore, *Body, Memory,* Ch. 4.

Page 177
Bloomer and Moore, *Body, Memory,* p. 1; Yudall, *ibid.,* p. 61.

Page 178
Blumenfeld, *Metropolis,* Ch. 23.

Bloomer and Moore, *Body, Memory,* p. 4.

CHAPTER 4

Page 179
Campbell, in John E. Pfeiffer, *The Emergence of Man* (Harper, 1972), p. 104.

Page 180
Hawley, *Human Ecology: A Theory of Community Structure* (N.Y.: Ronald Press, 1950).

Dubos, in Constantine Doxiadis, ed., *Anthropopolis* (Norton, 1974), pp. 259–60.

Page 181
Murdoch, *Social Structure* (Macmillan, 1949), pp. 79–80.

Alexander, in Helmer and Eddington, eds., *Urbanman,* p. 246.

Page 182
Pfeiffer, *Man,* pp. 376–77.

Primates: e.g., Michael Chance and Clifford Jolly, *Social Groups of Monkeys, Apes, and Men* (Dutton, 1970).

Page 182–83
Birdsell, in Pfeiffer, *Man,* pp. 376–77; Robert L. Carneiro, *Science,* 8/21/70; Clastres, *Society Against the State* (Urizen, 1977), p. 72; Adams, in Pfeiffer, *The Emergence of Society* (McGraw-Hill, 1977), pp. 155–56.

Page 183
Dubos, in *Anthropopolis,* ed., Doxiadis, p. 259.

Page 183–85
Neighborhood: Blumenfeld, *Metropolis;* Doxiadis, *Ekistics* (Oxford); Taylor, *Rethink,* pp. 151–52; Terrence Lee, *Human Relations,* 8/68; Charles Erasmus, *In Search of the Common Good* (Free Press, 1977), p. 130.

Page 184
Murdoch, in Pfeiffer, *Man,* p. 377.

Page 185
Sanders and Baker, and Rathje, in Pfeiffer, *Society,* pp. 467–69.

Sjoberg, in *Cities,* ed., *Scientific American* (Knopf, 1966), p. 28.

Doxiadis, *Science,* 10/10/68, and *Ekistics,* 6/77.

Page 186
Perry, *The Neighborhood Unit,* Vol. VII of *Regional Survey of New York and Its Environs,* 1929; and see Melville C. Branch, ed., *Urban Planning Theory* (Wiley, 1975).

Page 205
Morrison, *NYT,* 2/25/77.

Thompson, *Plenty of People* (Lancaster, Pa.: Jacques Cattel Press, 1944), p. 130.

De Torres, *NYT,* 8/6/77.

Page 206
Rand, in *Nation's Cities,* National League of Cities, 11/77.

Sundquist (Brookings, 1975), p. 281.

Page 207
Jacobs, *The Economy of Cities* (Vintage, 1970), p. 104.

Bookchin, *Limits,* p. 66.

CHAPTER 6

Page 209
Wellesley-Miller, in *Earth's Answer,* ed., Michael Katz, et al. (Harper, 1977).

Page 209–12
New Alchemists: Nancy Jack Todd, ed., *The Book of the New Alchemists* (Dutton, 1977); John Todd and Nancy Jack Todd, *Tomorrow Is Our Permanent Address* (Harper, 1980); *NYT Magazine,* 8/8/76; *People and Energy,* 4–5/79; *New Roots,* 11/79.

Page 212
John Todd, in *New Alchemists,* ed., N. J. Todd, p. 128.

Page 214
Miles, *Awakening from the American Dream* (Universe, 1976), p. 175.

Page 215
Plunkett, in Ray Reece, *The Sun Betrayed* (South End Press, 1979), pp. 79–80.

Page 215–27
Solar energy, generally: see Amory Lovins, *Soft Energy Paths* (Penguin, 1977); *Energy Strategies,* U.S. Senate Select Committee on Small Business, 2 vols. (GPO, 12/9/76); Barry Commoner, *The Poverty of Power* (Knopf, 1976); Denis Hayes, *Rays of Hope* (Norton, 1977); Alan Okagaki, et al., *Solar Energy: One Way to Citizen Control* (Washington, D.C.: Center for Science in the Public Interest, 1976); and *Solar Age, Alternate Sources of Energy, People and Energy, RAIN,* and *Science,* regularly.

Page 218
Lovins, *Soft Energy,* p. 45.

Page 219
NSF-NASA, *Solar Energy as a National Energy Resource* (GPO, 12/72).

Science, 9/2/77.

Page 220
Federal Power Commission, *Smithsonian,* 9/77.

Page 221
Lovins, *Soft Energy,* p. 135.

Page 223
Kapp, *The Social Costs of Private Enterprise* (Schocken, 1971), p. xi.

Page 224
Todd, in *New Alchemists,* ed., N. J. Todd, p. 66.

Page 225
Miles, *Awakening,* p. 90.

Lovins, *Soft Energy,* p. 101, and in *Energy Strategies,* p. 229; see also Portola Institute, *Energy Primer,* 1977; Lee Johnson, in *Stepping Stones,* eds., De Moll and Coe (Schocken, 1978).

People and Energy, 6/78.

Spectrolab, in Okagaki, *Solar Energy,* pp. 81–92.

Page 226
Institute for Local Self-Reliance, *Self-Reliance,* 9–10/78.

Page 227
Miles, *Awakening,* p. 135.

CHAPTER 7

Page 229–30
Food statistics: *SA* 1979, Nos. 1171–78, 1197; Richard Merrill, ed., *Radical Agriculture* (Harper, 1976); Joe Belden, *Toward a National Food Policy* (Washington, D.C.: Exploratory Project for Economic Alternatives, 1976); Frances Moore Lappé and Joseph Collins, *Food First* (San Francisco: Food First, 1977); Michael Perelman, *The Myth of Agricultural Efficiency* (Universe, 1977).

Page 230
Zwerdling, *New Times,* 5/29/78.

Page 231
Whiteside, *New Yorker,* 1/24/77.

Berry, *The Unsettling of America* (Sierra Club, 1977), p. 45.

Page 232
GAO, "Grain Reserves," 3/26/76, p. 11.

Berry, *Unsettling,* p. 43.

Page 233–35
Small farm: Belden, *National Food Policy,* Ch. 6; Merrill, ed., *Radical Agriculture,* Chs. 5, 6; J. D. Belanger, *The Case for the Family Farm* (Waterloo, Wis.: Countryside Publications, 1976); Philip M. Raup, "Economies and Diseconomies of Large-Scale Agriculture," U. Minnesota, 1969; USDA studies: *Agriculture and Economic Growth,* 1963; J. Patrick Madden, *Economics of Size in Farming,* 1967; Warren R. Bailey, *The One-Man Farm,* 8/73; Hearing, Senate Subcommittee on Migratory Labor, 1/11–13/72; Hearing, Senate Small Business and Interior Committees, 7/22/75.

Page 233
1973 report: Bailey, *One-Man Farm.*

Perelman, and government official, in Merrill, *Radical Agriculture,* Ch. 6.

Page 234
Economist, quoted in *Resurgence* (UK), 11–12/77, p. 27.

Agribusiness officials: Belden, *National Food Policy: RAIN,* 7/77.

Organic: Zwerdling, *New Times,* 5/29/78; *Organic Gardening* magazine, 1975–.

Page 235
Co-ops: *Business Week,* 2/7/77; *Co-op* (formerly *New Harbinger),* Summer 1978, 7–8/79, Winter 1979.

Page 238–39
World food: Lappé and Collins, *Food First,* and *Mother Jones,* 8/77; Michael Lipton, *Why Poor People Stay Poor* (Harvard UP, 1977); Barbara Ward, *Home of Man,* Ch. 19; *WIN,* 1/30/75; Belden, *National Food Policy.*

Page 239
Timmer, *NYT,* 12/4/76, p. 1.

McLaughlin, *NYT,* 6/14/77, Op Ed.

Page 240
Heady, *Scientific American,* 9/76.

Viability Project, *The Family Farm in California* (State of California), 11/77.

Page 241
Goldschmidt, *Small Business and the Community,* U.S. Senate report, Special Committee on Farm Problems, 12/46, and in *The People's Land,* ed., Peter Barnes (Rodale Press, 1975), and *As You Sow* (Free Press, 1947); update in Belden, *National Food Policy* and Viability Project, *Family Farm,* pp. 216 ff.

CHAPTER 8

Page 243
Love Canal: Michael Brown, *NYT Magazine,* 1/21/79; *NYT,* 8/3/78, p. 1.

Page 244–46
Sewage: Neil Seldman, *Garbage in America* (1975) and *New Directions in Solid Waste Planning* (1977), Institute for Local Self-Reliance; Leonard A. Stevens, *Clean Water* (Dutton, 1974); *NYT,* 7/9/76, p. A9.

Page 245
Singer, in Milton Moss, ed., *The Measurement of Economic and Social Performance,* (National Bureau of Economic Research, 1973), p. 536.

Page 246–47
Organic garbage: Seldman, *Garbage;* Merrill, *Radical Agriculture,* p. 312; *Energy in Solid Waste* (1975) and *Report to the President* (1972), Citizens Advisory Committee on Environmental Quality, GPO.

Page 247–49
Recycling: Jerome Goldstein, *Recycling* (Schocken, 1979); *Energy in Solid Waste; Self-Reliance,* No. 20, 1979, No. 21, 1980; Earl Cook, *Technology Review,* 6/75.

Page 248
Seldman and Bree, in Seldman, *New Directions.*

Page 250
EPA, *Wall Street Journal,* 5/22/79.

CHAPTER 9

Page 251
Ragtime (Random House, 1975), pp. 75–78, 80.

Page 252
GM: Bradford Snell, *American Ground Transportation,* Senate hearing (GPO, 2/74); Stephen Geisler, *WIN,* 8/17/78; Larry Sawyer, *Review of Radical Political Economics,* Spring 1975.

Page 254
Citizens for Clean Air, report by Brian Ketcham, *Societal Cost Accounting,* 1976.

Page 256
Ward, *Home of Man,* p. 142.

Page 258
Census, *SA* 1977, No. 1005; later estimates, *SA* 1979, No. 1113.

Ward, *Home of Man,* p. 150.

Page 259
Wilson, *RAIN,* 10/76, and *RAINBOOK,* p. 233.

Page 259–61
Bicycles: *NYT,* 1/29/77, Sec. 3, 10/7/79, Sec. 3; Papanek, *Design for the Real World,* pp. 201–6; Campbell bike, *Doing It!,* No. 4, 12/76, p. 19.

Page 260
Energy efficiency chart, originally by Vance E. Tucker, Duke University, from *Scientific American,* 3/73.

Page 261
Electric car: *Electric Vehicle News,* 5/77; *Consumer Reports,* 10/76; *People and Energy,* 3/78; *NYT,* 7/1/79, Sec. 3; *Symposium,* Edison Electric Institute advertising insert, Summer 1979.

Page 264
LTAs: *NYT,* 3/4/79, Sec. 4; *Lighter Than Air Society Newsletter* (Akron, Ohio).

British writer, Patrick Rivers, in *Radical Technology,* eds. Boyle and Harper, p. 227.

Page 265
Illich, *Energy and Equity* (Harper, 1974), pp. 59–61.

CHAPTER 10

Page 266
Thomas, in Michael Katz, et al., *Earth's Answer,* pp. 156 ff.

Page 267
President, *NYT,* 5/6/78, p. 11.

Illich, *Medical Nemesis* (Pantheon, 1976); and see Rick J. Carlson, *The End of Medicine* (Wiley, 1975); David Kotelchuck, ed., *Prognosis Negative* (Random House, 1976).

Surgery: *Newsweek,* 4/10/78; *NYT,* 5/24/77, 11/16/77.

Hospitals: Steven Jones, ed., *The Health Care Delivery System in the U.S.* (Springer, 1977); Ruth Roemer, *Planning Urban*

Health Systems (Springer, 1975); *Newsweek,* 5/9/77; *NYT,* 9/27/77.

Drugs: Barbara Ehrenreich, *Mother Jones,* 4/79; *NYT,* 8/17/79, p. A11.

McKinlays, in Drummon, *Mother Jones,* 1/78.

Page 268
Thomas, *NYT Magazine,* 7/4/76.

Page 269–70
Improved health: Edward A. Mortimer, *Science,* 5/26/78; McKinlays, in Drummon, *Mother Jones,* 4/79; Thomas McKeown, *The Modern Rise of Population* (Academic Press, 1976); *Science,* 5/12/77.

Page 270
Life expectancy: Illich, *Medical Nemesis; SA* 1979, No. 100; *NYT,* 5/30/78, p. B5; Whalen, *NYT,* 4/17/77, Op Ed.

Wildavsky and Somers, *NYT,* 5/30/78, p. B5; *NYT,* 1/20/80, p. 44.

Page 271
Williams, *New Times,* 11/78.

McNeill (Doubleday, 1976).

Page 272
Paramedics, *National Registry of Emergency Medical Technicians,* 10/78.

Page 272–73
Drugs: Ehrenreich, *Mother Jones,* 4/79; *Scientific American,* 9/73.

Page 273
JAGA, 11/76

Abel-Smith, *Value for Money in Health Services* (St. Martins, 1976).

Page 274–75
Optimal hospital size: *Hospitals,* 10/1/74; Leonard Ullman, *Institution and Outcome* (Pergamon, 1967); Lawrence Linn, *Archives of General Psychiatrics,* No. 23, 1970; M.S. Feldstein, in *Hospital Efficiency and Public Policy,* ed., Harry Greenfield (Praeger, 1973); Roemer, *Planning Urban Health Systems.*

Page 275
Chicago Regional Hospital Study, *Misused and Misplaced Hospitals and Doctors,* Association of American Geographers, Resource Paper No. 22, 1973.

Rutgers, *NYT,* 5/21/78, p. 47.

1969 survey, Greenfield, *Hospital Efficiency.*

Page 276–77
Ecotopia (Berkeley: Banyan Tree Books, 1975), pp. 142–44.

CHAPTER 11

Page 278–80
Barker and P. V. Gump (Stanford UP, 1964), esp. pp. 195–202.

Page 280
Statistics: *SA* 1978, No. 208.

Page 281
Illiteracy/achievement: Office of Education report, 10/29/75; Paul Copperman, *The Literary Hoax* (Morrow, 1980); Ivan Illich, *Deschooling Society* (Harper, 1972); *NYT Magazine,* 8/28/77; *Newsweek,* 11/6/78.

Page 282
Wicker and Baird, *Journal of Educational Psychology,* No. 60, 1969; see also Summers and Wolfe, *Business Review,* 2/75; Walberg, *Human Relations,* No. 22, 1969; Stanton Leggett, *Nation's Schools,* 9/70; E.P. Willens, *Child Development,* 12/67; Wicker, *Journal of Personal and Social Psychology,* No. 10, 1968, No. 13, 1969.

Page 283
WHO, in Arthur Morgan, *Community Comments* (Yellow Springs, Ohio: 10/70).

Page 284
Gallup poll, in Jonathan P. Sher, *Education in Rural America* (Boulder Colo.: Westview Press, 1977).

Coles, Foreword to Sher, *Education,* p. xiv.

Page 285
Morgan, *NYT,* 11/29/78, Op Ed.

Page 286
Large school: see Sher, *Education.*

Page 288
Footnote: *The Adults Learning Projects* (Ontario Institute for Studies in Education, 1971).

Page 289
Times, 12/7/78, Sec. 3.

Bernstein, in Vermilye, *Individualizing the System,* Ch. 2; and see Virginia V. Smith and Alison R. Bernstein, *The Impersonal Campus* (Jossey-Bass, 1979).

Page 290
Astin, *Four Critical Years* (Jossey-Bass, 1977), pp. 230–31, 244–4.

Gallant and Prothero, *Science,* 1/28/72.

Page 292
Russell, *NYT,* 12/1/77, p. 40.

Part Four: Economics on a Human Scale

CHAPTER 1

Page 297
NYT, 11/1/76, p. 65.

Page 298
Economic chaos: wealth—*SA* 1979, No. 741; Maurice Zeitlin, *Progressive,* 7/79; debt—*Newsweek,* 1/8/79; capital—*Business Week,* 9/79, *NYT,* 1/9/77, Sec. 3; Keyserling, *NYT,* 10/1/76, Letter.

Heilbroner, *Business Civilization in Decline* (Norton, 1976), p. 9.

Page 300
Science, 1/19/79.

Page 302
Tiger, *Newsweek,* 9/4/78, "My Turn."

Page 303
Henderson, *Planning Review,* 4–5/74; and

see Henderson, *Creating Alternative Futures* (Berkley, 1978).

Galbraith, *New York Review of Books,* 4/20/78.

Page 304
Kahn, *The Year 2000* (Macmillan, 1967), pp. 116, 74.

Page 305
Heilbroner, *New Yorker,* 8/28/78.

Page 306
Taylor, *Rethink,* p. 258.

CHAPTER 2

Page 307–09
Brewing industry: *NYT,* 2/24/79, 8/9/78, p. C1, 8/21/77, Sec. 3, 6/18 and 6/25/78, Sec. 3, 4/29/79, Sec. 4.

Page 310–11
Economies of scale: Stein, *Size, Efficiency and Community Enterprise,* Center for Community Economic Development, and testimony before Senate Small Business Committee, 12/2/75.

Page 311–13
Efficiency: Florman, *Harpers,* 8/77; Scherer, cited in Nader and Green, *NYT,* 4/17/79, Op Ed; Boulding and Union Carbide, in Stein, *Size, Efficiency,* pp. 5, 49; House Subcommittee, "Investigations of Conglomerate Corporations," 1971, p. 411; Treasury, in Stahrl Edmonds, *Futurist,* 2/79.

Page 313–15
Innovation: Galbraith, *American Capitalism* (Houghton-Mifflin, 1952), p. 91; Jacobs, *Economy of Cities,* pp. 52, 71 ff.; Norris, *NYT,* 3/4/77, Op Ed; Noyce, *NYT,* 12/10/76, p. D1; Quinn, in Kohr, *Breakdown,* p. 164; Schmookler, in Stein, *Size, Efficiency,* p. 34; Cooper, *HBR,* 5–6/64; Department of Commerce, *Technical Innovation,* Report of Panel on Invention and Innovation, 1967; Jewkes, *The Sources of Invention* (Norton, 1969); *NYT,* 1/18/80, p. D1.

Page 315–17
Cheaper prices: Borsodi, *Flight from the City* (Harper, 1933, 1972), pp. 10 ff.

Page 317
Profitability: Twentieth Century Fund and TNEC, Kohr, *Breakdown,* pp. 162–63; Stekler, in Stein, *Size, Efficiency,* p. 13; Antitrust Subcommittee, *Economic Concentration,* Part 4, pp. 1755 ff.; Stein, *Size, Efficiency,* p. 14.

CHAPTER 3

Page 320
Nordhaus and Tobin, in *The Measurement of Economic and Social Performance,* ed., Moss, pp. 509 ff.

Page 321
Consumer goods: *SA* 1978, Nos. 630–41. Cosmetic Association, "Bits and Pieces," *Mother Earth News,* 1/79.

Page 322
California workers: *NYT,* 1/18/79, p. A16, 7/3/79, Op Ed.

De Grazia, *Of Time, Work, and Leisure* (1962).

Sahlins, *Stone Age Economics* (Chicago: Aldine, 1972); and see Marvin Harris, *Cannibals and Kings* (Random House, 1977), pp. 9 ff.

Page 323
Vanek, *Scientific American,* 11/74.

Page 324
Model changes: Stein, *Size, Efficiency,* p. 41.

Packaging: *NYT,* 1/25/78, Sec. 3.

Page 325
Table: based generally on *SA* 1978, 1979, plus: fashion—figured at 25 percent of fashion-connected (clothing, auto, etc.) personal consumption, *SA* 1979, Nos. 723, 1465; shoddy work—Hart, in *The Center Magazine,* 1–2/72, p. 45, adjusted for inflation; advertising—*SA* 1979, No. 1000, packaging (above); business cheating—half of estimates of Beverly C. Moore, Jr., Corporate Accountability Research Group, in David Hapgood, *The Screwing of the Average Man* (Bantam, 1975), p. 265, plus data from Big Business Day (Washington, D.C.), 4/17/80; social unease—*SA* 1973, No. 249, adjusted for inflation; *SA* 1979, Nos. 405, 723; crime—*SA* 1973, No. 249; *SA* 1979, No. 313; accidents—*Science,* 8/21/70, pp. 723 ff.; government regulations—Weidenbaum, *Business Regulation,* Part II, Ch. 5; government waste—*NYT,* 4/16/78, p. 1, 12/14/78, p. A20, Common Cause in *NYT,* 5/7/78; tax loopholes—Philip M. Stern, *The Rape of the Taxpayer* (Random House, 1972), p. 5.

Page 327
Burns, *The Household Economy* (Beacon, 1977), p. 90.

Kohr, *The Overdeveloped Nations,* p. 45.

Page 328
Bender, in *Stepping Stones,* eds., De Moll and Coe, p. 196.

CHAPTER 4

Page 329
Mill, Book V, Ch. IV.

Page 330
Modern Sources: Boulding, e.g., in Herman E. Daly, *Essays Toward a Steady-State Economy* (Cuernavaca: Centro Intercultural de Documentacion, n.d.); Mishan (Praeger, 1967); Georgescu-Roegen, *The Entropy Law and the Economic Process* (Harvard UP, 1971); *The Limits to Growth* (Universe Books, 1972); Daly, *Essays Toward;* Schumacher, *Small Is Beautiful;* Kohr (Wales: Christopher Davies, 1973).

Page 331
Daly, *Steady-State Economics* (San Francisco: W. H. Freeman, 1977), p. 47; Boulding, in Daly, *Steady-State.*
Page 333
Hardin, *Science,* 12/13/68, and *Exploring New Ethics for Survival* (Viking, 1972).
Page 334
Ivins, *NYT,* 7/29/79, p. 1.
Page 335
Daly, *Essays Toward.*

CHAPTER 5

Page 336
Newsweek, 4/10/78; see also *Business Week,* 5/13/78; *Fortune,* 10/9/78.
Page 337
Gutmann, *Wall Street Journal,* 11/30/78, p. 1, and *U.S. News,* 10/22/79.

Household economy: Scott Burns, *Household Economy.*
Page 338
Nordhaus and Tobin, in *Measurement of Economic and Social Performance,* ed., Moss.
Page 339
Burns, *Household Economy,* pp. 117, 123.
Page 340
Times, 3/30/78.

Work in America (MIT, 1973), pp. 15–16.

SRI, Duane Elgin, Business Intelligence Program, Guidelines No. 100, 6/76, and *CoEvolution Quarterly,* Summer 1977.

Harris, 12/1/75, 12/4/75, 11/8/76, 5/23/77; and see Hazel Henderson, *Creating Alternative Futures,* p. 395.

Inglehart, *The Silent Revolution* (Princeton UP, 1977).

CHAPTER 6

Page 343–44
Five: A. Paul Hare, *Handbook of Small Group Research* (Macmillan, 1972), pp. 221 ff.; Handy, *Understanding Organizations* (Penguin, 1976), p. 152; Parkinson, *Parkinson's Law,* Ch. 4; *Small Groups* (Knopf, 1967); James, *American Sociological Review,* 8/51; Thayer (New Viewpoints/ Franklin Watts, 1973), pp. 8, 199–200.
Page 344–46
Small group studies: note above, and e.g., Rudolf Moos, ed., *Human Context* (Wiley, 1976); Barker and Gump, *Big School, Small School,* Part II, Ch. 11, Ch. 3; D. W. Johnson, *Joining Together* (Prentice-Hall, 1975); E. T. Reeves, *The Dynamics of Group Behavior* (American Management Association, 1970); B. Indik, *Human Relations,* No. 18, 1965; Lyman Porter and Edward Lawler III, *Psychological Bulletin,* No. 64, 1965.
Page 345
Hawthorne: Elton Mayo, *The Human Problems of an Industrial Civilisation* (Macmillan, 1933); and for crucial reanalysis, Paul Blumberg, *Industrial Democracy* (Schocken, 1973).
Page 346
Gilbert, *NYT,* 3/18/78, p. 29.
Page 346–47
Small plants: S. Talacchi, *Administration Science Quarterly,* No. 5, 1960; Moos, *Human Context,* esp. pp. 257 ff., 336 ff., 410.
Page 347
Stanford Business School, in Moos, *Human Context.*
Page 348–50
Mancur Olson, Jr., *The Logic of Collective Action* (Schocken, 1971), quotes from pp. 2, 28, 34–35, 44, 48, 54–55, 62, 166.
Page 350
Productivity: see Kohr, *Breakdown,* pp. 157 ff.

CHAPTER 7

Page 353
Brandeis, *The Curse of Bigness* (Viking, 1935).

Wood, quoted in *Nation,* 11/27/76, p. 564.

Quinn, *Nation,* 3/7/53.
Page 353–61
Workplace democracy: see generally, Jaroslav Vanek, *Self-Management* (Penguin, 1975); Vanek, *The General Theory of Labor-Managed Economies* (Cornell UP, 1970); Vanek, *The Labor-Managed Economy* (Cornell UP, 1977); Gerry Hunnius, et al., eds., *Workers' Control* (Vintage, 1973); George Benello and Dimitrios Roussopoulos, *The Case for Participatory Democracy* (Grossman, 1971); G. D. Garson, *On Democratic Administration and Socialist Self-Management* (Sage, 1975); *Journal of Social Issues,* 6/76; *Annals of American Academy of Political and Social Science,* 5/77; and below.
Page 355
Consultant, in Daniel Zwerdling, *Democracy at Work* (Washington, D.C.: Association for Self-Management, 1978), p. 56.

Whyte, NIH Summary Progress Report, 5/1/77; *Executive,* Spring 1977.

Hart poll, *Working Papers,* 5–6/79, p. 32.

EDA, TAP Report 99-6-09433, 1976.
Page 356
Michigan's ISR, in Zwerdling, *Working Papers,* 5–6/79.
Page 357
Footnote: Jones, "The Economics and Industrial Relations of Producer Cooperatives in the United States, 1790–1939" (Paper at ASM Conference, Washington, D.C., 1973).
Page 357–59
Plywood cooperatives: Katrina Berman,

Worker-Owned Plywood Cooperatives (Washington State UP, 1967), and in *Self-Management in North America,* Jaroslav Vanek, ed. (ASM, 1975) and in *Autogestion,* 6/75; Paul Berman, *Working Papers,* Summer 1974; Zwerdling, *Democracy,* pp. 91 ff.

Page 359
Ellerman, "On the Legal Structure of Workers' Cooperatives" (Cambridge: ICA, 1978); and *Self-Management,* Summer 1979.

Page 359–61
ESOPs: Zwerdling, *Democracy,* pp. 63 ff.; Katrina Berman, "ESOPs and Implementation of Worker Management," ASM, 5/76.

Page 360
Michigan's ISR, in Zwerdling, *Democracy,* pp. 63 ff.

Page 361–64
Mondragon: Alastair Campbell, et al., *Worker-Owners: The Mondragon Achievement* (London: Anglo-German Foundation for the Study of Industrial Society, 1977); Zwerdling, *Democracy,* pp. 151ff.; Anna Gutierrez Johnson and William Foote Whyte, *Industrial and Labor Relations Review,* 10/77; Robert Oakeshott, in *Self-Management,* ed., Vanek.

Page 362
Cornell researchers, Johnson and Whyte, *Industrial and Labor Relations Review,* 10/77.

Page 364
British observers, Campbell, *Worker-Owners.*

CHAPTER 8

Page 365–66
Saratoga: Zwerdling, *Democracy at Work,* pp. 68 ff.; *NYT,* 4/26/75.

Page 366
Swedish expert, Bertil Gardell, *Working Environment* (Stockholm: 1978), p. 15.

Page 367
Strongforce, *Democracy in the Workplace,* 1977, p. 39.

Page 368
Melman, *Review of Radical Political Economics,* Spring 1970, *Studies in Comparative International Development*, (Sage, 1969); and Hunnius, et al., eds. *Workers' Control,* pp. 252–54.

Page 369
Moberg and Thorsrud, *ITT,* 12/21–27/77.

Rushton: Zwerdling, *Democracy,* pp. 31 ff.

Page 370–71
Blumberg, *Industrial Democracy,* p.123.

Page 371
Phoenix, *WIN,* 11/2/78.

Page 372
Vocations for Social Change, *No Bosses Here,* 1976, p. 32.

Page 373
Burns, *Household Economy,* p. 96.

Sachs, in Strongforce, *Democracy,* p. 39, and Zwerdling, *Democracy,* p. 159.

Page 373–74
Kibbutzim: Keitha Sapsin Fine, in *Workers' Control,* ed., Hunnius, et al.; Haim Barkai, in Vanek, *Self-Management;* Tannenbaum, *Hierarchy in Organizations* (Jossey-Bass, 1974).

Page 374
Mondragon: Campbell, et al., *Worker-Owners,* p. 3.

Page 374–76
Cooperativa: Zwerdling, *Democracy,* p. 101 ff.

CHAPTER 9

Page 377
Dahl (Yale UP, 1970), p. 64.

Page 378
Conover, in Clement Bezold, ed., *Anticipatory Democracy,* ed. (Vintage, 1978), p. 208.

Page 379
McWhinney, in L. S. Stavrianos, *The Promise of the Coming Dark Age* (W. H. Freeman, 1976), p. 63.

Page 380
Yugoslavia: Hunnius, in *Workers' Control,* ed., Hunnius, et al., pp. 268 ff.; Ichak Adizes, *Industrial Democracy: Yugoslav Style* (Free Press, 1971); Blumberg, *Industrial Democracy,* Chs. 8, 9; Vanek, *Self-Management.*

Sachs, *Self-Management* (ASM), 6/78.

Page 381
Melman, in *Studies in Comparative Economic Development* (Sage, 1969).

Barkai, in Vanek, *Self-Management.*

Page 383
British team, Campbell, et al., *Worker-Owners,* p. 51.

Youngstown: NCEA news release, 4/11/78; *Mother Jones,* 4/78; *WIN,* 1/25/79.

Page 384
Schumacher, *Small Is Beautiful,* pp. 268 ff.

Vanek, *Labor-Managed Economy.*

Page 385
George, *Progress and Poverty* (1879; abridged reprint ed., N.Y.: Robert Schalkenbach Foundation, 1970), p. 159.

Page 386
Community land trust: National CLT Center (Cambridge); *Smithsonian,* 6/78; *Green Revolution,* 3/78; *Organic Gardening,* 8/77; *Communities,* 12/74; John McClaughry, *Harvard Journal on Legislation,* 6/75.

Page 387
Utopian experiments: Rosabeth Moss Kanter, *Commitment and Community* (Harvard UP, 1972); Charles Erasmus, *In Search of the Common Good.*

German/Austrian cities: Griscom Morgan, *Community Comments* (Yellow Springs), 3/69; *New Republic,* 8/10/62.

Kibbutz: Erasmus, *In Search of the Common Good,* pp. 183, 301.

Page 389–91

Spanish Civil War: Sam Dolgoff, *The Anarchist Collectives* (Free Life Editions, 1974); George Orwell, *Homage to Catalonia* (Beacon, 1955); George Woodcock, *Anarchism* (Meridian, 1962), Ch. 12.

Page 390

Laval, in Dolgoff, *Anarchist,* pp. 135 ff., 146 ff.

CHAPTER 10

Page 392–93

Paleolithics: Marshall Sahlins, *Stone Age Economics;* Marvin Harris, *Cannibals and Kings;* John E. Pfeiffer, *Emergence of Man;* René Dubos, *Beast or Angel?* (Scribners, 1974), Ch. 2.

Page 395

Jacobs, *The Economy of Cities,* Ch. 5.

Page 396

Darling, *American Scientist,* No. 39, 1951.

Schumacher, *Small Is Beautiful,* p. 55.

Page 397–403

Self-sufficiency: see Bookchin, *Post-Scarcity Anarchism,* pp. 83 ff., 141 ff; Kohr, *Breakdown,* and *Development Without Aid* (Wales: Christopher Davies, 1973); Barry Stein, *Size, Efficiency,* pp. 71 ff.; Edward Ullman, et al., *The Economic Base of American Cities* (Washington UP, 1969); Charles Tiebout, *Community Economic Base Study* (Campaign for Economic Democracy, 1962); Robert E. Dickinson, *City and Nation* (Routledge and Kegan Paul, 1964); Robert Goodman, *The Last Entrepreneurs* (Simon & Schuster, 1980), Chs. 8, 9.

Page 397

Manufacturing statistics: *SA* 1978, Nos. 1401, 1410.

Page 398

Table: *SA* 1978, p. 814 ff.

Page 400

Mumford, pp. 126 ff.

Braudel, *Capitalism and Material Life* (Harper, 1973), pp. 375, 376–79.

Page 401

Nineteenth-century communes, and Albertson quote: Kanter, *Commitment and Community,* p. 158; and Charles Nordhoff, *The Communistic Societies of the United States* (1875; reprint ed., Schocken, 1965).

Page 402

New Towns: Ebenezer Howard, *Garden Cities of Tomorrow* (1898; reprint ed., MIT, 1965); Frederick J. Osborn, *Green-Belt Cities* (Schocken, 1969), esp. pp. 144–45; J. E. Gibson, *Designing the New City;* Gideon Golany, *New-Town Planning* (Wiley, 1976), Chs. 4, 5.

CHAPTER 11

Page 404

Nielsen, in *LA Times* ad, *NYT,* 7/5/79, p. A18.

Page 405

Borsodi, *MEN,* 3/74.

Page 406

Papanek, *NYT,* 6/14/77, Sec. 3.

Page 407

Jojoba and guayule: *Mother Earth News,* 11–12/77; *NYT,* 3/30/77; "Jojoba," "Guayule," both National Academy of Science papers.

Page 409

Bookchin, *Post-Scarcity Anarchism,* p. 111.

Page 410

Land Economics, Charles G. Schmidt and Richard B. LeHeron, 11/76.

Page 411

Community economy: David Morris and Karl Hess, *Neighborhood Power* (Beacon, 1975), Ch. 4; Milton Kotler, *Neighborhood Government* (Bobbs-Merrill, 1969), and *Liberation,* Spring 1976; *Ecologist,* 8–9/77.

Page 412

Footnote: Shaw-Cardoza, Earl F. Mellor, *Public Goods and Services* (Washington, D.C.: Institute for Policy Studies, 10/69); Kotler, *Liberation;* national center, *Self-Reliance,* 11/77; Zion CDC, *Community Development Corporations* (Cambridge: Center for Community Economic Development, 1975).

Page 413

Lucca's Law: see, e.g., Kohr, *Overdeveloped Nations,* esp. Ch. 8, and *Development Without Aid,* esp. Chs. 5, 13, 14; Schumacher, *Small Is Beautiful,* pp. 59 ff.

Page 414–15

Gish, in *The Elements,* 7/77.

Part Five: Politics on a Human Scale

CHAPTER 1

Page 419–20

Carter, *NYT,* 7/16/79; polls, 7/18/79.

Page 421

Gans, *NYT,* 12/19/78, p. A13.

Hadley (Prentice-Hall, 1978).

Friedland, *Newsweek,* 5/22/78, p. 22; Greenfield, *Newsweek,* 4/24/78; Schorr, *NYT,* 1/7/79, Op Ed; Common Cause, mailing, 1979.

Caddell, *NYT,* 7/22/79, p. 1.

Page 422

Tuchman, *Newsweek,* 7/12/76, "My Turn."

Page 422–23
Congress: Tad Szulc, *Saturday Review,* 3/3/79; Elizabeth Drew, *New Yorker,* 6/26/78.
Page 423–24
Legislator, Patterson, Udall, in Drew, *New Yorker,* 6/26/78.
Page 423
Eleven minutes: *NYT,* 10/2/77, and see 8/28/77.
Page 424
Harrington, *Nation,* 12/23/78.
Page 425
Wolff, *In Defense of Anarchism* (Harper, 1970), pp. 30–31.
Page 427
Newsweek, and Field, *Newsweek* 11/6/78.

The Rebel, 2/1896, reprinted in *The Match* (Tucson), 1/72.

CHAPTER 2

Page 429
Phillips, *Harper's,* 5/78.
Page 430
Brustein, *NYT,* 11/19/78, Sec. 2.
Page 430–31
Nisbet (Oxford UP, 1975); quote, pp. 61–62.
Page 432
Newsweek, 3/6/78.

Health recommendations: *NYT,* 6/12/77, p. 55.

Recalls: *Reader's Digest,* 10/78.
Page 433
Macrae, *Economist,* 12/23/78, pp. 45 ff.

Hobsbawm, *New Left Review,* 9–10/77.

Zwerin, *A Case . . .* (London: Wildwood House, 1976), p. 4.

Sulzberger, *NYT,* 12/22/76, Op Ed.
Page 434
Nairn (London: New Left Books, 1977).
Page 435
Headlines, *NYT,* 5/26/79, 5/16/79.
Page 436
Letter to Brown: *Newsweek,* 9/17/79, p. 31.
Page 437
Lamm, *Newsweek,* 9/17/79, p. 31.

Herbers, *NYT,* 9/16/79.

Newsweek, 9/17/79, p. 31.

NYT, 3/18/79, p. 1.
Page 438
Schlesinger, in Phillips, *Harper's,* 5/78.

Localism: see generally, Morris and Hess, *Neighborhood Power* (Beacon, 1975); Bruce Stokes, "Local Responses to Global Problems," Worldwatch Institute, 1978; Milton Kotler, *Social Research,* 1975; *Social Policy,* 10/76 and 10/79; *NORG News Bulletin* (Bloomington, Ind.) and particularly its *Community Organization and Neighborhood Government* bibliographies; *Self-Reliance; Neighborhood Ideas; Community Service Newsletter.*

Monitor, Stewart Dill McBride, "A Nation of Neighborhoods," series, 9/9–12/23/77, reprinted 1/78, p. 40.
Page 439
NYT, 6/18/79.
Page 440
Regional groups: Paul Freundlich, et al., *A Guide to Cooperative Alternatives* (New Haven, Conn.; Louisa, Va.: Community Publications Cooperative, 1979); *Social Policy,* 10/79, pp. 17, 66 ff.

National Commission on Neighborhoods, report (GPO, 3/15/79).

National Housing Services, John McClaughry, *The Urban Lawyer,* Spring 1978.
Page 441
Herbers, *NYT,* 6/18/79.

CHAPTER 3

Page 443–53
Decentralism: see, generally, the corpus of Kohr, Schumacher, Mumford, Bookchin, Illich, Kropotkin, Ralph Borsodi, and Paul Goodman, and specifically, Herbert Agar, *Land of the Free* (Houghton-Mifflin, 1935); Richard Cornuelle, *De-Managing America* (Random House, 1975); Karl Hess, *Dear America* (Morrow, 1975); Thomas Hewes, *Decentralize for Liberty* (Appleton, 1947); Peter T. Manicus, *Death of the State* (Putnam's, 1974); Ioan B. Rees, *Government by Community* (London, Charles Knight, 1971); Mark Satin, *New Age Politics* (Dell, 1979); Colin Ward, *Anarchy in Action* (Harper, 1973); Ralph L. Woods, *America Reborn* (Longmans, Green, 1939). And see bibliographies in Satin, *New Age Politics;* Michael Marien, *Societal Directions and Alternatives* (Lafayette, NY: Information for Policy Design, 1976); and from Institute for Liberty and Community (Concord, Vt.).
Page 433–45
Arendt, *On Revolution* (Viking, 1965), pp. 254–85; quotes, pp. 260, 267, 252–53.
Page 446
Rocker, *Nationalism and Culture* (Los Angeles: Rocker Publications, 1937), Ch. 5 and p. 362.

Switzerland: Jonathan Steinberg, *Why Switzerland?* (Cambridge UP, 1976); quote, p. 73.

Kropotkin, *Mutual Aid,* (London: Heinemann, 1902), and "The State" in *Selected Writings,* ed. Miller.

Page 448–53
Decentralism, in America: see David DeLeon, *The American as Anarchist* (Johns Hopkins UP, 1979); Merrill Jensen, *The*

American Revolution Within America (New York UP, 1976), and *Articles of Confederation* (Wisconsin UP, 1940); Murray Rothbard, *Conceived in Liberty*, Vols. 1–4 (Arlington House, 1975–1979); Staughton Lynd, *Intellectual Origins of American Radicalism* (Pantheon, 1963); Lawrence Goodwyn, *Democratic Promise: The Populist Movement in America* (Oxford UP, 1976); Zinn, *A Peoples History of the United States.*

Page 449
Paine, in Colin Ward, *Anarchy in Action*, p. 65.

Shays, *Self-Reliance*, 7–8/77.

Page 450
Jefferson, Cartwright letter, 6/5/1824, Kercheval letters 7/12/1816 and 9/5/1816; Arendt, *On Revolution*, pp. 252–59.

Page 451
Jensen, *The New Nation* (Vintage, 1967), p. 120.

Populists: Goodwyn, *Democratic Promise.*

Page 452
1920s and 1930s: *I'll Take My Stand* (Harper, 1930, reprint, 1962); Herbert Agar and Allen Tate, eds., *Who Owns America?* (Houghton-Mifflin, 1936); Edward S. Shapiro, *Journal of American History*, No. 58, 1972; *Free America* (New York), 1937–47.

Page 453
Harrington, *Nation*, 12/23/78.

Woodward, *New York Review of Books*, 4/5/79.

Putnam County News, 8/22/79.

CHAPTER 4

Page 455
Pfeiffer, *Emergence of Man*, p. 290.

Walter, in Ward, *Anarchy in Action*, p. 50.

Page 456–61
Societies without states: Aidan Southall, "Stateless Societies," *International Encyclopedia of Social Sciences*, 1968; M. Fortes and E. E. Evans-Pritchard, eds., *African Political Systems* (Oxford UP, 1940); John Middleton and David Tait, eds., *Tribes Without Rulers* (Routledge and Kegan Paul, 1958); Pierre Clastres, *Society Against the State* (Urizen, 1977); Franz Oppenheimer, *The State* (Free Life, 1975); *Ecologist, Blueprint for Survival* (Penguin, 1973), pp. 106 ff.

Page 457
Southall, "Stateless Societies."

Farb, *Humankind* (Houghton Mifflin, 1978), p. 104.

Page 458
Clastres, *Society Against*, pp. 178 ff., p. 174.

Page 458–61
Dinka: Lienhardt, in Middleton and Tait, *Tribes;* quote, p. 114.

Page 462
Braudel, *Capitalism and Material Life*, p. 420.

Ireland: Joseph Peden, *The Libertarian Forum*, 4/71.

Page 463–65
Zuckerman, *Peaceable Kingdoms* (Vintage, 1972); quotes, pp. 32, 34, 37, 87, 46, 87, 93, 100–102; also see Robert E. Brown, *Middle-Class Democracy and the Revolution in Massachusetts, 1691–1780* (Cornell UP, 1955).

CHAPTER 5

Page 466–67
Dahl, *After the Revolution?*, pp. 146–47.

Page 468–71
Small-society peacefulness: Peter van Dresser, *Free America*, 7/38; Leopold Kohr, *Breakdown*, pp. 60 ff.

Page 469
Simons, *Economic Policy for a Free Society* (Chicago UP, 1948).

Hart and Pilling (Holt, 1960).

Page 474
Dahl, *After the Revolution?*, p. 86.

Dahl, in Dahl and Tufte, *Size and Democracy*, pp. 22–24.

Page 476
Taylor, *Anarchism and Cooperation* (Wiley, 1976), pp. 134–35.

Page 481
Weil, *The Need for Roots* (Putnam's, 1952), p. 43.

Kahn, debate with Steward Brand, 6/77.

CHAPTER 6

Page 482
Dunstable, in Zuckerman, *Peaceable Kingdoms*, p. 121.

Page 483
Switzerland: Steinberg, *Why Switzerland?*

Page 487
Rathje, *Ancient Civilization and Trade* (New Mexico UP, 1975).

Page 488
Pfeiffer, *Emergence of Society*, p. 467.

Page 489
Padavino, in Steinberg, *Why Switzerland?*, p. 23.

Spinellis, *NYT*, 8/19/79.

Page 490
Crime rates: *SA* 1978, No. 290.

CHAPTER 7

Page 493
Rousseau, *Social Contract*, Bk. III, Ch. XV; *Geneva Manuscript*, Bk. II, Ch. III; *Social Contract*, Bk. III, Ch. 1.

Aristotle, *Politics*, Bk. VII, Ch. 4.

Page 494
Madison, *Federalist 10.*

Mill, "The Ideal Best Form, *Representative Government*

Dahl, *After the Revolution?*, p. 85.

Kohr, *Breakdown*, pp. 99–100.

Page 495
Greece: see e.g., Kathleen Freeman, *Greek City-States* (Norton, 1950); M. I. Finley, *Democracy: Ancient and Modern* (Rutgers UP, 1973).

Page 495–96
Steinberg, *Why Switzerland?*, pp. 72–73, 190.

Page 497
Town meetings: Zuckerman, *Peaceable Kingdoms;* Merrill Jensen, *New Nation.*

Mansbridge, *Massachusetts Review*, Winter 1976.

Zimmerman, *NYT*, 3/15/79.

Page 498–99
Size and Democracy, pp. 57, 46–51, 87–88, 60; Swedish study, pp. 62 ff.; confirmative studies in Arthur Vidich and Joseph Bensman, *Small Town in Mass Society* (Doubleday, 1960), and Ritchie P. Lowry, in, Robert Rankin Ed. *Sociology*, (Lexington, Mass., 1977).

Page 500
Voting paradox: *Scientific American*, 6/76; Peter C. Fishburn, *The Theory of Social Choice* (Princeton UP, 1973).

Page 502
Hoffman, "Quaker Dialogue," in *The Civilization of the Dialogue*, from Peacemakers (Cincinnati, Ohio).

Page 503
Mansbridge, *Massachusetts Review*, Winter 1976.

MNS, Clamshell, see e.g., Paul Freundlich, et al., *A Guide to Cooperative Alternatives*, pp. 129–32.

Page 505
Kohr, *Breakdown*, pp. 98–99.

CHAPTER 8

Page 508
St. Denis: Horace Miner, *St. Denis, A French Canadian Parish* (Chicago UP, 1939), pp. 58–59.

Page 514
Roszak, in *Resurgence*, 1–2/79.

Part Six: Conclusion

Kohr, *Breakdown*, pp. 196–97.

Kolakowski, in Rudolf4Moos and Robert Brownstein, *Environment and Utopia* (Plenum, 1977), p. 262.

Kodner, *Progressive*, 9/77.

Page 520
Schelling, in Daly, *Steady-State Economics*, p. 145.

Index